DESIGNING HIGH AVAILABILITY SYSTEMS

IEEE Press
445 Hoes Lane
Piscataway, NJ 08854

DESIGNING HIGH AVAILABILITY SYSTEMS

DESIGN FOR SIX SIGMA AND CLASSICAL RELIABILITY TECHNIQUES WITH PRACTICAL REAL-LIFE EXAMPLES

Zachary Taylor
Subramanyam Ranganathan

IEEE
IEEE PRESS

WILEY

Library of Congress Cataloging-in-Publication Data:

Taylor, Zachary, 1959–
Designing high availability systems : design for Six Sigma and classical reliability techniques with practical real-life examples / Zachary Taylor, Subramanyam Ranganathan.
pages cm
ISBN 978-1-118-55112-7 (cloth)
1. Reliability (Engineering) 2. Systems engineering–Case studies. 3. Six sigma (Quality control standard) I. Ranganathan, Subramanyam. II. Title.
TA169.T39 2013
658.4'013–dc23

2013011388

ISBN: 9781118551127

10 9 8 7 6 5 4 3 2 1

To my wife Soheila, for her unwavering support

Zachary Taylor

To the Lotus Feet of Sringeri Jagadguru Sri. Abhinava Vidyatheertha Mahaswamigal, To the Lotus Feet of Sri. R. M. Umesh, whose divine grace brought me into the world of engineering, and to my parents Smt. Shantha Ranganathan and Sri. V. Ranganathan

Subramanyam Ranganathan

CONTENTS

PREFACE

Even as you begin to browse this book, you may be wondering, What's in it for me? Will it help solve my design challenges? Will it provide my product a more competitive edge? Will I have immediate takeaways to work? How much will it improve my capability to develop highly available and reliable systems? Can my team benefit practically from this book? Can I immediately adopt some of the techniques as best practices? Can it help deliver higher quality faster?

Whether you are a student, designer, architect, system engineer, mid-level manager, senior manager, director, VP, CEO, entrepreneur, or just someone with an intellectual curiosity, the techniques described in this book will provide you with a winning edge for designing world-class high-availability solutions.

The intent of this book is to bring to you a straight-forward, crisp, and practical approach to designing high-availability systems from the ground up, that is, systems in which high availability is an integral critical design element and differentiator, as well as a customer requirement. Typical business segments for these systems include telecom, automotive, medical, manufacturing, aerospace, financial, defense, and public safety. These systems typically consist of high reliability hardware, embedded and off-the-shelf software, multisite, multithreaded distributed processing environments, complex real-time applications, and high performance capabilities.

Though high availability and reliability are typically "must-haves" and taken for granted, designing such systems is usually complex and difficult for a variety of reasons. The design can take many iterations, involving significant time, cost, and effort. This book attempts to bring together different practical techniques used in the industry to successfully design, predict, and deploy high availability systems with maximum productivity and reduced total costs.

Our intent is to enable readers to quickly apply practical tools and techniques to projects within their organizations to realize potential benefits regardless of the current phase of their development project. Benefits include, but are not limited to, higher customer satisfaction, superior product differentiation, and delivery of high reliability and high performance systems to customers within shorter cycle times at lower overall cost.

Having worked in the telecommunication and aerospace industries for many years developing mission-critical and safety-critical embedded systems using these proven techniques, the authors strongly feel that practitioners will be able to employ this as a practical guide to designing high availability systems in a systematic and methodical fashion.

System engineers, architects, and designers who are driven to design high availability systems that are best in class can benefit immensely by employing the classical and Six Sigma tools in this book for predictive engineering. While the book focuses in-depth on the technical aspects, it is also sensitive to the underlying business frameworks, which demand designing a system right the first time in the face of market constraints. We use real-life examples throughout this book to explore predictive design methods, trade-offs, risk analysis, and "what-if" scenarios, so that the architect can realize the most effective impact to system design.

Designing high availability systems also requires a bit of skill from some of the more sophisticated arts of probability theory. A system is available unless something goes wrong. When something bad happens, we did not want it to happen and we certainly did not expect it to happen. So not only do we need to consider how to design a system that minimizes the probability of something going seriously wrong that disrupts operation, but we also need to quantify to some degree of confidence the likelihood of failure events occurring and how we can proactively minimize the impact of these failure events.

Reliability and availability are intimately intertwined with probability theory, and probability theory itself can be a difficult subject. But to fully understand how to design and manage highly available systems, we need to delve into the world of probability theory.

Many books on probability theory exist, and the reader is encouraged to explore some of the books recommended in the references. The authors believe that a firm understanding of the key aspects of theory is critical to understanding the application, as well as the limitations, of practical applications of theory. Therefore, we focus on those topics of probability and reliability theory that have the most influence on the practical applications of reliability theory. This includes exploring some concepts in depth, including proofs when needed for clarity.

Many approaches to presenting probability theory have been employed. Some make use of the typical dice, card, and coin problems that help the reader understand probability theory, but on the other hand these often fall short on the question of how to make the quantum leap from dice games to complex computer system redundancy strategies. Some texts take the approach that dice games are too simplistic and do not relate to the real world problems we are after. So dice and card games are dispensed with, and we are soon immersed in a set of complex equations that at first glance can be intimidating and difficult to follow for the uninitiated. In this case, we must somehow relate this theory back to the general problems we face in architecting a high available system. Another technique is to present the basic set of classical probability equations and ask the reader to accept these equations as a matter of faith and move on. Finally, some texts take a purely theoretical approach in which each equation is painstakingly derived based on a set of axioms and theorems previously laid out. Practical real-life examples are often left as an exercise for the reader. In addition, following the derivation process inevitably leads to a few steps being collapsed due to space constraints or assumptions of the knowledge of the reader.

This book takes a very different approach. We have uniquely blended classical and more recent Design for Six Sigma (DFSS) techniques, thus providing the practitioner with a broader repertoire of tools from which the designer or analyst can choose. We derive many of the equations that form the foundation of reliability theory. We believe these derivations are important for understanding the application of the theory, and we tackle the relevant mathematical foundation with some rigor. It is our firm belief that these derivations will be valuable to the reader by providing better foundational insight into the underpinnings of reliability theory. We then follow up with increasingly practical applications. The formalistic approach is avoided, since our goal is to help the practitioner not only apply a certain equation but also understand why it is valid. This is important, since the practitioner who wishes to become an expert in the field needs to know why a technique works, what the limitations are to this technique, what assumptions are necessary for this technique to be employed, and at what point a certain technique should be discarded for another approach.

Readers who have a strong background in reliability and probability theory may choose to skip those foundation chapters in which probability and reliability theory are introduced along with derivations of pertinent formulas, and instead jump straight to the practical applications and techniques. Practitioners will find many relevant examples of applications of classical reliability theory and DFSS techniques throughout the book.

The authors' goal is to deliver the following:

- *Application to a Broad Range of Industries:* Telecom, automotive, medical, manufacturing, aerospace, financial, information systems, such as fly-by-wire avionics, large telecommunication networks, life-saving medical devices, critical retail or financial transaction management servers, or other systems that involve innovative cutting-edge design and technology relevant to the everyday quality of life.
- *Practical Examples and Lucid Explanations:* Complex concepts are described in simple, easy-to-understand language, blending real-life examples and including step-by-step procedures and case studies.
- *Relevant Topics:* We bring together topics that are relevant for high availability design and at the same time have been sensitive to demands on the readers' time.
- *Comprehensive yet Focused Diagrams:* A wealth of illustrations and diagrams for in depth understanding are made available to the reader. We attempt to bridge theory with practice and have confined the derivations to key theoretical aspects most likely to be applicable in practical applications.
- *Immediate Takeaways with High Impact:* Readers can start applying techniques immediately to their projects at work. It will also enable them to quickly see the results, communicate success stories, and share best practices.

The authors hope that the book serves both the student and professional community to enrich their understanding as well as help them realize their objectives for designing and deploying superior high availability products.

MATLAB®

MATLAB is one of the most popular applications for mathematical calculations and plotting. The MATLAB programs used in several of the examples in this book are available on the book's website: http://booksupport.wiley.com. Enter the ISBN number 9781118551127 to access these files.

ACKNOWLEDGMENTS

The authors would like to thank the following reviewers who provided valuable comments on the manuscript: Dr. Thomas Garrity, Department of Mathematics at Williams College; Olli Salmela, Nokia Siemens Networks; and Michael D. Givot, Microsoft Corporation.

Numerous people, including present and past colleagues, have directly or indirectly contributed to enriching the content of this book. Greg Freeland worked on the session capacity design improvement illustrated in this book. Andy Moreno worked on the Paging Retries design improvement. Tim Klandrud worked on a complex critical parameter design prediction effort. Several teams have been part of the DFSS strategy and rollout, including DFMEA and Fault Tree Analysis. Thanks also to mentors and colleagues, including Eric Maass, Richard Riemer, and Cvetan Redzic, especially in the DFSS area.

Many late nights and weekends were spent working on this book; the authors would like to thank their families for adjusting their schedules and being there with a smile. Special thanks to Hema Ramaswamy for her persistent encouragement and support throughout the creation of the book.

Finally, thanks are due to Motorola Solutions Inc. and Nokia Siemens Networks.

Zachary Taylor
Subramanyam Ranganathan

LIST OF ABBREVIATIONS

ATCA	Advanced Telecommunications Computing Architecture
BIT	Built-In Test
BSC-DO	Base Station Controller—Data Only
BTS	Base Transceiver System
C_{pk}	Capability Metric
CDF	Cumulative Distribution Function
CDMA	Code Division Multiple Access
CPM	Critical Parameter Management
CTMC	Continuous Time Markov Chain
DF	Degrees of Freedom
DFMEA	Design Failure Modes and Effects Analysis
DFSS	Design for Six Sigma
DMAIC	Define Measure Analyze Improve Control
DOE	Design of Experiments
DTMC	Discrete Time Markov Chain
EVDO	EVolution Data Only
FIT	Failure in Time
FM	Fault Management
FMEA	Failure Modes and Effects Analysis
FRU	Field Replaceable Unit
FTA	Fault Tree Analysis
GOF	Goodness of Fit
GUI	Graphical User Interface
IEC	International Electrotechnical Commission
IP	Internet Protocol
KJ	Kawakito Jira
LSL	Lower Specification Limit
MDT	Mean Downtime
MOL	Maintenance on Line
MRM	Markov Reward Model
MTBF	Mean Time between Failures
MTBV	Mean Time between Visits
MTFF	Mean Time to First Failure
MTTF	Mean Time to Failure
MTTR	Mean Time to Repair
NASA	National Aeronautics and Space Administration

NUD	New, Unique, Difficult
O&M	Operations and Maintenance
ODE	Ordinary Differential Equation
OOS	Out of Service
OS	Operating System
PAM	Process Advanced Mezzanine
PAND	Priority AND Gate
PDF	Probability Density Function
PMF	Probability Mass Function
QFD	Quality Function Deployment
RAM	Random Access Memory
RBAC	Role-Based Access Control
RBD	Reliability Block Diagram
RCA	Root Cause Analysis
RF	Radio Frequency
ROI	Return on Investment
RPN	Risk Priority Number
SCI	Slot Cycle Index
SSPD	Six Sigma Product Development
USL	Upper Specification Limit
VOC	Voice of Customer

CHAPTER 1

Introduction

We live in a complex and uncertain world. Need we say more? However, we can say quite a bit about some aspects of randomness that govern behavior of systems—in particular, failure events. How can we predict failures? When will they occur? How will the system we are designing react to unexpected failures? Our task is to help identify possible failure modes, predict failure frequencies and system behavior when failures occur, and prevent the failures from occurring in the future. Determining how to model failures and build the model that represents our system can be a daunting task. If our model becomes too complex as we attempt to capture a variety of behaviors and failure modes, we risk making the model difficult to understand, difficult to maintain, and we may be modeling certain aspects of the system that provide only minimal useful information. On the other hand, if our model becomes too simple, we may leave out critical system behavior that dramatically reduces its effectiveness. A model of a real system or natural process represents only certain aspects of reality and cannot capture the complete behavior of the real physical system. A good model should reflect key aspects of the system we are analyzing when constrained to certain conditions. The information extracted from a good model can be applied to making the design of the system more robust and reliable.

No easy solutions exist for modeling uncertainty. We must make simplifying assumptions to make the solutions we obtain tractable. These assumptions and simplifications should be identified and documented since any model will be useful only for those constrained scenarios. Used outside of these constraints, the model will tend to degrade and provides us with less usable information. That being the case, what type of model is best suited for our project?

When designing a high availability system, we should carefully analyze the system for critical failure modes and attempt to prevent these failures by incorporating specific high availability features directly in the system architecture and design.

Designing High Availability Systems: Design for Six Sigma and Classical Reliability Techniques with Practical Real-Life Examples, First Edition. Zachary Taylor and Subramanyam Ranganathan.

However, from a practical standpoint, we know unexpected failures can and will occur at any time despite our best intentions. Given that, we add a layer of defense, known as fault management, that mitigates the impacts of a failure mode on the system functionality. Multiple failures and/or failure modes not previously identified may cause system performance degradation or complete system failure. It is important to characterize these failures and determine the expected overall availability of the system over its lifetime of operation.

Stochastic models are used to capture and constrain randomness inherent in all physical processes. The more we know about the underlying stochastic process, the better we will be able to model that process and constrain the impacts of the random failures on the system we are analyzing. For example, if we can assume that certain system components have constant failure rates, a wealth of tools and techniques are available to assist us in this analysis. This will allow us to design a system with a known confidence level of meeting our reliability and availability goals. Unfortunately, two major impediments stand in our way: (1) The failure rate of many of the components that comprise our system are not constant, that is, independent of time over the life of the system being built or analyzed, but rather these failure rates follow a more complicated trajectory over the lifetime of the system; and (2) exact component failure rates—especially for new hardware and software—are not known and cannot be exactly determined until after all built and deployed systems reach the end of their useful lives.

So, where do we start? What model can we use for high availability design and analysis? How useful will this model be? Where will it fail to correctly predict system behavior? Fortunately, many techniques have already been successfully used to model system behavior. In this book, we will cover several of the more useful and practical models. We will explore techniques that will address reliability concerns, identify their limitations and assumptions that are inherent in any model, and provide methods that in spite of the significant hurdles we face, will allow us to effectively design systems that meet high availability requirements.

Our first step in this seemingly unpredictable world of failures is to understand and characterize the nature of randomness itself. We will begin our journey by reviewing important concepts in probability. These concepts are the building blocks for understanding reliability engineering. Once we have a firm grasp on key probability concepts, we will be ready to explore a wide variety of classical reliability and Design for Six Sigma (DFSS) tools and models that will enable us to design and analyze high availability systems, as well as to predict the behavior of these systems.

CHAPTER 2

Initial Considerations for Reliability Design

2.1 THE CHALLENGE

One of the biggest challenges we face is to predict the reliability and availability of a particular system or design with incomplete information. Incomplete information includes lack of reliability data, partial historical data, inaccuracies with data obtained from third parties, and uncertainties concerning what to model. Inaccuracies with data can also stem from internal organizational measurement errors or reporting issues. Although well-developed techniques can be applied, reliability attributes, such as predictive product or component MTBF (*Mean Time between Failures*), cannot be precisely predicted—it can only be estimated. Even if the MTBF of a system is accurately estimated, we will still not be able to predict when *any* particular system will fail. The application of reliability theory works well when scaled to a large number of systems over a long period of time relative to the MTBF. The smaller the sample and the smaller the time frame, the less precise the predictions will be. The challenge is to use appropriate information and tools to accomplish two goals: (1) predict the availability and reliability of the end product to ensure customer requirements are met, and (2) determine the weak points in the product architecture so that these problem areas can be addressed *prior* to production and deployment of the product.

A model is typically created to predict the problems we encounter in the field, such as return rates, and to identify weak areas of system design that need to be improved. A good model can be continually updated and refined based on new information and field data to improve its predictive accuracy.

2.2 INITIAL DATA COLLECTION

How do we get started? Typically, for the initial availability or reliability analysis, we should have access to (1) the initial system architecture, (2) the availability and reliability requirements for our product, and (3) reliability data for individual components (albeit in many cases data are incomplete or nonexistent).

Designing High Availability Systems: Design for Six Sigma and Classical Reliability Techniques with Practical Real-Life Examples, First Edition. Zachary Taylor and Subramanyam Ranganathan.

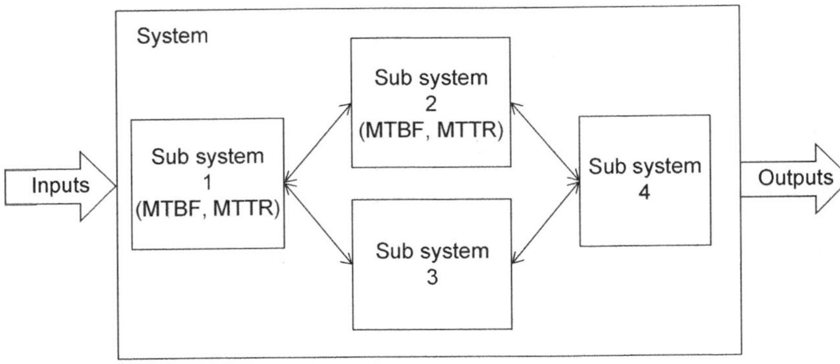

Figure 2.1 System Block Diagram

For reliability purposes, a system or product is decomposed into several components and reliability information is associated with these components. How do we determine these components? Many components can be extracted from the system architecture block diagram (Fig. 2.1). For hardware components, one natural division is to identify the Field Replaceable Units (FRU) and their reliability data estimates or measurements. FRUs are components, such as power supplies, fan trays, processors, memory, controllers, server blades, and routers, that can be replaced at the customer site by the customer or contracted field support staff. For software components, several factors need to be considered, such as system architecture, application layers, third party vendors, fault zones, and existing reliability information.

Let us say you have a system with several cages and lots of cards and have implemented fault tolerance mechanisms. If the design of the system is such that the customer's maintenance plan includes procedures for the repair of the system by replacing failed cards (FRUs), then at a minimum, those individual cards should be uniquely identified along with certain reliability data associated with them—in particular, MTBF and MTTR (*Mean Time to Repair*) data. This basic information is required to calculate system availability.

The MTTR is dependent upon how quickly a problem is identified and how quickly it can be repaired. In the telecom industry, typically a maintenance window exists at low traffic peak times of the day, during which maintenance activities take place. If a failed component does not affect the system significantly, then the repair of that component may be delayed until the maintenance window, depending on the customer's maintenance plan, ease of access to equipment, and so on. However, if the system functionality is significantly impacted due to a failure event, then the system will require immediate repair to recover full service.

In addition to hardware and software component failures, we should also take into consideration other possible failures and events that impact the availability of our system. These may include operator-initiated planned events (e.g., software upgrade), environment failures, security failures, external customer network failures, and operator errors. The objective is to identify those events and failures that can affect the ability of our system to provide full service, and create a model that incorporates these potential failures. We need to determine which of these events are significant (characterized by severity and likelihood of occurrence) and are within the scope of the system we are modeling. If we group certain failure modes into a fault zone, we can modularize the model for better analysis and maintainability.

2.3 WHERE DO WE GET MTBF INFORMATION?

For hardware, we may be able to obtain MTBF information from industry standard data or from a manufacturer-published reliability data. If a particular component does not have a published or known MTBF, then the next step is to look for parts or components that are similar and estimate the MTBF based on that data under similar operating conditions. Another method is to extrapolate the MTBF based on data mining from past similar projects. In the worst case scenario, if a totally new component or technology is to be employed with little reliability data available, then a good rule of thumb is to look at a previous generation of products of a similar nature that do have some reliability data and use that information to estimate the MTBF. Use engineering judgment to decrease the MTBF by a reasonable factor, for example, $x/2$ to account for uncertainty and typical early failures in the hardware cycle. It is better to err on the side of caution.

Once we have made this initial assessment, the MTBF becomes the baseline for our initial model.

Let us consider a hypothetical communication system that consists of a chassis with several slots in which processing cards can be inserted. One of these cards is a new RF (radio frequency) carrier card. We start with the original data provided from the manufacturer, which indicate an MTBF of 550,000 hours. We then sanity-check these data by looking at MTBFs of similar cards successfully deployed over a number of years from different manufacturers. This analysis reveals an average MTBF of 75,000 hours. How might we reconcile the difference? Which estimate do we use? It turns out that the right answer may be both! In Chapter 13, we will explore techniques for combining reliability data to obtain an updated MTBF estimate that may be more accurate than any single source of MTBF data.

On the software side, if the software is being built in-house, we can derive the MTBF from similar projects that have been done in the past and at the same level of maturity. Release 1 always has the most bugs! We can also extrapolate information from field data in the current system release or previous releases. This can get more complicated if the software is written by multiple vendors (which is typical). We may also consider software complexity, risk, schedules, maturity of the organization developing the software, nonconstant failure rates, and so on as factors that affect MTBF.

Are you reusing existing software that has already been working in the field? For example, if we build on software from a previous project, and then we add additional functionality on top of this, we can extract software failure rate information from the previous project. We can also identify failure rates from industry information for common software, such as the Linux operating system. It is also quite possible that we identify MTBF information from third-party suppliers of off-the-shelf standard software.

We need to take into account as many relevant factors as possible when assigning MTBFs, MTTRs, and other reliability data. We should also note that MTBFs of products generally tend to improve over time—as the system gets more field exposure and software fixes are made, the MTBF numbers will generally increase.

2.4 MTTR AND IDENTIFYING FAILURES

An important part of our architectural considerations is the detectability of the problem. How likely is it that when we have a problem, we will be able to automatically detect it? Fault Management includes designing detection mechanisms that are capable of picking up those failures.

Mechanisms, such as heartbeat messaging, checkpointing, process monitoring, checksums, and watchdog timers, help identify problems and potential failures. If a particular failure is automatically detected and isolated to a card, then recovery mechanisms can be invoked, for example, failover to a standby card.

Autorecovery is a useful technique for transient software failures, such as buffer overflows, memory leaks, locks that have not been released, and unstable or absorbing states that cause the system to hang. For these failures, a reboot of the card in which the problem was detected may repair the problem. If the failure reoccurs, more sophisticated repair actions may need to be employed. The bottom line is that we want to maximize the time the system is available leveraging the simplest recovery options for providing the required service.

In addition to detectable failures, a portion of the failures will be undetected or "silent" failures that should be accounted for as part of reliability calculations. These undetected failures can eventually manifest themselves as an impact to the functionality of the system. Since these failures may remain undetected by the system, the last line of defense is manual detection and resolution of the problem. Our goal is to reduce the number of undetected failures to an absolute minimum since these undetected failures potentially have a much larger impact on system functionality due to the length of time the problem remains undiscovered and uncorrected.

In situations where the problem cannot be recovered by an automatic card reset, for example, the MTTR becomes much larger. Instead of an MTTR of a few minutes to account for the reboot of a card, the MTTR could be on the order of several hours if we need to manually replace the card or system, or revert to a previous software version. So the more robust the system and fault management architecture is, the more successfully we can quickly identify and repair a failure. This is part of controllability—once we know the nature of the problem, we know the method that can be employed to recover from the problem.

There are several ways to reduce the MTTR. In situations where the software fails and is not responsive to external commands, we look at independent paths to increase the chances of recovering that card. In ATCA (Advanced Telecommunications Computing Architecture) architectures, a dedicated hardware line can trigger a reboot independent of the software. Other methods include internal watchdog time-outs that will trigger a card reboot. The more repair actions we have built into the system, the more robust it becomes. There is of course a trade-off on the number and complexity of such mechanisms that we build into a system. We must be careful that our fault management mechanisms are not so complex that the fault management software itself begins to negatively impact the reliability and response time of the system! In the authors' experience, redundancy management tends to be the most problematic of fault management techniques due to their inherent complexity necessary to minimize outages or downtimes for a large number of possible known (and unknown) failures.

2.5 SUMMARY

Prior to designing a high availability system, we must have a set of availability requirements or goals for our system. We then set out to design a system that meets these requirements. By decomposing the system into appropriate components, we can create a system reliability model and mechanisms for ensuring high availability. For each component in this model, we allocate specific MTBF, MTTR, and other reliability information. We described a few general methods can be used to estimate and improve this reliability information. The more knowledge we have regarding the reliability of these components, the maintenance plan for the system, and the number of systems we expect to deploy, the more accurate our model will be in predicting actual system reliability and availability once deployed to the field.

Now that we have introduced the mechanics of obtaining initial reliability information, the next several chapters will dwell on basic mathematical concepts that will set the foundation for more advanced techniques and applications to build, predict, and optimize high availability systems.

CHAPTER 3

A Game of Dice: An Introduction to Probability

3.1 INTRODUCTION

All systems are subject to failures. We need powerful techniques to analyze and predict the reliability of high availability systems. Standard techniques have been developed and put into practice that can assist us in designing these systems. Many of the techniques necessary for this analysis are rooted in probability theory. This chapter introduces important concepts of probability theory that will be applied to the practical application of reliability in later chapters. We begin our discussion by posing the following question:

"How can an exponential failure distribution represent a constant failure rate?"

The vast majority of analysis techniques we employ are based on the concept of constant random failure rates. That is, the probability of a failure occurring at any time remains the same over the lifetime of the system. In other words, the rate of these random failures is independent of time. With this assumption regarding the nature of the failures, a wealth of techniques become available—including the exponential failure probability distribution.

An exponential function is not constant, right? In fact, it is not even a simple linear equation of the form $y = mx + b$. If we know anything about the exponential function e, we know that either e^x gets bigger faster when x is positive and the magnitude of x becomes larger, or gets smaller quicker when x is negative and the magnitude of x becomes larger.

Observe the graphs of e^x in Figure 3.1 and Figure 3.2. We see that the slope of the function become larger as x becomes larger in the first figure and the slope becomes smaller as x becomes larger in the latter figure.

Now, what does a constant failure rate mean? It means that a component, system, or device that we are analyzing can fail randomly at any time, and the random nature of this failure does not depend on time, that is, it does not depend on how long the system has been in operation. Plenty of physical systems exhibit this property, including electrical components, and in certain cases, software. Constant failure rate models

Designing High Availability Systems: Design for Six Sigma and Classical Reliability Techniques with Practical Real-Life Examples, First Edition. Zachary Taylor and Subramanyam Ranganathan.

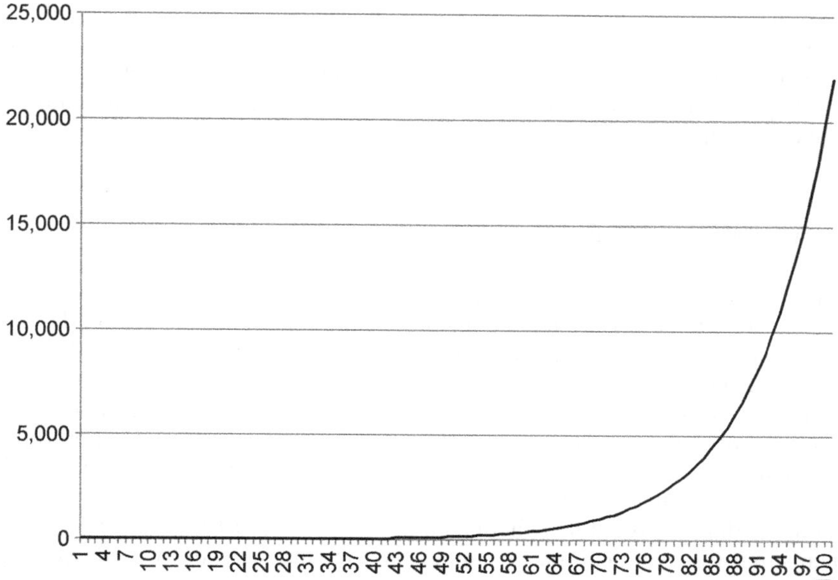

Figure 3.1 Exponential Function (Increasing)

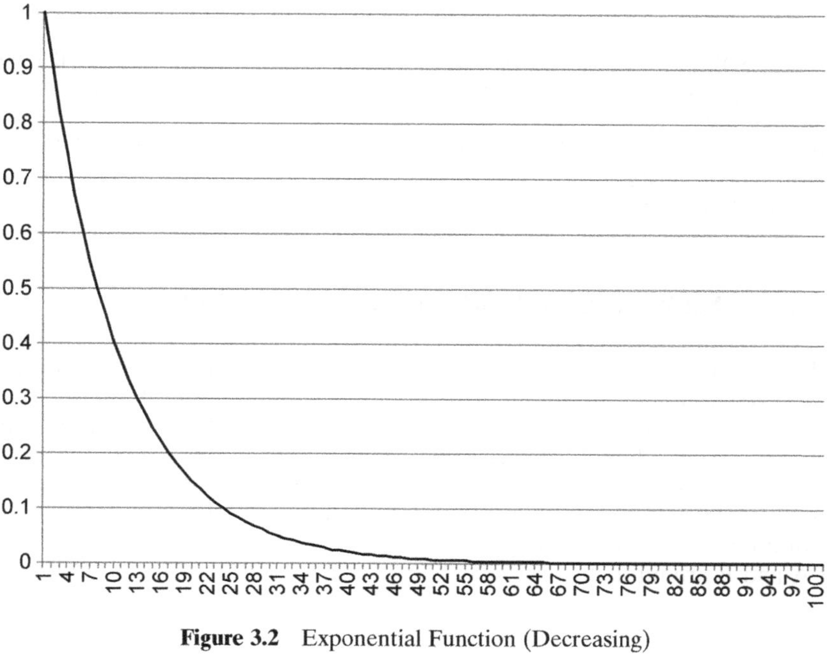

Figure 3.2 Exponential Function (Decreasing)

are popular since they are easier to analyze and actually provide very valuable information. Generally, the device, component, or system under consideration can be described by the constant failure rate model during its useful operational life.

We begin with a game of dice, which will help illustrate some fundamental properties that govern reliability engineering.

3.2 A GAME OF DICE

Using only one die, your task is to predict whether a roll of the die will result in a particular number. Since we have dots representing the numbers 1–6 on the die, one of those values will be displayed on the top face of the die after a roll. And yes, we are considering a perfect die—no hidden defects to make the die land more frequently on one number or another. So given all six numbers are equally likely, we ask you the probability of rolling a '5', for example. You may reason upon reflection since we have 6 choices with each of them being equally probable, then the '5' will on average show up one out of six times. Thus, the probability of a '5' will be 1/6. Remember that in probability theory, all probabilities must fall between 0 and 1. A 0 means that the event can never happen. A 1 means the event will always happen. All possible events that can happen are assigned a probability, and the sum of all of these probabilities must equal 1. Let us check that. The probability of rolling a *'1'* is 1/6, rolling a *'2'* is 1/6, and so on. Now add all these probabilities up:

$$1/6\,(\text{for a `}\mathit{1}\text{'}) + 1/6\,(\text{for a `}\mathit{2}\text{'}) + 1/6\,(\text{for a `}\mathit{3}\text{'}) + 1/6\,(\text{for a `}\mathit{4}\text{'}) + 1/6\,(\text{for a `}\mathit{5}\text{'}) + 1/6\,(\text{for a `}\mathit{6}\text{'}) = 1. \tag{3.1}$$

All 6 possible values of *'1'* through *'6'* sums to the probability of 1. So if I ask you what is the probability of rolling either a *'1'* or a *'2'* or a *'3'* or a *'4'* or a *'5'* or a *'6'*, you immediately can say confidently "probability of 1."

Now I ask you what is the probability of rolling either a *'5'* or a *'4'*. Oh, that is easy, too, you say. Just add 1/6 (for a *'5'*) and 1/6 (for a *'4'*) and we get 1/6 + 1/6 = 2/6 = 1/3. So one-third of the time, we can expect to roll a *'4'* or a *'5'*.

3.3 MUTUALLY EXCLUSIVE AND INDEPENDENT EVENTS

We digress briefly from our game to review two important probability concepts.

Two events *'A'* and *'B'* are *mutually exclusive* if when event *'A'* occurs, event *'B'* cannot occur and vice versa. When we roll a single die, if the event *'5'* occurs, this event excludes all other events occurring. Only one event can occur: either a *'1'* or a *'2'* or a *'3'* or a *'4'* or a *'5'* or a *'6'* occurs. The occurrence of this event precludes all other events from occurring. For example, throwing a die once can yield a *'3'* or *'4'*, but not both, in the same toss.

If A and B are mutually exclusive,

$$P(A \cup B) = P(A) + P(B). \tag{3.2}$$

Comparing Equation (3.1) with Equation (3.2), we note that Equation (3.1) is a specific example of the mutually exclusive rule.

So far so good—our intuition appears to be in good working order. Now let us extend the game by having you roll the same die twice. We ask you this question: If you roll the die twice (keeping track of the values after each roll), what is the probability of rolling two *'4'*s in a row? Ok, let us reason this out. We know that the first roll of the die will give a probability of 1/6 for any number, so for the first roll, we have a 1/6 probability of getting a *'4'*. Now we take the same die and roll it again. We reason that the roll of the die again should not be influenced by the value we obtained from the first roll. If we assume that the value of the first roll has no effect on the value of the second roll, we have 1/6 probability of getting a *'4'* on the first roll and we will have a 1/6 probability of getting a *'4'* on the second roll. We have 1/6th of 1/6th chance of obtaining two '4's in a row which is equivalent to multiplying these probabilities together (1/6) (1/6) = 1/36. We have one chance out of 36 (or ~2.8%) of obtaining two *'4'*s in a row.

We say the rolls of the die are *independent*; the roll of the die is not influenced by previous rolls of the die. When two events are *independent*, we multiply the probabilities together to obtain the joint probability that *both* events will occur.

If A and B are independent,

$$P(A \cap B) = P(A)P(B). \tag{3.3}$$

Note that if A and B are mutually exclusive, they cannot be independent since the occurrence of event A affects the occurrence of event B and vice versa. Conversely if A and B are independent, they cannot be mutually exclusive, since the event A has no impact on the occurrence of event B and vice versa.

Now let us pose a slightly different question. What is the probability that after two rolls, *'4'* appears at least once? Ok, thinking about this, we may say: For the first roll of the die we have a 1/6 chance of a *'4,'* and since the roll of the die the second time is not influenced by the value of the die we obtained after the first roll, we assign a 1/6 chance of a *'4'* for the second roll as well. Now since we have two chances of getting a *'4'* our odds should double then. So we add the two probabilities together. 1/6 + 1/6 = 1/3. We expect a 1/3 probability of seeing at least one *'4'* after rolling the die twice. Now we ask, "What is the probability that after 10 rolls of the die, a *'4'* appears at least once?" Ok, simple, let us add them up. 1/6 probability for each roll of the die times the number of times we roll the die, that is, 10 should give us the number. 1/6 + 1/6 + 1/6 + 1/6 + 1/6 + /1/6 + 1/6 + 1/6 + 1/6 + 1/6 = 10/6. "10/6 probability of getting a *'4'*", we proudly say.

Wait a minute! How can we have a probability that exceeds 1? After all, a probability value can only range from 0 (absolutely impossible) to 1 (happens always every time). Does obtaining 10/6 or 167% must mean that we are absolutely certain this will happen after 10 rolls and just to make sure we have added 67% padding on top for good measure? Well, something is not right. We know a probability of any event or series of events can never exceed 1. Where did we go wrong?

Start with the first roll. We did that part correctly: 1/6 probability of a *'4'* after the first roll. But now we roll again. Let us take a look at all of the possibilities (Fig. 3.3) where the row values represent the first roll and the column values represent the second roll of the die.

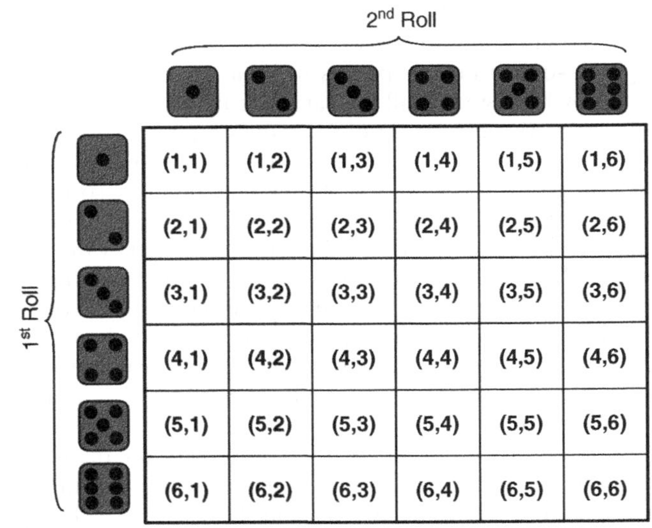

Figure 3.3 Possible Combinations of Rolling a Die Twice

We can see 36 possible combinations. Now count how many of these combinations have at least one *'4'*. We have highlighted those in Figure 3.4.

The number of combinations in which at least one *'4'* appears is 11. 11?! It should be 12, you say. Since the roll of the die are independent, the probability is 1/6 (first roll) and 1/6 (second roll), we should have 1/6 (die 1) + 1/6 (die 2) = 2/6 or equivalently 12 out of 36 possible combinations. So we are missing one combination. Where did it go?

This is where our naive intuition and probability theory can conflict. How can the roll of the die the second time be affected by the value shown after the first roll? Impossible you say. You are correct. The probability of rolling a *'4'* on the second roll is independent of the value obtained after the first roll. Then how did we get 11 out of 36?

Study Figure 3.3 again. We can confirm that a *'4'* is as equally probable on the first die roll as on the second die roll by first counting how many times a *'4'* shows up on the first die roll. Let us highlight those in Figure 3.5.

We have 6 of the 36 combinations in which the first die shows a *'4'*, which is 1/6 and in agreement with the probability of rolling a die once as we previously discussed.

Now count how many times a *'4'* shows up on the second die roll—highlighted in Figure 3.6.

For the second roll, we also have 6 of the 36 combinations in which a *'4'* appears corresponding to 1/6 and in agreement with the probability obtained by rolling the die a second time as we previously discussed.

But as we noted, they do not add up. Six out of 36 for the first die roll and 6 out of 36 for the second die roll does not equal 12 out of 36, but instead equals 11 out of

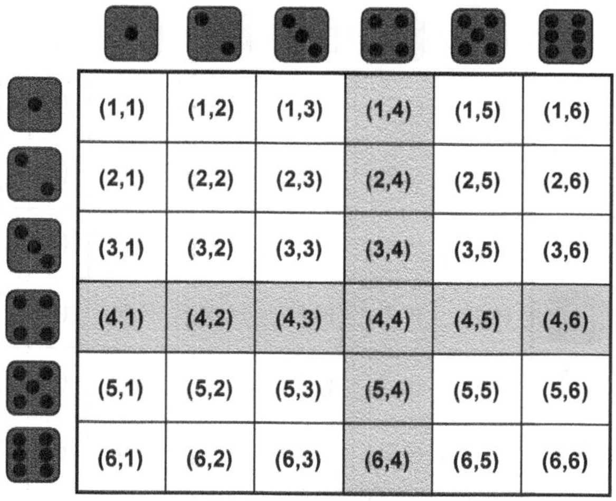

Figure 3.4 Combinations with at Least One *'4'*

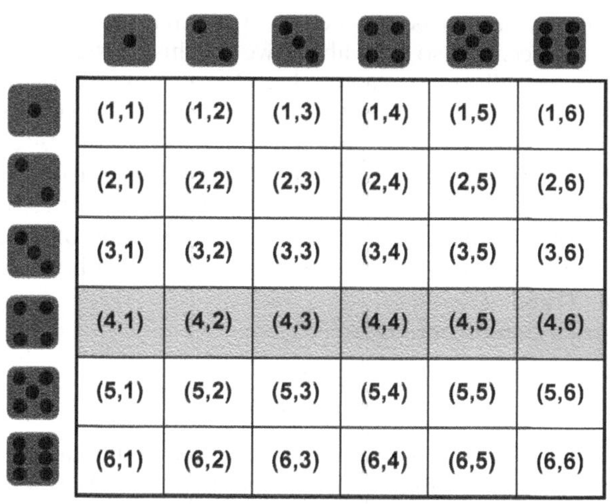

Figure 3.5 Combinations in which the First Die Is a *'4'*

36. Examine Figure 3.6 again. You will notice that one of the combinations is shared by both rolls of the die. That is the combination (4,4) highlighted in Figure 3.6.

This combination is shared by both events and we can only count it *once*. Or you might say, throw it in the count for the first roll or throw it in the count for the second roll, but not for both—do not double count it. So let us arbitrarily throw it in the count for the first roll and exclude it from the count for the second roll. That means the count for the first roll is 6 and the count for the second roll is 5 (cannot count the (4,4) shared event twice). The total is 11.

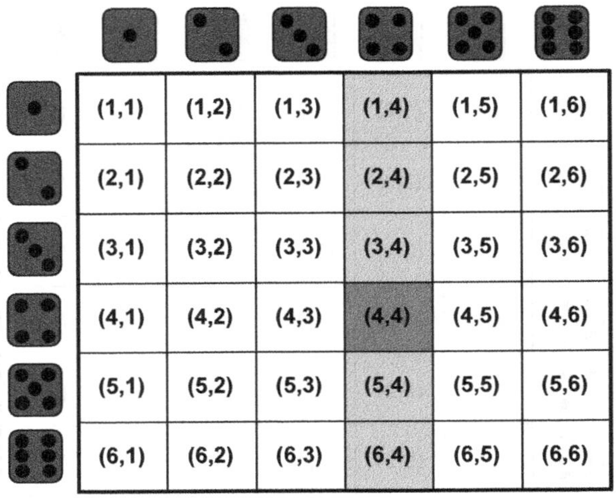

Figure 3.6 Combinations in which the Second Die Is a *'4'*

Now at this point you may be inclined to argue that we should indeed count the (4,4) combination twice. Ok, so how about we do this? Instead of rolling the same die twice, let us roll two different colored dice. What would our figure look like then?

Figure 3.7 shows the (4,4) combination. It should be clear that a second combination of (4,4) cannot be added. But perhaps you are still not quite convinced. How about we add the combination of the second die showing *'4'* followed by the first die showing *'4'*, you say. Ok let us do that. Now what is the difference between (**4**,4) and (4,**4**)? There is none. Both represent the exact same pair of dice with the same number. That *combination* can occur in only one way. Then how come we have two combinations of *'3'* and *'4'*?

Ok, let us extract them from Figure 3.7:

(**4**,3)
(3,**4**)

Is there a difference between them? Yes. In the first combination, the first die is a *'4'*, whereas in the second combination, the first die is a *'3'*. We can tell these combinations apart from the color of the die. But in the case of (**4**,4), there really is only one combination of 4 and 4 taken together whenever we roll the dice and double 4s show up. In dice games, that's why doubles are so valuable. There is only one way to throw a double *'4'* with two dice. However, there are two ways to throw a *'3'* and a *'4'* (first die *'4'*, second die *'3'*) or (first die *'3'*, second die *'4'*). The probability of a pair of dice showing a (3,4) (or vice versa) is 2 out of 36, double the probability of rolling double *'4'*.

Now that we hopefully have convinced ourselves that the probability of rolling *at least* one *'4'* after 2 rolls of a die is indeed 11/36, let us expand our investigation to other possibilities. As an exercise, take a look at Figure 3.3 and prove for yourself that the probability of rolling at least one *n* where *n* can be 1, 2, 3, 4, 5, or 6 twice in a row using a single die is 11/36.

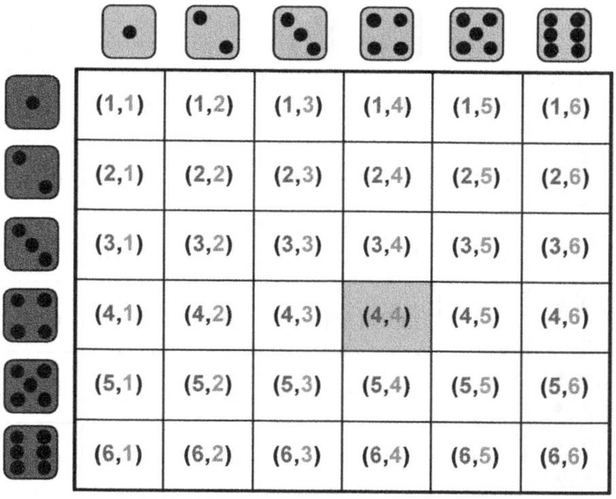

Figure 3.7 The Double *'4'* Combination

Nonmutually exclusive events are events that share one or more of the same outcomes. Rolling a *'4'* on the first die does not preclude us from rolling a *'4'* on the second die, thus the probability of rolling any number twice in a row is a nonmutually exclusive probability.

For the double combinations, when event *'A'* occurs (at least one *'4'*), event *'B'* also occurs (at least one *'4'*). Since we cannot double count these, that is, we can only count this combination once, we need to subtract one of the double-event probabilities from the probability of event A occurring and the event B occurring.

We summarize the general case in the rule for nonmutually exclusive events. For two events *'A'* and *'B'* that are not mutually exclusive, the probability of at least one of these events occurring is captured in the following probability theorem:

If A and B are nonmutually exclusive,

$$P(A \cup B) = P(A) + P(B) - P(A \cap B). \tag{3.4}$$

Let us test this using our dice game and determine the probability of rolling a *'4'* at least once in two rolls of a single die. Inserting values for A and B into Equation (3.4):

$$P(A \cup B) = 1/6 + 1/6 - 1/36 = 11/36. \tag{3.5}$$

This is the same result as we got from counting combinations in Figure 3.4!

3.4 DICE PARADOX PROBLEM AND CONDITIONAL PROBABILITY

A paradox is simply something that is counterintuitive. Let us explore another example where we describe how to increase the odds of winning if we bet on a roll of 7.

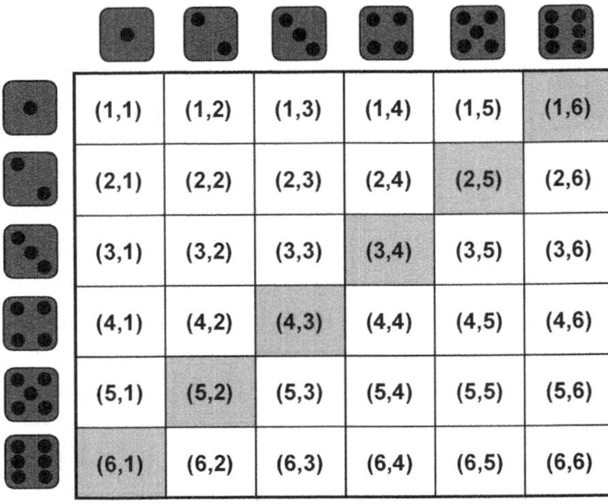

Figure 3.8 Dice Roll of Seven

The probability of rolling a sum of 7 with 2 dice is 1/6. We can verify this by counting the combinations highlighted in Figure 3.8. We see six events out of a total of 36 possible events in which the sum is 7. If we bet on 7 for any random roll of two dice, we would have a chance of winning 1/6 of the time. Now let us say we want to improve our odds of winning. We have an impartial observer take a look at the dice after they are rolled, but we are not allowed to see the result. Let us say our favorite number is *'4'*, so anytime our assistant sees a *'4'* appear on at least one of the die, she shouts out "I see a four!" With this extra bit of information provided before we place our bet, we have improved our odds of winning from 1 in 6 to 2 in 11!

How can that be? Let us go back to our faithful figure (Fig. 3.9). We have shaded the combinations containing at least one *'4'*.

We have 11 cells that indicate that 11 out of 36 times the value *'4'* appears in a roll of two dice. Of those shaded cells, how many have a sum equal to 7? Exactly 2. Since we only consider betting on those dice rolls in which the number *'4'* appears on at least one of the die, we have only 11 possibilities. Of those 11 possibilities, only 2 can sum up to 7, thus our probability of winning when we bet on 7, given that our trusty assistant has seen a *'4'*, is now 2/11.

This extra information is captured in the conditional probability theorem:

Conditional probability:

$$P(A \mid B) = \frac{P(A \cap B)}{P(B)}. \tag{3.6}$$

And can be read as: Given that event B has occurred, the probability of event A occurring is equal to the probability of both A and B occurring divided by the probability of B occurring.

(1,1)	(1,2)	(1,3)	(1,4)	(1,5)	(1,6)
(2,1)	(2,2)	(2,3)	(2,4)	(2,5)	(2,6)
(3,1)	(3,2)	(3,3)	(3,4)	(3,5)	(3,6)
(4,1)	(4,2)	(4,3)	(4,4)	(4,5)	(4,6)
(5,1)	(5,2)	(5,3)	(5,4)	(5,5)	(5,6)
(6,1)	(6,2)	(6,3)	(6,4)	(6,5)	(6,6)

Figure 3.9 Dice Combinations of *'4'* and *'3'*

For our dice problem, let event B represent the probability that a *'4'* appears on at least one of the die. Let event A represent the probability of the sum of the two dice is 7.

$P(B)$ = Probability of a *'4'* appearing on at least one of the die = 11/36 (highlighted in Fig. 3.9).

$P(A \cap B)$ = Probability of both A and B occurring = 2/36 (underlined in Fig. 3.9). Thus $P(A|B) = (11/36)/(2/36) = 2/11$.

The conditional probability we obtained is the same as we got by carefully considering each combination in Figure 3.9.

We can extend this problem to one of a more general nature. If we choose any number from 1 to 6 beforehand and call that number out when we see it appear on at least one of the two dice, the probability will remain the same. Pick a number and check the list.

Now to make this even more strange and counter intuitive, how about we randomly pick any number from 1 to 6 and check our probabilities again? Picking a random number will not make any difference to the probabilities. So even if we somehow accept that a specific number may influence events, how can choosing a random number before each roll of the dice and calling out that number if it appears, increase the probability that the sum of 7 will occur?

We can check these various scenarios by simulation. The tool we used for the simulation is Matlab® (refer to Section 14.16.5). The Matlab program (DiceRoll.m) shown in Figure 3.10 was executed using a simulation based on 100,000 rolls of the dice.

The simulation results after one particular run are shown in Figure 3.11. Actually, picking a random number should be no different than picking a specific number. If we pick *'4'*, we expect the number to show up 1 out of 6 times for the first die and 1 out of 6 times for the second die. If we pick a random number, it, too,

```
% Simulation to determine probability of rolling a 7 with 2 dice when one of
% the die shows the number 4

simsize = 100000; %number of dice rolls
dieDist = [];
count7 = 0; %count of total number of 7's rolled during the simulation
Num4countcond = 0;
anyNumcountcond = 0; %count of total number of times a randomly chosen
                     %number 1-6 shows up after each roll of the dice
Num4Firstcountcond = 0; %count of total number of times a '4' appears on the
                        % first die.
bet7cond4 = 0; % count of total number of times we bet on a 7 given that a
               % a '4' is shown on one of the die
bet7condAny = 0; %count of total number of times we bet on a 7 given that
                 % anyNumcountcond occurs
bet7cond4First = 0; %count of total number of times we bet on a 7 givent
                    %that the first die shows a '4'

dieRolls = ceil(rand(3,simsize)*6); %simulate the roll of 3 dice over the
                                    %simulation period.
dieSum = [];

i = 1:simsize;
dieSum(1,i) = dieRolls(1,i) + dieRolls(2,i);
for i = 1:1:simsize
    if dieSum(i) ==7
        count7 = count7 + 1;
        if ((dieRolls(1,i) == dieRolls(3,i)) || (dieRolls(2,i)==dieRolls(3,i)))
            anyNumcountcond = anyNumcountcond+1;
        end
         if((dieRolls(1,i) == 4) || (dieRolls(2,i)== 4))
           Num4countcond = Num4countcond+1;
         end

        if(dieRolls(1,i) == 4)
           Num4Firstcountcond = Num4Firstcountcond+1;
        end

    end

    % Assistant says at least one of the die shows the rand number I just picked.
    if ((dieRolls(1,i) == dieRolls(3,i)) || (dieRolls(2,i)==dieRolls(3,i)))
        bet7condAny = bet7condAny+1;
    end

    % Assistant says at least one of the die shows a '4'
    if((dieRolls(1,i) == 4) || (dieRolls(2,i)== 4))
          bet7cond4 = bet7cond4+1;
    end

    % Assistant says the first die shows a '4'
    if(dieRolls(1,i) == 4)
          bet7cond4First = bet7cond4First+1;
    end
end

bet7CondProb4 = Num4countcond/bet7cond4;
bet7CondProbAny = anyNumcountcond/bet7condAny;
bet7CondProb4First = Num4Firstcountcond/bet7cond4First;
bet7Prob = count7/simsize;
display (6/36);
display (2/11);
display (bet7Prob);
display (bet7CondProb4);
display (bet7CondProbAny);
display (bet7CondProb4First);
```

Figure 3.10 Matlab Dice Simulation Program

```
                                           Simulation  Theory

Probability of rolling a sum of 7:         0.1661      0.1667 (1/6)
Probability of rolling a sum of 7
     if at least one die is a four:        0.1800      0.1818 (2/11)
Probability of rolling a sum of 7
     if at least one die is the
     value of a random number:             0.1850      0.1818 (2/11)
Probability of rolling a sum of 7
    if the first die is a four:            0.1649      0.1667 (1/6)
```

Figure 3.11 Dice Simulation Results

should show up one out of six times for the first die and one out of six times for the second die.

From the simulation results in Figure 3.11, we can see that the probability of rolling a sum of 7 if at least one die is a four and the probability of rolling a sum of seven if at least one die is the value of a random number are fairly close to the theoretical probability of 2/11.

Now that we have convinced ourselves that the probability in this scenario is indeed 2/11, let us instead have our assistant tell us only when the value of the first die thrown is a *'4'*. The probability of correctly predicting a sum of seven drops from 2/11 to 1/6. Why?

Heuristically, we can say that the observer is prevented from providing any information about the second die, so we have reduced the amount of information available to us. The probability of a *'4'* appearing on the first die is 1/6. With that information, we restrict the possible value of the sample set to 6 values: the six possible values of the second die. These available values are shaded in Figure 3.12.

From this reduced sample set, only one of the events results in a sum of seven. Plugging in the numbers using our conditional probability rule Equation (3.6):

$$P(A \mid B) = \frac{1/36}{6/36} = 1/6. \tag{3.7}$$

So in this case, the information provided on the first die did not help improve our odds.

Now, to help understand why the probability of a sum of seven increases when we are given the event that a *'4'* appears on at least one of the dice, consider the following: We asked the question if a *'4'* appears on *at least* one of the dice, so the *'4'* showing up once or twice satisfies the condition equally and the probabilities of any of the events in the sample space of the two dice are the same.

We can conclude that although the roll of each die is independent, the events of interest are not mutually exclusive.

For our dice, as previously discussed, any throw of a double is a shared event between die 1 and die 2. If the dice were mutually exclusive, then any time we rolled a *'4'* for example, the other die would have a zero probability of being a *'4'*. We could construct mutually exclusive dice. Let us keep die 1 the same with value 1–6 on the 6 faces, but now let us change the values of die 2 to have the values 7–12. Now the dice are mutually exclusive. No matter what we roll with die 1, die 2 can never have the same value.

(1,1)	(1,2)	(1,3)	(1,4)	(1,5)	(1,6)
(2,1)	(2,2)	(2,3)	(2,4)	(2,5)	(2,6)
(3,1)	(3,2)	(3,3)	(3,4)	(3,5)	(3,6)
(4,1)	(4,2)	(4,3)	(4,4)	(4,5)	(4,6)
(5,1)	(5,2)	(5,3)	(5,4)	(5,5)	(5,6)
(6,1)	(6,2)	(6,3)	(6,4)	(6,5)	(6,6)

Figure 3.12 Combinations in which the First Die Is a *'4'*

We have hopefully shown to the reader's satisfaction that a four will occur 11 times out of the 36 possible combinations of throws in general. From a probability perspective, a double-four is not any more valuable than, let us say, a (6,4), when we are searching for the probability of finding a four. That is because of the way the problem is phrased. Since the event (4,4) occurs with equal probability as (1,4), the fact that the value we are looking for shows up in both dice is irrelevant to the problem.

Consider that when we look for a *'4'* and call it out when it appears, it overlaps with the events when we look for numbers other than *'4'* for all possibilities except for double-four. The double-four shows up both in die 1 and die 2, but is only counted once, thus reducing the frequency of the event "at least one die shows a four."

What this means is that the chances of at least one *'4'* showing up when we roll the dice is 11/36, a slightly more rare event than the "intuitive" but incorrect answer of 1/6. Since this event is rarer, our sample space for conditional probability is smaller, that is, our sample space is now 11 events. Of these 11 events, 2 pairs have a sum of 7, thus the probability is 2/11.

What if instead of picking a random number before each dice throw to see if it shows up, let us instead just look at the pair of dice and tell the gambler one of the values shown. We can arbitrarily pick either of the dice and shout out its value. Using our conditional probability equation, let us see how this affects the odds.

Let event B be the probability of the number we see on one of the dice being correct. Since we see the die and report its value, the probability is exactly 1 every time. $P(A \cap B)$ then reduces to $P(A)$, which is the probability of getting a sum of seven.

$$P(A \mid B) = \frac{P(A \cap B)}{P(B)}$$

$$P(A \mid B) = \frac{1/6}{1}$$

$$P(A|B) = 1/6. \tag{3.8}$$

So even if we call out the value of on every roll, our chances of getting a 7 have not improved.

Another way to envision the problem is to note when we indicate a *'4'* has occurred, we immediately remove all possible events in which a *'4' cannot* occur, dramatically reducing our sample set from 36 to 11. And since any number 1–6, on any die can be used to create the sum 7, we get two chances out of the 11 possibilities to make the sum 7.

Okay, enough about dice. Let us flip a coin instead!

3.5 FLIP A COIN

It might help in understanding the dice paradox if we take a look at a simpler but similar situation. Instead of two six-sided dice, let us toss two coins. Now we arbitrarily assign a value of one for *"heads"* and a value of two for "*tails*." Then we pose the question: What is the probability on any given toss of both coins that the sum of the two coins will be three?

Let us create a matrix of possible events (Fig. 3.13). We see that 50% of the time, the sum will be 3.

Now suppose the coins are hidden from view so that we cannot see the results of the coin toss. We ask our assistant to look at the coins after they are tossed and let us know that if at least one of the coins is a head. Now what are the odds that the sum is three?

Given that at least one of the coins is heads, the probability has now magically increased to 67%. This is because our sample space reduces to 3.

We should *always* bet on sum = 3 whenever our assistant calls out at least one of the coins is a head! We could instead ask our assistant to call out whenever at least one of the coins is a tail. The odds of getting a sum of 3 is the same. Wait a minute, you say. If the chances of getting a head on any single coin toss is 50% and the chances of getting a tail on any single coin toss is 50%, then why if our assistant calls out *"heads!"*, we win 2/3 of the time, and when our assistant calls out *"tails!"*, we win 2/3 of the time? They both can't be right, can they?

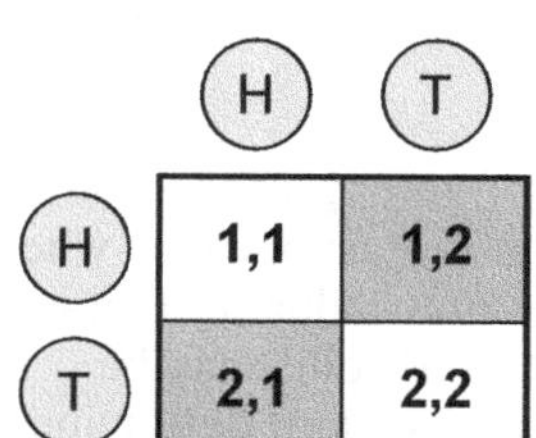

Figure 3.13 Possible Outcomes of Flipping Two Coins

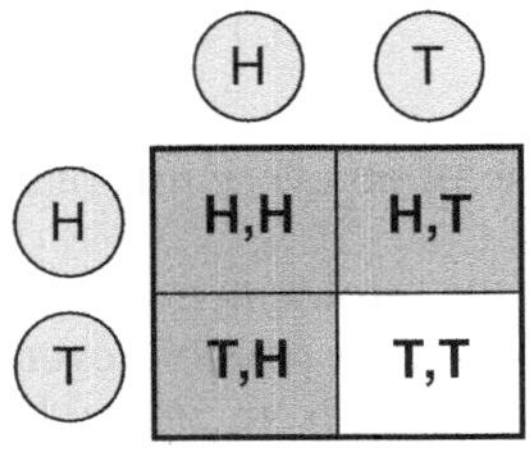

Figure 3.14 Combinations in which at Least One "H" Occurs

We can help explain the mystery by considering the case of how often our assistant will say that one of the coins is a head. From Figure 3.14, we see that three out four possibilities contain at least one head, so 75% of the time our assistant will call out "heads" and only then will we place our bet, because we know that the double tails scenario cannot be included in the set of possibilities we consider.

Similarly, if we ask our assistant to call out *"tails!"* whenever at least one of the coins is a tail, then 75% of the time she will call out *"tails!"* and 25% of the time our assistant will be mute, that is, one in four chance both coins are a head.

Now we can see why. Since the values shown on the two coins are not mutually exclusive, all combinations are possible for seeing at least one head except for the event in which both coins are tails. If both coins are tails, they are excluded from the sample space, thus increasing the odds of the sum of 3 occurring in the reduced sample space. The knowledge of at least one head means we don't need to consider the case of both tails, which sums up to 4. But you say a head or a tail occurs with equal frequency, so how can we get rid of both tails or both heads?

The probability of the four events is based on the assumption that heads or tails occur with equal likelihood. The key is the events that occur, which allow us to indicate a head has occurred. The event in which the assistant calls out one of the coins is a head overlaps with the event one of the coins is a tail:

Event	Watch for Heads	Watch for Tails
(H,H)	*"heads!"*	
(H,T)	*"heads!"*	*"tails!"*
(T,H)	*"heads!"*	*"tails!"*
(T,T)		*"tails!"*

Although the probability of getting a head on any given coin toss is 50%, the probability of getting a head on at least one coin when two coins are tossed increases to 75%. Thus, the latter case is more valuable, that is, provides more information that helps us determine what the tossed coins may be.

What happens if we have our assistant determine whether to call out *"heads!"* or *"tails!"* by flipping a third coin prior to the toss of the two coins under examination? If coins 1 and 2 are not identical, call out the state of the third coin; also, if coins 1 and 2 are identical and coin 3 is the same, call out the state of the third coin. Since flipping a single coin results in a 50% chance of either a head or tail, we will modify our event table to include the third coin:

Event	First, Second Coins	Third Coin	Assistant Reports
1	(H,H)	(H)	*"heads!"*
2	(H,H)	(T)	
3	(H,T)	(H)	*"heads!"*
4	(H,T)	(T)	*"tails!"*
5	(T,H)	(H)	*"heads!"*
6	(T,H)	(T)	*"tails!"*
7	(T,T)	(H)	
8	(T,T)	(T)	*"tails!"*

For these eight possible combinations, if we bet only when our assistant calls out heads or tails (events 1, 3, 4, 5, 6, and 8), 1 and 8 are loser bets (recall we were betting on a sum of 3; we gave the value of 1 to a head and 2 to a tail). Thus, we win 6 out of 8 times (2/3), which is the same for the nonrandom scenario.

What if we change the rule for our assistant and have her tell us *"heads!"* when both coins are heads, *"tails!"* when both coins are tails, and when one of the coins is a head and the other is a tail, 50% of the time tell us *"heads!"* and the other 50% of the time, tell us *"tails!"*

Now our table looks like this:

Event	First, Second Coins	Assistant Reports		
1	(H,H)	*"heads!"*		
2	(H,T)	*"heads!"*	-OR-	*"tails!"*
3	(T,H)	*"heads!"*	-OR-	*"tails!"*
4	(T,T)			*"tails!"*

Now what information can we extract to make our bet? You can see from the table above that if the assistant calls out *"heads!"* and we bet on heads, we win only 50% of the time. If our assistant calls out *"tails!"* and we bet on tails, we also win only 50% of the time. This is because what the assistant says is not adding any extra information.

3.6 DICE PARADOX REVISITED

From the discussion on two-coin toss example, the two-dice paradox is simply a 6×6 version of the 2×2 coin toss. For both examples, the information provided after the

(1,1)	(1,2)	(1,3)	(1,4)	(1,5)	(1,6)
(2,1)	(2,2)	(2,3)	(2,4)	(2,5)	(2,6)
(3,1)	(3,2)	(3,3)	(3,4)	(3,5)	(3,6)
(4,1)	(4,2)	(4,3)	(4,4)	(4,5)	(4,6)
(5,1)	(5,2)	(5,3)	(5,4)	(5,5)	(5,6)
(6,1)	(6,2)	(6,3)	(6,4)	(6,5)	(6,6)

Figure 3.15 Combinations of Only One '4' Showing Up

event reduces the number of possible events that need to be considered, thus increasing our chances of guessing correctly that a particular event (or events) has occurred.

Can we improve our chances even more? Sure we can! Recall earlier we were betting on the sum of the dice to be 7. The probability of getting 7 increases to 1/5 if we change the condition to call out the number four only when either die shows a '4', but not when both dice show a '4'. Now we have 10 events in which this new condition is true (Fig. 3.15).

Plugging into our conditional probability formula:

$$P(A \mid B) = \frac{P(A \cap B)}{P(B)}$$

$$P(A \mid B) = \frac{2/36}{10/36}$$

$$P(A|B) = 1/5. \tag{3.9}$$

So, place your bets!

3.7 PROBABILITIES FOR MULTIPLE DICE THROWS

Now that we have hopefully satisfied ourselves that the probability of rolling at least one '4' on two rolls of a single die is 11/36, let us extend this to n throws of a single die.

Another way to look at the two die throw scenario is as follows:

$$P(\text{first throw is '5'}) = 1/6. \tag{3.10}$$

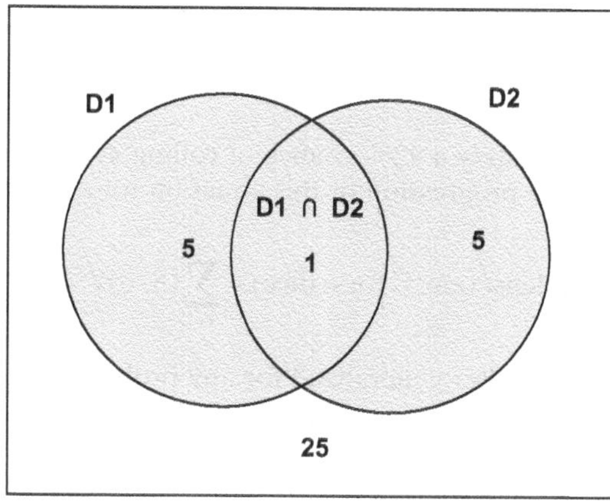

Figure 3.16 Rolling a '5' of Die 1 or Die 2

Now the probability of throwing at least one '5' after two throws is the same as 1 minus the probability of *not* rolling at least one 5 after two throws, that is, *P(at least one '5' after two throws) = 1 – P('5' not appearing after two throws)*.

P(not rolling a '5' after two throws) = *(5/6)(5/6)* = 25/36, that is, five out of six times you will not roll a '5' on the first roll and five out of six times you will not roll a '5' after the second roll. These probabilities are independent and thus multiplied together. Thus, the probability of rolling at least one '5' after two throws is 1 – (25/36) = 11/36.

Consider the probability of rolling a '5' on the first roll is 1/6, then the probability of rolling a '5' on the second roll is also 1/6. The roll of the first die is independent of the roll of the second die. However, we cannot therefore say that the probability of rolling at least one '5' after the die is rolled twice is: 1/6 + 1/6 = 1/3. How come? This is a very common mistake, and as we will continue to find out, probability can be counterintuitive.

A good way to picture this problem is by using a Venn diagram (Fig. 3.16). D1 represents the number of combinations a '5' shows up on the first die. Similarly, D2 represents the number of times a '5' shows up on the second die. D1 $\cap$ D1 represents the number of times both dice show '5' simultaneously.

So we have 5/36 (probability a '5' is rolled on the first die but not on the second) plus 1/36 (probability a '5' is rolled on both dice), plus 5/36 (probability a '5' is rolled on the second die but not the first die.

Written another way, the probability of rolling a '5' on the first die (or first roll) is 6/36 *P*(at least one '5' after two throws) = 1/6 + 1/6 – 1/36, which can be also be written as:

$$P(\text{at least one '5' after two throws}) = 1/6 + (5/6)(1/6) = 11/36. \qquad (3.11)$$

The possibility of getting a '5' on the first die and a '5' on the second die does not affect the calculation since we are considering *P*(at least one '5'), which is satisfied if we get one '5' or two '5's.

Let us extend this to rolling three dice at once. Now we get:

$$P(\text{at least one '5' for 3 dice}) = 1/6 + (5/6)(1/6) + (5/6)(5/6)(1/6) = 0.4213, \quad (3.12)$$

that is, we have approximately a 42% chance of rolling at least one '5'.

We can generalize the progression of this equation for n dice:

$$P(\text{at least one '5' for } n \text{ dice}) = \sum_{x=1}^{n} (5/6)^{x-1}(1/6). \quad (3.13)$$

Equation (3.13) can be further generalized for any probability:

$$P(X \le x) = P(\text{at least one event occurs after } n \text{ trials}) = \sum_{x=1}^{n} (1-p)^{x-1} p, \quad (3.14)$$

where $P(X \le x)$ is the probability that event x occurs within n attempts and p is the probability of the event x occurring during a single attempt.

The above equation is the cumulative geometric distribution.

Now, let us say the probability of having a "success" on the first attempt is p. In general, to succeed for the first time on the nth attempt, we would have had $n - 1$ previous failures. If at first you do not succeed, try again! Since each subsequent event is independent, any sequence having $n - 1$ failures followed by a success occurs with probability $(1 - p)^{n-1}p$. Thus, the *probability* of succeeding for the first time on the nth attempt is given by:

$$P(X = n) = P(\text{event with prob } p \text{ occurs after the } n\text{th trial}) = (1-p)^{n-1}(p). \quad (3.15)$$

The distribution for the probability represented by Equation (3.15) is a *geometric distribution.*

The probability of rolling the first '5' after n throws of a single die:

$$P(\text{rolling a '5' on the } n\text{th throw}) = (5/6)^{n-1}(1/6). \quad (3.16)$$

What is the probability of *not* rolling a '5' after n trials?

$$P(\text{not rolling a '5' after the } n\text{th throw attempt}) = (5/6)^{n}. \quad (3.17)$$

Looking at it in a different way, the probability of rolling at least one '5' after three rolls is the same as 1 minus the probability of not rolling a '5' after three rolls:

$$P(\text{at least one '5' after 3 rolls}) = 1 - (5/6)(5/6)(5/6) = 0.4213. \quad (3.18)$$

Generalizing this equation for n attempts:

$$\begin{aligned} &P(\text{rolling at least one '5' after the } n\text{th throw attempt}) \\ &= 1 - P(\text{not rolling a '5' after the } n\text{th throw attempt}) \\ &= 1 - (5/6)^{n}. \end{aligned} \quad (3.19)$$

Or considering any general success event after n trials:

$$\begin{aligned}P(\text{at least one success in } n \text{ trials}) &= 1 - P(\text{event } x \text{ occurs at least once in } n \text{ trials})\\ &= 1-(1-p)^n.\end{aligned} \tag{3.20}$$

This is the closed form equivalent *cumulative distribution function* (*cdf*) of Equation (3.14). Equations (3.14) and (3.20) are equivalent:

$$\sum_{x=1}^{n}(1-p)^{x-1}p = 1-(1-p)^n. \tag{3.21}$$

3.8 CONDITIONAL PROBABILITY REVISITED

Let us start with a sample space S, representing one of several possible sample spaces. Now let us define the probability of an event A occurring by $P(A)$. Now what is the probability of event A occurring within this sample set S? In other words, we define $P(A|S)$ as the probability of event A, given that event S has occurred.

EXAMPLE 3.1

After manufacture, 1000 flashlights are inspected and tested prior to shipment. Twenty of these fail the electrical test and 17 of the 1000 are missing a screw.

Our sample space, S, is the set of 1000 flashlights. Let $N(S)$ = total number of flashlights being evaluated, $N(D)$ = the number of defective flashlights in the sample, and $N(M)$ = the number of flashlights in the sample missing a screw. Let us further define $P(D|S)$ as the probability that a flashlight does not work properly given the sample space of 1000 flashlights, and $P(M|S)$ is the probability that a flashlight is missing a screw given the sample space of 1000 flashlights.

The Venn diagram in Figure 3.17 shows the relationship between the sample spaces and events. Using the diagram as a guide, and referring to conditional probability Equation (3.6), if we randomly choose a flashlight, the probability that it does not work is:

$$P(D \mid S) = \frac{N(D)}{N(S)} = \frac{20}{1000} = 0.02. \tag{3.22}$$

And the probability that the flashlight is missing a screw is:

$$P(M \mid S) = \frac{N(M)}{N(S)} = \frac{17}{1000} = 0.017. \tag{3.23}$$

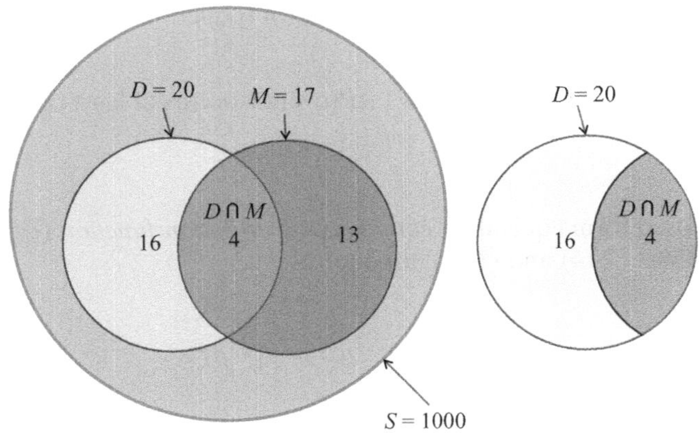

Figure 3.17 Flashlights Venn Diagrams

Now let us consider the following. Given that we have selected a flashlight that does not work, what is the probability that it is also missing a screw? Our sample space is now reduced from 1000 to the 20 we know do not work. Referring back to the Venn diagram, four flashlights are both defective and have a missing screw.

If we divide the numerator and denominator by the size of the sample set, we get $P(M|D)$ in terms of probability relationships:

$$P(M \mid D) = \frac{\frac{N(D \cap M)}{N(S)}}{\frac{N(D)}{N(S)}}$$

$$P(M \mid D) = \frac{P(D \cap M)}{P(D)}.$$

We have derived the conditional probability Equation (3.6). This equation applies generally to any event M and any event D within a sample space S.

Continuing with the example, we substitute the values into Equation (3.6) and obtain:

$$P(M \mid D) = \frac{4}{20} = 0.2. \tag{3.24}$$

We read the above equation as follows: Given that we have selected a defective electrical switch, the probability that this flashlight is also missing a screw is 0.2.

3.9 SUMMARY

This chapter introduced some of the fundamentals of probability theory illustrated by our analysis of a game of dice and coin flip experiments. By analyzing these familiar examples in greater depth, using familiar illustrations, we have hopefully been able to shed some light on some fundamental concepts of probability theory. These concepts are the building blocks for more practical and varied applications of reliability analysis.

CHAPTER 4

Discrete Random Variables

4.1 INTRODUCTION

Probability concepts are fundamental to the study of high availability systems. In this chapter, we will build on the concepts introduced in the previous chapter, define random variables, and then introduce several discrete probability distributions of random variables that play an important role in the analysis of high availability systems.

If we flip on a light switch, we expect the light to turn on. In the exceedingly vast majority of times we attempt this, the light will turn on. If it fails to turn on, then we search for reasons for the failure such as the light bulb burned out, the light bulb is missing, the switch is defective, and so on. Barring these extraordinary circumstances, we expect that the light will turn on.

This type of experiment is *deterministic*. When a certain set of conditions are valid (power available, light bulb working, switch working, and switch flipped on), the light turns on.

Another type of experiment is *probabilistic*. We cannot ascertain the result of any one particular experiment, but we may be able to draw some useful conclusions when the experiment is repeated a large number of times. As is typically the case, given the light was previously working, what is the chance that it will not work the next time I flip the switch? Any one experiment may not provide much information. But if the probability of failure remains the same for each attempt, we can repeat the experiment (light on/off) many times and perhaps be able to draw some conclusions on the likelihood of failure and certain failure attributes for the general case of any electrical system with the same type of switch and light bulb.

Suppose we are interested in finding the probability of a device failure during a period of time. This time could be the duration of the test time in a laboratory environment or the mission time of a deployed device. We can empirically calculate the probability of failure during this time interval by counting the number of devices that failed and divide that by the number of devices under test. We test N devices, and of

Designing High Availability Systems: Design for Six Sigma and Classical Reliability Techniques with Practical Real-Life Examples, First Edition. Zachary Taylor and Subramanyam Ranganathan.

these, n devices fail. Let us define F as the event that a device has failed. Then we can estimate the probability of failure of the device by:

$$P_{est}(F) = \frac{n}{N}. \tag{4.1}$$

As we increase the number of devices under test, our confidence that we find the true probability of failure of the device increases. Formally, the probability of failure is defined as:

$$P(F) = \lim_{N \to \infty} \frac{n}{N}. \tag{4.2}$$

Taking another example, if we flip a coin repeatedly and count the number of heads that appear, we would expect that after an infinite number of trials, N, that the probability of heads appearing would converge to 0.5.

4.2 RANDOM VARIABLES

We can extend the basic probability concept by defining multiple events for a particular experiment. For example, we may be interested in the probability that one device has failed during testing, or two of them have failed during testing, and so on. Or we may be interested in the probability that a device survives (continues to function) after a certain time in operation. For a coin toss experiment, we may want to know the probability that three heads in a row appear during a coin toss.

A *random variable* $X(\varphi)$ is a real function that maps each sample event φ that is in a sample space, S, of a random process, to a real number. We classify random variables as discrete or continuous. A discrete random variable has a finite or countable infinite number of values. A continuous random variable has an uncountable infinite number of values. Let us consider discrete random variables first. A *discrete random variable* X is a variable whose value is taken from a set of possible discrete values we have defined. For the finite case, those values are: $(x_0, x_1, x_2, \ldots, x_n, \ldots, x_N)$. These values must be real numbers in order to characterize the resulting probability distributions. We will use a lower case letter to denote the value of a random variable and an upper case letter to denote the random variable. The set of events we are interested in $(\varphi_1, \varphi_2, \ldots, \varphi_M)$ must be mapped to these discrete values $(x_0, x_1, x_2, \ldots, x_n, \ldots, x_N)$ by the random variable X.

The set of discrete values associated with this mapping is referred to as the value of $X(\varphi)$. The random variable represents a set of random values that are possible for a particular sample space. Typically, $X(\varphi)$ is abbreviated to X (the sample point parameter is omitted for brevity), and X represents all possible values in the sample set. The sample space S is called the *domain* of the random variable, and the set of all values of X is called the *range* of the random variable. Note that two or more different sample values may give the same random value of $X(\varphi)$, but two different numbers in the range cannot be assigned to the same sample value. We can write the general relation:

$$X(\varphi) = x. \tag{4.3}$$

Once we have this relationship, we are interested in the probability that the events ($\varphi_1, \varphi_2, \ldots, \varphi_M$) occur. For each x_n, we associate a probability that the random variable X is equal to x_n. that is:

$$P(\varphi_j \in S: X(\varphi_j) = x_n). \tag{4.4}$$

As an example, when a coin is flipped, we have two possibilities: it lands on one side of the coin, heads (H) or the other side of the coin, tails (T). The total sample space we have to choose from is $\{H, T\}$. Our random variable X thus maps these two sample values to a real number. Let us choose:

$$X(H) = 1 \text{ and } X(T) = 0.$$

Note: Other assignments can be used; for example, we could have chosen $X(H) = 0$ and $X(T) = 1$.

The events of interest are set $\varphi = \{H, T\}$, and the values assigned to these is the set $x = \{0,1\}$.

Since random variables have probability distributions associated with them, we would like to characterize the likelihood or probability of various events occurring. Let us start with the event $X(\varphi) = x$. This event can be abbreviated as:

$$(X = x).$$

We can also define events in terms of other relationships:

$$(X < x), (X > x), (X \geq x), (X \leq x), (x_1 < X \leq x_2), \text{ and so on.}$$

Once we have determined the events we are interested in, the next step is to assign probabilities of these events happening based on the probability distribution information associated with the random variable X. These probabilities are denoted as:

$$P(X = x), P(X < x), P(X > x), P(X \geq x), P(X \leq x), P(x_1 < X \leq x_2), \text{ and so on.}$$

EXAMPLE 4.1

We toss two coins simultaneously and want to analyze the probability of obtaining two heads after the experiment.

We first consider our sample space. It is composed of the various possibilities that can occur: $S = \{HH, HT, TH, TT\}$. Next, we identify the event we are interested in: $X(HH) = 2$, where 2 is the number of heads. For $X = 2$, the subset R of the sample space S in which two heads occur is: $R = (X = 2) = \{HH\}$. Since all sample values are equally likely, we can calculate the probability of two heads:

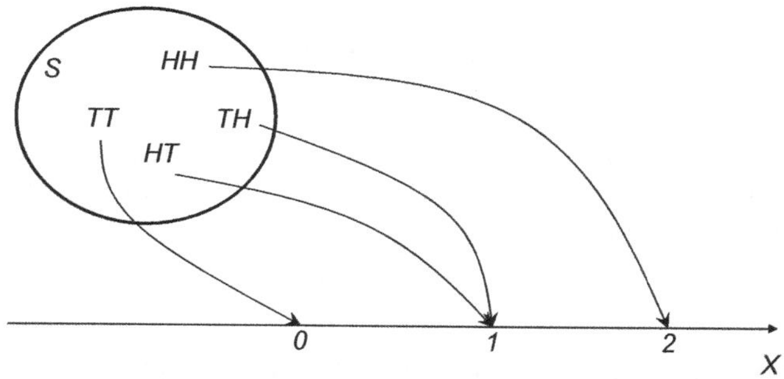

Figure 4.1 Random Variable X as a Function of Two Coin Tosses

$$P(R) = P(X = 2) = (1/4).$$

Now let us determine the probability that at least one of the coins will be a head, where we consider a head to have a value of 1 and a tail to have a value of 0.

Since our sample set has four values, Figure 4.1 shows how the random variable X maps these four sample values in the domain to the following real values in the range

$$X(HH) = 2, X(HT) = 1, X(TH) = 1, X(TT) = 0.$$

The subset T of the sample set S in which the event of at least one head occurs is:

$$T = X \geq 1 = \{HH, HT, TH\}.$$

Since all sample values in the sample set are equally likely, the probability of at least one head occurring is:

$$P(R) = P(X \geq 1) = (3/4).$$

4.3 DISCRETE PROBABILITY DISTRIBUTIONS

A *discrete* random variable is a random variable that can have a *countable* number of possible values. Since X must have one of these values, we have:

$$\sum_{0}^{\infty} P(X = x) = 1. \tag{4.5}$$

To find the probability that the random variable X takes on a value between a and b, we can write:

$$P(a \le X \le b) = \sum_{a}^{b} P(X = x). \tag{4.6}$$

A *probability mass function (pmf)* is a function that gives the probability that a discrete random variable is exactly equal to some value. For a discrete random variable X, the *pmf* is defined as:

$$f_X(a) = P(X = a). \tag{4.7}$$

Equation (4.7) gives the probability that the event random variable X has the value a.

Discrete probability functions are referred to as probability mass functions, and continuous probability functions are referred to as *probability density functions* (*pdf*), as we will see in the next chapter. The term probability function covers both discrete and continuous distributions. When we are referring to probability functions in generic terms, we may use the term probability density functions to mean both discrete and continuous probability functions.

A complete description of a random variable is provided by the *cumulative distribution function* (*cdf*) $F_X(x)$ defined as:

$$F_X(x) = P(X \le x), \quad -\infty < x < \infty. \tag{4.8}$$

The mean of a discrete random value X, identified by μ_x or $E(X)$, is defined as:

$$\mu_X = E(X) = \sum_{i} x_i f_X(x_i). \tag{4.9}$$

The variance of a discrete random variable X, identified by σ_x^2 or $\mathrm{Var}(X)$ is defined as:

$$\sigma_x^2 = \mathrm{Var}(X) = E\{[X - E(X)]^2\} = \sum_{i} (x_i - \mu_X)^2 f_X(x_i). \tag{4.10}$$

Now let us review common, useful discrete probability distributions.

4.4 BERNOULLI DISTRIBUTION

If an experiment has two outcomes, success or failure, the independent repetition of this experiment is called a Bernoulli trial. A Bernoulli trial satisfies the following assumptions:

1. Only two possible outcomes exist for each trial. We call these outcomes "success" or "failure."
2. The probability of success is the same for each trial.
3. Each trial is independent of the other trials.

Let us denote s as the probability of success and $1 - s$ as the probability of failure. Since the probability of a success or a failure in any trail is independent, each probability is multiplied together (Eq. (3.3)). The probability of x successes and $n - x$ failures after n trials is $s^x(1-s)^{n-x}$. For an individual trial, we will have a value s if $k = 1$ (a success occurs) and $1 - s$ if $k = 0$ (a failure occurs). k represents the outcome of the trial, mathematically expressed as:

$$P(X = k) = P(\text{event } k \text{ occurs in a single trial}) = s^k(1-s)^{1-k}, k = 0, 1. \quad (4.11)$$

Equation (4.11) is referred to as a **Bernoulli distribution**. The Bernoulli distribution is the simplest distribution and serves as the building block for other distributions.

For the cumulative distribution, the probability for $k = 0$ is $1 - s$. For $k = 1$, the probability is s. Since this is a cumulative distribution, the probability that X is ≤ 1 is the sum of $1 - s$ and s, which is 1.

Replacing k with x, the cumulative Bernoulli distribution is:

$$\begin{aligned} F(x) &= P(X \leq x) = 0 \; for \; x < 0 \\ &= 1 - s \; for \; 0 \leq x < 1 \\ &= 1 \; for \; x \geq 1. \end{aligned} \quad (4.12)$$

4.5 GEOMETRIC DISTRIBUTION

If we are interested in the first success case, (e.g., how many trials occur before the first "head" occurs), the number of trials required is not known in advance, and this number becomes our random variable. If X is the random variable representing the number of Bernoulli trials required for the first success (e.g., the first "head" occurs), then X is a geometric random variable.

Suppose we perform n tests or trials on a single component. For x successes, we multiply the probability of success in each trial in which a success occurred and obtain s^x. Then we multiply all of the trials, in which a failure occurred ($n - x$ failures) to obtain $(1 - s)^{n-x}$. Then these probabilities are multiplied together to get the overall probability of obtaining x successes and $n - x$ failures.

The probability of obtaining x successes and $n - x$ failures after n trials in a specific order is:

$$P(X = x) = s^x(1-s)^{n-x}. \quad (4.13)$$

Although we have not specified an order, we could. One way in which three successes in five trials could occur in a specific sequence is:

Trial 1: f
Trial 2: s
Trial 3: f
Trial 4: s
Trial 5: s.

We previously explored the geometric distribution in Chapter 3. This is a distribution of the number x representing the number of independent Bernoulli trials required to obtain a success after $x - 1$ failures:

$$P(X = x) = s(1 - s)^{x-1}. \tag{4.14}$$

The probability of Equation (4.14) follows a **geometric distribution**.

Let us say we have n components and we are interested in quantifying the reliability of these components. We will now define f as the probability of failure. f is the probability of failure of a component or system, and $s = 1 - f$ is the probability of survival (or success) of a component or system.

If we are interested in the probability of one failure in n trials, then we must have $n - 1$ successes before a failure at the nth trial. If we have two trials, the first trial is a success and the second trial is a failure. That is, the probability of one failure in two trials is $(1-f)f$. Applying Equation (4.14) for n trials:

$$f(1 - f)^{n-1}. \tag{4.15}$$

Each trial is a success until the last trial (which is a failure); thus, this combination can occur in one order only.

From Chapter 3, Equation (3.21), the cumulative geometric distribution is:

$$P(X \le x) = P(\text{at least one event occurs after } n \text{ trials}) = 1 - (1 - f)^n. \tag{4.16}$$

EXAMPLE 4.2

What is the probability that after a tossing a coin six times, five heads in a row will occur (success) until the first tail (failure) appears? Assuming an unbiased coin where tails or heads are equally likely during a coin toss:

$$\begin{aligned} P(6 \text{ tosses to get the first tail}) &= fs^{n-1} = (1 - 0.5)(0.5)^{6-1} \\ &= 0.015625. \end{aligned} \tag{4.17}$$

EXAMPLE 4.3

The probability of on-time arrival of a flight at a particular destination is 85%. Assume that the probability of success (arriving on time) remains constant irrespective of other factors, such as the weather, air traffic, and so on.

(a) What is the probability that the first nine flights are on time and the last flight is late? From Equation (4.15):

$$P(X = 10) = 0.85^9(1 - 0.85) = 0.034743.$$

(b) What is the probability that at least one of the flights is late after 10 flights? For this question, we make use of the cumulative geometric distribution, Equation (4.16).

$$\begin{aligned} P(\text{at least one flight arrives late after 10 flights}) &= 1-(1-f)^n \\ &= 1-(1-0.15)^{10} \\ &= 0.8031. \end{aligned}$$

Although the chance that only the last flight will be late is ~3.5%, the probability that at least one of the 10 flights arrives late is greater than 80%.

Figure 4.2 shows the *pmf* and *cdf* for values of total number of flights ranging from 1 through 30. For the left figure, the probabilities become increasingly small. We must be careful how to interpret this. More fully qualified, the graph on the left represents the probability that *only* the last flight is late given that the previous $n-1$ flights were all on time. Intuitively, the probability of having all flights on time with only the last one late must become smaller.

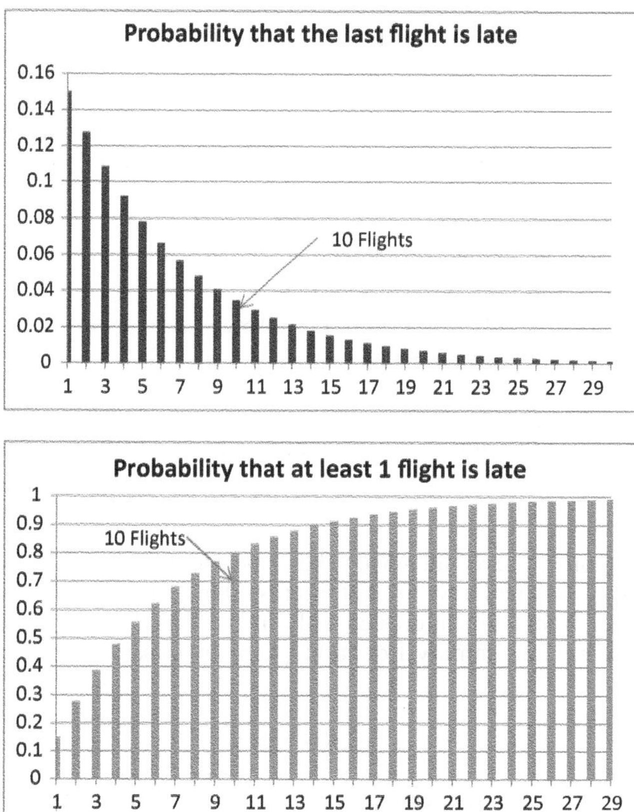

Figure 4.2 *pmf* and *cdf* for Delayed Flight Example

For the bottom part of the figure, the probabilities become larger and asymptotically approach one as the number of flights increase. In this case, we are only interested in the probability that *at least* one of the flights is late. More than one flight could be late as well and still satisfy the equation.

The geometric distribution is the discrete counterpart to the continuous exponential distribution (described in the following chapter). If we take a continuous exponential distribution and slice it up into equal discrete units, we get a geometric distribution.

4.6 BINOMIAL COEFFICIENTS

Assume you have a number of components and we want to predict the components average failure probability after testing them. From this sample, we can determine the probability of failure of an individual component.

Let us define the following probabilities:

$p(0)$: *The probability of no components failing*

$p(1)$: *The probability of 1 component failing*

. . . .

$p(N)$: *The probability of all N components failing.*

Let us first consider a test of two units of identical design and construction and assume each component can fail independent of the other component. For each test or trial, we have four possible outcomes:

s_1s_2: Probability that neither unit fails

f_1s_2: Probability that only the first unit fails

s_1f_2: Probability that only the second unit fails

f_1f_2: Probability that both units fail.

Here the subscripts identify the two components. Since these are the only possible outcomes of the test, the sum of the probabilities must equal one.

$$s_1s_2 + f_1s_2 + s_1f_2 + f_1f_2 = 1. \tag{4.18}$$

If the components are identical and have identical failure characteristics (i.e., $f_1 = f_2$, and $s_1 = s_2$), and if we are not interested in *which* particular unit has failed (the permutations), but are interested only in the number of components failing (combinations), we can simplify Equation (4.18) by considering only the possible combinations:

$$s^2 + 2fs + f^2 = 1. \tag{4.19}$$

We note that the last equation can be written in the following form:

$$(s + f)^2. \tag{4.20}$$

Thus we have the following probabilities of component failures:

$$p(0) = s^2 \tag{4.21}$$

$$p(1) = 2fs \tag{4.22}$$

$$p(2) = f^2. \tag{4.23}$$

If we are interested in the test results for three units that can fail independently, we will have the following possibilities which must sum to one:

$$s_1s_2s_3 + s_1s_2f_3 + s_1f_2s_3 + s_1f_2f_3 + f_1s_2s_3 + f_1s_2f_3 + f_1f_2s_3 + f_1f_2f_3 = 1. \tag{4.24}$$

Now assume the units are identical and thus have identical failure probabilities. If we are interested only in the number of components failing, Equation (4.24) simplifies to:

$$s^3 + 3sf^2 + 3s^2f + f^3 = 1 \tag{4.25}$$

We have the following probabilities of component failures:

$$p(0) = s^3 \tag{4.26}$$

$$p(1) = 3s^2f \tag{4.27}$$

$$p(2) = 3sf^2 \tag{4.28}$$

$$p(3) = f^3. \tag{4.29}$$

Equation (4.25) is the binomial expansion of:

$$(s + f)^3. \tag{4.30}$$

Extending this concept to the general case of n components, the binomial coefficients are obtained from the polynomial expansion of the binomial power $(a + b)^n$.

Each binomial coefficient can be written as ${}^nC_k$, where n represents the number of components and k represents the number of combinations k things can be chosen from the set of n components, or more specifically in terms of reliability, k represents the number of components that can fail.

The general binomial coefficient formula is:

$${}^nC_k = \binom{n}{k} = \frac{n!}{k!(n-k)!}. \tag{4.31}$$

Equation (4.31) is also the general formula for determining the values of Pascal's triangle. A Pascal's triangle is an array of binomial coefficients arranged in a triangle (Fig. 4.3).

For three components, the coefficients are

$${}^3C_0 = 1$$

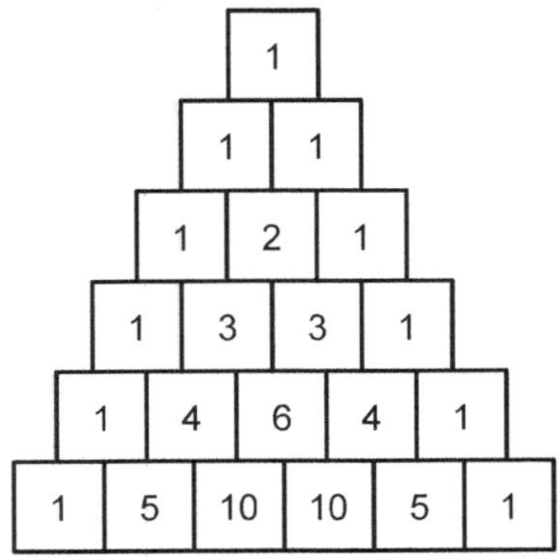

Figure 4.3 Pascal's Triangle

$$^3C_1 = 3$$

$$^3C_2 = 3$$

$$^3C_3 = 1.$$

These agree with the coefficients we previously determined in Equations (4.26)–(4.29).

These coefficients correspond to the number of units that have failed. As an example, to calculate the probability that any two units fail after the test, we need to first determine the probability of two specific units failing and multiply that by the number of combinations in which any two units that can fail:

$$p(2) = 3sf^2. \tag{4.32}$$

4.7 BINOMIAL DISTRIBUTION

Since there are many ways in which x successes can occur, if we want to know the probability of x successes and $n - x$ failures *in any order*, then we need to determine the number of combinations of x successes selected from n trials. Recall that Equation (4.31) is the number of combinations of n objects taken k at a time.

So the probability of x successes in any order for n trials is determined by multiplying the number of possible combinations in which x successes can occur in n trials by the probability of obtaining x successes in a given order. This distribution is called the **binomial distribution**. If x represents the number of successes that occur in n independent trials where the success of any given trial has a probability s, then X is a binomial random variable with parameters (n, s). Its probability mass function is:

$$f_X(x) = P(X = x) = P(x \text{ successes in } n \text{ trials}) = \frac{n!}{x!(n-x)!} s^x (1-s)^{n-x}. \tag{4.33}$$

The binomial distribution must satisfy the following conditions:

1. The number of trials is fixed.
2. The outcome must be either a success or a failure.
3. The probability of success is the same for each trial.
4. The trials are independent.

4.7.1 Relationship between Bernoulli and Binomial Random Variables

A binomial random variable can be expressed in terms of n Bernoulli random variables. If $X_1, X_2, \ldots, X_n$ are independent Bernoulli random variables with success probability s, then the sum of the random variables is distributed as a binomial distribution with parameters (n, s).

If we want to know the probability that r or fewer events occur in n trials, we sum the evaluation of Equation (4.33) for each possible value of x ranging from 0 to r:

$$F_X(x) = P(X \leq x) = \sum_{x=0}^{r} \frac{n!}{x!(n-x)!} s^x (1-s)^{n-x}. \tag{4.34}$$

Equation (4.34) is defined as the **cumulative binomial distribution**. An example of a binomial distribution is shown in Figure 4.4.

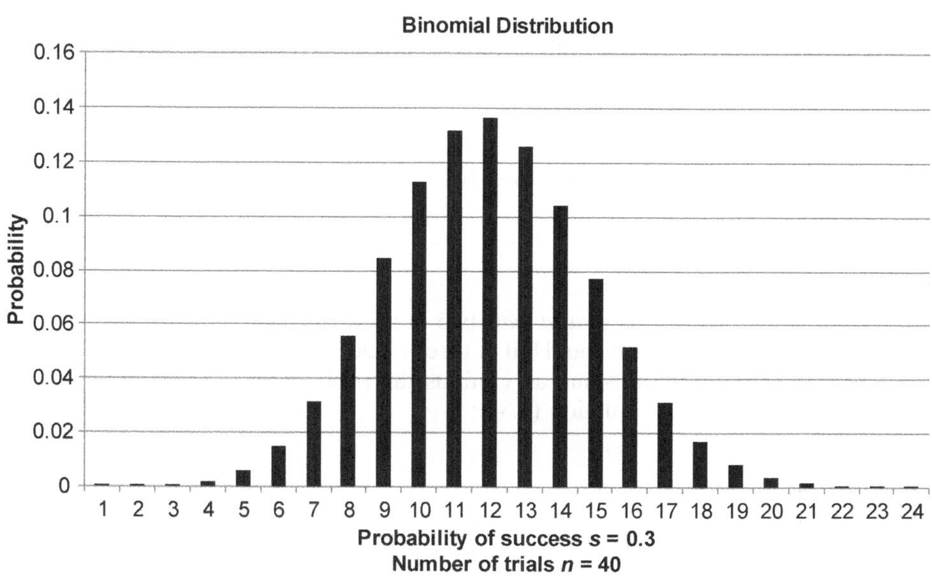

Figure 4.4 Binomial Distribution Example

4.7.2 Relationship between Geometric and Binomial Random Variables

The binomial random variable represents the number of successes in a fixed number of trials (e.g., how many "heads" occur in n trials). In a geometric distribution, the number of successes is fixed to 1 and the number of trials is the random variable. However, in a binomial experiment, the number of trials is fixed in advance and the total number of successes is the random variable of interest.

EXAMPLE 4.4

What is the probability that a five-letter password contains three Gs?

We make the simplifying assumption that each letter in the alphabet is equally likely and thus each letter has a probability of 1/26. Using Equation (4.33):

$$P(3\text{ Gs in an five-letter word}) = \frac{5!}{3!(5-3)!}\left(\frac{1}{26}\right)^3\left(1-\frac{1}{26}\right)^{5-3}$$
$$= 0.000526.$$

EXAMPLE 4.5

(a) A certain model microwave oven is known to have a failure rate of 0.05 units per year. If a dealer sells 10 of these microwave ovens, what is the probability that two of these will fail within a year?

In this case, the number of trials n is represented by the number of microwave ovens. From Equation (4.33):

$$P(2\text{ failures within a group of }10) = \frac{10!}{2!(10-2)!}(0.05)^2(1-0.05)^{10-2}$$
$$= 0.074635.$$

(b) What is the probability that *at least* two of these microwave ovens fail?

For this problem, two could fail or three could fail, and so on up to all 10 failing. This probability is the same as 1 minus the probability of no failures (or one failure). Employing Equation (4.34):

$$P(\text{at least two failures}) = 1 - \sum_{x=0}^{1}\frac{10!}{x!(10-x)!}(0.05)^x(1-0.05)^{10-x}$$

$$P(\text{at least two failures}) = 1 - 0.5987 - 0.3151 = 0.0862.$$

EXAMPLE 4.6

What is the probability of rolling *at least* one '5' after three rolls of a single die?
Substituting the values into Equation (4.34):

$$P(\text{at least one '5'}) = \sum_{x=1}^{3} \frac{3!}{x!(3-x)!}(1/6)^x(1-1/6)^{3-x}$$

$$= 3(1/6)(5/6)^2 + 3(1/6)^2(5/6) + (1/6)^3$$

$$P(\text{at least one '5'}) = 0.4213$$

Note this is the same answer we obtained in Equations (3.12) and (3.18).
For this example, let us summarize three methods we have explored so far that provide the same result.

Method 1. Cumulative geometric distribution. The probability of a success in one trial plus the probability of a success in the second trial plus the probability of success in the third trial is calculated.

Method 2. Multiple independent events. The probability of at least one success is the probability of 1 minus the probability of *not* obtaining a successful result in the first trial multiplied by the probability of *not* obtaining a success in the second trial multiplied by the probability of *not* obtaining a success in the third trial.

Method 3. Binomial distribution. Using Equation (4.33), calculate the probability of exactly one '5' in three trials in any order, then calculate the probability of exactly two '5's in three trials in any order and then finally calculate the probability of exactly three '5's in a row and then add these probabilities.

4.8 POISSON DISTRIBUTION

The Poisson distribution is a discrete probability distribution that represents the probability of a number of events occurring in a fixed interval given that each of these events occurs with a known average probability and that the events are independent.

The Poisson distribution is used to model the number of events occurring within a given interval. The term "interval" can refer to a time interval, a geometric area, and so on, depending on the context of the problem.

Some examples are:

- The number of new cars recalled for warranty repair during the first year.
- The number of books shipped from a manufacturer in a year that have defective bindings.
- The number of policyholders that submit a claim in any given year.
- The number of white blood cells found in a cubic centimeter of blood.

- The number of defects found in a batch of silicon wafer chips.
- The number of deaths per year due to automobile accidents in a certain state.

The Poisson distribution is based on four assumptions.

1. The probability of observing a single event over a small interval is approximately proportional to the size of that interval.
2. The probability of two events occurring in the same narrow interval is negligible.
3. The probability of an event within any interval is the same.
4. The probability of an event in one interval is independent of the probability of an event in any other nonoverlapping interval.

A Poisson experiment has the following properties:

1. The experiment results in outcomes that can be classified as either successes or failures.
2. The average number of successes that occurs in a specified region or time is known.
3. The probability that a success will occur is proportional to the size of the region or time interval.

The Poisson distribution is sometimes called the law of small numbers because it represents the probability distribution of an event that has many opportunities to occur, but the event rarely occurs. The Poisson distribution can be applied to systems with a large number of possible events, each of which is rare.

For example, the number of calls received at a base station controller in 1 hour follows a Poisson distribution with the events appearing frequent to the controller, but are rare from the point of view of the average caller, who is very unlikely to make a call routed to that controller in that hour.

4.8.1 Poisson Random Variable

A random variable X is a Poisson random variable with parameter λ if its *pmf* is:

$$f_X(i) = P(X = i) = \frac{e^{-\lambda}\lambda^i}{i!}, \quad i = 0, 1, \ldots, \tag{4.35}$$

where $P(X = i)$ is the probability of i events occurring in some interval and λ is the expected average number of events in a given interval.

4.8.2 Poisson Distribution Derivation

The Poisson distribution can be derived from the binomial distribution (Eq. (4.34)).

Starting with the binomial distribution, we allow the number of Bernoulli trials to become very large, and the probability of success for each trial to become very small. The mean of the binomial distribution, np, is constrained to be finitely large. Let λ be a constant equal to np. Equation (4.33) becomes:

$$P(x=i)=\frac{n!}{i!(n-i)!}\left(\frac{\lambda}{n}\right)^{i}\left(1-\frac{\lambda}{n}\right)^{n-i}$$

$$=\frac{n(n-1)\cdots(n-i+1)}{n^{i}}\frac{\lambda^{i}}{i!}\frac{\left(1-\frac{\lambda}{n}\right)^{n}}{\left(1-\frac{\lambda}{n}\right)^{i}}. \tag{4.36}$$

As $n \to \infty$:

$$\frac{n(n-1)\cdots(n-i+1)}{n^{i}} \approx 1,$$

and

$$\left(1-\frac{\lambda}{n}\right)^{n}=e^{-\lambda}, \quad \text{as } n \to \infty.$$

Equation (4.36) simplifies to:

$$P(X=i)=\frac{e^{-\lambda}\lambda^{i}}{i!}. \tag{4.37}$$

The Poisson distribution is shown in Figure 4.5. Note the similarities with the binomial distribution example, Figure 4.4.

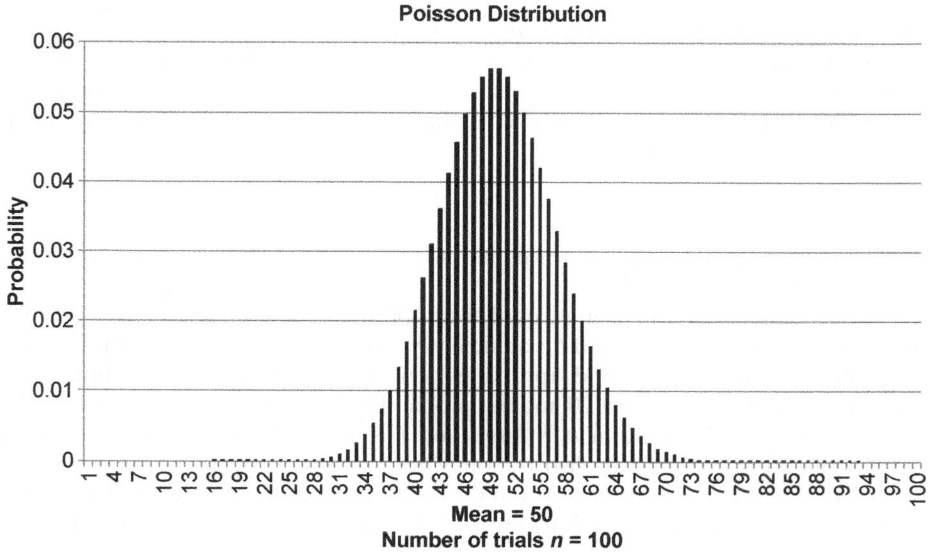

Figure 4.5 Poisson Distribution Example

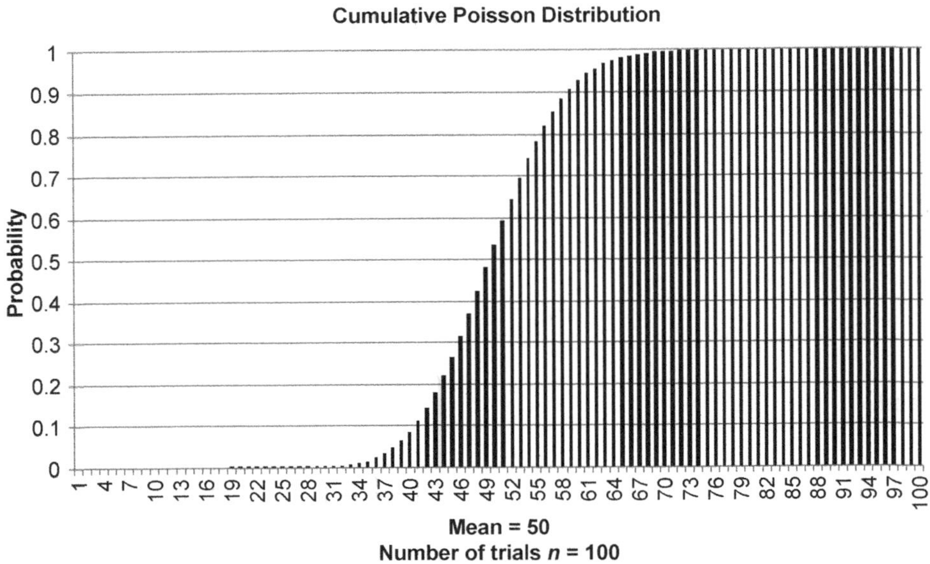

Figure 4.6 Poisson Cumulative Distribution Function Example

The corresponding *cdf* is:

$$F_X(x) = P(X \leq i) = e^{-\lambda} \sum_{i=0}^{n} \frac{\lambda^i}{i!}, \quad i \geq 0. \tag{4.38}$$

An example of the Poisson *cdf* is shown in Figure 4.6.

The parameter i that characterizes the Poisson distribution is just the average number of events expected if we repeat the counting experiment many times. If we know the average rate r for these events, the expected average number of events in a given time interval T is:

$$i = rT. \tag{4.39}$$

Conversely, if the rate is unknown, then we can estimate i by counting the number of events in a time interval T.

The Poisson distribution can be used as an approximation of the binomial distribution if n is sufficiently large and f is sufficiently small. As a general rule of thumb, the Poisson distribution is a good approximation of the binomial distribution if the number of events n is at least 20 and the probability f of any failure event is less than or equal to 0.05. Thus, when n is large and f is small, binomial probabilities are approximated by the Poisson distribution, where $\lambda = nf$.

As a side note, the Poisson distribution itself can be approximated by the normal distribution with mean λ and variance λ (standard deviation) when λ is large ($\lambda > 1000$).

Since reliability tests define a maximum number of failures for acceptance, the cumulative Poisson distribution can be used, where:

$N(T)$ = number of failures in the interval T
λ = expected average failure rate
i = number of failure events
f = probability of failure in a small time interval.

Let us observe a reliability test starting at $t = 0$. If we monitor the test during the interval $(0,T)$, then the number of faults observed, $X(T)$, could be $0, 1, 2, \ldots, n$.

Another way to describe the test is to break it into a sequence of small intervals of length Δt. In the time span $(0, T)$, we create small intervals of size Δt, with the number of intervals n equal to $T/\Delta t$. If Δt is small, we would expect zero faults in most of the intervals, no interval would have more than one fault, and consequently at most one fault in one interval. If you think of each interval's result (0 or 1) as a Bernoulli trial, then the number of events that occur during $(0,T)$, $Y(T)$, should have a binomial distribution. In other words, the binomial random variable $Y(T)$ and Poisson random variable $X(T)$ should have very similar probabilities. At the same time, $X(T)$ has a Poisson distribution with mean λt, while $Y(T)$ has a binomial distribution with parameters $n = T/\Delta t$ and $p = \lambda \Delta t$ (and mean $nf = \lambda T$).

If Δt is small enough that at most one event can occur in an interval, we have two ways of describing the same process: binomial and Poisson. Assuming the constraints are met, if faults occur at the rate of λ per second, we would expect the number of faults $X(T)$ in the interval $(0,T)$ to have a Poisson distribution with mean λT.

EXAMPLE 4.7

Refer back to Example 4.5. A certain model microwave oven is known to have a probability of failure of 0.05. If a dealer sells 10 of these microwave ovens, what is the probability that two of these will fail within a year?

We previously solved this problem by employing the binomial distribution equation. Now let us solve it using the Poisson distribution.

$$i = 2$$

$$\lambda = nf = (10)(.05) = 0.5$$

$$f_X(i) = \frac{\lambda^i e^{-\lambda}}{i!}$$

$$f_X(i) = (0.5)^2 e^{-0.5}/2!$$

$$f_X(i) = 0.0758.$$

The answer we obtained from the binomial distribution is 0.0862, giving an error of 12.1%. Note that if either the probability of an event becomes smaller (satisfying the rare event assumption of the Poisson distribution) or the event count becomes larger (satisfying the assumption of a large number of events), then the inaccuracies due to the use of the Poisson equation will be increasingly smaller until negligible. As a practical matter, for a large number of events, using the binomial formula becomes unwieldy when large factorial terms are calculated (e.g., calculating the factorial of 200 microwave ovens)! In these cases, the Poisson distribution equation is recommended.

As we explore high availability system components, individual component failure rates are typically many orders of magnitudes smaller than the dice and microwave oven problems we have encountered; thus, the Poisson distribution accurately captures the probability distribution and is commonly used in reliability theory.

4.9 NEGATIVE BINOMIAL RANDOM VARIABLE

A negative binomial distribution represents the number of Bernoulli trials to obtain a specified number of successes or failures.

Let us assume a row of 100 light bulbs and define the experiment as the act of turning on and off the light bulbs sequentially. We define "success" as the event that the light bulb illuminates. Let us then define "failure" as the event when a light bulb fails to illuminate. In this case, the negative binomial distribution represents the number of light bulbs that we should turn on before two bulbs fail to illuminate.

Suppose a sequence of independent trials are performed where the probability of success of each trial is s. Instead of fixing the number of trials, n, suppose the trials will continue until exactly k successes have occurred. In this case, the random variable is n—the number of trials necessary for exactly k successes. If the independent trials continue until the kth success, then the last trial must have been a success. Prior to the last trial, there must have been $k-1$ successes in $n-1$ trials. Substituting into Equation (4.31), the number of distinct ways $k-1$ successes can be observed in $n-1$ trials is:

$$\binom{n-1}{k-1} = \frac{(n-1)!}{(k-1)!(n-k)!}. \tag{4.40}$$

From the binomial distribution Equation (4.33), replacing x with $k-1$, the probability that $k-1$ successes occur is:

$$f_X(k) = P(X = k-1) = \frac{(n-1)!}{(k-1)!(n-k)!} s^{k-1}(1-s)^{n-k}.$$

If a series of n trials are performed until exactly k successes occur, with the last trial being a success, then the random variable X is known as the negative binomial random variable with parameter s, where s is the probability of a success at the end of each trial. The *pmf* is obtained from the above equation using the same binomial coefficients, since the $k-1$ failures can occur in any order, but the last failure must occur at the end of the trials:

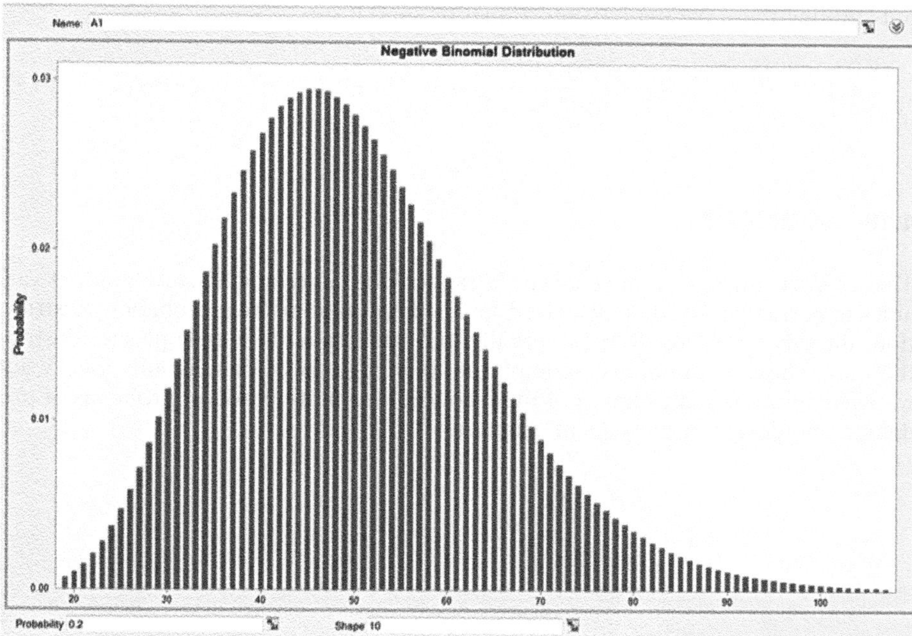

Figure 4.7 Negative Binomial Distribution Example from Crystal Ball

$$f_X(k) = P(X = k) = \frac{(n-1)!}{(k-1)!(n-k)!} s^k (1-s)^{n-k}. \tag{4.41}$$

Equation (4.41) is the negative binomial distribution. An example of a negative binomial distribution is shown in Figure 4.7.

EXAMPLE 4.8

A single light bulb is turned on every day to illuminate a room and turned off every night. If the light bulb fails, it is immediately replaced. The probability of a light bulb failing when it is switched on is 0.01.

What is the probability that we have two light bulb failures after 100 days given with the second failure occurring at the end of the trials?

The parameters are:

$$n = 100$$

$$k = 2$$

$$s = 0.99.$$

Substituting these parameters into Equation (4.41):

$$f_X(2) = P(X = 2) = \frac{(100-1)!}{(2-1)!(100-2)!}(0.99)^{98}(1-0.99)^{100-98} = 0.0036973. \tag{4.42}$$

4.10 SUMMARY

This chapter introduced the concept of random variables and discrete random variables in particular. We then discussed several important discrete probability distributions directly applicable to reliability theory, including a set of examples to illustrate their use. These distributions are important building blocks for reliability theory, and we will explore the application of these concepts to more practical problems of prediction and design in subsequent chapters.

CHAPTER 5

Continuous Random Variables

5.1 INTRODUCTION

X is a continuous random variable if it can have a range of possible values that are infinite and *uncountable*. The probability density function $f(x)$ is used for continuous random variables.

The probability of a discrete random variable having any particular value can be a zero or a nonzero value. However, for a continuous random variable (as opposed to discrete random variables), the probability of it having any particular value is *always* zero. For example, consider a target as illustrated in Figure 5.1. The probability of an archer hitting any particular spot on a target with an arrow is zero—assuming for the sake of argument the point of the arrow is infinitesimally small! However, the probability of hitting a marked area on the target (the red bull's eye for example) is nonzero.

For continuous random variables, a range of values must be considered and not a particular value as was the case for discrete random variables. The *probability distribution function* (*pdf*) represents the distribution of probabilities over a particular range of values between some value a at the lower limit and some value b at the upper limit.

The integral of a *pdf* over this range (a, b), gives the probability of the random variable falling within that range. For a continuous random variable, we must specify an interval over which the function is evaluated.

$$P(a \le X \le b) = \int_a^b f(x)dx. \tag{5.1}$$

Note that if we let $a = b$, (one value), then the above equation will evaluate to 0. Referring to Figure 5.1, this equates to the zero probability of striking a particular point.

Designing High Availability Systems: Design for Six Sigma and Classical Reliability Techniques with Practical Real-Life Examples, First Edition. Zachary Taylor and Subramanyam Ranganathan.

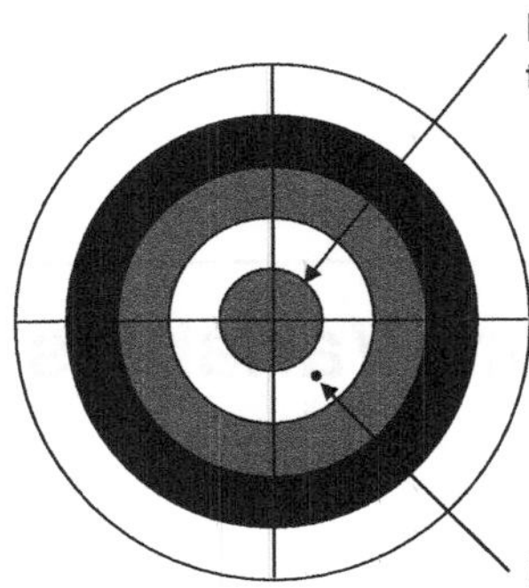

Figure 5.1 Probability of Hitting a Bull's Eye

The relationship between the cumulative distribution function $F(x)$ and the probability density function $f(x)$ is:

$$F(a) = P(-\infty \le X \le a) = \int_{-\infty}^{a} f(x)dx, \tag{5.2}$$

and

$$\frac{d}{dx}F(x) = f(x). \tag{5.3}$$

The mean or expected value of a continuous random variable X is defined as:

$$\mu_X = E(X) = \int_{-\infty}^{\infty} x f_X(x)dx. \tag{5.4}$$

The variance of a continuous random variable X is defined as:

$$\sigma_X^2 = \text{Var}(X) = E\{[X - E(X)]^2\} = \int_{-\infty}^{\infty} (x - \mu_X)^2 f_X(x)dx. \tag{5.5}$$

Continuous random variables play an important role in reliability. We will now consider some of the more common distributions applied in reliability theory.

5.2 UNIFORM RANDOM VARIABLES

A continuous random variable is *uniformly* distributed over an interval (a,b) if its *pdf* and *cdf* are:

$$f(x) = \frac{1}{b-a}, \quad \text{for } a < x < b, \tag{5.6}$$

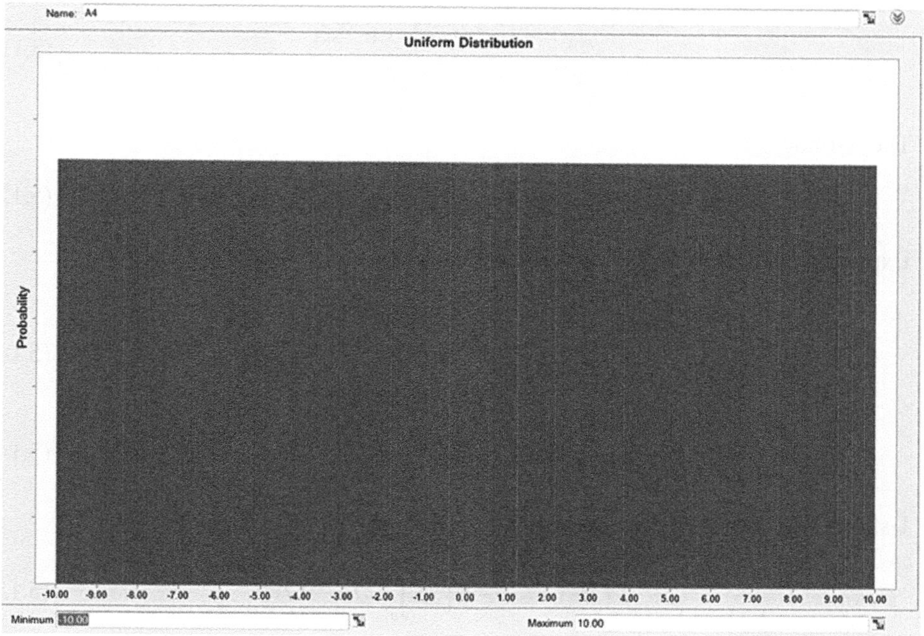

Figure 5.2 Uniform Distribution Example from Crystal Ball

$$F(x) = P(X \le x) = \int_0^x f(x)dx = 0 \quad \text{if } x \le a, \quad \frac{x-a}{b-a} \quad \text{if } a < x < b, \quad x = 1 \quad \text{if } x \ge b. \tag{5.7}$$

The *pdf* of a uniform distribution is shown in Figure 5.2.

5.3 EXPONENTIAL RANDOM VARIABLES

A continuous random variable is exponentially distributed for a given parameter λ, if its *pdf* and *cdf* are:

$$f(x) = \lambda e^{-\lambda x}, \quad \text{for } x \ge 0 \tag{5.8}$$

$$F(x) = P(X \le x) = \int_0^x f_X(x)dx = \int_0^x \lambda e^{-\lambda x}dx$$

$$F(x) = 1 - e^{\lambda x}, \quad \text{for } x \ge 0. \tag{5.9}$$

The mean is:

$$\mu_X = E(X) = \int_0^\infty x\lambda e^{-\lambda x}dx. \tag{5.10}$$

Using integration by parts, where $u = x$, and $dv = \lambda e^{-\lambda x}dx$,

$$\mu_X = -xe^{-\lambda x}\Big|_0^\infty - \int_0^\infty -e^{-\lambda x}dx$$
$$\mu_X = 1/\lambda. \tag{5.11}$$

The variance is:

$$\sigma_X^2 = \text{Var}(X) = E\{[X - E(X)]^2\} = E(X^2) - [E(X)]^2. \tag{5.12}$$

Expanding the term $E\{[X - E(X)]^2\}$ in the above equation, we get:

$$\begin{aligned} E\{[X - E(X)]^2\} &= E\{X^2 - 2XE(X) + [E(X)]^2\} \\ &= E(X^2) - 2E(X)E(X) + [E(X)]^2 \\ &= E(X^2) - [E(X)]^2 \end{aligned} \tag{5.13}$$

$$E(X^2) = \int_{-\infty}^{\infty} x^2 \lambda e^{-\lambda x} dx. \tag{5.14}$$

Using integration by parts again, where $u = x^2$, and $dv = \lambda e^{-\lambda x}dx$

$$E(X^2) = -x^2 e^{-\lambda x}\Big|_0^\infty - \int_0^\infty -2xe^{-\lambda x}dx. \tag{5.15}$$

The first term evaluates to zero. For the integral in the second term, we use integration by parts once again, where $u = 2x$, and $dv = e^{-\lambda x}dx$

$$E(X^2) = 2x(1/\lambda)e^{-\lambda x}\Big|_0^\infty - \int_0^\infty -\frac{2}{\lambda}e^{-\lambda x}dx. \tag{5.16}$$

The first term evaluates to zero, and the second term becomes:

$$E(X^2) = \frac{2}{\lambda^2}. \tag{5.17}$$

We previously showed $E(x) = \mu_x = 1/\lambda$. Thus, the variance of an exponential random variable is:

$$\sigma_X^2 = E(X^2) - [E(X)]^2 = \frac{2}{\lambda^2} - \frac{1}{\lambda^2}$$
$$\sigma_X^2 = \frac{1}{\lambda^2}. \tag{5.18}$$

The *pdf* of an exponential distribution is shown in (Fig. 5.3).

5.4 WEIBULL RANDOM VARIABLES

The Weibull distribution is used in reliability analysis for describing the distribution of times to failure. For example, the characteristics of a hardware failure with respect

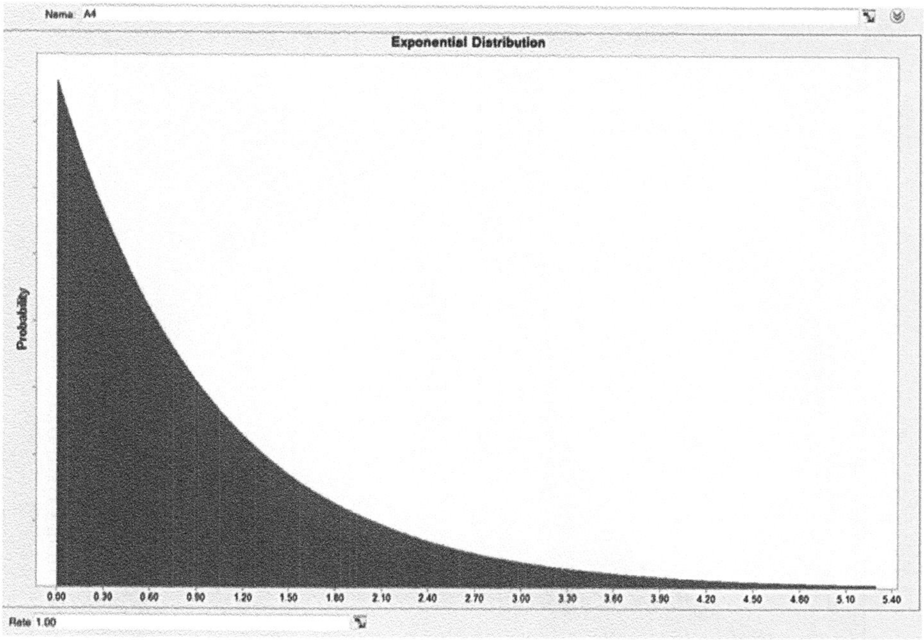

Figure 5.3 Exponential Distribution Example from Crystal Ball

to time is illustrated by the bathtub curve (Fig. 7.9), wherein we expect a high rate of failures early on. However, those failures decrease dramatically when it reaches its useful lifespan where the failure rate becomes almost a constant. After the useful lifespan when the hardware wears out and reaches obsolescence, the failure rate increases dramatically. The *cdf* is given by:

$$F(x) = 1 - e^{-(x/a)^m}, \text{ for } x \geq 0, \tag{5.19}$$

where a is the scale parameter and m is the shape parameter. Taking the derivative of Equation (5.19), we obtain the *pdf*:

$$f(x) = \frac{m}{a}(x/a)^{m-1} e^{-(x/a)^m}, \text{ for } x \geq 0. \tag{5.20}$$

The *pdf* of the Weibull distribution for different scale parameters and shape parameters are shown in Figure 5.4.

5.5 GAMMA RANDOM VARIABLES

The gamma distribution, $G(\alpha, \beta)$, is a two-parameter distribution (both parameters are positive real numbers). α is the shape parameter and β is the scale parameter.

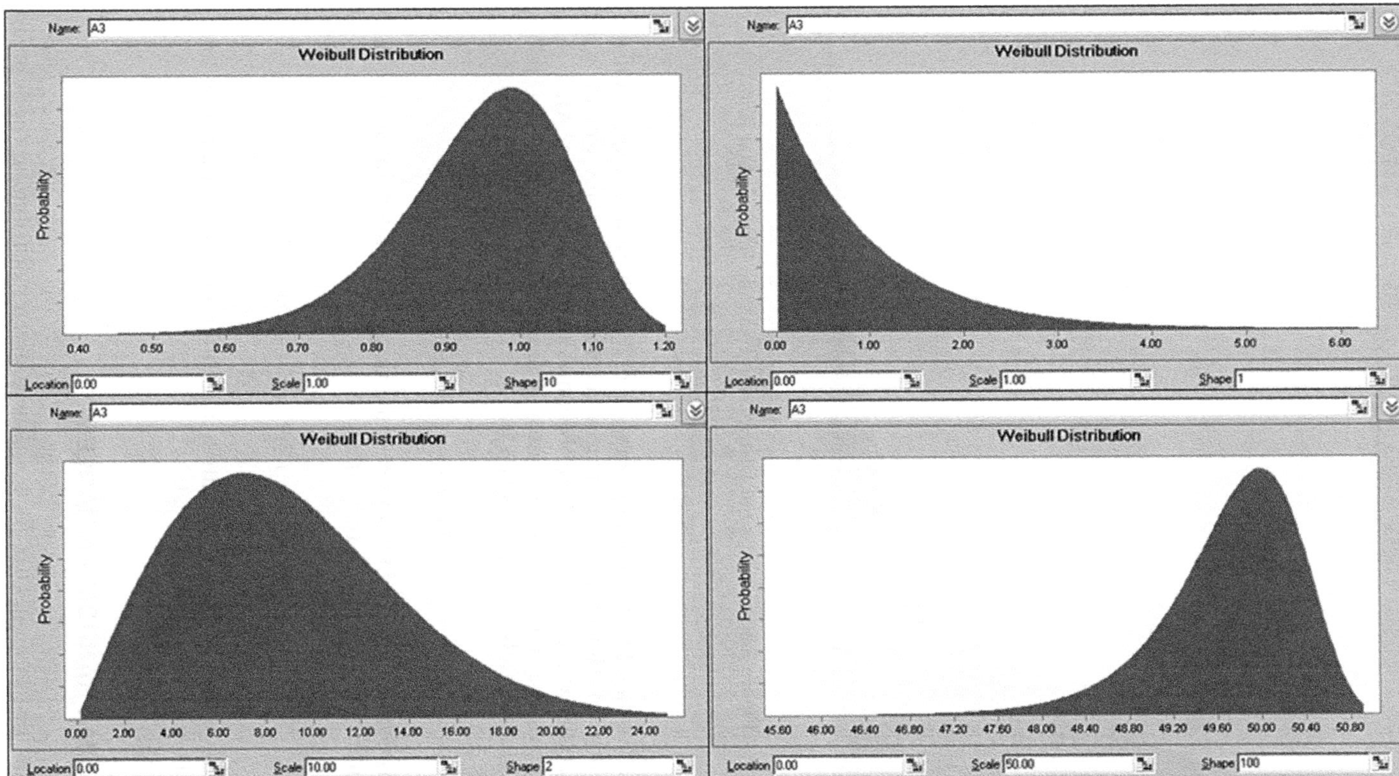

Figure 5.4 Weibull Distribution Examples from Crystal Ball

The gamma probability density is defined as:

$$f(x)=\frac{x^{\alpha-1}e^{-x/\beta}}{\beta^{\alpha}\Gamma(\alpha)} \quad \text{for } x\geq 0. \tag{5.21}$$

where $\Gamma(\alpha)$, called the gamma function, is the constant that makes the integral of the density sum to 1:

$$\Gamma(\alpha)=\int_0^{\infty} e^{-x}x^{\alpha-1}dx. \tag{5.22}$$

If α is an integer, we can evaluate $\Gamma(\alpha)$ using integration by parts:

$$\Gamma(\alpha)=(\alpha-1)\Gamma(\alpha-1). \tag{5.23}$$

And since:

$$\Gamma(1)=\int_0^{\infty} e^{-x}dx=1 \tag{5.24}$$

$$\Gamma(\alpha)=(\alpha-1)(\alpha-2)(\alpha-3)\cdots\Gamma(1).$$

We obtain:

$$\Gamma(\alpha)=(\alpha-1)! \tag{5.25}$$

When used to describe the sum of a series of exponentially distributed variables, the shape factor represents the number of variables, and the scale factor is the mean of the exponential distribution. When the shape parameter α is set to one, and the scale parameter β is set to the mean interval between events, the gamma distribution simplifies to the exponential distribution:

$$f(x)=\frac{e^{-x/\beta}}{\beta} \quad \text{for } x\geq 0. \tag{5.26}$$

The cumulative gamma distribution is defined as:

$$P(\text{gamma}(\alpha,\beta)\leq x)=\frac{\int_0^x t^{\alpha-1}e^{-t/\beta}dt}{\beta^{\alpha}\Gamma(\alpha)}. \tag{5.27}$$

The *cdf* for a gamma random variable for integral α is:

$$F(x)=\frac{\int_0^x x^{\alpha-1}e^{-x/\beta}dx}{\beta^{\alpha}(\alpha-1)!}. \tag{5.28}$$

The *pdf* of the gamma distribution for different scale parameters and shape parameters is shown in Figure 5.5.

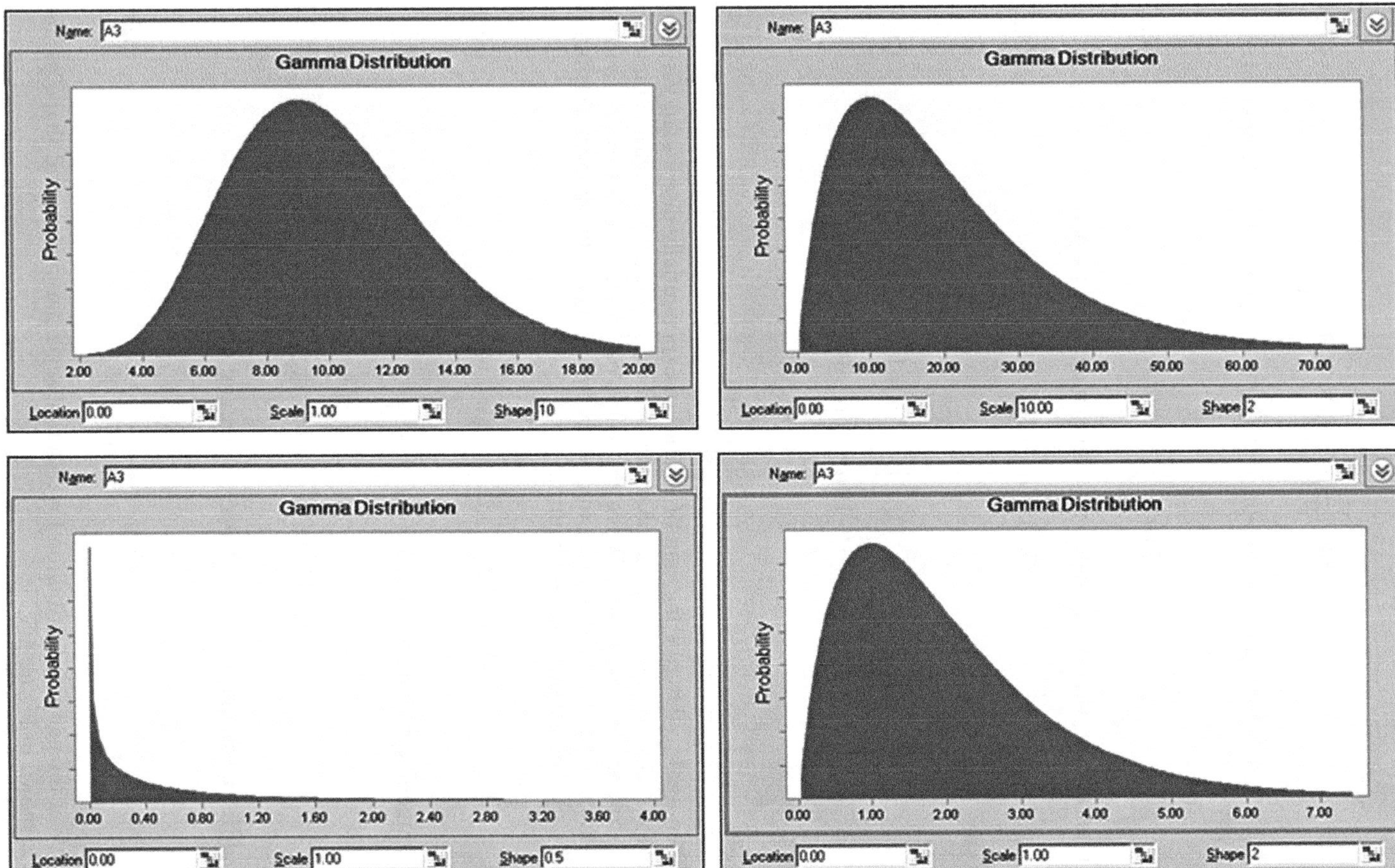

Figure 5.5 Gamma Distribution Examples from Crystal Ball

The gamma random variable also has the following important property that we will derive in Chapter 12:

Let $X_1 \ldots X_n$ be exponentially distributed and independent random variables with parameter λ, and let $Y = \Sigma_{i=1}^{n} X_i$, then Y is a gamma (n,λ) random variable.

5.6 CHI-SQUARE RANDOM VARIABLES

The chi-square distribution is a special case of the gamma distribution. If the shape parameter α is $n/2$, where n is the degrees of freedom, and the scale parameter β is equal to 2, then the gamma probability density is known as the chi-square distribution with n degrees of freedom.

The degrees of freedom n is the number of independent standard normal random variables that make up the chi-square distribution. The chi-square distribution with n degrees of freedom represents the distribution of the sum of the squares of n independent standard normal random variables.

The chi-square probability density function equation is defined in Equation (5.29), where $\Gamma(n/2)$ is the gamma function.

$$f(x) = \frac{x^{(n/2)-1}e^{-x/2}}{2^{n/2}\Gamma(n/2)}. \tag{5.29}$$

For integral multiples of $n/2$ (from Eq. (5.25)):

$$\Gamma(\alpha) = (\alpha - 1)! \tag{5.30}$$

Thus, Equation (5.28) becomes:

$$f(x) = \frac{x^{(n/2)-1}e^{-x/2}}{2^{n/2}(n/2-1)!}. \tag{5.31}$$

We obtain the cumulative chi-square distribution by substituting parameter $\alpha = n/2$ and $\beta = 2$ into Equation (5.27), getting:

$$P(\text{gamma}(\alpha,\beta) \le x) = \frac{\int_0^x t^{n/2-1}e^{-t/2}dt}{2^{n/2}\Gamma(n/2)}. \tag{5.32}$$

The *pdf* of the chi-square distribution for different degrees of freedom is shown in Figure 12.1.

5.7 NORMAL RANDOM VARIABLES

X is a normal random variable with parameters μ and σ^2 if the *pdf* of X is:

$$f(x) = \frac{1}{\sigma\sqrt{2\pi}}e^{-(x-\mu)^2/2\sigma^2}. \tag{5.33}$$

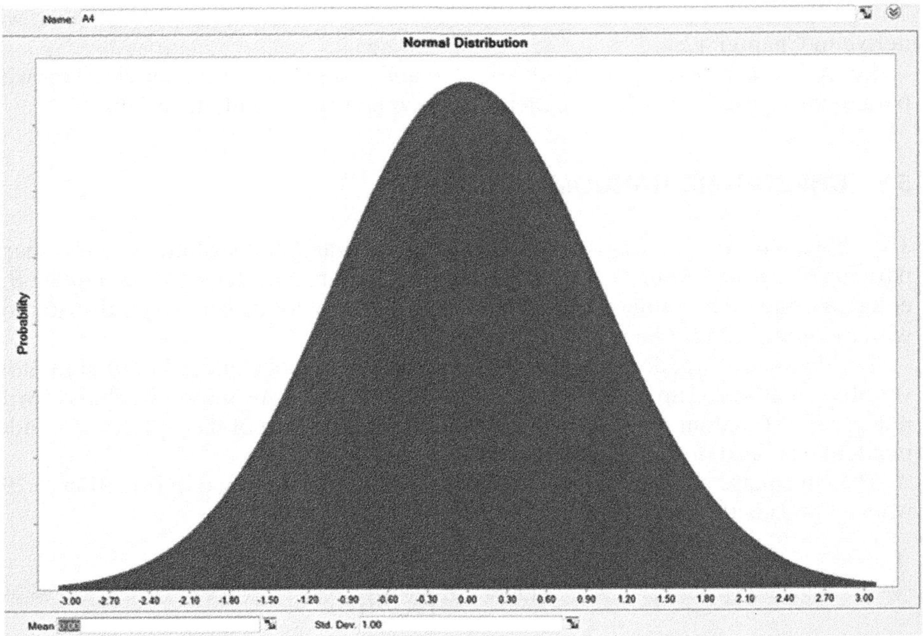

Figure 5.6 Normal Distribution Example from Crystal Ball

The *pdf* of a uniform distribution is shown in Figure 5.6.

5.8 RELATIONSHIP BETWEEN RANDOM VARIABLES

Understanding the relationships between random distributions allows us to apply a variety of techniques to reliability problems. We will summarize a few of the more important relationships.

An exponential distribution is a special case of a gamma distribution with the shape parameter $\alpha = 1$ (previously shown).

The geometric distribution models the number of Bernoulli trials that were failures before the first success occurs. To find the number of Bernoulli trials that were failures until the rth success occurs, we need to wait for r geometric distributions. If Y is a negative binomial with parameters r and p, and X is a geometric distribution with parameter p, then:

$$Y = \sum_{i=1}^{r} X_i. \tag{5.34}$$

The sum of r geometric distributions (with the same parameter p) is the negative binomial distribution.

The negative binomial and gamma are *identically independently distributed* (IID) sums. Based on the central limit theorem, as r gets large, these two distributions approach a normal distribution.

The binomial random variable models probabilities for the number of successes in n Bernoulli trials. In the discrete approximation to the Poisson process, we have a series of Bernoulli trials in which p is the probability of a "success" in each time interval of size Δt. How many time intervals would we have to wait before the first "success" occurs? This number of time intervals must have a geometric distribution (reasoning from the Bernoulli trials setting).

The continuous analog of the discrete geometric distribution is the exponential distribution. As the units of time become smaller, the two distributions converge (Chapter 7).

The time until the first success in the sequence of Bernoulli trials has a very similar distribution as the time until the first event in the Poisson process. This analogy extends to the negative binomial–gamma relationship. The time until the rth success in a sequence of Bernoulli trials has a very similar distribution to the time until the rth event in the Poisson process. The time until the rth event (sometimes called the waiting time until the rth event) has a gamma distribution.

Just as the geometric distribution is a special case of the negative binomial distribution, the Exponential is a special case of the gamma. If you substitute $r = 1$ in the negative binomial, you get the geometric. If you substitute $\alpha = 1$ in the gamma, you get the exponential.

5.9 SUMMARY

This chapter provided a brief introduction to the continuous random variables in particular and contrasting this concept with discrete random variables. Several important continuous random variables are applicable to reliability theory. We will see in later chapters how these concepts are applied to more practical reliability problems.

We will now turn our attention to random processes.

CHAPTER 6

Random Processes

6.1 INTRODUCTION

A random process (also known as a stochastic process) is a set of random variables evolving over time. In other words, a random process is a dynamic system whose output of interest varies randomly over time, based on a probability distribution model. Each random variable is indexed by the time parameter t, where t is a member of the sample space T. Previously, we indicated that a random variable, $X(\varphi)$, is a function that maps each sample point φ in a sample space S. Thus, a random process is really a function of two arguments: t and φ: $X(t, \varphi)$. For a fixed time, t_k, $X(t_k, \varphi) = X(\varphi)$ is a random variable that varies over the sample space S. Conversely, for a fixed sample point φ_j, $X(t, \varphi_j)$ evolves as a function of t for the selected random sample point and is called a sample function of the process. Generally, we represent a random process as $X(t)$ (the sample point φ assumed but not explicitly shown).

A discrete random process has a countable finite or infinite number of time intervals or values within the given time space T, also known as a random sequence, whereas a continuous random process has an uncountable set of infinite time values for the given time space T (in other words continuous time as opposed to discrete time intervals). The values assumed by $X(t)$ are called states, and the set of all possible values is the state space. Note that a discrete-time process (or discrete-state process also referred to as a *chain*) is represented by $X(n)$, $n = 1, 2, 3 \ldots$, where n represents the states of the process.

The mapping from the sample space S to a set of time functions representing the random process is shown in Figure 6.1.

Various special random processes have been classified. We will explore a few of the more important processes that we will apply to reliability models and techniques.

Designing High Availability Systems: Design for Six Sigma and Classical Reliability Techniques with Practical Real-Life Examples, First Edition. Zachary Taylor and Subramanyam Ranganathan.

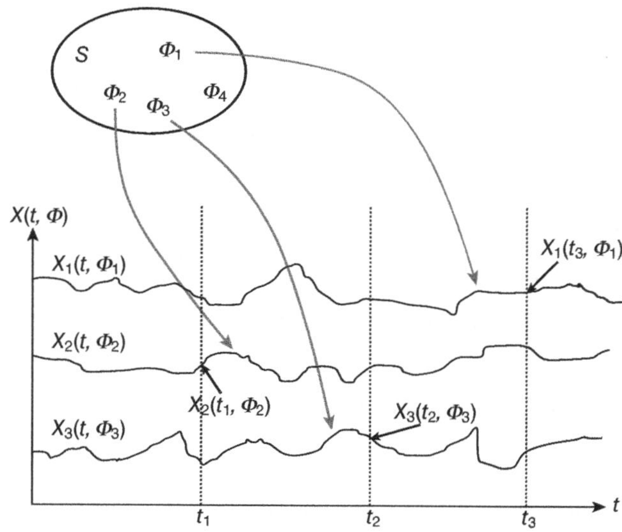

Figure 6.1 Random Process Mapping

6.2 MARKOV PROCESS

The Markov process has wide applicability in reliability theory. We will explore Markov processes in great depth beginning with Chapter 8.

6.3 POISSON PROCESS

A counting process is a stochastic process where $X(t)$ represents the total number of events that have occurred in an interval $(0, t)$. From this definition, the following must be true:

1. $X(t) \geq 0$ and $X(0) = 0$.
2. $X(t)$ is an integer representing the countable number of events.
3. $X(s) \leq X(t)$, if $s < t$.
4. $X(t) - X(s)$ is the number of events that have occurred in the interval (s, t).

A Poisson process is a counting process that has the following additional characteristics:

1. $X(t)$ has independent increments, that is, the expected rate of occurrence of random events in the interval (t_1, t_2) is the same as in the nonoverlapping interval (t_3, t_4).

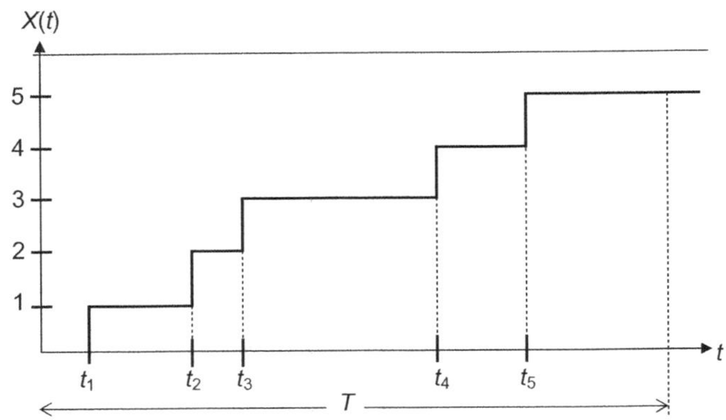

Figure 6.2 Poisson Process

2. The probability of a single event occurring in a short interval t approaches λt as t approaches 0, where λ represents the expected rate of occurrence of these events:

$$\lim_{t \to 0} P(X(t) = 1) = \lambda t. \tag{6.1}$$

3. The probability of more than one event occurring in a short interval t approaches 0 as t approaches 0:

$$\lim_{t \to 0} P(X(t) \geq 2) = 0. \tag{6.2}$$

Intuitively, 2 and 3 make sense if we consider that in this counting process, the rate of occurrence always has the same probability regardless of the time interval we consider. If we make the time interval sufficiently small, noting that each event is sequential and multiple events cannot occur simultaneously, we reduce the probability of multiple events occurring in the interval to zero. However, we still have a nonzero, albeit small probability of a single event occurring in the small interval.

The Poisson process is the basis of modeling diverse scenarios, including the number of cell phone calls received at a base station, the number of people waiting in line at the post office, automobile traffic, radioactive decay, and service requests on a web server, to name a few.

Figure 6.2 illustrates a Poisson process.

6.4 DERIVING THE POISSON DISTRIBUTION

From Equations (6.1) and (6.2), we obtain:

$$\lim_{t \to 0} P(X(t) = 0) = 1 - \lambda t. \tag{6.3}$$

Defining $P_n(t) = P(X(t) = n)$, where n is the number of events, then the probability that n events have occurred by time $t + \Delta t$ is equal to the probability that n events have occurred at time t and no new events occur in the interval Δt plus the probability that $n - 1$ events have occurred at time t and one event occurs during the interval Δt:

$$P_n(t+\Delta t) = P_n(t)(1-\lambda\Delta t) + P_{n-1}(t)\lambda\Delta t + o(\Delta t). \tag{6.4}$$

where $o(\Delta t)$ is a function that approaches zero faster than Δt.

Also, the probability that no events have occurred by time $t + \Delta t$ is equal to the probability that no events have occurred at time t and no new events occur in the interval Δt:

$$P_0(t+\Delta t) = P_0(t)(1-\lambda\Delta t) + o(\Delta t). \tag{6.5}$$

Equations (6.4) and (6.5) can be rewritten as:

$$\frac{P_n(t+\Delta t) - P_n(t)}{\Delta t} = \frac{-P_n(t)\lambda\Delta t + P_{n-1}(t)\lambda\Delta t + o(\Delta t)}{\Delta t} \tag{6.6}$$

$$\frac{P_0(t+\Delta t) - P_0(t)}{\Delta t} = \frac{-P_0(t)\lambda\Delta t + o(\Delta t)}{\Delta t}. \tag{6.7}$$

Now taking the limit as $\Delta t \to 0$, we obtain:

$$\frac{d}{dt}P_n(t) = -\lambda P_n(t) + \lambda P_{n-1}(t) \tag{6.8}$$

$$\frac{d}{dt}P_0(t) = -\lambda P_0(t). \tag{6.9}$$

The solution to Equation (6.9) is:

$$P_0(t) = Ce^{-\lambda t}. \tag{6.10}$$

The constant C is determined from the boundary condition in which $X(0) = 0$ and thus the probability of 0 events at time $t = 0$ is 1, that is, $P_0(0) = 1$.

Thus,

$$P_0(t) = e^{-\lambda t}. \tag{6.11}$$

Substituting $n = 1$ into Equation (6.8), we obtain:

$$\frac{d}{dt}P_1(t) = -\lambda P_1(t) + \lambda P_0(t) = -\lambda P_1(t) + \lambda e^{-\lambda t}. \tag{6.12}$$

To solve this equation, we can use the integrating factor technique (Section 6.4.1).

6.4.1 Using an Integrating Factor to Solve a Linear Ordinary Differential Equation (ODE)

If a first-order ODE can be written in normal linear form:

$$y' + p(t)y = q(t), \tag{6.13}$$

then the ODE can be solved using the integrating factor technique.

If we define $\mu(t) = exp(\int p(t)dt)$, then we can multiply both sides of the equation by the factor and obtain:

$$\mu(t)[y' + p(t)y] = \mu(t)q(t). \tag{6.14}$$

This equation can be rewritten as:

$$(\mu(t)y)' = \mu(t)q(t), \tag{6.15}$$

which follows from the product rule:

$$(\mu(t)y)' = \mu(t)y' + \mu'(t)y. \tag{6.16}$$

And

$$\mu'(t) = p(t)\mu(t), \tag{6.17}$$

based on using the chain rule to differentiate

$$\mu(t) = e^{\int p(t)dt}. \tag{6.18}$$

Equation (6.15) can be solved by integrating both sides:

$$\mu(t)y = \int \mu(t)q(t)dt + C. \tag{6.19}$$

Let us try the technique on Equation (6.12). First, we put the equation into standard form:

$$\frac{d}{dt}P_1(t) + \lambda P_1(t) = \lambda e^{-\lambda t}.$$

Next, we define the integrating factor:

$$\mu(t) = e^{\int \lambda dt} = e^{\lambda t}.$$

Next, multiply each side of the equation by the integrating factor:

$$e^{\lambda t}\frac{d}{dt}P_1(t) + e^{\lambda t}\lambda P_1(t) = e^{\lambda t}\lambda e^{-\lambda t} = \lambda,$$

which can be written (using the product rule for derivatives) as:

$$\frac{d}{dt}e^{\lambda t}P_1(t) = \lambda.$$

Integrate both sides to get:

$$e^{\lambda t}P_1(t) = \int \lambda dt$$

$$e^{\lambda t}P_1(t) = \lambda t + C.$$

Finally, divide by $e^{\lambda t}$ to get the solution for $P_1(t)$:

$$P_1(t) = \lambda t e^{-\lambda t} + Ce^{-\lambda t}. \tag{6.20}$$

C is determined from the boundary condition in which $X(0) = 0$ and thus the probability of 1 event at time $t = 0$ is 0: $P_1(0) = 0$, and thus $C = 0$. Our solution becomes:

$$P_1(t) = \lambda t e^{-\lambda t}. \tag{6.21}$$

This solution can be checked by substituting into Equation (6.12).

Continuing for $n = 2$, Equation (6.12) becomes:

$$\frac{d}{dt}P_2(t) = -\lambda P_2(t) + \lambda P_1(t) = -\lambda P_2(t) + \lambda t e^{-\lambda t}. \tag{6.22}$$

Using the same integrating factor technique, the solution to this equation is:

$$P_2(t) = \frac{(\lambda t)^2}{2!}e^{-\lambda t}. \tag{6.23}$$

For $n = 3$, we get the following solution:

$$P_3(t) = \frac{(\lambda t)^3}{3!}e^{-\lambda t}. \tag{6.24}$$

In general, we see a pattern emerging, and the solution for any n value is:

$$P[X(t) = n] = P_n(t) = \frac{(\lambda t)^n}{n!}e^{-\lambda t}. \tag{6.25}$$

From the Poisson conditions, we know that the arrivals in any disjoint time period are independent, thus:

$$P[X(t+\tau) - X(t) = n] = P[X(\tau) = n] = P_n(\tau) = \frac{(\lambda \tau)^n}{n!}e^{-\lambda \tau}. \tag{6.26}$$

We conclude that the number of events in any interval of length t is Poisson distributed with a mean of λt, that is:

$$f_{X(t)}(n)=\frac{(\lambda t)^n}{n!}e^{-\lambda t},\quad n=0,1,\ldots. \tag{6.27}$$

Note how the Poisson process is uniquely derived from the three simple conditions listed at the beginning. Compare this equation with the derivation of the Poisson approximation to the binomial distribution (Eq. (4.37)).

EXAMPLE 6.1

The number of calls to a doctor's office in a given business day follows a Poisson distribution.

(a) If the average number of calls per hour is five, then what is the probability that there are exactly four calls in 1 hour?

$\lambda = 5$ calls per hour
$n = 4$
$t = 1$ hour.

Substituting into Equation (4.37):

$$f_{X(t)}(n=4)=\frac{((5)(1))^4}{4!}e^{-(5)(1)}=0.1755.$$

(b) What is the probability that less than five calls are received in 1 hour?

$$f_{X(t)}(n\le 4)=\frac{((5)(1))^0}{0!}e^{-(5)(1)}+\frac{((5)(1))^1}{1!}e^{-(5)(1)}+\frac{((5)(1))^2}{2!}e^{-(5)(1)}$$
$$+\frac{((5)(1))^3}{3!}e^{-(5)(1)}+\frac{((5)(1))^4}{4!}e^{-(5)(1)}=0.4405.$$

From condition 2, the expected value of the Poisson distribution is:

$$E[X(t)]=\lambda t,$$

or calculated directly from Equation (6.27) and Equation (4.9):

$$E[X(t)]=\sum_{n=0}^{\infty}nP[X(t)=n]=\sum_{n=0}^{\infty}ne^{-\lambda t}\frac{(\lambda t)^n}{n!}$$

$$E[X(t)]=\sum_{n=1}^{\infty}e^{-\lambda t}\frac{(\lambda t)^{n-1}}{(n-1)!}$$

$$E[X(t)]=\lambda t\sum_{n=0}^{\infty}e^{-\lambda t}\frac{(\lambda t)^n}{n!}.$$

Since the probability that zero or more events will occur for all possible n is equal to one, the expectation becomes:

$$E[X(t)] = \lambda t. \tag{6.28}$$

To calculate the variance, first calculate:

$$E[X^2(t)] = \sum_{n=0}^{\infty} n^2 P[X(t) = n] = \sum_{n=0}^{\infty} n^2 e^{-\lambda t} \frac{(\lambda t)^n}{n!}$$

$$E[X^2(t)] = \lambda t \sum_{n=1}^{\infty} n e^{-\lambda t} \frac{(\lambda t)^{n-1}}{(n-1)!}$$

$$E[X^2(t)] = \lambda t \sum_{n=1}^{\infty} n e^{-\lambda t} \frac{(\lambda t)^{n-1}}{(n-1)!}$$

$$E[X^2(t)] = \lambda t \sum_{n=0}^{\infty} (n+1) e^{-\lambda t} \frac{(\lambda t)^n}{n!}$$

$$E[X^2(t)] = \lambda t \sum_{n=0}^{\infty} \left[n e^{-\lambda t} \frac{(\lambda t)^n}{n!} + e^{-\lambda t} \frac{(\lambda t)^n}{n!} \right]$$

$$E[X^2(t)] = (\lambda t)^2 + \lambda t.$$

From Equation (5.12), the variance of a Poisson distribution is therefore:

$$\text{Var}[X(t)] = \sigma^2 = E[X^2(t)] - E[X(t)]^2 = \lambda t. \tag{6.29}$$

6.5 POISSON INTERARRIVAL TIMES

The Poisson distribution provides the probability density of the number of arrivals in a large interval t with the expected number of arrivals of λt.

Now let us ask a different question. For a Poisson process, what is the distribution of the interarrival times?

Let us start with finding the probability of zero arrivals in time t where T represents the time of the event. To have zero arrivals in time t, T must be greater than t:

$$P(\text{zero arrivals in time } t) = P(T \geq t) = P_0(t).$$

In a small interval dt, the probability of finding one arrival is λdt, thus the probability of finding no arrivals in this interval is $1 - \lambda dt$. Since the interarrival times are independent, the probability of finding no arrivals in the time $t + dt$ is equal to the probability of finding no arrivals in time t and finding no arrivals in the interval dt.

$$P_0(t+dt) = P_0(t)(1 - \lambda dt).$$

Subtracting $P_0(t)$ from both sides we obtain:

$$\begin{aligned} P_0(t+dt)-P_0(t) &= P_0(t)(1-\lambda dt)-P_0(t) \\ &= P_0(t)(1-\lambda dt-1) = -P_0(t)\lambda dt. \end{aligned}$$

Dividing each side by dt:

$$\frac{P_0(t+dt)-P_0(t)}{dt} = -\lambda P_0(t).$$

In the limit as dt goes to zero, we obtain:

$$\dot{P}_0(t) = -\lambda P_0(t).$$

The only function that satisfies this differential equation is:

$$P_0(t) = Ce^{-\lambda t}.$$

Using the boundary condition $P_0(0) = 1$, we get $C = 1$.

Thus, the probability of no arrivals in a time interval t is

$$P(T \geq t) = e^{-\lambda t}.$$

And the interarrival times is therefore:

$$P(T < t) = 1 - e^{-\lambda t}. \tag{6.30}$$

For the exponential distribution, we previously showed in Section 5.3 that the mean is $1/\lambda$, which in this case is the average time between arrivals.

Note that Equation (6.30) is a *continuous* distribution in which the random variable $X(t)$ represents the interarrival time of a failure, whereas Equation (6.27) is a *discrete* distribution in which the random variable $X(t)$ represents the number of failure events that occur in a specified time interval.

Referring back to Figure 6.2, the number of events in a time span T is a Poisson process determined by Equation (6.27), and the time from one event to the next event occurring, that is, the interinterval time, is determined by Equation (6.30). For example, given that event 3 at time t_3, has occurred, the time to the next event t_4 is given by Equation (6.30).

Let us say that a test team finds bugs in a software release at the rate of 10 defects per week. When should the team stop testing given that they have already found 90 defects? We do not know beforehand the number of defects remaining and need to make a decision on when to stop testing and release the product given cost–benefit tradeoffs.

We will consider in later chapters techniques to help answer these questions by quantifying the confidence level that testing has been adequate, that is, a sufficient number of bugs have been fixed.

6.6 SUMMARY

A random process is a set of discrete or continuous random variables that evolve over time. Reliability theory makes use of several important random processes. One of the most important of these processes is the Poisson process. The Poisson process is a counting process in which the rate of occurrence of any random event is the same over any interval of time and as this time interval becomes very short, the expected rate of occurrence becomes a constant. This process is a building block for many reliability concepts. We derived the Poisson process and provided an example of its use.

CHAPTER 7

Modeling and Reliability Basics

7.1 INTRODUCTION

Successful reliability engineering is the implementation of prioritized improvements early in the project development cycle that result in increased ROI (Fig. 7.1).

Reliability considerations:

1. *Complexity.* Reliability depends on the complexity and the number of components that comprise the system. As an initial approximation for a series system, the failure rate is proportional to the number of components. In hardware models, typically the failure rates of each component on a board, (e.g., capacitors, integrated circuits, and connectors) are summed to get an approximate estimate of the board failure rate—each component failure on the board is assumed to cause the board to fail.
2. *Reliability Improves with Redundancy.* Types of redundancy, their application, and limitations will need to be considered.
3. *Reliability Improves with Product Maturity.* For software, the first release typically has the most bugs. When we fix the bugs, generally software reliability improves—although new bugs can be introduced by fixing problems.
4. *Maintainability.* A system should be periodically repaired and maintained. Identifying problems and quickly correcting them either automatically or manually has a dramatic impact on availability.
5. *Operability.* Reliability is enhanced by superior system operability. If the operator is well trained and the system is well designed, problems due to operator error or inability to quickly solve problems or system behavior issues are reduced, thus improving availability.
6. *System Effectiveness.* This is a measurement of the system's ability to meet the specific mission requirements. System effectiveness is a function of availability, dependability, and capability.

Designing High Availability Systems: Design for Six Sigma and Classical Reliability Techniques with Practical Real-Life Examples, First Edition. Zachary Taylor and Subramanyam Ranganathan.

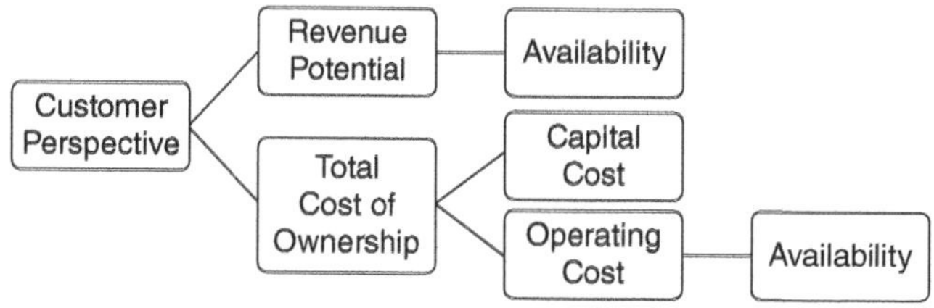

Figure 7.1 Impact of Availability on Revenue and Cost

7. *Dependability (Reliability).* The probability that the system will enter a specific operating mode and perform the functions associated with the operating mode, that is, a measure of dependability.
8. *Capability (Performance).* Capability is the ability of the system to achieve its mission objectives.

The IEEE Reliability Society definition of reliability: "Reliability is a design engineering discipline which applies scientific knowledge to assure a product will perform its intended function for the required duration within a given environment."

Reliability engineering includes designing-in the ability to maintain, test, and support the product throughout its total life cycle. To achieve the specified reliability, we often require rigorous reliability improvements in the product design phase. Numerous methods are used for increasing system reliability. Redundancy and the use of highly reliable components are classical reliability design approaches. But these techniques must be balanced against the increased cost and complexity often required to achieve a highly reliable system.

Robust design and modeling are powerful techniques for improving reliability at low cost up-front. The goal of reliability design is to establish reliability into products at the design stage through the use of DFSS techniques such as Fault Tree Analysis, DFMEA, Capability Analysis, Design Margins, and others. Reliability design is perhaps the most important part in a reliability program in terms of the cost versus the benefits it brings. The ordering of these tasks is optimized using the DFSS approach so that the effectiveness of the reliability design is maximized.

The goal of classical reliability engineering is to ensure that the system delivered to the customer meets or exceeds its reliability and availability requirements. DFSS, on the other hand, is a methodology that controls the process for satisfying the customer's expectations early in the product development life cycle. In this chapter, we will review reliability basics that will be applied in subsequent chapters. In the latter part of the book, we will see how DFSS techniques work together with classical reliability techniques to meet and exceed customer expectations. The integration of classical reliability and DFSS techniques to create high availability systems is shown Figure 7.2.

With random failures, we cannot predict exactly when a particular system will fail, but we can gather statistical information that allows us to predict to some degree of confidence when a system on average will fail and what percentage of the systems

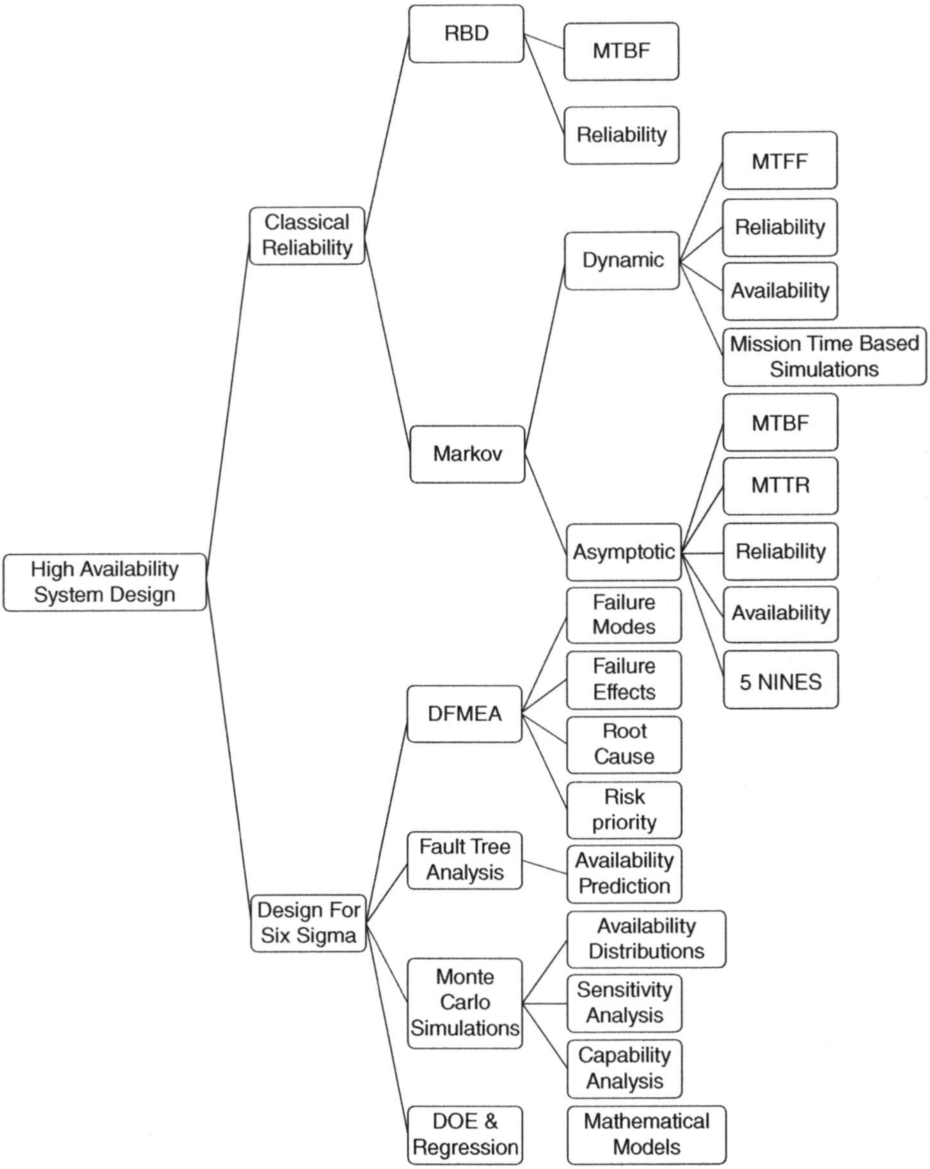

Figure 7.2 Integrated View of High Availability System Design

will still be operational or fault free at certain times. We will also be interested in the return rates and repair times. For example, we might design our system with the following reliability goals.

1. After 2 years of deployment, we require greater than 99% of the units to be failure-free.

2. We predict less than 0.5% of the delivered units to be returned each year.
3. The repair time for each returned unit is 2 days.

7.2 MODELING

Modeling is the art and science of capturing some useful aspect of the functionality and the relevant characteristics of real-world systems and presenting such systems in a form that facilitates understanding system behavior, analyses, improvement, and optimization.

Two types of models can be created for simulation and behavioral study: the static model and the dynamic model.

For each of these models, we can have deterministic or probabilistic simulations. The models we discuss in this book are probabilistic simulations. For example, we are interested in the reliability of a system for which the exact time to failure is not known beforehand. We cannot predict when a particular system will fail, but we can ascribe certain attributes to the system reliability and use this information to predict the behavior of the system when many systems or multiple trials are considered.

The static model describes and characterizes the steady-state behavior of the system, that is, the behavior of the system is independent of time.

The dynamic model describes how the system evolves over time. A good example of this is the behavior of a system when it is first deployed. The initial system behavior may be radically different from its steady-state behavior. In most cases, for repairable systems, because of the large Mean Time between Failures (MTBFs) that are part of the system components, the time frame of the dynamic component will be very small and could range on the order of days or weeks. However, this behavior could be different for software-intensive systems where the quality of the software will have a great influence on the time frame of the transient response. This is also typically seen during and after system upgrades.

A simplifying assumption is to make the model static over a certain period of time (either the mission time or narrower slice of time of interest). Especially when the variances in the estimation of the various MTBFs and other probabilities can be quite wide or have a large range, the level of inaccuracies introduced by employing a constant failure rate model may be overshadowed by the greater inaccuracies introduced by our lack of knowledge of the "true" MTBF. It is interesting to observe how availability is impacted during the early stages of field deployment by product immaturity and new maintenance policies. An unusually low availability can reflect on both the quality and quantity of performance and system testing we perform in the labs prior to field deployment.

A huge advantage of using the static model is that our equations characterizing the system behavior become much simpler and more tractable. One of the dangers of making a model too complicated is that we may run into human errors either in creating the model, understanding the results, or managing the model down the road either for subsequent product iterations or evolutions. In general, it is better to have the simplest of models that convey the most accurate information that we can obtain—diametrically opposing goals. You definitely don't want to overanalyze information where the payback on the data you get from that analysis becomes smaller as your model becomes more complicated.

A wealth of software support and tools exist for taming even the most complicated model and these tools should be used whenever possible to aid in the analysis. For example, Matlab and Relex are good tools for dynamic models and simulation, whereas Crystal Ball is a good tool for static models and Monte Carlo simulations.

7.2.1 Reliability Models

A variety of reliability models exist. These can be broadly classified as:

1. Combinatorial models
2. Hierarchical models
3. State-space models.

These models show structural relationships between idealized system components and how failure events affect the operation of the system.

The following model types are discussed in this book. Reliability block diagrams, which are combinatorial, fault tree analysis, which can be classified as hierarchical, and Markov models, which belong to state-space models. Other model types, such as reliability graphs, are not discussed in this book. The interested reader is referred to the literature.

RBDs are discussed in this chapter. Fault trees and Markov models are discussed in later chapters.

7.2.2 Modeling Approaches

Large-scale computer systems, such as telecommunication systems and servers, have become increasingly complex and a total failure of these systems could potentially affect millions of customers. Redundancy and fault management schemes are put in place to reduce the chance of a total failure. Partial failures may be more common, allowing the system to function with reduced redundancy, capacity, or functionality over a short period of time. This degradation is designed to be temporary, as the system automatically restores full functionality, or the maintenance team restores full system operation. The identification of these operational, degraded, and failed states begins with the initial architecture of the system. A variety of failures and operational scenarios are analyzed to determine the impacts of these events on system performance, operation, capacity, and so on. The time spent in each of these states is also important to determine, since these times affect overall system availability and functionality. This type of system is referred to as a multistate system.

A number of well-known reliability techniques have been employed to address binary-state systems, where the system is either fully functioning or completely fails. Reliability block diagrams and fault trees have been used with much success to characterize the reliability of binary systems. However, extending these models to the analysis of multistate systems with partial and degraded functionality as well as different modes of operation is not straightforward, can be more complex, and may not clearly show the relationships between events and each of the states. Other tools and models, such as Markov analysis, address multistate systems directly.

Several approaches can be used for reliability assessment:

1. *The Structure Function Approach.* Boolean models, such as fault trees and reliability block diagrams, are extended to encompass multivalued states.
2. *Markov Analysis.* States are identified and probabilities are assigned to each state.
3. *Static Monte Carlo Simulation.* A variety of scenarios are simulated using random sampling from probability distributions for each trial.
4. *Mission Time-Based Monte Carlo Simulation.* Component failure rates and repair rates are sampled from exponential distributions with mean values equal to the component MTBFs and MTTRs. The system behavior for multiple trials is observed over the mission time.

Each of these techniques will be examined in depth along with practical examples. Before we apply modeling techniques, let us explore the basics of reliability theory.

7.3 FAILURE PROBABILITY AND FAILURE DENSITY

The failure state of a system at a given time t may be described by a state variable $X(t)$:

$$X(t) = 1 \quad \text{(if the system is functioning at time } t\text{)}$$

$$X(t) = 0 \quad \text{(if the system has failed at time } t\text{)}.$$

T is the time at which the failure occurs from the time the system is put into operation ($t = 0$) until it fails for the first time. Since the time to failure is not known beforehand, T is a random variable with some probability distribution, and $X(t)$ is the state variable that represents the state of the system (not failed, failed) at time t. This relationship is illustrated in Figure 7.3.

We will now consider important reliability parameters and will describe the relationship between them.

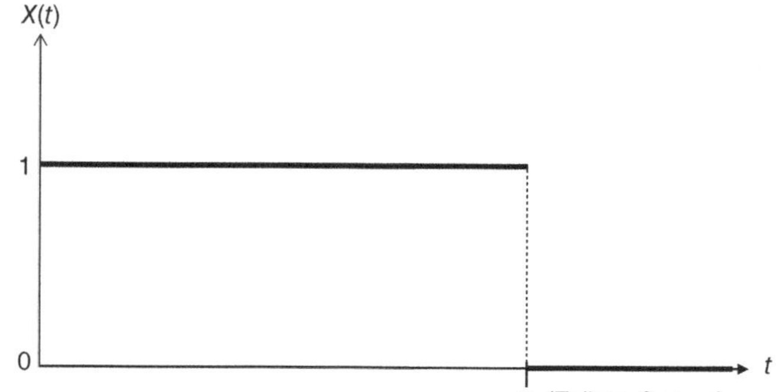

Figure 7.3 State Variable for System Failure

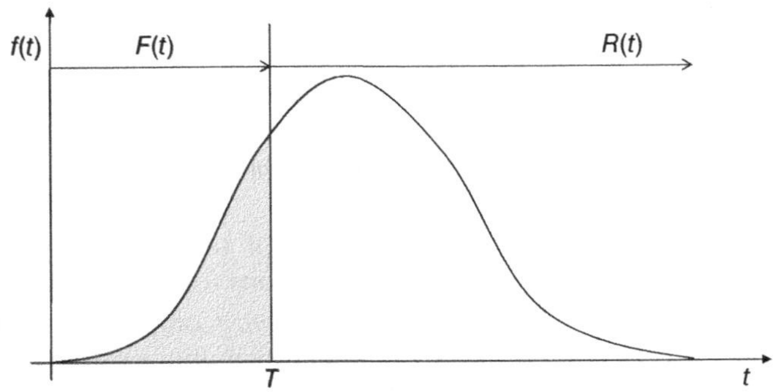

Figure 7.4 Failure Probability Density Function

7.4 UNRELIABILITY, $F(t)$

Unreliability of a component or system is defined as the probability that the component or system fails during the time interval (0, t), given that it was operating or repaired to a like new condition at time 0.

Generally we make some assumption for a probability distribution for failures of the item of interest. Figure 7.4 shows a general failure probability density function. At any point in time, the probability of failure may vary based upon this density function. If we want to know how likely it is that a failure occurs between $t = 0$ and $t = T$, we integrate the area under the curve from 0 to T to obtain the failure distribution function $F(t)$. As t goes to ∞, the probability of failure approaches 1.

The reliability of a component is $R(t) = 1 - F(t)$, that is, $R(t)$ is the probability of *not* failing, which is 1 minus the probability of failing given by $F(t)$. $R(t)$ is the probability that at any given time t the system will still be operational.

The time to failure T is a continuously distributed random variable with a probability density function $f(t)$ and distribution function $F(t)$:

$$F(t) = P(T \le t) = \int_0^t f(x)dx, \tag{7.1}$$

that is, $F(t)$ is the probability that the component will fail in the time interval (0, t).

The failure probability density function $f(t)$ is defined as:

$$f(t) = \frac{dF(t)}{dt} = \lim_{\Delta t \to 0} \frac{[F(t+\Delta t) - F(t)]}{\Delta t} = \lim_{\Delta t \to 0} \frac{P\{t < T \le t + \Delta t\}}{\Delta t}. \tag{7.2}$$

This implies that when Δt is small, the probability that a failure occurs between time t and $t + \Delta t$ is:

$$P(t < T \le t + \Delta t) \approx f(t)\Delta t. \tag{7.3}$$

7.5 RELIABILITY, *R*(*t*)

Reliability is defined as the probability that a system or component will operate without a failure for a given period of time (often called Mission Time) and under given operating and environmental conditions. It can also be defined as the probability that a system or component survives for some specified period of time (probability of nonfailure):

$$R(t) = P(T > t) = 1 - F(t); \tag{7.4}$$

that is, $R(t)$ is the probability that the system operates without failure up to time T. In other words, $R(t)$ is the probability that the system will *not* fail in the time interval $(0,T)$. If the mission time is arbitrary, then $R(t)$ is called the survival function of t.

Initially, we assume the system is operational, so at $t = 0$, $R(0) = 1$. Eventually, the system will fail, so $R(\infty) = 0$.

7.6 MTTF

The *Mean Time to Failure* (MTTF) is the expected value $E(t)$, or mean value (usually expressed in hours) of time to failure for a component or system where t is a random variable. The mean time is measured from the time a component or system begins operation until the time it fails. This terminology applies to repairable and non-repairable components and systems.

To calculate the MTTF, we substitute the random variable t for x into Equation (5.4):

$$\text{MTTF} = \int_0^{\infty} tf(t)dt. \tag{7.5}$$

Taking the derivative of Equation (7.4) and then solving Equation (7.5) by employing integration by parts:

$$\text{MTTF} = -\int_0^{\infty} \frac{d}{dt} R(t)dt = -tR(t)\Big|_0^{\infty} + \int_0^{\infty} R(t)dt.$$

The term $tR(t)$ vanishes at $t = 0$ and $t = \infty$. Thus, the MTTF becomes:

$$\text{MTTF} = \int_0^{\infty} R(t)dt. \tag{7.6}$$

7.7 MTBF

The MTBF represents the expected time (usually expressed in hours) between consecutive failures of a component or system. The MTBF is calculated by taking the total operating time of all systems and dividing it by the total number of failures from all systems. The MTBF is the inverse of the failure rate. This terminology applies to repairable components and systems. It is easy to confuse MTBF with MTTF. When the MTTR is very short in comparison with the MTTF, then the MTBF and the MTTF can be used interchangeably. Vendors do not necessarily know what the maintenance practices of a customer are or how their components will be integrated into a larger system, so MTTR as a practical matter is not known by parts and subsystem suppliers.

7.8 REPAIRABLE SYSTEM

A repairable system is one which upon failure, can be automatically or manually repaired. For example, if a card or a component within a system is replaced on site, the system is still considered to be repairable. For high availability systems, great emphasis is placed on automatic repair, including redundancy protection as applicable so that the system is available at all times and outages associated with the system are very short since the vast majority of the failure modes are designed to be addressed automatically.

7.9 NONREPAIRABLE SYSTEM

A nonrepairable system is one which upon failure cannot be repaired and is generally replaced. For example, a light bulb is not repaired after it burns out.

Another example is from the aerospace industry. If a flight control system fails in flight, it cannot be manually repaired by replacing the failed system or failed board within the system. The system must be designed to operate in a full or degraded mode until the flight terminates—by successfully landing! Redundancies are a very important part of ensuring the critical flight systems remain operational, so we initially approach the design of these systems with nonrepairable models. We will also add that the definition of a repairable system can be quite broad. For example, if we have a software failure, after we recover from the failure by switching to a redundant system, automatically rebooting the subsystem or card where the failure occurred may repair the system.

For systems such as satellites and interplanetary spacecraft, we may be able to download new software that corrects the problem and reboot the system remotely to repair a software bug.

7.10 MTTR

The *Mean Time to Repair* is the expected time it takes to repair a system or component after it fails. This terminology applies to repairable components and systems. The repair time can vary significantly depending on the action to be taken. For example replacing a failed card may require travel to the location of the failed equipment which may take several hours. In other cases, such as restarting a failed software task,

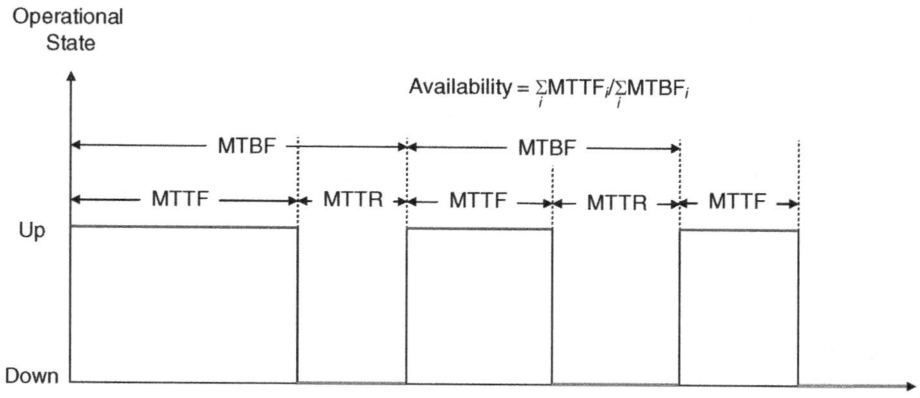

Figure 7.5 MTBF, MTTF, MTTR, and Availability Relationships

it may only take a few seconds. Repairing a failed card on an airplane must occur after the airplane has landed and is brought to a maintenance hangar. The repair time encompasses the time from which the component fails to the time the component is fully functional within the system.

Note the MTBF is the sum of MTTF and MTTR. The relationships between availability, MTTF, MTBF, and MTTR are shown in Figure 7.5.

The mean time to repair is the mean value of the repair distribution function $r(t)$:

$$\text{MTTR} = \int_0^{\infty} tr(t)dt.$$

7.11 FAILURE RATE

The failure rate is the number of failures of an item per unit time measurement. Failure rates are commonly written using FIT (Failure in Time) which corresponds to the number of failures per billion hours.

7.12 MAINTAINABILITY

Maintainability is the probability that a system or component that has failed will be restored to operational effectiveness within a given time (probability of repair).

7.13 OPERABILITY

Operability is the ease and accuracy with which users can operate the system and react to and resolve problems that occur with the system.

7.14 AVAILABILITY

Availability is the probability that a system performs its intended function under specified operational and environmental conditions. Availability provides information about the time in which the system is operational. Contrast this with reliability, which provides information about the failure-free interval. Both are described in percentage values. *Availability is driven by time lost over a time interval, whereas reliability is driven by the number of failures over a time interval.* Availability incorporates system maintenance and repair information after a system failure has occurred, whereas reliability does not.

In its simplest form, availability is the ratio of the system uptime to the total system time:

$$A_{\text{avg}} = \frac{\text{Uptime}}{\text{Uptime} + \text{Downtime}}. \tag{7.7}$$

Using MTTF (Uptime) and MTTR (Downtime) information, the average availability can be calculated by:

$$A_{\text{avg}} = \frac{\text{MTTF}_{\text{avg}}}{\text{MTTF}_{\text{avg}} + \text{MTTR}_{\text{avg}}}. \tag{7.8}$$

If the MTBF is much larger than MTTR, the average availability can be calculated by:

$$A_{\text{avg}} = \frac{\text{MTBF}_{\text{avg}}}{\text{MTBF}_{\text{avg}} + \text{MTTR}_{\text{avg}}}. \tag{7.9}$$

We see that the availability is a function not only of the failure time but also of the repair time which is determined by maintenance policies and equipment self-repair design. If MTBF (or MTTF) is very large compared with MTTR, the system availability will become very high. Conversely, if the MTBF is not large but the MTTR is very small, high availability can also be achieved. From a practical standpoint, as average reliability decreases (i.e., MTTF becomes smaller), better maintainability (i.e., shorter MTTR) will achieve the same availability. Trade-offs between MTTF and MTTR can be made to meet the availability goals or requirements of the system.

The MTTR is dependent on a variety of factors, not all of which are directly controllable by the system vendor. Since different facilities or customers may have different maintenance and logistics procedures (maintenance can be preventive or corrective), and the number and skill level of the maintenance staff may be different, the MTTR will vary. Also, the level and responsiveness of the customer support team can influence the MTTR. Assumptions are made based on discussions with the customer regarding their maintenance policies.

The MTTR variations can have a very dramatic impact on availability; thus, we can consider two types of availability:

1. *Operational availability* provides the actual availability measured after system deployment.

2. *Baseline availability*, where availability is calculated using a baseline MTTR independent of the customers and facilities to allow system architecture comparisons.

The MTBF of a particular component, let us say a telecommunications system call processing board, can be improved by improved design, more reliable parts, and so on. However, even greater system MTBF values can be achieved using redundancy design techniques. (The system MTBF and component MTBF are usually different).

No matter which distribution models are used for the failure rates and the repair rates, the steady-state availability depends on the system MTBF and the system MTTR.

7.14.1 Availability as a Function of Time

As we will learn in Markov modeling, Chapter 8, availability is a function of time. The instantaneous availability $A(t)$ is defined as the probability that a component or system is operating at time t given that is was operating at time 0.

$A(t)$ can be calculated from the solution to the dynamic Markov equations. However, in many cases, we will be interested in the asymptotic or steady-state behavior of the system, which is calculated using Equations (7.7)–(7.9).

We can also compute the mean availability over a time interval $[0,t]$:

$$\bar{A} = E[A] = \frac{1}{t}\int_0^t A(x)dx. \tag{7.10}$$

The steady-state availability is the limit of the instantaneous availability as $t \to \infty$:

$$A_{ss} = \lim_{t\to\infty} A(t). \tag{7.11}$$

7.14.2 Service Level Availability

Service level availability is typically negotiated between the customer and the vendor. The definition is dependent on this agreement but typically takes into consideration those outages that technically may affect the availability of the system, but may not be counted as a contributing event to the availability if certain criteria have been met.

Examples include the following:

1. The duration of the outage is less than 5 seconds. For example, if we have a call processing design in which data connection service remains intact for outages less than 5 seconds, then a system outage of less than 5 seconds does not impact call service and is not counted.
2. The outage only impacts Operations and Maintenance (O&M) of the system, that is, does not impact end-user service. We decompose the system into modular subsystems, such that the end-user service is unaffected by downtime of the O&M portion of the system. However, we may have separate outage requirements associated with the O&M portion, since the ability to monitor and maintain the system is impaired.

3. The percentage of service impacted is less than 10%. This reflects TL9000 requirements.
4. The physical capacity of the system has been affected but the percentage of current service impacted is 0%.
5. The outage is attributable to an external environment event. This is usually negotiated between the vendor and the customer. Outages due to the equipment vendor are counted, but outages due to the customer environment may not be counted depending on the effectiveness of the monitoring that may be part of the delivered equipment.
6. The outage is caused by other equipment. Isolating the problem to a particular system is important for system outage calculations, as well as root cause analysis and equipment repair.
7. The outage is caused by operator error. The equipment vendor should take great care that the system has robust operability functionality. A system that is difficult to operate will lead to more errors on the part of the operator that can cause inadvertent outages. Thus, operability requirements are an important consideration.
8. The outage is expected as part of normal system maintenance, for example, software upgrade. Depending on the design of the system, an outage window of varying sizes will occur during software upgrade, board upgrades, and certain reconfiguration activities. Generally, these outage times are captured and monitored separately than the unexpected outages associated with failures.

Once the set of exceptions that do not contribute to the system availability are identified, the service availability calculation is modified to incorporate these expectations.

The Telecom Management Standard TL9000, defines an outage to have occurred when the system is operating at less than 90% of full capacity for a period of more than 15 seconds.

$$A_{SL} = \frac{\text{Uptime}}{\text{Uptime} + \text{Svc Outage Time}}, \tag{7.12}$$

where service outage time is defined uniquely, using criteria such as those listed above.

One method to achieve 90% of full capacity is to have the service functionality distributed across 10 or more boards or systems. If only one board fails, and redistribution of the service occurs across the remaining boards seamlessly (albeit at a lower capacity), we may not have an outage.

7.15 UNAVAILABILITY

Unavailability, $Q(t)$, is the probability that a component or system is not operating at time t, given that is was operating at time zero. The sum of availability and unavailability by definition must equal 1:

$$A(t) + Q(t) = 1.$$

Unavailability provides the probability of failure on demand, that is, the probability that when a request for a system service is made, the system is unable to fulfill that request.

The steady-state or asymptotic unavailability is:

$$Q_{lt} = 1 - A_{lt}$$

7.16 FIVE 9S AVAILABILITY

Another form of availability is the concept of Five 9s. For a high availability system, the availability is typically defined to be at least 99.999% which corresponds to Five 9s availability. As systems or components exhibit even higher levels of availability, the availability differences among them become very small and not as easy to discern. By counting the number of "9"s in the availability, we have a more intuitive method to compare relative availabilities. The number of nines is calculated using the following formula:

$$\text{Five 9s} = -\log(10, 1 - A), \tag{7.13}$$

where A is the availability.

7.17 DOWNTIME

Downtime (usually expressed in outage minutes per year) is another way of expressing availability.

$$\text{Downtime} = (525,960)(1 - A). \tag{7.14}$$

Comparisons of different levels of availability are shown in Table 7.1.

TABLE 7.1 Availability Comparisons

Availability	Unavailability	Five 9s	Downtime per Year
0.9	0.1	1	36 days
0.99	0.01	2	87.7 hours
0.999	0.001	3	8 hours, 46 minutes
0.9999	0.0001	4	52.5 minutes
0.99999	0.00001	5	5 minutes, 16 seconds
0.999999	0.000001	6	32 seconds

7.18 CONSTANT FAILURE RATE MODEL

The dice throw scenario (Chapter 3) can be generalized to apply to any event that is equally likely after each successive trial.

Let us say we have a hardware component that has a constant failure rate, λ. *A constant failure rate means that the chances of a failure do not change over any given time interval.* The failure can randomly occur by chance at any time, but the average failure rate approaches a constant λ.

If we consider a time t, we can slice the time into intervals Δt such that $t = n\Delta t$. Since the failure rate is constant, the probability of a device failing in any particular time interval Δt is $\lambda t/n$. In the first time interval Δt, the probability of the device failing after time Δt is equal to 1 minus the probability of *not* failing within the interval Δt:

$$P(\text{device failure after time } \Delta t) = 1 - (1 - \lambda t/n). \tag{7.15}$$

Given that the device has *not* failed in the first interval, then the probability that the device will fail after the second interval is equal to 1 minus the probability of not failing within the first interval Δt and not failing in the second interval Δt:

$$\begin{aligned} P(\text{device failure after time } 2\Delta t) &= 1 - (1 - \lambda t/n)(1 - \lambda t/n) \\ &= 1 - (1 - \lambda t/n)^2. \end{aligned} \tag{7.16}$$

Extending this to $n\Delta t$ intervals, the probability of failing after time $t = n\Delta t$ is equal to one minus the probability of not failing in any of the previous n intervals:

$$P(\text{device failure after time } t) = 1 - (1 - \lambda t/n)^n. \tag{7.17}$$

Conversely, the probability of the device not failing after time t is

$$R(t) = P(\text{no device failure after time } t) = (1 - \lambda t/n)^n. \tag{7.18}$$

Compare this equation with the geometric distribution Equation (4.15), where $(1 - f)^n$ is equal to $(1 - \lambda t/n)^n$. The geometric distribution then multiplies this probability by f (the probability of one failure occurring).

If the time intervals are small and n is very large, Equation (7.18) approaches an exponential distribution since by definition:

$$e^x = \lim_{n \to \infty}(1 + x/n)^n.$$

We can now see that the discrete geometric distribution converges to the continuous exponential distribution when n becomes very large.

From the above relationship, Equation (7.17) becomes:

$$P(\text{device failure after time } t) = F(t) = 1 - e^{-\lambda t}. \tag{7.19}$$

Figure 7.6 shows the failure probability.

The reliability or survivability of the device is the probability that the device will not fail, thus:

$$R(t) = 1 - F(t) = e^{-\lambda t}. \tag{7.20}$$

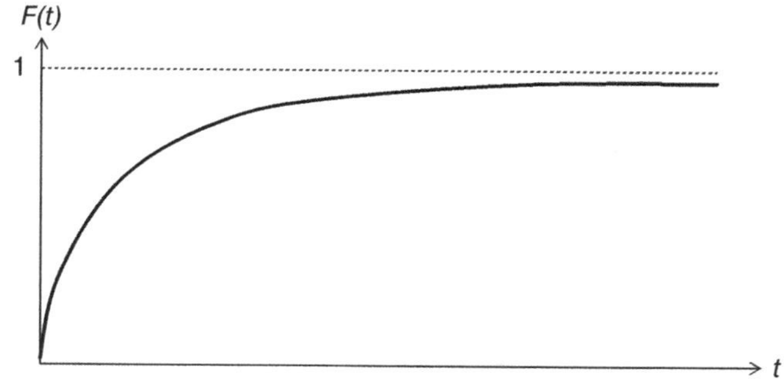

Figure 7.6 Exponential Failure Probability

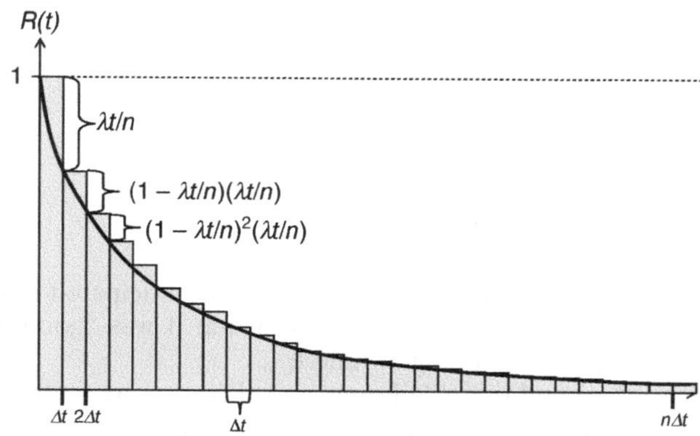

Figure 7.7 Constant Failure Rate Reliability

The reliability function for constant failure rate as a function of time for both finite n (geometric distribution) and the exponential limit is shown in Figure 7.7. This shows that initially, when the component is first put into service, the reliability is 100%, and then begins to decrease with time, which would be expected.

If we take the derivative of this function with respect to time, we obtain the failure density, that is, the probability per unit time that the system experiences its first failure at time t.

$$f(t) = \frac{dF(t)}{dt} = \frac{d(1-e^{-\lambda t})}{dt}$$
$$f(t) = \lambda e^{-\lambda t}. \tag{7.21}$$

This probability density function is shown in Figure 7.8.

This figure illustrates that the probability that the device fails at $t = 0$, is equal to the failure rate λ, and that the probability that the device will fail at any time t becomes smaller as t becomes larger. How do we interpret this, since we already

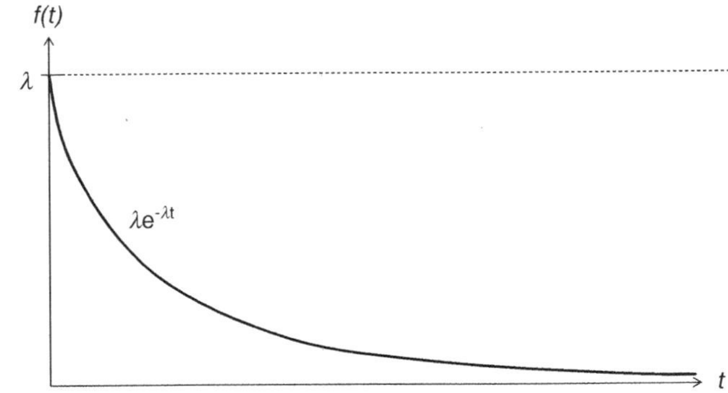

Figure 7.8 Exponential Failure Density

assumed the failure rate is constant (i.e., does not change with time)? The next section addresses this question.

7.19 CONDITIONAL FAILURE RATE

Conditional failure rate, or failure intensity, $\lambda(t)$, is the anticipated number of times an item will fail in a specified time period, given that it was "good as new" at an observed time zero and is functioning at time t.

Let us determine the conditional failure probability at some time $t + \Delta t$ given that the component has already survived up to time t.

Let A denote the event "the component will survive time $t + \Delta t$." The probability that the component will survive to time $t + \Delta t$ is $P(A) = P(T > t + \Delta t)$, where T is a random variable representing the survival time. Let B denote the event "the component has survived time t." The probability that the component will survive to time t, is $P(B) = P(T > t)$. $P(A|B)$ is then the probability that the component will survive time $t + \Delta t$ given that it has survived to time t, written more succinctly as $P(T > t + \Delta t|(T > t))$. From the conditional probability formula (Eq. (3.6) repeated here for convenience):

$$P(A \mid B) = \frac{P(A \cap B)}{P(B)}, \text{ where } P(A \cap B) = P(A)P(B|A). \tag{7.22}$$

$P(B|A)$ is the probability that the event B will occur given that event A has occurred as previously discussed in Chapter 3.

Since B represents the survival probability after A has already occurred, then event B will always be true when A is true, thus $P(B|A) = 1$ and $P(A \cap B)$ becomes $P(A)$:

$$P(T > t + \Delta t \mid (T > t)) = \frac{P(T > t + \Delta t)}{P(T > t)}. \tag{7.23}$$

$P(T > t)$ is equivalent to the reliability function $R(t)$ (the probability that the component will survive time t) and $P(T > t + \Delta t)$ is equivalent to the reliability function $R(t + \Delta t)$ (the probability that the component will survive time $(t + \Delta t)$). Furthermore, let us call $P(T > t + \Delta t|(T > t))$, the conditional reliability or $R(T|t)$.

$$R(T \mid t) = \frac{R(t + \Delta t)}{R(t)}. \tag{7.24}$$

Substituting in the exponential reliability Equation (7.20):

$$R(T \mid t) = \frac{e^{-\lambda(t+\Delta t)}}{e^{-\lambda t}} = e^{-\lambda \Delta t}. \tag{7.25}$$

The exponential conditional reliability equation expresses the reliability for a component within the time interval Δt, having already successfully accumulated T hours of operation.

Referring to Figure 7.9, we can observe that at any given time t in which the component has not already failed, the failure density is given by:

$$f(t + \Delta t) = \lambda e^{-\lambda \Delta t} \tag{7.26}$$

Equation (7.26) indicates that the reliability for a component in the time interval Δt after the component or equipment has already accumulated T hours of operation from time zero is only a function of the time interval Δt, and ***not*** a function of the time the component has been in service. The result is the same exponential distribution independent of the time the component has been in service. This is referred to as the *memoryless* property.

In general, a random variable has the memoryless property if the probability that an event occurs after time $t + \Delta t$ given that it has not occurred after time t is dependent only on the time interval Δt (i.e., the event is independent of the time t):

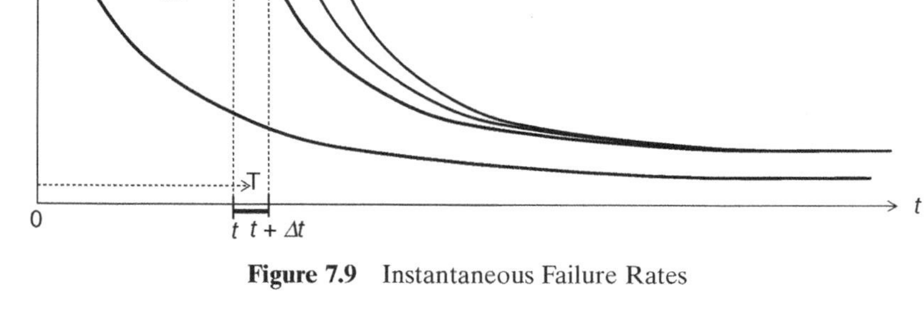

Figure 7.9 Instantaneous Failure Rates

$$P(X > t + \Delta t | (X > t)) = P(X > \Delta t). \tag{7.27}$$

Substituting Equation (7.20) into Equation (7.23), we obtain:

$$\begin{aligned} P(X > t + \Delta t \mid X > t) &= \frac{e^{-\lambda(t+\Delta t)}}{e^{-\lambda t}} \\ P(X > t + \Delta t \mid X > t)) &= e^{-\lambda \Delta t} = P(X > \Delta t). \end{aligned} \tag{7.28}$$

Thus the exponential distribution satisfies the memoryless property and in fact is the *only* distribution that has the memoryless property.

From Equation (7.19), the probability that a device will fail at any time t is equal to 1 minus the probability that the device will survive at any time t. Practically, this means that the probability that a device will fail given that it has not yet failed remains the same regardless of how long the device has been in service.

A random variable has the memoryless property if the probability that the event occurs after time t is independent of the probability that the event occurs within the interval Δt. Thus, using the independence rule, the probability that the random event occurs after time $t + \Delta t$ is equal to the probability that the random event occurs after time t and the probability that the random event occurs after the interval Δt:

$$P(X > t + \Delta t) = P(X > t)P(X > \Delta t). \tag{7.29}$$

Substituting Equation (7.20) into Equation (7.29), we obtain:

$$\begin{aligned} P(X > t + \Delta t) &= e^{-\lambda t} e^{-\Delta t} \\ P(X > t + \Delta t) &= e^{-\lambda(t+\Delta t)}. \end{aligned} \tag{7.30}$$

If we consider the time to failure as a waiting time, the occurrence of failures is a Poisson process, assuming the component that fails is immediately replaced with a new component having the same constant failure rate.

7.19.1 Hazard Rates and Failure Rates

For a stated period in the life of a system, the ratio of the total number of observed failures, k, to the total cumulative observed time T is the observed failure rate λ:

$$\lambda = \frac{k}{T}. \tag{7.31}$$

As previously discussed, $f(t)$ is defined as the probability density function of the time to failure of a given component. The probability that a component will fail between times t and $t + \Delta t$ is given by: $f(t)\Delta t$. Since the probability that a component will fail on the interval 0 to t is given by:

$$F(t) = \int_0^t f(x)d(x). \tag{7.32}$$

The probability that the component will fail in the interval from t to Δt is $F(t + \Delta t) - F(t)$.

Since $R(t) = 1 - F(t)$, and from Equation (7.24), the conditional probability of *survivability* in the interval $t + \Delta t$, given that the component has survived to time t, is given by:

$$P(T > t + \Delta t \mid T > t) = \frac{1 - \{F(t+\Delta t) - F(t)\}}{1 - F(t)}. \tag{7.33}$$

The conditional probability of *failure* is thus:

$$P(T < t + \Delta t \mid T > t) = \frac{F(t+\Delta t) - F(t)}{R(t)}. \tag{7.34}$$

Dividing by Δt we get the average rate of failure in this interval:

$$\frac{F(t+\Delta t) - F(t)}{\Delta t R(t)}.$$

Taking the limit as $\Delta t \to 0$, we then get the instantaneous (conditional) failure rate, also known as the *hazard rate*, $h(t)$:

$$\begin{aligned} h(t) &= \lim_{\Delta t \to 0} \frac{F(t+\Delta t) - F(t)}{\Delta t R(t)} \\ h(t) &= \frac{F'(t)}{R(t)}. \end{aligned} \tag{7.35}$$

Since $F'(t) = f(t)$, we get the general equation for the hazard rate function:

$$h(t) = \frac{f(t)}{R(t)} = \frac{-\frac{d}{dt}R(t)}{R(t)} = \frac{f(t)}{1 - F(t)}. \tag{7.36}$$

The hazard rate is sometimes called a *conditional failure rate*, since the denominator $1 - F(t)$ converts the expression into a conditional rate, given the component has survived up to time t.

$h(t)$ can be explained as the conditional probability that a device will fail in a unit time interval after t, given that it was working at time t. In other words, the instantaneous failure rate at time t equals the failure density function divided by the reliability evaluated at time t. Note that hazard rates are equivalent to failure rates only when the interval over which the failure rate is calculated approaches zero.

The hazard rate measures the instantaneous failure rate at a given time. In the next instant, the failure rate may change, and the units that have already failed have no impact on the hazard rate, since only components that have survived will count.

Knowledge of the hazard rate uniquely determines the failure density function, the unreliability function, and the reliability function.

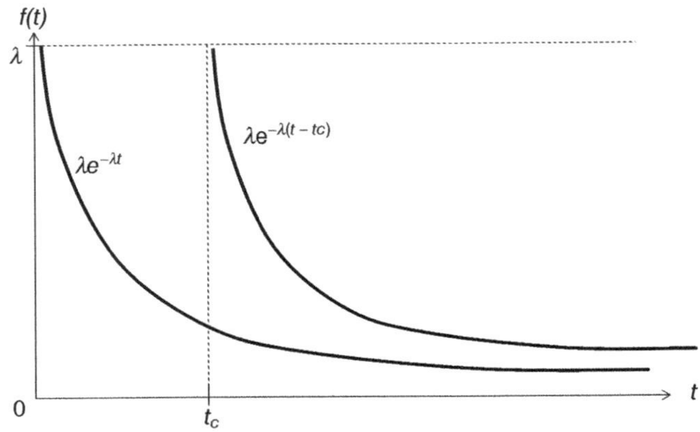

Figure 7.10 "Good as New" Failure Rate

For a component with a *constant* failure rate, substituting for Equations (7.20) and (7.21) into Equation (7.36), the hazard rate is:

$$h(t) = \frac{f(t)}{R(t)} = \frac{\lambda e^{-\lambda t}}{e^{-\lambda t}} = \lambda. \tag{7.37}$$

As Δt approaches zero, the instantaneous failure rate approaches λ. If we plot the reliability function $f(T)$ and $f(T + \Delta t)$ as shown in Figure 7.9, we see that the exponential distribution is identical for the two times. We can interpret this as follows: Whenever we observe a component and we know that the component has not failed, then the failure distribution will remain the same.

From Figure 7.10, we can see graphically that in the limit, as Δt approaches zero, the failure rate approaches the constant λ, which is the instantaneous (conditional) failure rate $h(t)$ from Equation (7.37).

Another way to look at this phenomenon is to consider any arbitrary time t_c (the time we observe the device). If the device is still functioning, then the failure density at $t = t_c$ is the same as the failure density at time $t = 0$ "Good as New" (Fig. 7.10).

7.19.2 General Reliability Equation

The failure rate $\lambda(t)$ is the probability of failure in a period t to $t + \Delta t$ under the condition that no failure occurred before t, divided by Δt and Δt going to 0.

$$\lambda(t) = \lim_{\Delta t \to 0} \frac{F(t + \Delta t) - F(t)}{\Delta t}.$$

The probability of a system failing in the interval between t and $t + dt$ given that it has survived until time t is:

$$\lambda(t)dt,$$

where $\lambda(t)$ is the failure rate.

The probability of failure in the interval t to $t + dt$ unconditionally is:

$$f(t)dt.$$

Now let us determine the most general form of the reliability equation. Rearranging Equation (7.36), we obtain:

$$h(t) = \frac{-\frac{d}{dt}R(t)}{R(t)} = -\frac{d(\ln R(t))}{dt}. \tag{7.38}$$

When integrating the ln form of $h(t)$, the general reliability function is obtained:

$$R(t) = e^{-\int_0^x h(x)dx}. \tag{7.39}$$

This is a mathematical description of reliability in the most general form. It is independent of the failure distribution used.

Many physical systems exhibit failure rates similar to the classic bathtub curve shown in Figure 7.11. The system has a large number of failures in the early stage that decrease over time. The system's useful life is then entered where failures are random but occur at a constant failure rate. As the system begins to wear out, it enters the end-of-life phase, and the rate of failures begins to increase over time. The Weibull distribution can be used to model these three phases. From Chapter 5, Equation (5.20), substituting the shape parameter β for m and t for x, we obtain:

$$h(t) = \frac{\beta}{a}(t/a)^{\beta-1}e^{-(t/a)^{\beta}}. \tag{7.40}$$

From Equation (7.36), $f(t) = R(t)h(t)$, thus

$$f(t) = h(t)e^{-\int_0^t h(x)dx}. \tag{7.41}$$

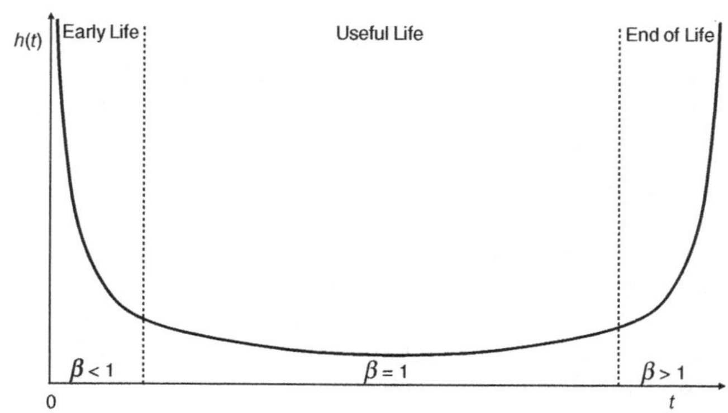

Figure 7.11 Reliability "Bathtub" Curve

If the failure rate is constant during the period of useful life of a component, we can substitute a constant failure rate of λ for $h(t)$ into the general failure equation:

$$f(t) = \lambda e^{-\lambda t}. \tag{7.42}$$

This matches the failure density function for a constant failure rate system—Equation (7.21). Note also that this equation matches the Weibull distribution Equation (7.40) for $\beta = 1$ and $a = 1/\lambda$.

7.20 BAYES'S THEOREM

We derive Bayes's theorem from conditional probability.

The conditional probability, Equation (3.6) is:

$$P(A \mid B) = \frac{P(A \cap B)}{P(B)}, \tag{7.43}$$

or equivalently, the conditional probability is:

$$P(B \mid A) = \frac{P(A \cap B)}{P(A)}. \tag{7.44}$$

We can combine these two equations to obtain:

$$P(A \mid B)P(B) = P(A \cap B) = P(A)P(B \mid A) \tag{7.45}$$

$$P(A \mid B) = \frac{P(B \mid A)P(A)}{P(B)}. \tag{7.46}$$

Equation (7.46) is Bayes's theorem. This formula determines the probability that event A occurs, given event B occurs.

Each term is interpreted as follows:

- $P(A)$: Probability of A prior to taking into account information about event B.
- $P(A|B)$: Probability of A given that event B has occurred. This conditional probability is a posterior probability since it is dependent on the event B.
- $P(B)$: Probability of event B.
- $P(B|A)$: Probability of event B given that event A occurred.

From a reliability point of view, before the collection of data, the failure rate is described by the prior probability, $P(A)$. After the collection of data, the new information is incorporated to modify the probability by calculating a posterior probability $P(A|B)$. In other words, the initial assumptions regarding the failure rate is updated (but not discarded!) by incorporating new data.

Bayes's rule can be extended if we partition the event space B into n mutually exclusive partitions $A_1 \ldots A_n$ (Fig. 7.12).

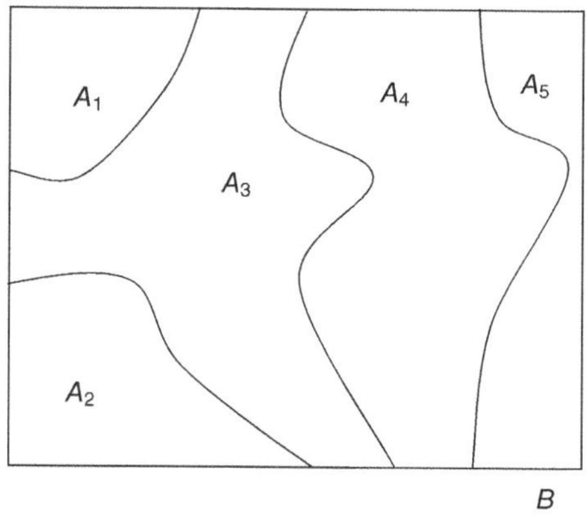

Figure 7.12 Mutually Exclusive Partitions Venn Diagram

$P(B)$ is then the sum of the probability of each partition multiplied by the probability of B:

$$P(B) = \sum_i P(B \cap A_i). \tag{7.47}$$

From Equation (7.44):

$$P(A \cap B) = P(B \mid A)P(A). \tag{7.48}$$

Thus:

$$P(B) = \sum_i P(B \cap A_i) = \sum_i P(B \mid A_i)P(A_i) \tag{7.49}$$

Equation (7.49) is the rule for total probability for mutually exclusive events.

EXAMPLE 7.1

To illustrate this rule in action, consider a processor board made by three manufacturers. We receive 30% of the boards from supplier A_1, 50% from supplier A_2, and the remaining 20% from supplier A_3. Ninety-eight percent of the boards from supplier A_1, 80% of the boards from supplier A_2, and 95% of the boards from supplier A_3 pass the acceptance test.

(a) What is the probability that any board randomly selected for testing will fail?
(b) What is the probability that the failed board is from supplier A_2?

(a) Substituting the data into Equation (7.49), we obtain:

$$\begin{aligned} P(B) &= P(B|A_1)P(A_1) + P(B|A_2)P(A_2) + P(B|A_3)P(A_3) \\ P(B) &= (0.02)(0.3) + (0.2)(0.5) + (0.05)(0.2) \\ P(B) &= 0.116. \end{aligned} \tag{7.50}$$

(b) From Figure 7.12, to determine the probability of any mutually exclusive event A_i given B, we have:

$$P(A_i \mid B) = \frac{P(B \mid A_i)P(A_i)}{P(B)}. \tag{7.51}$$

Now, we can substitute Equation (7.51) into Equation (7.49) to obtain Bayes's equation for mutually exclusive events:

$$P(A \mid B) = \frac{P(B \mid A)P(A)}{\sum_i P(B \mid A_i)P(A_i)}. \tag{7.52}$$

Using Bayes's theorem and substituting into the equation above:

$$P(A_2 \mid B) = \frac{(0.5)(0.2)}{0.116} = 0.862.$$

Although supplier $A2$ supplied only 50% of the boards, over 86% of the failed boards came from this supplier.

EXAMPLE 7.2

A fun illustration of Bayes's theorem in action is the Monty Hall problem. This problem was made famous by Marilyn vos Savant in 1990 when she posed the problem to readers of *Parade* magazine.

Here is the problem. You are on a game show in which three doors are shown. Two of the doors when opened will reveal a goat and the remaining door when opened reveals an expensive sports car.

You choose one of the three doors, but this door is not opened. After you announce your choice, the game show host opens one of the two remaining doors, which reveals a goat and offers you a chance to switch your choice to the other remaining door that has not been opened. Should you switch? Most respondents to *Parade* magazine—including math professors—indicated that it is equally as likely that the door you originally chose has the same probability of hiding a car as the remaining unopened door. Surprisingly, the odds are 2-to-1 in your favor if you switch your choice. How can that be?

Assume you choose door 1. For the two remaining doors, doors 2 and 3, our host must open one of these two remaining doors.

(a) What is the probability that the car is behind door 3, if the host opens door 2? Let us apply Bayes's theorem:

$$P(\text{Car}_{D=3} \mid \text{Host}_{D=2}) = \frac{P(\text{Host}_{D=2} \mid \text{Car}_{D=3}) P(\text{Car}_{D=3})}{P(\text{Host}_{D=2})}.$$

We read the above relationship as: The probability that the car is behind door 3 given that the host selected door 2 is equal to the probability that the host selected door 2 given that the car is behind door 3 times the probability that the car is behind door 3 all divided by the probability the host selects door 2.

Since the host knows which door has the car, and will not open a door that reveals a car, we can assign probabilities to the right-hand side of the equation and calculate the probability that the car is behind door 3 if the host selects door 2.

The probability that the car is behind any door is 1 out of 3:

$$P(\text{Car}_{D=3}) = 1/3.$$

The probability that the host picks either of the two remaining doors is 50%:

$$P(\text{Host}_{D=2}) = 1/2.$$

The probability that the host selects door 2 if the car is behind door 3 is 100%. The host must select door 2:

$$P(\text{Host}_{D=2} \mid \text{Car}_{D=3}) = 1.$$

Putting the probabilities into the equation, we obtain:

$$P(\text{Car}_{D=3} \mid \text{Host}_{D=2}) = \frac{(1)(1/3)}{1/2} = 2/3.$$

(b) Similarly, for the remaining possibility, what is the probability that the car is behind door 2, given the host opens door 3?

$$P(\text{Car}_{D=2} \mid \text{Host}_{D=3}) = \frac{P(\text{Host}_{D=3} \mid \text{Car}_{D=2}) P(\text{Car}_{D=2})}{P(\text{Host}_{D=3})}$$

$$P(\text{Car}_{D=2} \mid \text{Host}_{D=3}) = \frac{(1)(1/3)}{1/2} = 2/3.$$

(c) Finally, what is the probability that the car is behind door #1, the original choice by our contestant?

$$P(\text{Car}_{D=1} \mid \text{Host}_{D=3}) = \frac{P(\text{Host}_{D=3} \mid \text{Car}_{D=1}) P(\text{Car}_{D=1})}{P(\text{Host}_{D=3})}$$

$$P(\text{Car}_{D=1} \mid \text{Host}_{D=3}) = \frac{(1/2)(1/3)}{1/2} = 1/3.$$

So why did the probability go to 2/3 when you switch? It is because the host knows which door conceals the car and that knowledge is reflected in the decision she makes on which door to open. In particular, the host cannot open the door that contains the prize, and so that knowledge can be used to improve the odds for the contestant.

Bayes's theorem and conditional probability gives you more information and that information affects the probabilities.

We will use the same technique in Chapter 13 for updating our initial estimate of MTBF with new data, although we are unlikely to win a new sports car!

Bayes's theorem will be extended to included estimation of reliability parameters in Chapter 13.

7.21 RELIABILITY BLOCK DIAGRAMS

Reliability block diagrams (RBD) are arguably the simplest technique both in terms of understandability and calculation. RBDs show blocks representing a logical or physical component of the system. The relationships between these blocks are shown by lines that connect them and indicate how a failure of block components affects system reliability. These blocks are typically arranged in a series-parallel fashion to indicate redundant and nonredundant relationships. In a series relationship, failure of any block component results in a failure of all functionality associated with each block in series with the failed block. Blocks in parallel indicate that a failure of a single block will not result in failure of the system. For example, two blocks connected in parallel can represent a 2N redundant system. If one component fails, the redundant component is able to provide full system functionality.

Each component has an MTTF or a random failure distribution associated with it, and the probability of failure of these block components are independent of other components in the reliability block diagram. Armed with this information, quantitative system reliability is quickly calculated from the constituent parts and part relationships.

Disadvantages of RBDs include the following:

1. System architectures that do not reflect a series/parallel structure generally cannot be modeled.
2. Only nonrepairable components and systems can be modeled. Repair times cannot be captured separately. If the system is nonrepairable, then the system availability and reliability are identical.
3. RBDs can represent only two states per component: functioning and failed. Dependencies between components and multiple states within a component cannot be captured.
4. Time dependency is not captured. Reliability can be dramatically affected by Mission Time, for example.
5. The components in an RBD are considered to have independent failure probabilities. Dependent failures cannot be represented.

We will now consider several common RBDs.

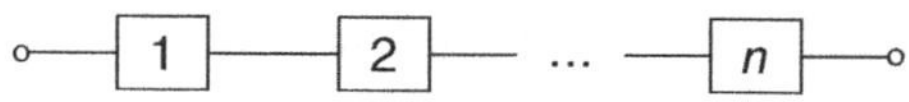

Figure 7.13 *n*-Component Series RBD

7.21.1 RBD for a Series System

Figure 7.13 shows the RBD for an n-component series system.

The series RBD indicates that if any component fails, the system fails. We are interested in the probability that the system survives during the mission time T. This is equal to the probability that *all* of the components survive during the mission time T. Since for an RBD, each component is independent, applying the probability of independent events (Eq. (3.3)), we have:

$$R_S(t) = P(T_S > t) = \prod_{i=1}^{n} P(T_{c_i} > t) = \prod_{i=1}^{n} R_{c_i}, \tag{7.53}$$

where S represents the system and c_i represents each component i.

If the reliability of the components is high, the system reliability can be approximated by:

$$R_S(t) = 1 - \sum_{i=1}^{n} (1 - R_{c_i}(t)). \tag{7.54}$$

We verify Equation (7.54) by noting that the component failure, $F(t)$, is $1 - R(t)$. Thus,

$$R_S(t) = \prod_{i=1}^{n} (1 - F_{c_i}(t)).$$

If the failure probabilities are very small, we can neglect the $F^2(t)$ terms and higher terms, leaving us with:

$$R_S(t) \approx \sum_{i=1}^{n} (1 - F_{c_i}(t)) = 1 - \sum_{i=1}^{n} (1 - R_{c_i}(t)).$$

Let us consider a two-component series system. The system is functioning if both components are functioning and the system is failed if either component fails.

We are interested in the probability that the system survives during the mission time T. This is equal to the probability that component 1 survives during the mission time T and the probability that component 2 survives during the mission time T.

$$R_S(t) = P(T_S > t) = P(T_{c1} > t)P(T_{c2} > t) \tag{7.55}$$

$$R_S(t) = R_{c1}(t)R_{c2}(t). \tag{7.56}$$

If components 1 and 2 exhibit constant failure rates, then:

$$R_S(t) = e^{-\lambda_1 t} e^{-\lambda_2 t} = e^{-(\lambda_1 + \lambda_2)t}. \tag{7.57}$$

And if both components have identical failure rates Equation (7.57) becomes:

$$R_S(t) = e^{-2\lambda t}. \tag{7.58}$$

We can extend this concept to n components in series to get the following general reliability equation:

$$R_S(t) = e^{-\sum_{i=1}^{n} \lambda_i t}. \tag{7.59}$$

The system failure rate λ_S is:

$$\lambda_S = \frac{-\ln(R_S(t))}{t} = \frac{-\ln\left(e^{-\sum_{i=1}^{n} \lambda_i t}\right)}{t} \tag{7.60}$$

$$\lambda_S = \sum_{i=1}^{n} \lambda_i.$$

We can also calculate the system failure rate using the hazard function (Eq. (7.36)):

$$\lambda_S(t) = h(t) = \frac{-\frac{d}{dt}R(t)}{R(t)}$$

$$\lambda_S(t) = \frac{\sum_{i=1}^{n} \lambda_i e^{-\sum_{i=1}^{n} \lambda_i t}}{e^{-\sum_{i=1}^{n} \lambda_i t}} \tag{7.61}$$

$$\lambda_S = \sum_{i=1}^{n} \lambda_i,$$

which gives the same result as Equation (7.60).

The failure rate of a system of n components in series is a constant failure rate equal to the sum of the individual component failure rates.

For constant failure rate models, the MTTF is the reciprocal of the failure rate. The system MTTF of an n-component system is calculated as follows:

$$\text{MTTF}_S = \frac{1}{\frac{1}{\text{MTTF}_{C1}} + \frac{1}{\text{MTTF}_{C2}} + \cdots + \frac{1}{\text{MTTF}_{Cn}}}. \tag{7.62}$$

For n identical components, the system reliability shown in Equation (7.59) reduces to:

$$R_S(t) = (R_c)^n = e^{-n\lambda_c t}. \tag{7.63}$$

The system failure rate for n identical components becomes:

$$\begin{aligned} \lambda_S &= \frac{-\ln(R_S(t))}{t} = \frac{-\ln(e^{-n\lambda_c t})}{t} \\ \lambda_S &= n\lambda_c. \end{aligned} \tag{7.64}$$

EXAMPLE 7.3

We have a two-component series system with component 1 MTTF = 10,000 hours and component 2 MTTF = 25,000 hours. We want to calculate the following after 1000 hours of operation:

(a) System MTTF
(b) System failure rate (failures per million hours)
(c) System reliability
(d) System availability.

Solution

(a) From Equation (7.62):

$$\text{MTTF}_S = 1/(1/10{,}000 + 1/25{,}000) = 7142.857 \text{ hours.}$$

(b) From Equation (7.61):

$$\lambda_S = (1/10{,}000 + 1/25{,}000)(1{,}000{,}000) = 140.$$

(c) From Equation (7.57):

$$R_S = e^{-(0.0001+0.00004)1000} = 0.8694.$$

(d) For nonrepairable systems, reliability and availability are the same.

$$A_S = R_S = 0.8694.$$

RBDs may be created using commercial reliability software. Figure 7.14 and Figure 7.15, respectively, show the RBD and the calculation results of this example using Relex.

The Relex tool can also graph the reliability of the system over the mission time of 100,000 hours (Fig. 7.16).

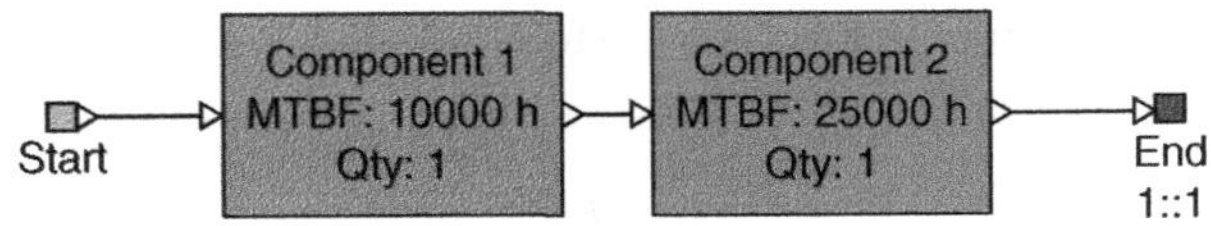

Figure 7.14 Two-Component Series RBD Using Relex Tool (Windchill)

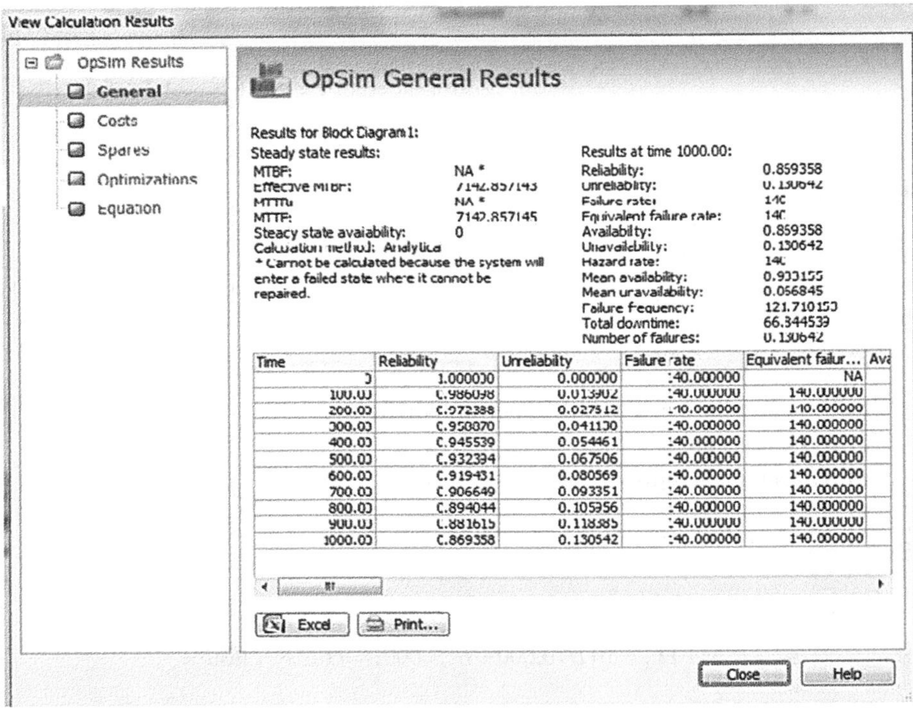

Figure 7.15 Two-Component Series RBD Calculations Using Relex Tool (Windchill)

7.21.2 RBD for a Parallel System

The RBD for an n-component parallel system is shown in Figure 7.17.

The parallel RBD indicates the system is functioning if *at least* one of the components in parallel is functioning.

The probability that the system survives during the mission time T is equal to the probability that *any* component survives during the mission time T:

$$R_S(t) = P\{T_S > t\} = P\{(T_{c1} > t) \cup (T_{c2} > t) \cdots \cup (T_{cn} > t)\}. \tag{7.65}$$

To express Equation (7.65) in terms of component reliabilities, we note that *all* components must fail for the system to fail, and since the probability of failure for each component is independent, we have:

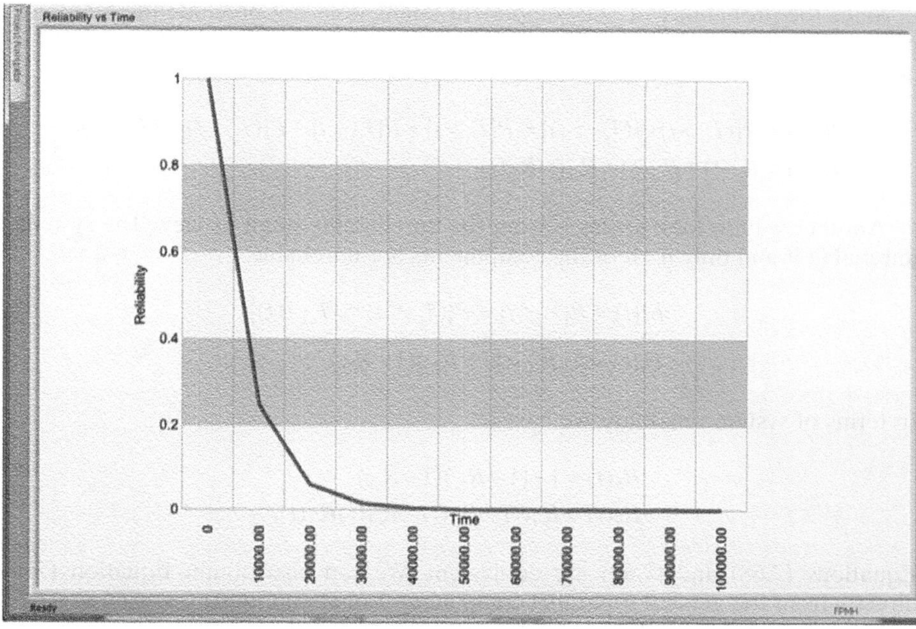

Figure 7.16 Two-Component Series RBD Reliability versus Time Using Relex Tool (Windchill)

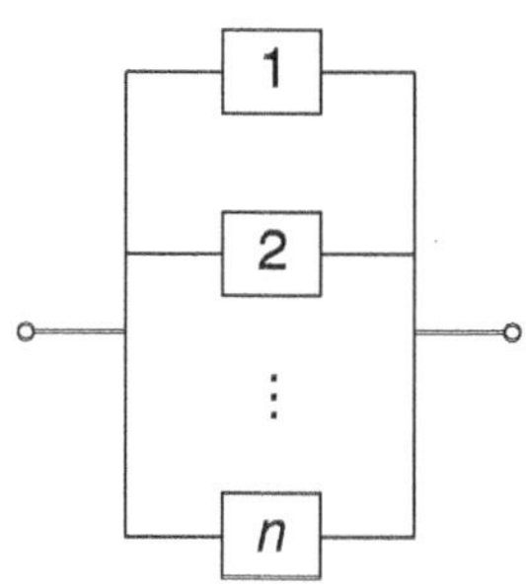

Figure 7.17 n-Component Parallel RBD

$$R_S(t) = 1 - \prod_{i=1}^{n} F_{c_i}(t) = \prod_{i=1}^{n}(1 - R_{c_i}(t)). \tag{7.66}$$

Let us consider a two-component parallel system. The probability that the system survives during the mission time T is equal to the probability that either component 1 or component 2 survives during the mission time T or the probability that both components survive during the mission time T:

$$R_S(t) = P\{T_S > t\} = P\{(T_{c_1} > t) \cup (T_{c2} > t)\}. \tag{7.67}$$

Since the probability of one component failing does not preclude the other component from failing, the components are nonmutually exclusive, thus Equation (3.4) applies:

$$
\begin{aligned}
R_S(t) &= P\{(T_{c1} > t) \cup (T_{c2} > t)\} = P\{T_{c1} > t\} + P\{T_{c2} > t\} - P\{(T_{c1} > t) \cap (T_{c2} > t)\} \\
R_S(t) &= R_{c1}(t) + R_{c2}(t) - R_{c1}(t) R_{c2}(t).
\end{aligned}
\tag{7.68}
$$

Another way to look at this is from the unreliability point of view. The system is unreliable if and only if all of the components are unreliable:

$$
\begin{aligned}
F_S(t) &= P\{T_S \le t\} = P\{(T_{c1} \le t) \cap (T_{c2} \le t)\} \\
F_S(t) &= F_{c1} F_{c2} = (1 - R_{c1})(1 - R_{c2}).
\end{aligned}
$$

In terms of system reliability, we have:

$$
\begin{aligned}
R_S(t) &= 1 - (1 - R_{c1})(1 - R_{c2}) \\
R_S(t) &= R_{c1}(t) + R_{c2}(t) - R_{c1}(t) R_{c2}(t).
\end{aligned}
\tag{7.69}
$$

Equations (7.68) and (7.69) are equivalent. We can also obtain Equation (7.68) directly from the general n-parallel component, Equation (7.66).

If components 1 and 2 exhibit constant failure rates, then:

$$
\begin{aligned}
R_S(t) &= 1 - (1 - e^{-\lambda_1 t})(1 - e^{-\lambda_2 t}) \\
R_S(t) &= e^{-\lambda_1 t} + e^{-\lambda_2 t} - e^{-(\lambda_1 + \lambda_2)t}.
\end{aligned}
\tag{7.70}
$$

From Equation (7.6):

$$
\text{MTTF}_S = \int_0^{\infty} (e^{-\lambda_1 t} + e^{-\lambda_2 t} - e^{-(\lambda_1 + \lambda_2)t})\,dt \tag{7.71}
$$

$$
\text{MTTF}_S = \left.\frac{-e^{-\lambda_1 t}}{\lambda_1}\right|_0^{\infty} - \left.\frac{e^{-\lambda_2 t}}{\lambda_2}\right|_0^{\infty} + \left.\frac{e^{-(\lambda_1 t + \lambda_2)t}}{(\lambda_1 + \lambda_2)}\right|_0^{\infty}
$$

$$
\text{MTTF}_S = \frac{1}{\lambda_1} + \frac{1}{\lambda_2} - \frac{1}{(\lambda_1 + \lambda_2)} \tag{7.72}
$$

$$
\text{MTTF}_S = \text{MTTF}_1 + \text{MTTF}_2 - \frac{1}{\dfrac{1}{\text{MTTF}_1} + \dfrac{1}{\text{MTTF}_2}}. \tag{7.73}
$$

If both components have identical failure rates, Equation (7.70) above becomes:

$$
R_S(t) = 2e^{-\lambda t} - e^{-2\lambda t}. \tag{7.74}
$$

Another method to calculate the reliability is by using Bayes's theorem, Equation (7.49). Restating reliability in terms of Bayes's theorem, the reliability of a system is equal to the sum of (the reliability of the system when component x is working) multiplied by (the probability that component x is working) plus (the reliability of the system when component x is *not* working) multiplied by (the probability that component x is not working).

$$R_S = (R_s|\text{one component functioning})R_c + (R_s|\text{one component failed})(1 - R_c).$$

For the identical two-component parallel system, we have:

$$R_S = (1)R_c + R_c(1 - R_c) = 2R_c - R_c^2.$$

Substituting R_c, we have:

$$R_S(t) = 2e^{-\lambda t} - e^{-2\lambda t},$$

which agrees with Equation (7.74).

For this simple system, the Bayesian approach is trivial. However, when analyzing much more complex RBDs, this method can simplify the calculations.

To calculate the system conditional failure rate for a two-component parallel system, we need to employ Equation (7.38):

$$\lambda_S(t) = h(t) = \frac{-\frac{d}{dt}R(t)}{R(t)}$$

$$\lambda_S(t) = \frac{\lambda_1 e^{-\lambda_1 t} + \lambda_2 e^{-\lambda_2 t} - (\lambda_1 + \lambda_2)e^{-(\lambda_1+\lambda_2)t}}{e^{-\lambda_1 t} + e^{-\lambda_2 t} - e^{-(\lambda_1+\lambda_2)t}}. \tag{7.75}$$

From this equation, we note that the system failure rate is no longer constant even though the two components that compose the system have constant failure rates. This nonconstant system failure rate holds true for the general case of n components.

For constant failure rate models, the MTTF is the reciprocal of the failure rate. The system MTTF of an identical n-component parallel system is calculated as follows:

$$\text{MTTF}_S = \sum_{i=1}^{n} \text{MTTF}_c - \frac{1}{\sum_{i=1}^{n} \frac{1}{\text{MTTF}_c}} = \text{MTTF}_c\left(1 + \frac{1}{2} + \frac{1}{3} + \cdots + \frac{1}{n}\right). \tag{7.76}$$

The reliability of an n-component system is:

$$R_S(t) = 1 - (1 - e^{-\lambda_1 t})(1 - e^{-\lambda_2 t})\cdots(1 - e^{-\lambda_2 t})$$

$$R_S(t) = \sum_{i=1}^{n} \binom{n}{i} R_c^i (1 - R_c)^{n-i}.$$

EXAMPLE 7.4

A two-component parallel system has component 1 MTTF of 10,000 hours and component 2 MTTF of 25,000 hours. We want to calculate the following after 1000 hours of operation:

(a) System MTTF
(b) System failure rate (failures per million hours)
(c) System reliability
(d) System availability.

Solution

We know that the failure rates are:

$$\lambda_1 = 1/10{,}000 = 0.0001$$

$$\lambda_2 = 1/25{,}000 = 0.00004$$

(a) From Equation (7.73):

$$\mathrm{MTTF}_s = 1/0.0001 + 1/0.00004 - 1/(0.0001 + 0.00004) = 27{,}857.14 \text{ hours.}$$

(b) From Equation (7.75):

$$\lambda_S = 0.00000723.$$

(c) From Equation (7.70):

$$R_S = 0.9963.$$

(d) For nonrepairable systems reliability and availability are the same.

$$A_S = R_s = 0.9963.$$

Figure 7.18 shows the RBD model in Relex, and Figure 7.19 and Figure 7.20 show the reliability calculation results.

We will consider more examples of employing parallel and series RBDs during Markov analysis is Chapters 10 and 11.

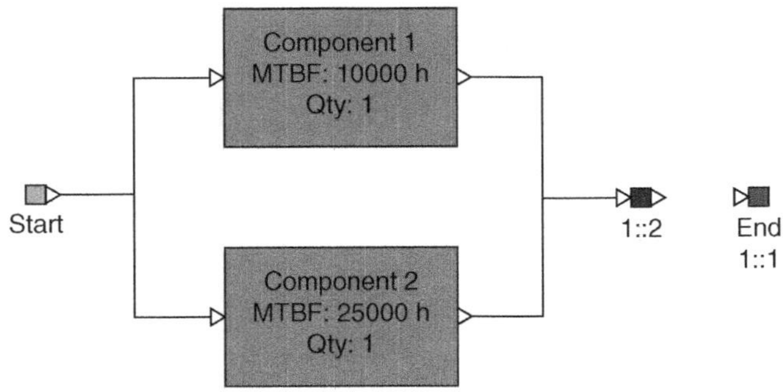

Figure 7.18 Two-Component Parallel RBD Using Relex Tool (Windchill)

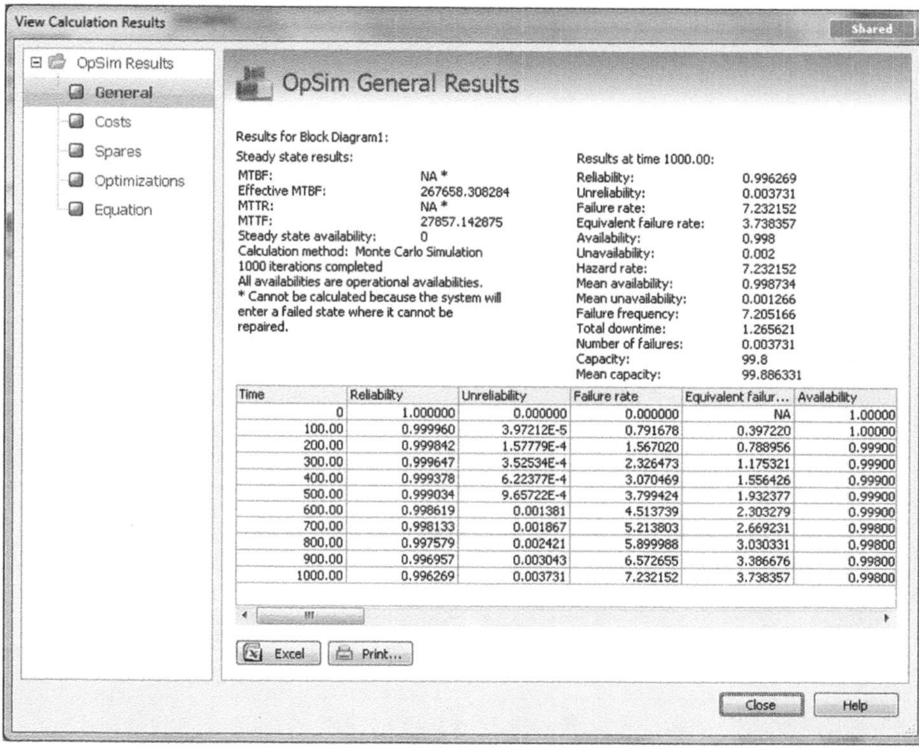

Time	Reliability	Unreliability	Failure rate	Equivalent failur...	Availability
0	1.000000	0.000000	0.000000	NA	1.00000
100.00	0.999960	3.97212E-5	0.791678	0.397220	1.00000
200.00	0.999842	1.57779E-4	1.567020	0.788956	0.99900
300.00	0.999647	3.52534E-4	2.326473	1.175321	0.99900
400.00	0.999378	6.22377E-4	3.070469	1.556426	0.99900
500.00	0.999034	9.65722E-4	3.799424	1.932377	0.99900
600.00	0.998619	0.001381	4.513739	2.303279	0.99900
700.00	0.998133	0.001867	5.213803	2.669231	0.99800
800.00	0.997579	0.002421	5.899988	3.030331	0.99800
900.00	0.996957	0.003043	6.572655	3.386676	0.99800
1000.00	0.996269	0.003731	7.232152	3.738357	0.99800

Figure 7.19 Two-Component Parallel RBD Calculations Using Relex Tool (Windchill)

7.21.3 Complex RBDs

More elaborate configurations, such as complex communication networks, exist that combine the system components in a variety of ways (Fig. 7.21). These RBDs will not be analyzed in this text. Refer to the literature for calculation methods.

7.22 SUMMARY

The important relationships between reliability distributions are captured in Table 7.2.

We defined reliability engineering and design and showed the relationship with other related terminology. Reliability modeling is a tool used for reliability prediction, and a variety of these models are available. The modeling approaches for reliability assessment were described. Failure probability and failure density were introduced, followed by a description of the most important reliability terminology along with fundamental reliability equations and several examples.

We also introduced Bayes's theorem in the context of reliability theory followed by a discussion of RBDs.

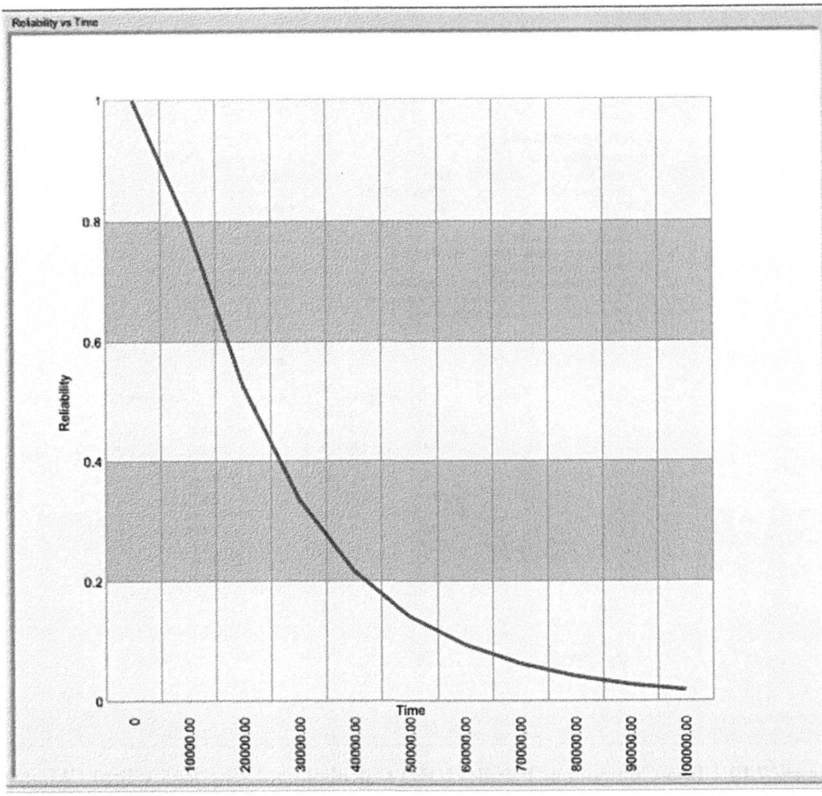

Figure 7.20 Two-Component Parallel RBD Reliability versus Time Using Relex Tool (Windchill)

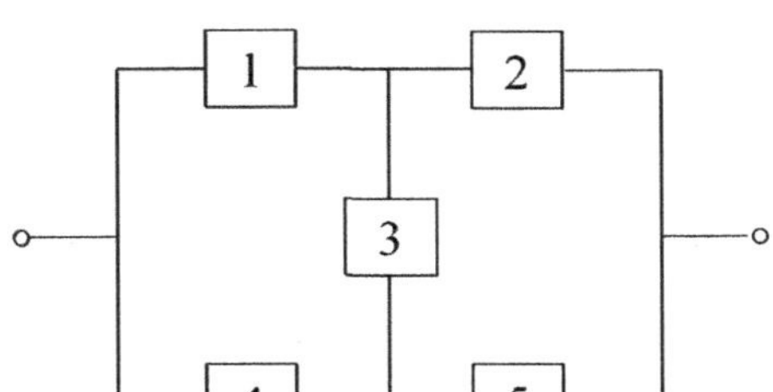

Figure 7.21 Complex RBD

TABLE 7.2 Relationship between Various Reliability Functions

$Y = f(x)$ Y: Row $F(x)$: Column	$f(t)$	$F(t)$	$R(t)$	$h(t)$
$f(t)$	–	$\frac{dF(t)}{dt}$	–	$h(t)e^{-\int_0^t h(x)dx}$
$F(t)$	$\int_0^t f(x)dx$		$1 - R(t)$	$1-e^{-\int_0^t h(x)dx}$
$R(t)$	$\int_t^\infty f(x)dx$	$1 - F(t)$	–	$e^{-\int_0^t h(x)dx}$
$h(t)$	–	$\frac{\frac{dF(t)}{dt}}{1-F(t)}$	$-\frac{d}{dt}\ln R(t)$	–
MTTF	$\int_0^\infty tf(t)dt$	–	$\int_0^\infty R(t)dt$	–

CHAPTER 8

Discrete-Time Markov Analysis

8.1 INTRODUCTION

Markov models are frequently used in reliability theory to model and understand the behavior of complex dynamic systems. We will explore Markov models in some depth over the next several chapters and consider further design applications in later chapters. Markov models, when applicable, provides a wealth of information on the reliability of specific systems. If we can describe a system and its components by creating an architecture block diagram, we can just as easily describe the reliability of the system and its components by creating a Markov state diagram.

A Markov process is a stochastic process in which the history of the process has no influence on the future evolution of the process; the future is *solely* influenced by the current state of the process. The next random variable is dependent *only* on the current random variable, irrespective of the past history of the process. The current random variable is referred to as the current state of the process, and the next random variable is the next state of the process.

This chapter provides the fundamental definition and application of discrete-time Markov modeling.

Chapter 9 will introduce the concept of continuous-time Markov modeling as applied to reliability. We will derive some important attributes of reliability from the model.

Chapter 10 will describe two types of systems: (1) nonrepairable systems along with illustrative examples, and (2) partially repairable systems in which some but not all components are repairable; thus, once the process enters a particular failure state, the system will remain in that state forever.

From a practical standpoint, a nonrepairable system applies to one of the following scenarios:

1. The system has failed and cannot be repaired due to unavailability of a repair process.
2. The system has failed and is not repaired due to the infeasibility of repair.

Designing High Availability Systems: Design for Six Sigma and Classical Reliability Techniques with Practical Real-Life Examples, First Edition. Zachary Taylor and Subramanyam Ranganathan.

3. The system may be repaired or replaced outside the mission time and hence is outside the scope of the Markov analysis.

Chapter 11 will then explore several examples of repairable systems. In terms of reliability, this implies that once any component has failed, it can be repaired, thus allowing a transition from any failed or degraded state.

A Markov chain is a discrete-time Markov process that can occupy a finite or countably infinite number of states. A discrete-time Markov chain is a discrete-time stochastic process $\{X_n | n = 0, 1, 2, \ldots\}$ that has a set of finite or countably infinite values. If $X_n = j$, then the process is in state S_j at time step n. The process changes only occur at fixed-time intervals represented by the index. Each time interval can represent a day, an hour, a year, or remain nondescript depending on the application of the Markov chain to a particular problem. Given the historical sequence of all random variables up to the current random variable X_n, the next random variable X_{n+1} depends only on X_n and the constant transition probabilities:

$$\begin{aligned} &p[X_{n+1} = S_j \mid X_n = S_i, X_{n-1} = S_{i_{n-1}}, \ldots, X_0 = S_{i_0}] = \\ &p[X_{n+1} = S_j \mid X_n = S_i] = p_{ij}. \end{aligned} \tag{8.1}$$

p_{ij} is the constant probability that the next state will be S_j given that the current state is S_i.

In a Markov model, the system is modeled as a state machine with each state representing some characteristic of interest for the system. The system begins with an initial state and can transition to other states. These state machines are defined by a set of probabilities. For each state, a probability is defined for leaving that state and transitioning to another state. The transition, if it occurs, happens instantaneously.

Recall from Chapter 6 that a random variable maps a set of random events that evolve over time from a sample space to a set of real values (Fig. 6.1). We will now apply this general mapping to a one-component reliability model. The one-component system has two states: S_1 is the operational state and state S_2 is defined as the failed state. The system is assumed to be initially operational at $t = 0$. Based on the failure rate of the component, the system will eventually fail. Upon failure, the system enters State 2 and will remain in this state until repaired. The system will then transition back to State 1. This repairable system is cyclical, that is, it will cycle between the operational and failed states for the defined mission time.

Figure 8.1 illustrates this reliability model with five consecutive events. The random process maps the events within the sample space S to the real values $\{0,1\}$, where 0 is the failed state and 1 is the operational state. The time-dependent system operational event (identified by S_1) is mapped to the real value 1 at the time of the event occurrence. The random process will remain at the value 1 until the failure event occurs. Once the failure event occurs, the event is mapped to the real value 0. Each time a transition occurs, a change in the value of the random process occurs.

These random events are governed by the underlying probability distributions. Recall that the constant failure rate and repair rate systems have an exponential probability distribution (see Fig. 3.2). The failure probability is equal to the failure rate at the point in time when the system is operational and the probability that the system fails decreases exponentially with time. This failure distribution determines

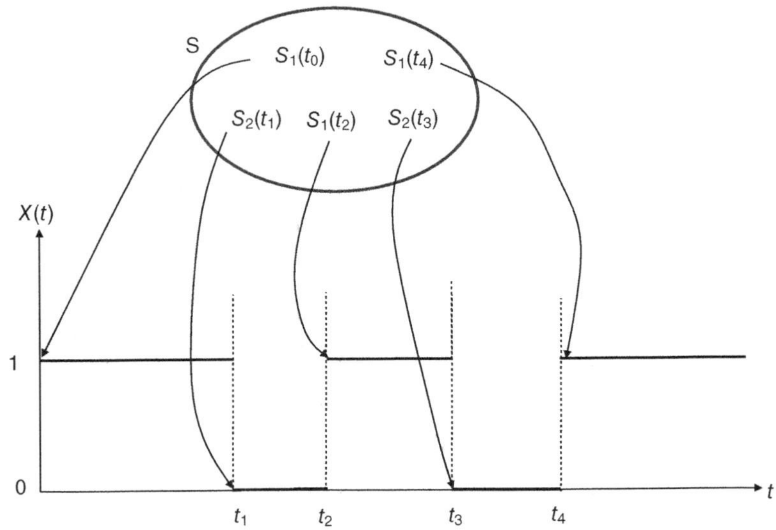

Figure 8.1 Random Process Mapping of a Repairable System

how likely the device will fail at different times, that is, determines how likely the system will enter the failed state.

We will consider two classes of Markov models applied to reliability:

1. *Irreducible.* Every state can be reached from every other state. This model corresponds to a repairable system. No matter what state you are in, you can leave the state by one or more paths.
2. *Absorbing.* One or more states once entered, cannot be exited—you can check in but you can never checkout! The process becomes "stuck" once it enters an absorbing state. This model corresponds to a nonrepairable system.

8.2 MARKOV PROCESS DEFINED

A Markov process is a stochastic process that moves from state to state with the following properties:

1. The probability of transitioning to a particular state depends only on the current state of the process.
2. The transition probabilities between states are constant over time.
3. The sum of all transition probabilities from a given state to any other state (including the current state) must be 1.

In this text, we will consider only those processes in which the number of possible states is finite.

A Markov process may be represented graphically by a *state transition diagram* that shows the states and the possible transitions from each state of the Markov process. In the state transition diagram, each state is represented as a labeled bubble or circle, and directed arcs represent transition rate probabilities between the states. The states are mutually exclusive, that is, the system can only be in one state at a time.

EXAMPLE 8.1

To illustrate a simple Markov process, let us go back to our dice example and roll a die every 10 minutes. Each time we roll a '*1*', we open the window if it is shut and close the window if it is open. Create a Markov state transition diagram for the system.

Solution

Since the probability of rolling a '*1*' is 1/6, we can expect to take an action only 1/6 of the time. Thus if the window is open, we can conclude that it is more likely that the window will remain open after the next toss of the die (probability of 5/6). If the window is closed, we can then conclude that the probability that the window will remain closed after the next toss of the die is 5/6 as well. Thus, the system has memory. The current state affects the probability of what state the window will be in after the next roll of the die, and we can predict the probabilistic evolution of the states for all future times. However, keep in mind that the history of the state transitions has no impact on the next state of the system. The opening and closing on the window is a Markov process since it meets all three criteria for the Markov process listed above.

1. The number of states is 2: Window open and window closed.
2. The probability of the window being open depends on whether the window is currently opened or closed,
3. The transition probabilities are constant (based on the random throw of the die).

From this information, the state transition diagram for the system is created (Fig. 8.2), where State 1 corresponds to window open and State 2 corresponds to window closed:

Conversely, the roll of the die itself is *not* a Markov process, since the second criteria is violated: The probability of rolling any particular value on the next throw is independent of the current state of the die. The die has no memory. Similarly, each time you flip a coin, it has no memory of what happened on the previous coin toss. Each toss is an independent event, that is, not dependent on the previous sequence of heads and tails events.

A set of state transition equations can be extracted from this diagram and solved using the techniques described in this chapter and in the following chapters.

A Markov model as applied to reliability and availability analysis is a set of defined states for a system being modeled, and a set of directed transition paths from one state to another state. For each of these paths, a constant rate parameter is assigned,

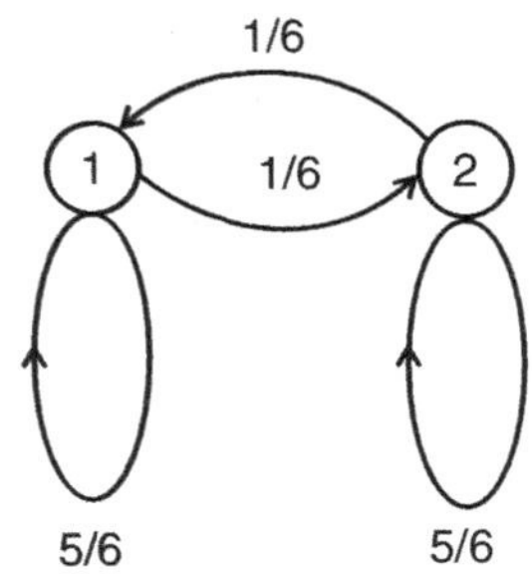

Figure 8.2 Open/Close Window State Machine

which defines the rate of transition from one state to the other. The probability of remaining in a state or transitioning out of that state is dependent solely on the current state of the system and the rate probabilities for the state transitions.

In comparison with combinational techniques, such as RBD (Chapter 7) and fault tree analysis (FTA, described in Chapter 16), partial failures, capacity loss, and repair strategies can be easily modeled and identified.

Although the number of states in a Markov model can grow to be very large, these models are relatively easy to solve, even for complex systems, using commonly available tools, such as Microsoft Excel.

EXAMPLE 8.2

Let us consider a weather prediction model. It is predicted that after a sunny day, there is a 75% chance that the next day will be sunny. And after a cloudy day, there is a 60% chance that the next day will be cloudy.

Solution

We first create the state transition diagram with the two states and the probabilities of remaining in a state or transitioning to another state based on the probabilities given in the problem (Fig. 8.3).

State 1 represents the sunny day state and State 2 represents the cloudy day state. We will remain in State 1 with a 75% probability and transition to State 2 with a 25% probability, and so on.

Writing the state transition matrix P, based on the state transition diagram, we have:

$$P = \begin{bmatrix} p_{11} & p_{21} \\ p_{12} & p_{22} \end{bmatrix}$$

$$P = \begin{bmatrix} 0.75 & 0.4 \\ 0.25 & 0.6 \end{bmatrix}.$$

Observe the following: The sum of the probabilities in each column is equal to 1.

If we assume the initial condition of the system $\mathbf{p}_1$ on day 1 is

$$\begin{bmatrix} 1 \\ 0 \end{bmatrix},$$

that is, we look out the window and see with 100% certainty that it is sunny outside.

If we want to predict the state of the next day, $\mathbf{p}_2$, we take the transition probability matrix P and multiply it by the initial state $\mathbf{p}_1$:

$$\mathbf{p}_2 = P\mathbf{p}_1 = \begin{bmatrix} 0.75 & 0.4 \\ 0.25 & 0.6 \end{bmatrix}\begin{bmatrix} 1 \\ 0 \end{bmatrix} = \begin{bmatrix} 0.75 \\ 0.25 \end{bmatrix}.$$

Thus, the next day, we have a 75% chance of being sunny and a 25% chance of being cloudy.

To predict the state of the third day, $\mathbf{p}_3$, we take the transition probability matrix P and multiply it by State $\mathbf{p}_2$:

$$\mathbf{p}_3 = P\mathbf{p}_2 = \begin{bmatrix} 0.75 & 0.4 \\ 0.25 & 0.6 \end{bmatrix}\begin{bmatrix} 0.75 \\ 0.25 \end{bmatrix} = \begin{bmatrix} 0.6625 \\ 0.3375 \end{bmatrix}.$$

Or we can find $\mathbf{p}_3$ directly from $\mathbf{p}_1$:

$$\mathbf{p}_3 = P(P\mathbf{p}_1) = P^2\mathbf{p}_1 = \begin{bmatrix} 0.75 & 0.4 \\ 0.25 & 0.6 \end{bmatrix}^2\begin{bmatrix} 1 \\ 0 \end{bmatrix} = \begin{bmatrix} 0.6625 \\ 0.3375 \end{bmatrix}.$$

Thus, the third day, we have a 63.2% chance of being sunny and a 36.8% chance of being cloudy, given that the first day was sunny.

For our weather example as shown in the state diagram in Figure 8.3, no absorbing state exists, that is, there is no state in which after entering we cannot exit.

In general, the probability that the chain is in state S_i after n steps is the ith entry in the vector.

$$\mathbf{p}_n = P^{n-1}\mathbf{p}_1. \tag{8.2}$$

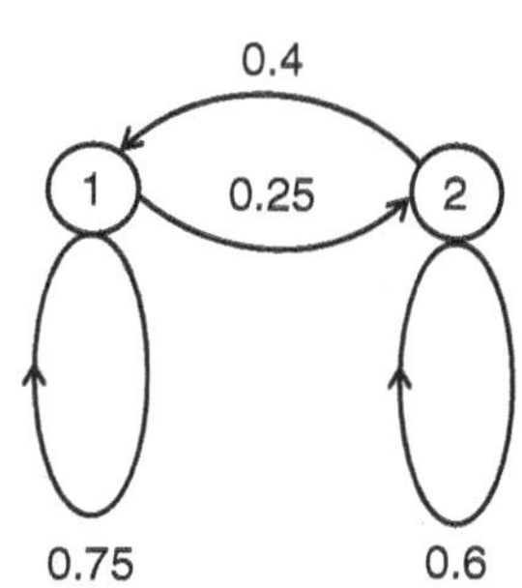

Figure 8.3 Example of a Weather Prediction State Machine

8.3 DYNAMIC MODELING

Dynamic modeling is used to model the evolution of processes over time. In order for some change to occur in the system, some time must have passed. Two types of time representations can be employed: continuous time and discrete time. Discrete time is a series of time points with a fixed or arbitrary constant time interval between each point. This time progression can be represented as a sequence of numbers, $1, 2, 3, \ldots$ with an arbitrary or specified fixed time step between each sequence number.

Discrete time requires a distinction between time points and time intervals. The time axis is divided into a number of adjacent time segments (typically of fixed length), whereas continuous time is a continuum of an infinite number of points.

Discrete time points or time intervals:

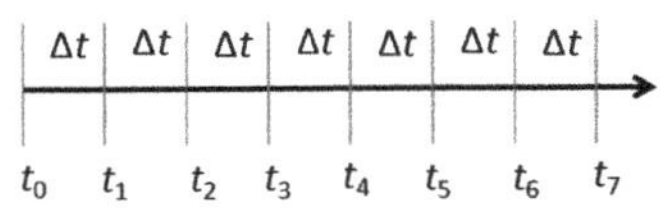

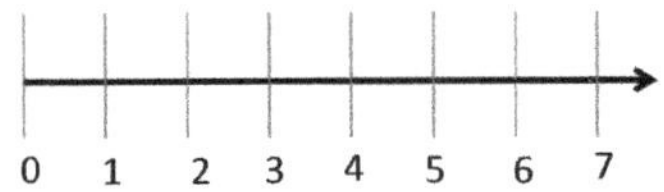

A *Continuous-Time Markov Chain* (CTMC) is one in which changes to the system can happen at any time along a continuous interval. An example is the number of phone calls received at the computer system help desk at a company throughout the day. A computer user can face an unexpected problem and call the help desk at any time t rather than at discrete time intervals. If we assume call arrivals are a Markov process, then if you know the number of calls received by noon on a work day, what happened before noon does not give you any additional information that would be useful in predicting the number of calls that will be received by the end of the business day.

For continuous time modeling, no series of time steps is required. The change in time for any point in time is indicated by an infinitesimally small time step, dt. Instead of a series of time-points, in the continuous model, we express changes in system behavior using infinitesimal time intervals. Since all changes are based on momentary rates of change to time points, fixed time intervals are no longer required.

Continuous sequence of points in time:

8.4 DISCRETE TIME MARKOV CHAINS

Which model should we use? If the data are quantized over specific fixed intervals, then the discrete model can be applied, whereas if the data are continuous and we are interested in a change at any arbitrary time, the continuous model can be applied. For example, if we are not interested in the weather at any arbitrary point in time during the day, but are interested in the general weather description on any given day (days are countable discrete units of time), then a *Discrete-Time Markov Chain* (DTMC) can be applied. The discrete Markov chain has a transition matrix that provides the probabilities of changing from one state to another state during one iteration step. Given an initial weather condition on the first day, the weather prediction for the next day is simply the product of the probability transition matrix times the current state of the weather. All that is needed is the current state of the system and the probability transition matrix. Note that once we denote the meaning of the time step (in this case, days), we cannot arbitrarily change it to represent a different time interval, that is, the weather at each hour of the day. To do so, we would need to reformulate the given problem.

If we define a finite set of states $S = \{S_1, S_2, \ldots, S_r\}$, then we can define a *probability vector*, the set of probabilities of occupying the states:

$$\mathbf{p} = \begin{bmatrix} p_1 \\ p_2 \\ \vdots \\ p_r \end{bmatrix}. \tag{8.3}$$

$\mathbf{p}$ represents the set of probabilities of being in a particular state.

Since all probabilities added together should be equal to one, we have $p_1 + p_2 + \cdots + p_r = 1$.

Note that the system must be in one of the defined states at any time step n, and the sum of the probabilities across all states must be equal to 1.

$$\sum_{i=1}^{r} p_i = 1, \tag{8.4}$$

where r is the number of states.

We can then define the transition probability matrix P for all possible state transitions with the condition that the probability of transitioning from any state to any other state is dependent only on the current state.

$$\begin{array}{ccccccl} & & S_1 & S_2 & & S_r & \leftarrow \text{State } n \\ & S_1 & P_{11} & P_{21} & \cdots & P_{r1} & \\ P = & S_2 & P_{12} & P_{22} & \cdots & P_{r2} & \\ & \vdots & \vdots & \vdots & \vdots & \vdots & \\ & S_r & P_{1r} & P_{2r} & \cdots & P_{rr} & \\ & \uparrow & & & & & \\ & \text{State } n+1 & & & & & \end{array}$$

Each element in the array represents a transition probability from the current state to the next state.

$$P_{ij} = P(\text{transitioning to state } j | \text{current state is } i).$$

From the Markov process definition, the above transition probabilities are constant. From the matrix, observe that:

$$\mathbf{p}_{n+1} = P.\mathbf{p}_n, \tag{8.5}$$

where $\mathbf{p}_n$ is the probability vector of being in one of the states at the time stamp n, and $\mathbf{p}_{n+1}$ is the probability vector of being in one of the states at time stamp $n + 1$.

Given an initial probability vector P_0, we can determine the probability vector at any step by iterating through Equation (8.5).

8.4.1 Limiting State Probabilities

The steady-state behavior of Markov processes when the number of steps becomes very large is important to understand for the analysis of most systems.

For a regular Markov chain, the limiting probability distribution as $n \to \infty$ is:

$$\boldsymbol{w} = \lim_{n \to \infty} P^n,$$

that is, if the transition from State i to State $i + 1$ returns the same probability, we have reached the steady state. Thus, the probability vector $\mathbf{p}$ approaches $\boldsymbol{w}$ as $n \to \infty$.

This implies a unique probability vector $\boldsymbol{w}$ exists, such that:

$$\boldsymbol{w} = P\boldsymbol{w}. \tag{8.6}$$

Proof:

$$(wP)_i = \sum_{j \in S} w_j p_{ji}$$

$$(wP)_i = \sum_{j \in S} \lim_{n \to \infty} w_j^{(n)} p_{ji}$$

$$(wP)_i = \lim_{n \to \infty} \sum_{j \in S} w_j^{(n)} p_{ji}$$

$$(wP)_i = \lim_{n \to \infty} w_j^{(n+1)}$$

$$(wP)_i = w_i$$

$$wP = w.$$

A Markov process that satisfies Equation (8.6) for any time step, is a *stationary* process.

For a finite Markov chain, we have three possibilities for steady-state behavior:

1. The probability vector $\boldsymbol{w}$ exists and is independent of the initial state and transition probabilities of the system.

   OR

2. The probability vector $\boldsymbol{w}$ exists but is dependent on the initial state.

OR

3. The probability vector $\boldsymbol{w}$ does not exist.

To solve for $\boldsymbol{w}$, we can rewrite Equation (8.6) as:

$$(P - I)\boldsymbol{w} = \mathbf{0}. \tag{8.7}$$

where I is the identity matrix.

$$\boldsymbol{w} = \begin{bmatrix} w_1 \\ w_2 \\ \vdots \\ w_j \end{bmatrix}. \tag{8.8}$$

Steps to obtain the limiting probability distribution:

1. Create the state transition diagram.
2. Define and number each state.
3. For each state at time = i, identify all possible states that can be transitioned into at the next observed time = $i + 1$.
4. Create the probability matrix P.
5. Using Equations (8.4), (8.7), and (8.8) and solve for $\boldsymbol{w}$.

EXAMPLE 8.3

Going back to our previous example, we ask: what is the long-term probability of a cloudy or sunny day?

Solution

We calculate the matrix $(P - I)$ and solve for $\boldsymbol{w}$ using Equation (8.7).

$$\begin{bmatrix} -0.25 & 0.4 \\ 0.25 & -0.4 \end{bmatrix} \boldsymbol{w} = \begin{bmatrix} 0 \\ 0 \end{bmatrix}.$$

And since for any probability vector $\mathbf{p}$, the sum of the probabilities must be 1, so

$$w_1 + w_2 = 1.$$

Solving for $\boldsymbol{w}$, we get:

$$-0.25w_1 + 0.4w_2 = 0.$$

and since $w_1 + w_2 = 1$,

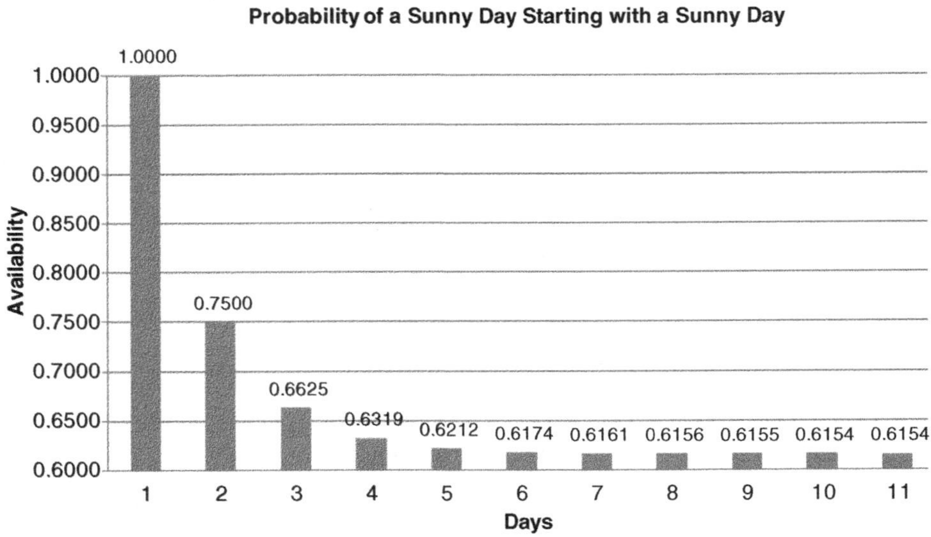

Figure 8.4 Weather Prediction State Machine Starting with a Sunny Day

$$-0.25w_1 + 0.4(1 - w_1) = 0 \Rightarrow w_1 = 0.6154$$

$$0.25w_1 - 0.4w_2 = 0 \Rightarrow w_2 = 0.3846.$$

So:

$$\boldsymbol{w} = \begin{bmatrix} 0.6154 \\ 0.3846 \end{bmatrix}.$$

In the long run, that is, in the steady state, we have a 61.54% chance of having a sunny day and a 38.46% chance of having a cloudy day, irrespective of whether the first day was sunny or cloudy. The long-term probabilities of the system are constant and independent of the initial state on the system. We graph these probabilities of a sunny day over 11 days beginning with a sunny day (Fig. 8.4) and a cloudy day (Fig. 8.5). The probabilities begin to converge by the 11th day.

The probability of being in a sunny day or a cloudy day is the same after 20 days or 200 days (steady state) irrespective of whether we start with a sunny day or a cloudy day.

EXAMPLE 8.4

We have two redundant servers at our office. Only one server needs to be operational to meet our requirements. At the end of each week, we track the following server operational service:

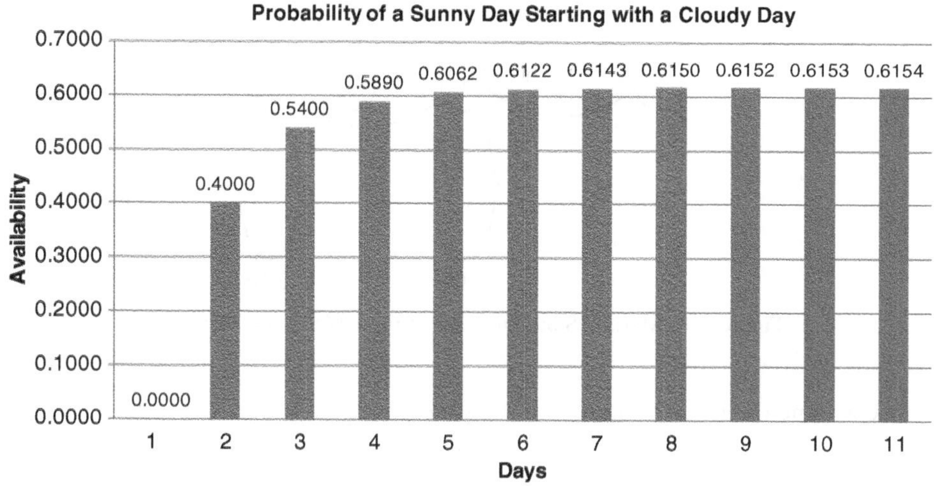

Figure 8.5 Weather Prediction State Machine Starting with a Cloudy Day

1. Both servers are operational.
2. One server is operational and the other server is failed and being repaired.
3. Both servers have failed and are both under repair.

When both are working at the beginning of the week, there is a 25% chance one of them will fail by the end of the week and a 10% chance both will fail by the end of the week. Servers under repair are fixed and repaired prior to the beginning of the next week. If a server fails at any time during the week, it is repaired only during the weekend. If only one server has failed, there is an 85% chance it will be repaired by the beginning of the next week.

(a) What is the steady-state probability of no servers failed, 1 server failed, and both servers failed at the end of the week?

Solution

We first create the state transition diagram (Fig. 8.6).

- *State 1: Both servers are working.*
- *State 2: One server has failed.*
- *State 3: Both servers have failed.*

From this diagram, we can create the state transition probability matrix:

$$P = \begin{array}{c} \\ 1 \\ 2 \\ 3 \end{array} \begin{array}{c} \begin{matrix} 1 & 2 & 3 \end{matrix} \\ \begin{pmatrix} 0.65 & 0.85 & 1 \\ 0.25 & 0.15 & 0 \\ 0.1 & 0 & 0 \end{pmatrix} \end{array}$$

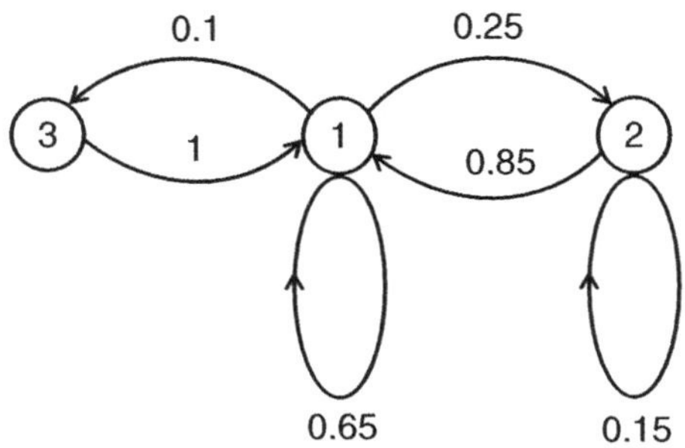

Figure 8.6 Markov State Diagram of the Office Server Example

If we assume the initial condition of the system $\mathbf{p}_1$, is

$$\begin{bmatrix} 1 \\ 0 \\ 0 \end{bmatrix},$$

This means that the two servers are operational at the beginning of the week.

If we want to predict the state of the servers at the end of the first work week, $\mathbf{p}_2$, we take the transition probability matrix P and multiply it by the initial state $\mathbf{p}_1$:

$$\mathbf{p}_2 = P\mathbf{p}_1 \begin{bmatrix} 0.65 & 0.85 & 1 \\ 0.25 & 0.15 & 0 \\ 0.1 & 0 & 0 \end{bmatrix} \begin{bmatrix} 1 \\ 0 \\ 0 \end{bmatrix} = \begin{bmatrix} 0.65 \\ 0.25 \\ 0.1 \end{bmatrix}.$$

Thus, at the end of the first work week, we have a 65% chance that no servers has failed, a 25% chance that one of the servers has failed, and a 10% chance that both servers have failed.

(b) What is the probability of both servers being down at the end of week 4?

$$\mathbf{p}_5 = P^4\mathbf{p}_1 = \begin{bmatrix} 0.65 & 0.85 & 1 \\ 0.25 & 0.15 & 0 \\ 0.1 & 0 & 0 \end{bmatrix}^4 \begin{bmatrix} 1 \\ 0 \\ 0 \end{bmatrix} = \begin{bmatrix} 0.718 \\ 0.210 \\ 0.071 \end{bmatrix}.$$

(c) What is the steady-state probability of 1 server being failed?

We know that

$$(P-I)\boldsymbol{w} = \mathbf{0}$$

$$\boldsymbol{w} = \begin{bmatrix} w_1 \\ w_2 \\ w_3 \end{bmatrix}, \tag{8.9}$$

where I is the identity matrix.

And since for any probability vector **p**, the sum of the probabilities must be 1, so

$$w_1 + w_2 + w_3 = 1.$$

We calculate the matrix $(P - I)$ and solve for $\boldsymbol{w}$.

$$\begin{bmatrix} -0.35 & 0.85 & 1 \\ 0.25 & -0.85 & 0 \\ 0.1 & 0 & -1 \end{bmatrix} \boldsymbol{w} = \begin{bmatrix} 0 \\ 0 \\ 0 \end{bmatrix}.$$

These three equations are not linearly independent. To solve this set of equations, we must replace one of the equations with the conservation of probability equation ($w_1 + w_2 + w_3 = 1$). For a discussion on why this is necessary, refer to Chapter 9, Section 9.5. Replace the last equation with the conservation of probability equation:

$$\begin{bmatrix} -0.35 & 0.85 & 1 \\ 0.25 & -0.85 & 0 \\ 1 & 1 & 1 \end{bmatrix} \boldsymbol{w} = \begin{bmatrix} 0 \\ 0 \\ 1 \end{bmatrix}.$$

Solving for $\boldsymbol{w}$, we get:

$$\boldsymbol{w} = \begin{bmatrix} 0.7173 \\ 0.2110 \\ 0.0717 \end{bmatrix}.$$

In the long run, that is, in the steady state, we have a 71.73% chance of no servers failing, a 21.1% chance of one server failing, and a 7.17% chance of both servers failing.

The evolving office server reliability is captured in Figure 8.7.

8.5 ABSORBING MARKOV CHAINS

A state S_k is called an absorbing state if once that state is entered, it remains in that state forever, that is, the probability of leaving that state is zero.

A Markov chain is said to be absorbing if it has at least one absorbing state, and for every state, the probability of reaching an absorbing state in a finite number of steps is not 0.

The nonabsorbing states are called transient states.

If we arrange the probability matrix P such that the absorbing states are listed first, then the transition matrix with i absorbing states and j transient states has the following form:

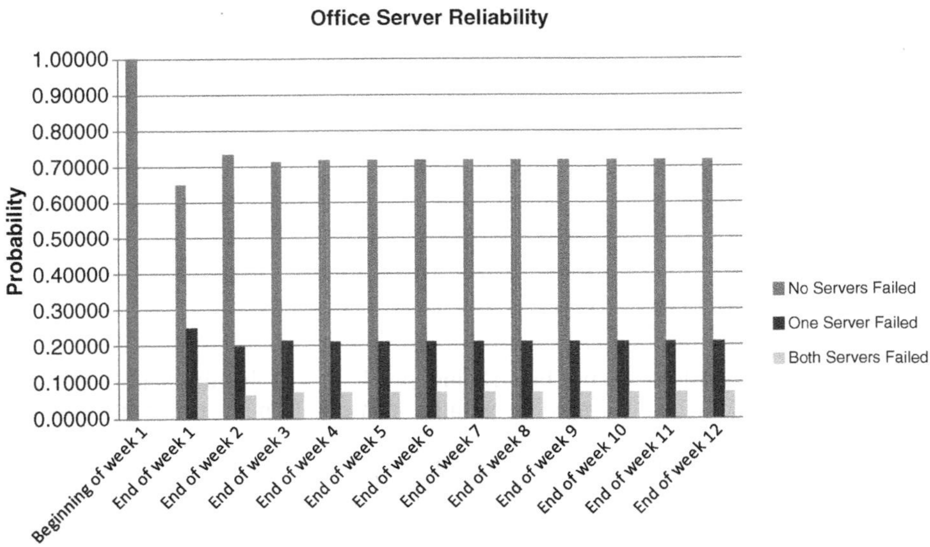

Figure 8.7 Office Server Availability Prediction

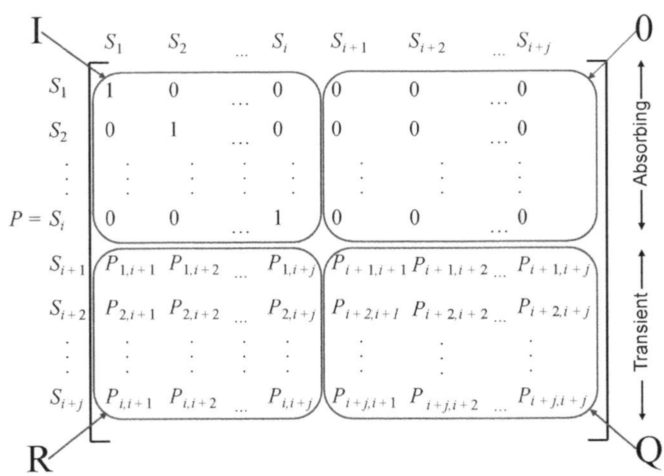

From the matrix $\boldsymbol{P}$ above, we can partition the matrix in the following canonical form:

$$P = \begin{bmatrix} I & 0 \\ R & Q \end{bmatrix}, \tag{8.10}$$

where I is the $i \times i$ identity matrix, 0 is the $i \times j$ zero matrix, R is an $i \times j$ matrix, and Q is a $j \times j$ matrix. R contains the set of probabilities of transitions from transient states to absorbing states while Q contains the set of probabilities of transitions from transient states to transient states.

To show the limiting form of the matrix P^n, let us start with the initial powers of P:

$$P^2 = P \cdot P = \begin{bmatrix} I & 0 \\ R & Q \end{bmatrix} \cdot \begin{bmatrix} I & 0 \\ R & Q \end{bmatrix} = \begin{bmatrix} I & 0 \\ R(I+Q) & Q^2 \end{bmatrix}$$

$$P^3 = P^2 \cdot P = \begin{bmatrix} I & 0 \\ R(I+Q+Q^2) & Q^3 \end{bmatrix}$$

$$P^n = \begin{bmatrix} I & 0 \\ R(I+Q+Q^2+\cdots Q^{n-1}) & Q^n \end{bmatrix}$$

$$P^n = \begin{bmatrix} I & 0 \\ \sum_{k=0}^{n-1} Q^k R & Q^n \end{bmatrix}.$$

As $n \to \infty$, the state transition matrix becomes:

$$\lim_{n\to\infty} P^n = \begin{bmatrix} I & 0 \\ R(I-Q)^{-1} & 0 \end{bmatrix}. \tag{8.11}$$

Since in the long run, the probability of the process being in a transient state is zero,

$$Q^n \to 0 \text{ and as } n \to \infty, R(I+Q+Q^2+\cdots) = R(I-Q)^{-1}.$$

Proof:

Show that $\Sigma Q^i = (I - Q)^{-1}$:
Let $N = \Sigma Q^i = I + Q + Q^2 + \cdots$
Then $QN = Q\Sigma Q^i = Q + Q^2 + Q^3 + \cdots = (I + Q + Q^2 + Q^3 + \cdots) - I = N - I$.
Then:

$$QN = N - I$$

$$N - QN = I \tag{8.12}$$

$$N(I-Q) = I$$

$$N = (I-Q)^{-1}. \tag{8.13}$$

Another way to show that as $n \to \infty$, $R(I + Q + Q^2 + \cdots) = R(I - Q)^{-1}$ is to note that:

$$N_n = I + Q + Q^2 + \cdots + Q^n.$$

Now multiply each side by $(I - Q)$:

$$(I-Q)N_n = I - Q^{n+1}.$$

Since Q is the part of matrix P that corresponds to the transient state, Q^n converges to a matrix of zeros as $n \to \infty$. Thus, $N_n \to N$ as $n \to \infty$, and we have

$$(I-Q)N = I \to N = (I-Q)^{-1}.$$

The matrix $N = (I - Q)^{-1}$ is called the *fundamental matrix* of P:

$$P = \begin{bmatrix} n_{11} & n_{12} & \cdots & n_{1j} \\ n_{21} & n_{22} & \cdots & n_{2j} \\ \vdots & \vdots & \ddots & \vdots \\ n_{i1} & n_{i2} & \cdots & n_{ij} \end{bmatrix}, \tag{8.14}$$

where:

$$n_{ij} = E(\text{time in transient state } j|\text{start at transient state } i).$$

Each entry of N provides the time of the system entering the absorbing states given that the process started in the ith transient state.

The matrix $B = R(I-Q)^{-1} = RN$ provides the long-term probabilities of the system entering the absorbing states given that the process started in the ith transient state. This matrix is called the *absorption probability matrix* of P.

$$B = \begin{bmatrix} b_{11} & b_{12} & \cdots & b_{1j} \\ b_{21} & b_{22} & \cdots & b_{2j} \\ \vdots & \vdots & \ddots & \vdots \\ b_{i1} & b_{i2} & \cdots & b_{ij} \end{bmatrix}, \tag{8.15}$$

where

$$b_{ij} = \text{Prob}(\text{process absorbed in absorption state } j|\text{start at transient state } i).$$

Let M_i be the expected number of steps until absorption starting in transient State i and let M be the $M \times 1$ column vector with elements M_i. The total number of steps until absorption is the sum of the number of steps spent in each of the transient states before absorption given that we start in transient State i.

Thus,

$$M_i = N_{i,1} + N_{i,2} + \cdots + N_{i,m}.$$

So

$$M = \begin{bmatrix} 1 \\ 2 \\ \vdots \\ m \end{bmatrix}.$$

Let B_{i1} be the probability of being absorbed in the absorbing State 1 starting in transient State i. Breaking up the overall probability into the sum of the probabilities of being absorbed in State 1 in each of the possible steps, we get:

$$B_{i1} = R_{i1} + (QR)_{il} + (Q^2R)_{il} + \cdots,$$

so that:

$$\begin{aligned} B &= R + QR + Q^2R + Q^3R + \cdots \\ &= (1 + Q + Q^2 + \cdots)R \\ &= NR. \end{aligned}$$

Thus, $B = NR$.

N, M, and B describe the evolution of the absorbing Markov chain, given the matrices Q and R.

Let s_i and s_j be two transient states, and assume i and j are fixed. Let $X(k)$ be a random variable that equals 1 if the chain is in State s_j after k steps, and equals 0 otherwise. For each k, this random variable depends on both i and j.

$$\begin{aligned} P(X(k) = 1) &= q_{ij}^{(k)} \\ P(X(k) = 0) &= 1 - q_{ij}^{(k)}, \end{aligned}$$

where $q_{ij}^{(k)}$ is the ijth entry of Q^k. These equations hold for $k = 0$ since $Q^0 = I$. Therefore, since $X^{(k)}$ is a 0,1 random variable $E(X^{(k)}) = q_{ij}^{(k)}$.

The expected number of times the chain is in State s_j in the first n steps, given that it starts in State s_i, is:

$$E(X^{(0)} + X^{(1)} + X^{(2)} + \cdots + X^{(n)}) = q_{ij}^{(0)} + q_{ij}^{(1)} + \cdots + q_{ij}^{(n)}.$$

Letting n tend to infinity, we have:

$$E(X^{(0)} + X^{(1)} + X^{(2)} + \cdots) = q_{ij}^{(0)} + q_{ij}^{(1)} + \cdots = n_{ij}.$$

We can also approach this intuitively. Suppose that we want to calculate the expected number of times the chain spends in a transient State j starting in transient State i. The expected value equals the probability of being in state $Q_{i,j}$ at time = 0 plus the probability of being in state $Q_{i,j}$ at time = 1, and so on:

$$\begin{aligned} E(Q_{i,j}) &= Q_{i,j}^{(0)} + Q_{i,j}^{(1)} + Q_{i,j}^{(2)} + \cdots + Q_{i,j}^{(n)} \\ E(Q_{i,j}) &= I + Q + Q^2 + \cdots + Q^n. \end{aligned}$$

The first term is I since the probability of being in the given initial state at $t = 0$ is 1 for the matrix entries in which $i = j$ and 0 otherwise.

We note that $E(Q_{i,j})$ is the same as the entries for the fundamental matrix N.

The expected time for the system to enter an absorption state given that the process started in a transient State i is given by the sum of the entries n_{ij} for a given State i:

$$E(\text{time to absorb}|\text{start at transient state } S_i) = (I-Q)^{-1}\begin{bmatrix}1\\1\\\vdots\\1\end{bmatrix}. \tag{8.16}$$

If we define the transient states as $T_1, T_2, \ldots, T_j$ and recall that the matrix Q provides the probabilities of transitions from transient states to transient states, then:

$$q_{ij} = \text{Prob}(T_i|T_j).$$

Furthermore, let us define $\text{Num}(T_i, T_j)$ as the expected number of times the process will be in State T_i, given that the process started in State T_j. Thus $\text{Num}(T_1,T_j)$ is the expected number of times the process will be in T_1. Since q_{i1} is the probability of transitioning from T_1 to T_i, the product of $q_{il}\text{Num}(T_1,T_j)$ is the expected number of transitions from T_1 to T_i. Therefore, the total number of expected transitions to T_i is $q_{i1}\text{Num}(T_1, T_j) + q_{i2}\text{Num}(T_2, T_j) + \cdots + q_{il}\text{Num}(T_l, T_j)$.

The expected number of transitions into a given state is the same as the expected number of times the system is in this state (except for the case in which the process started since there is no transition into the initial state). Thus, we need to add 1 to the expected number of transitions into T_i to get the correct number of times that the process was in T_i. So,

$$\text{Num}(T_i, T_j) = 1(\text{if } i=j) + q_{i1}\text{Num}(T_1, T_j) + q_{i2}\text{Num}(T_2, T_j) + \cdots q_{ij}\text{Num}(T_i, T_j). \tag{8.17}$$

Equation (8.17) is the (i,j) entry in the matrix equation:

$$N = I + QN \tag{8.18}$$

$$N = \begin{bmatrix}\text{Num}(T_1,T_1) & \text{Num}(T_1,T_2) & \cdots & \text{Num}(T_1,T_j)\\ \text{Num}(T_2,T_1) & \text{Num}(T_2,T_2) & \cdots & \text{Num}(T_2,T_j)\\ \vdots & \vdots & \ddots & \vdots\\ \text{Num}(T_i,T_1) & \text{Num}(T_i,T_2) & \cdots & \text{Num}(T_i,T_j)\end{bmatrix}. \tag{8.19}$$

Thus, the (i,j) entry of $(I - Q)^{-1}$ is the expected number of times that the process is in the ith transient state, given that it started in the jth transient state. The sum of the ith column of N is the expected number of times that the process will be in some transient state, given that the process started in the jth transient state.

Let t_i be the expected number of steps before the chain is absorbed, given the state the chain starts in State s_i, then:

$$t = Nc,$$

where c is a column vector whose entries are 1.

Now, let us put some of these equations to work!

8.6 NONREPAIRABLE RELIABILITY MODELS

If we define a system consisting of components that have constant failure rates and has at least one absorbing state (i.e., system has failed with no chance of recovery), we can determine the Mean Time between Failures (MTBF) by summing the expected number of times that the process will be in some transient state given that the process started in the jth transient state. Steps to calculate MTBF:

1. Define the set of states describing the system.
2. Draw the state transition diagram.
3. Identify the state transition matrix from the diagram.
4. Partition the matrix into canonical form.
5. Derive the fundamental matrix N.
6. Sum the elements of the last row of the fundamental matrix N.

8.6.1 Two-State Nonmaintained System

A *nonmaintained system* contains components which are not repaired upon failure.

Step 1:
A simple system that has two states:

- *State 1: System fully functional.*
- *State 2: System has failed.*

The probability of the system failing at any time is dependent on the failure rate of the system. We assume a constant failure rate λ.

Step 2:
Create the state transition diagram from the information provided (Fig. 8.8).

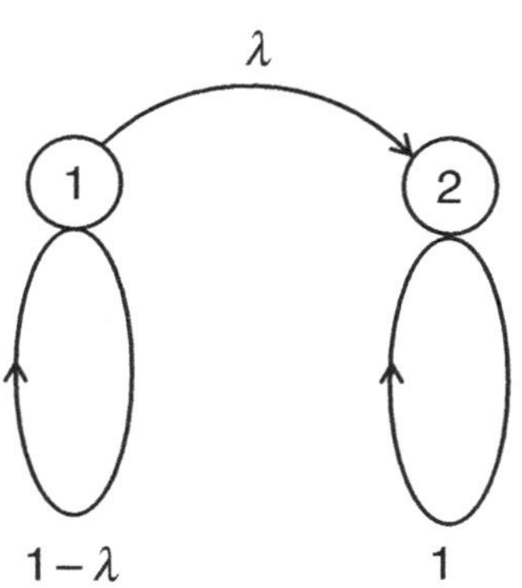

Figure 8.8 A Two-State Nonmaintained System

Step 3:
P_{11} is the probability that the system has not failed:

$$P_{11} = 1 - \lambda.$$

P_{21} is the probability that the failed system returns to operational conditions:

$$P_{21} = 0.$$

P_{12} is the probability that the system has failed:

$$P_{12} = \lambda.$$

And finally, P_{22} is the probability that the failed system remains failed:

$$P_{22} = 1.$$

Putting these probabilities into the state transition matrix P:

$$\begin{array}{c} \quad\quad S_1 \quad\quad S_2 \leftarrow \text{State } n \\ P = \begin{matrix} S_1 \\ S_2 \end{matrix}\begin{pmatrix} 1-\lambda & 0 \\ \lambda & 1 \end{pmatrix} \\ \uparrow \\ \text{State } n+1 \end{array}$$

Step 4:
Rearrange P so that it is in the following canonical form

$$P = \begin{bmatrix} I & 0 \\ R & Q \end{bmatrix}.$$

P becomes:

$$P = \begin{pmatrix} 1 & 0 \\ \lambda & 1-\lambda \end{pmatrix}$$

where $R = \lambda$ and $Q = 1 - \lambda$.

Steps 5 and 6:
Now we can calculate $N = (I - Q)^{-1}$. Since N is a 1×1 matrix, $N = 1/\lambda$.

Since the sum of each entry of N provides the time to absorption given that the process started in the ith transient state, the expected time for the system to enter the absorbing State 2 given that it started in the transient State 1, is $1/\lambda$. This is the MTBF of the system.

We can also calculate the absorption probability B:

$$B = R(I - Q)^{-1} = RN = \lambda(1/\lambda) = 1.$$

Thus, the long-term probability of the system entering the absorbing State 2, that is, failing, is 1.

8.6.2 Three-State Nonmaintained System

Let us next consider a redundant system in which two components are connected in parallel. One component is active and the other component is a hot standby, that is, it is powered and takes over from the active component upon active component failure.

The system is not repairable and the failure rate of components 1 and 2 are λ_1 and λ_2, respectively. We further assume both components have the same failure rate ($\lambda_1 = \lambda_2$) and the components are independent as shown in the reliability block diagram of Figure 8.9.

To derive the transition probabilities for a hot-standby system, we first define the following states:

- *State 1: No components have failed.*
- *State 2: One component has failed.*
- *State 3: Both components have failed.*

The state diagram for this system is shown in Figure 8.10. This model does not take into account *common mode failures* (i.e., failures that affect both components simultaneously). These failure modes are *not* usually negligible and should be considered when modeling a complex system. If a common mode failure is added, we would show an additional directed transition from State 1 to State 3 labeled λ_c, and we would need to assign a failure rate for the common mode failure. We will discuss these failure modes and several other configurations in subsequent chapters.

Following the transition diagram above, we can calculate the following nine possible state transition probabilities for the three states.

P_{11} is the probability that component 1 has not failed and the probability that component 2 has not failed:

$$P_{11} = (1-\lambda_1)(1-\lambda_2) = 1-2\lambda+\lambda^2.$$

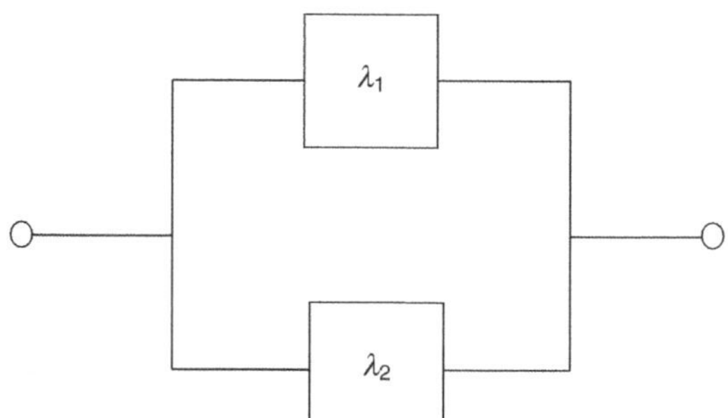

Figure 8.9 Parallel Components Block Diagram

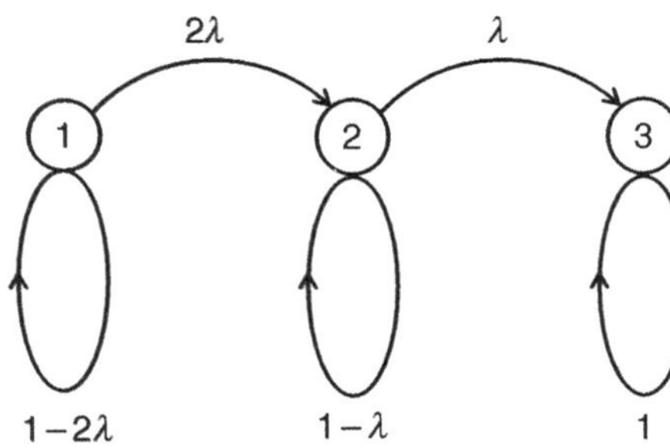

Figure 8.10 Nonmaintained 2N Redundant Markov State Diagram

Since we assume both components cannot fail simultaneously ($\lambda^2 = 0$), then

$$P_{11} = 1 - 2\lambda.$$

P_{12} is the probability that component 1 has failed and the probability that component 2 has not failed plus the probability that component 2 has failed and component 1 has not failed:

$$P_{12} = \lambda_1(1-\lambda_2) + \lambda_2(1-\lambda_1) = 2\lambda(1-\lambda)$$
$$P_{12} = 2\lambda, \quad \text{if } \lambda^2 = 0.$$

P_{21} is the probability that once a component has failed, it can be repaired. Since we are analyzing a nonmaintained system, the component will remain failed:

$$P_{21} = 0.$$

P_{22} is the probability that the second component has not failed, given that one of the components has already failed:

$$P_{22} = 1 - \lambda.$$

P_{13} is the probability that both components fail simultaneously (no transition from State 1 to State 3):

$$P_{13} = 0.$$

P_{23} is the probability that the second component has failed, given that one of the components has already failed:

$$P_{23} = \lambda.$$

P_{31} is the probability that given both components have failed, both will be repaired:

$$P_{31} = 0.$$

P_{32} is the probability that given both components have failed, at least one of the failed components will be repaired:

$$P_{32} = 0.$$

P_{33} is the probability that given both components have failed, the system will always remain failed:

$$P_{33} = 1.$$

From Figure 8.10 and the transition probabilities, the state transition probability matrix is:

$$P = \begin{matrix} & \begin{matrix} 1 & 2 & 3 \end{matrix} \\ \begin{matrix} 1 \\ 2 \\ 3 \end{matrix} & \begin{pmatrix} 1-2\lambda & 0 & 0 \\ 2\lambda & 1-\lambda & 0 \\ 0 & \lambda & 1 \end{pmatrix} \end{matrix}.$$

Now, identify the Q matrix:

$$P = \begin{matrix} & \begin{matrix} 1 & 2 & 3 \end{matrix} \\ \begin{matrix} 1 \\ 2 \\ 3 \end{matrix} & \left(\begin{array}{cc|c} 1-2\lambda & 0 & 0 \\ 2\lambda & 1-\lambda & 0 \\ \hline 0 & \lambda & 1 \end{array}\right) \end{matrix}.$$

(Q is the upper-left 2×2 block enclosed by the dashed box.)

Next, calculate the fundamental matrix $N = (I - Q)^{-1}$:

$$I - Q = \begin{bmatrix} 2\lambda & 0 \\ -2\lambda & \lambda \end{bmatrix}.$$

Recall from matrix theory that for a 2×2 matrix,

$$A = \begin{bmatrix} a & b \\ c & d \end{bmatrix},$$

the inverse is:

$$A^{-1} = \frac{1}{ad - bc} \begin{bmatrix} d & -b \\ -c & a \end{bmatrix}.$$

Thus,

$$(I - Q)^{-1} = \frac{1}{2\lambda^2} \begin{bmatrix} \lambda & 0 \\ 2\lambda & 2\lambda \end{bmatrix} = \begin{bmatrix} 1/(2\lambda) & 0 \\ 1/\lambda & 1/\lambda \end{bmatrix}.$$

The Mean Time to First Failure (MTTF) of the system can be found by summing the elements of the column corresponding to State 1.

$$\mathrm{MTTF}_S = \sum_{k=1}^{\infty} N_{1,k}$$

$$\mathrm{MTTF}_S = \frac{1}{\lambda} + \frac{1}{2\lambda}$$

$$\mathrm{MTTF}_S = (1.5)\frac{1}{\lambda}$$

$$\mathrm{MTTF}_S = 1.5(\mathrm{MTTF}_c),$$

where MTTF_c = the MTTF of an individual component of the system.

We conclude that the MTTF of the redundant system is 50% better than the non-redundant system.

8.6.3 Maintained Systems: Two-Component Parallel System with Single-Component Repair

Now let us consider a maintained system.

Maintained systems are systems in which failed units are repaired or replaced.

Assumptions:

- As with the failure rate, the repair rate $\mu(t)$ is also assumed to be memoryless and is a constant, $\mu(t) = \mu$.
- The failure rate is much less than the repair rate, $\lambda \ll \mu$.

Using the previous example with the same three state definitions, let us assume that each time a component fails, it will be repaired with the time to repair a component is $1/\mu$, thus, the repair rate is μ. We also assume as in the previous example that the probability of simultaneous failures is 0. The state transition diagram for this maintained system is as shown in Figure 8.11.

The state transition probabilities are the following:

P_{11} = probability that component 1 has not failed and the probability that component 2 has not failed:

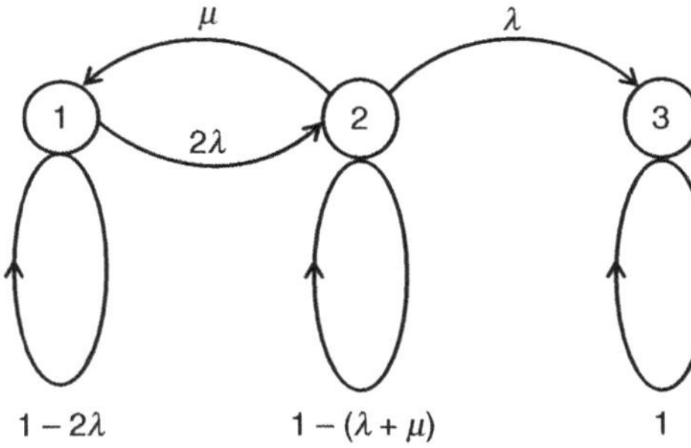

Figure 8.11 Maintained 2N Redundant Markov State Diagram

$$P_{11} = (1-\lambda)(1-\lambda) = 1 - 2\lambda + \lambda^2$$

$$P_{11} = 1 - 2\lambda, \quad \text{if } \lambda^2 = 0.$$

P_{12} = probability that component 1 has failed and the probability that component 2 has not failed, plus the probability that component 2 has failed and component 1 has not failed:

$$P_{12} = \lambda_1(1-\lambda_2) + \lambda_2(1-\lambda_1) = 2\lambda(1-\lambda)$$

$$P_{12} = 2\lambda, \quad \text{if } \lambda^2 = 0.$$

P_{21} is the probability that the failed component is repaired and the probability the other component does not fail:

$$P_{21} = (1-\lambda)\mu$$

$P_{21} = \mu$, if the probability of $\mu\lambda = 0$ (repair of one component and failure of another component during any time interval is negligible).

P_{22} is the probability that the second component has not failed and the probability that the first component has not been repaired:

$$P_{22} = (1-\lambda)(1-\mu) = 1 - \lambda - \mu + \mu\lambda.$$

$$P_{22} = 1 - (\lambda + \mu), \quad \text{if } \lambda_\mu = 0.$$

P_{13} is the probability that both components fail simultaneously:

$$P_{13} = 0.$$

P_{23} is the probability that the second component has failed and the probability that the first component has not been repaired, given that one of the components has already failed:

$$P_{23} = \lambda(1-\mu).$$

$$P_{23} = \lambda, \text{ if } \lambda\mu = 0.$$

P_{31} is the probability that both failed components are simultaneously repaired:

$$P_{31} = 0.$$

P_{32} is the probability that at least one of the two failed components will be repaired:

$$P_{32} = 0.$$

P_{33} is the probability that given both components have failed, the system will always remain failed:

$$P_{33} = 1.$$

The state transition probability matrix is:

$$P = \begin{array}{c} \\ 1 \\ 2 \\ 3 \end{array}\!\!\begin{array}{c} \begin{array}{ccc} 1 & 2 & 3 \end{array} \\ \begin{pmatrix} 1-2\lambda & \mu & 0 \\ 2\lambda & 1-(\mu+\lambda) & 0 \\ 0 & \lambda & 1 \end{pmatrix} \end{array}.$$

Now, identify the Q matrix:

$$P = \begin{array}{c} \\ 1 \\ 2 \\ 3 \end{array}\!\!\begin{array}{c} \begin{array}{ccc} 1 & 2 & 3 \end{array} \\ \left(\begin{array}{cc|c} 1-2\lambda & \mu & 0 \\ 2\lambda & 1-(\mu+\lambda) & 0 \\ \hline 0 & \lambda & 1 \end{array}\right) \end{array}$$

(Q = the dashed upper-left 2×2 block.)

Next, calculate the fundamental matrix $N = (I - Q)^{-1}$:

$$I - Q = \begin{bmatrix} 2\lambda & -\lambda\mu \\ -2\lambda & \mu+\lambda \end{bmatrix}.$$

Calculate the matrix inverse:

$$(I-Q)^{-1} = \frac{1}{2\lambda^2}\begin{bmatrix} \lambda+\mu & \mu \\ 2\lambda & 2\lambda \end{bmatrix}.$$

Sum the elements of the column corresponding to State 1 (the first column) to obtain the MTTF:

$$\mathrm{MTTF}_S = \sum_{k=1}^{n} N_{1,k}$$

$$\mathrm{MTTF}_S = \frac{\lambda+\mu+2\lambda}{2\lambda^2} = \frac{1}{\lambda} + \frac{1}{2\lambda} + \frac{\mu}{2\lambda^2}$$

$$\mathrm{MTTF}_S = 1.5(\mathrm{MTTF}_c) + \frac{\mu}{2\lambda^2}$$

$$\mathrm{MTTF}_S = \mathrm{MTTF}_{\mathrm{parallel}} + \frac{\mu}{2\lambda^2}.$$

where MTTF_c = the MTTF of an individual component and $\mathrm{MTTF}_{\mathrm{parallel}}$ = the MTTF of the two identical component parallel system.

Thus, the MTTF of the parallel system with single component repair is increased over the single component system by $= \mu/(2\lambda^2)$.

Note that the MTTF increases with the decrease in the failure rate and also increases with an increase in the repair rate. If we had an instantaneous repair rate, the MTTF would be infinite (i.e., the system would never fail). Conversely, if the component failure rate increases to a very large value, the MTTF will approach zero.

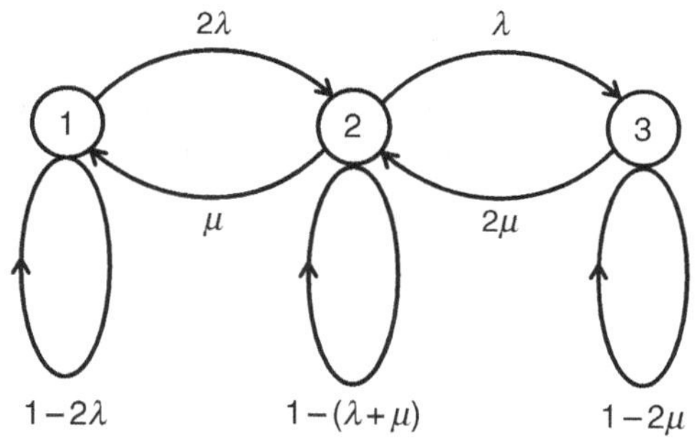

Figure 8.12 Maintained 2N Redundant Markov State Diagram

8.6.4 Maintained Systems: Two-Component Parallel System, Multiple Component Repair

Using the assumptions from the previous example with the same three state definitions, let us also assume that each time a component fails, it will be repaired with a repair rate of μ, and if both components fail, the system will be brought back into service by repairing one of the failed components. The state transition diagram for this maintained system is shown in Figure 8.12.

The state transition probabilities are:

P_{11} = probability that component 1 has not failed and the probability that component 2 has not failed:

$$P_{11} = (1-\lambda_1)(1-\lambda_2) = 1-2\lambda+\lambda^2$$

$$P_{11} = 1-2\lambda, \quad \text{if } \lambda^2 = 0.$$

P_{12} is the probability that component 1 has failed and the probability that component 2 has not failed plus the probability that component 2 has failed and component 1 has not failed:

$$P_{12} = \lambda_1(1-\lambda_2)+\lambda_2(1-\lambda_1) = 2\lambda(1-\lambda)$$

$$P_{12} = 2\lambda, \quad \text{if } \lambda^2 = 0.$$

P_{21} is the probability that the failed component is repaired and the probability that the other component does not fail:

$$P_{21} = (1-\lambda)\mu$$

$$P_{21} = \mu, \quad \text{if } \lambda\mu = 0.$$

P_{22} is the probability that the second component has not failed and the probability that the first component has not been repaired:

$$P_{22} = (1-\lambda)(1-\mu) = 1-\lambda-\mu+\mu\lambda.$$

$P_{22} = 1 - (\lambda + \mu)$, if $\lambda\mu = 0$ (repair of one component and failure of another component occurring during any time interval is negligible).

P_{13} is the probability that both components fail simultaneously:

$$P_{13} = 0.$$

Note: We assume both components cannot fail within the same time step.

P_{23} is the probability that the second component has failed and the probability that the first component has not been repaired, given that one of the components has already failed:

$$P_{23} = \lambda(1-\mu).$$

$$P_{23} = \lambda, \quad \text{if } \lambda\mu = 0.$$

P_{31} is the probability that both failed components are simultaneously repaired:

$$P_{31} = 0.$$

Note: We assume both components cannot fail within the same time step.

P_{32} is the probability that at least one of the two failed components will be repaired:

$$P_{32} = \mu + \mu$$

$$P_{32} = 2\mu.$$

P_{33} is the probability that given both components have failed, the system will remain failed:

$$P_{33} = (1-\mu_1)(1-\mu_2) = 1-2\mu+\mu^2.$$

$P_{33} = 1 - 2\mu$, if the probability of μ^2 (simultaneous repair of both components) occurring during any time interval is negligible.

The state transition probability matrix is:

$$P = \begin{bmatrix} 1-2\lambda & \mu & 0 \\ 2\lambda & 1-(\lambda+\mu) & 2\mu \\ 0 & \lambda & 1-2\mu \end{bmatrix}.$$

Note that in this example, no absorbing state exists.

EXAMPLE 8.5

Applying this general solution, consider a system that is monitored every hour and each component has a failure rate of 0.05 failures per hour and a repair rate of 0.2

repairs per hour. Given the system is in a good state at time step 0, what is the probability that the system fails after 2 hours?

Using Equation (8.3),

$$\mathbf{p}_2 = P^2\mathbf{p}_1 \begin{bmatrix} 0.9 & 0.2 & 0 \\ 0.1 & 0.75 & 0.4 \\ 0 & 0.05 & 0.6 \end{bmatrix}^2 \begin{bmatrix} 1 \\ 0 \\ 0 \end{bmatrix} = \begin{bmatrix} 0.830 \\ 0.165 \\ 0.005 \end{bmatrix}.$$

The system fails when a transition to State 3 occurs. The probability that the system will fail in 2 hours is 0.005.

The reliability of this system is equal to the probability of the system being in either State 1 or State 2 given that State 3 is transformed into an absorbing state. Using the previous example, in which State 3 is an absorbing state, we can calculate the Mean Time to First Failure, MTTFF_S:

$$\text{MTTFF}_S = \frac{1}{\lambda} + \frac{1}{2\lambda} + \frac{\mu}{2\lambda^2}$$
$$\text{MTTFF}_S = 70 \text{ hours.}$$

What is the steady-state probability vector? We calculate the matrix $(P - I)$ and solve for $\boldsymbol{w}$:

$$\begin{bmatrix} -0.1 & 0.2 & 0 \\ 0.1 & -0.25 & 0.4 \\ 0 & 0.05 & -0.4 \end{bmatrix} \boldsymbol{w} = \begin{bmatrix} 0 \\ 0 \\ 0 \end{bmatrix}.$$

Since $w_1 + w_2 + w_3 = 1$, replace the last equation with the conservation of probability equation:

$$\begin{bmatrix} -0.1 & 0.2 & 0 \\ 0.1 & -0.25 & 0.4 \\ 1 & 1 & 1 \end{bmatrix} \boldsymbol{w} = \begin{bmatrix} 0 \\ 0 \\ 1 \end{bmatrix}.$$

Solving for $\boldsymbol{w}$, we get:

$$\boldsymbol{w} = \begin{bmatrix} 0.64 \\ 0.32 \\ 0.04 \end{bmatrix}.$$

The availability of the system is the probability of being in State 1 or State 2, thus

$$A_{lt} = 0.96.$$

The MTBF is the mean time between transitions from a functioning state into a failed state. The system can transition into a failed state from State 2. The frequency

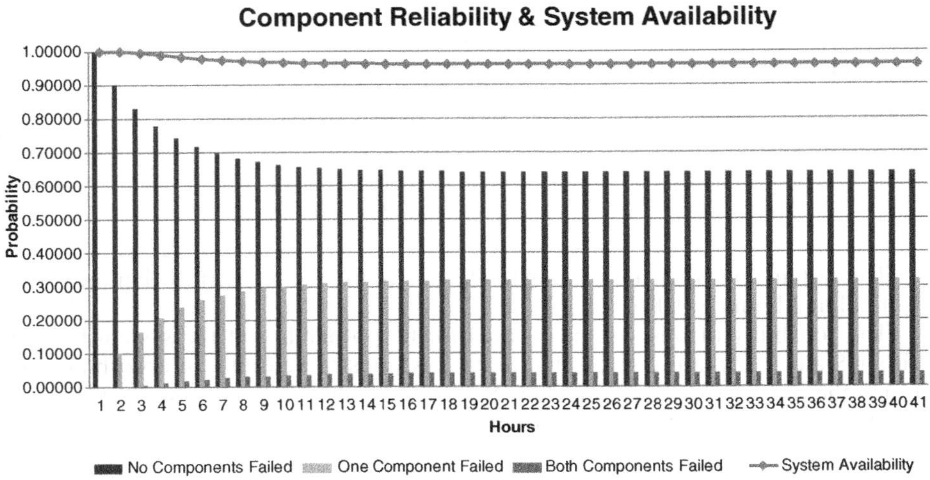

Figure 8.13 Two-Component Reliability and System Availability Prediction

of transitioning from a good state to a bad state is the probability of being in State 2 multiplied by the transition rate from State 2 to State 3. Thus, the $MTBF_S$ is:

$$\text{MTBF}_S = \frac{1}{P_2\lambda} = \frac{1}{(0.32)(0.05)} = 62.5 \text{ hours.}$$

We can calculate the probabilities of being in each of the three states for any discrete time interval (say hours) using Equation (8.2), where n represents the number of hours and plot these over 41 hours (Fig. 8.13).

8.7 SUMMARY

In this chapter, we introduced the concept of a Markov process. We considered two types of Markov processes: DTMC and CTMC. These two processes are differentiated by the time scales employed. A DTMC is based on discrete equally spaced time intervals. These time intervals can be any arbitrary or fixed values of time increments, whereas a CTMC is based on continuous time. The continuous time Markov processes are analyzed in much more detail in the following two chapters.

The state transition matrix and probability vector were introduced and the relevant equations derived. We then considered several examples of the application of DTMCs, including several practical reliability problems, in which these equations were applied.

DTMCs are a useful tool that can be applied to many practical reliability problems.

CHAPTER 9

Continuous-Time Markov Systems

9.1 INTRODUCTION

In the previous chapter, we provided a brief overview on the discrete time Markov chains and continuous-time Markov processes. We then discussed at length Discrete Time Markov Chain (DTMC) along with several application examples. In this chapter, we will describe the continuous Markov process and compare it with its discrete counterpart. The following two chapters will then describe in some detail examples of common continuous-time Markov process models that serve as the building blocks for more complicated systems.

9.2 CONTINUOUS-TIME MARKOV PROCESSES

The random variable $X(t)$ denotes the state of a process at time t. The collection of all possible states is called the state space, and we will denote it by $\boldsymbol{X}$. The state space $\boldsymbol{X}$ we discuss here is finite and corresponds to real states of a system. The states are $\{1, 2, \ldots, n\}$, such that $\boldsymbol{X}$ contains n different states.

Consider a Markov process $\{X(t), t \geq 0\}$ with the states $X = \{1, 2, 3, \ldots, n\}$ and constant transition probabilities:

$$P_{ij}(t) = P(X(t) = j | X(0) = i) \text{ for all } i, j \in X. \tag{9.1}$$

$P_{ij}(t)$ is the probability that a process currently in State i will transition to State j.

Equation (9.1) can be read as: The probability that a process will transition to a new State j over a time interval t given that the process is in State i at time 0.

A continuous-time Markov process (CTMC) is a continuous-time stochastic process such that for the infinitesimal time increment Δt, the following holds:

$$p[X(t+\Delta t) = j \mid X(t) = i] = p_{ij}\Delta t, \tag{9.2}$$

where p_{ij} is the transition rate probability of transitioning from State i to State j.

Designing High Availability Systems: Design for Six Sigma and Classical Reliability Techniques with Practical Real-Life Examples, First Edition. Zachary Taylor and Subramanyam Ranganathan.

This equation indicates that the probability of transitioning from the current State i to the next State j is dependent only on the current state and the transition probabilities.

We develop Markov differential equations by describing the probability of being in each system state at time $t + dt$ as a function of the state of the system at time t.

Comparing Equation (9.2) with Equation (8.1), note the similarities. The time steps we used for the DTMC can be any time increment, whereas for CTMC, the time step is an infinitesimally small value. Note: p_{ij} in this chapter refers to the transition *rate* probability, whereas in the previous chapter, p_{ij} refers to a constant transition probability. The DTMC is used for discrete time system models where transitions between states occurs at regular time intervals based on the time steps, $i = 0, 1, \ldots, n$. In the previous chapter, we used DTMC to model repairable and nonrepairable systems with constant failure rates and repair rates. Although practically, failures could occur at any time irrespective of the time stamp, the DTMC model was sufficient for our requirements since we were only concerned with the state of the system at regular intervals (based on a time step of 1 hour for example).

The CTMC model assumes that only a single state transition can occur in the small interval Δt, whereas for DTMC, we have no such restriction. In fact, for the DTMC reliability models, we made some simplifications in the transition equations that discounted the contribution of multiple state transitions occurring simultaneously within a given time stamp.

Note also that since the transition rates p_{ij} are constant, the time until the next transition is an exponential random variable. The probability that no transition occurs at some time t where $s > t > r$ is:

$$p[X(s) = i \mid X(r) = i] = e^{p_{ii}t}. \tag{9.3}$$

Refer to Poisson and exponential discussions in Chapter 4 for more in-depth discussion of this relationship. Since the transition rates are constant, the time spent in the current state does not affect the transition time to the next state.

We discussed the constant P matrix in the previous chapter. We can extend this constant matrix to a dynamic matrix. The state transition probabilities for n states may be arranged as an $n \times n$ matrix of all possible probabilities of transitions from any State i to any State j.

$$P(t) = \begin{bmatrix} P_{11}(t) & P_{21}(t) & \cdots & P_{n1}(t) \\ P_{12}(t) & P_{22}(t) & \cdots & P_{n2}(t) \\ \vdots & \vdots & \ddots & \vdots \\ P_{1n}(t) & P_{2n}(t) & \cdots & P_{nn}(t) \end{bmatrix}. \tag{9.4}$$

Since all entries in $P(t)$ are probabilities, for any state probability

$$0 \le P_{ij}(t) \le 1 \quad \text{for all } t \ge 0, i, j \in X. \tag{9.5}$$

We must also have the sum of the probabilities of transitioning out of any State i plus the probability of remaining in the same state i always equals 1:

$$\sum_{j=1}^{n} P_{ij}(t) = 1. \tag{9.6}$$

The sum of each column in the matrix $P(t)$ is therefore equal to 1. Note that the entries in column i represent the transitions out of State i (for $j \neq i$) plus the probability of not transitioning out of State i. Also, the entries in each row j represent the transition into State j (for $i \neq j$) plus the probability of not transitioning into State j.

When the process leaves State i, it will next enter State j with some probability $P_{ij}(t)$. Let us define α_i as the rate at which the process leaves State i. If $\alpha_i = 0$, then State i is called *absorbing* since once the process enters that state, it can never leave. We also define β_{ij} as the constant rate of transition of the system from State i to State j:

The rate at which the process remains in State i (α_i) is equal to the negative sum of the rates at which the process transitions from State i to State j (β_{ij}):

$$\alpha_i = -\sum_{j=1, j\neq i}^{n} \beta_{ij}. \tag{9.7}$$

We can arrange the constant transition rates β_{ij} as a matrix:

$$\mathbf{A} = \begin{bmatrix} \beta_{11} & \beta_{21} & \cdots & \beta_{n1} \\ \beta_{12} & \beta_{22} & \cdots & \beta_{n2} \\ \vdots & \vdots & \ddots & \vdots \\ \beta_{1n} & \beta_{2n} & \cdots & \beta_{nn} \end{bmatrix}. \tag{9.8}$$

$\mathbf{A}$ is the *transition rate matrix* of the Markov process.

For the diagonal elements, we can define the following:

$$\beta_{ii} = -\alpha_i = -\sum_{j=1, j\neq i}^{r} \beta_{ij}. \tag{9.9}$$

Observe that the entries of column i are the transition rates out of State i (for $j \neq i$). We will call these entries *departure rates* from State i. The sum of the departure rates from State i, is α_i, the *total departure rate* from State i. The entries of row i are transition rates into State i (for $j \neq i$). The sum of the entries in column i is equal to 0, for all $i \in X$.

With this canonical form of the continuous Markov process, we will be able to determine the dynamic probabilities of each state over time.

9.3 TWO-STATE DERIVATION

We begin with a single-component repairable system.

The system begins in the operational state and is assumed to be fully functional. Eventually, the system fails with the random time to failure is determined by the

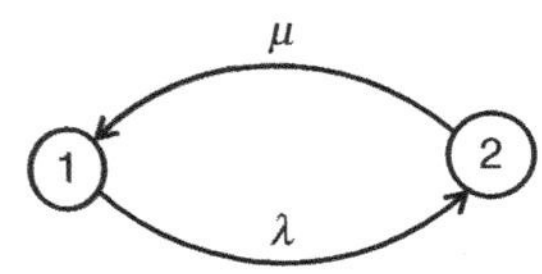

Figure 9.1 Single-Component System with Repair

constant failure rate λ. Once failed, the system remains failed (remains in the failed state) until repaired with the random time to repair determined by the repair rate μ. Once repaired, the system transitions back to the fully operational state.

We model the system using two states. We define State 1 to be the fully functional state and State 2 to be the failed state. A directed path from State 1 to State 2 represents the failure rate of the system and the directed path from State 2 to State 1 is the repair rate.

The following states are defined:

- *State 1: System fully functional.*
- *State 2: System has failed.*

Figure 9.1 shows the state transition diagram for this two state repairable system with a system failure rate of λ and a system repair rate of μ.

Let us consider a small increment of time Δt. The state transition diagram shows a transition out of State 1 at a rate of λ, thus the probability of transitioning out of State 1 at time $t+\Delta t$ given that the system is in State 1 at time t is equal to the transition rate λ multiplied by the increment Δt. The change in the probability of being in State 1 is reduced by $\lambda\Delta t$, that is, $\Delta P_1(t) = -\lambda\Delta t$. The diagram also shows a transition out of State 2 at a rate of μ. Given that the system is in State 2 at time t, the probability of transitioning out of State 2 is equal to the transition rate μ multiplied by the increment Δt, that is, $\Delta P_2(t) = -\mu\Delta t$.

What is the probability of being in State 1 at time $t + \Delta t$? From the diagram, we see that the total probability of being in State 1 at time $t+\Delta t$ is equal to the probability of being in State 1 at time t, and not transitioning out of State 1 at time $t + \Delta t$ plus the probability of being in State 2 at time t and transitioning out of State 2 at time $= t + \Delta t$.

We can write this in equation form:

$$P_1(t+\Delta t) = P_1(t)(1-\lambda\Delta t) + P_2(t)\mu\Delta t. \tag{9.10}$$

Now consider the probability of being in State 2 at time $t + \Delta t$. This is equal to the probability of being in State 1 and not transitioning out of State 2 plus the probability of being in State 1 and transitioning out of State 1.

In equation form, we obtain:

$$P_2(t+\Delta t) = \lambda P_1(t)\Delta t + (1-\mu\Delta t)P_2(t). \tag{9.11}$$

Rearranging Equations (9.10) and (9.11):

$$\frac{P_1(t+\Delta t)-P_1(t)}{\Delta t}=-\lambda P_1(t)+\mu P_2(t)$$

$$\frac{P_2(t+\Delta t)-P_2(t)}{\Delta t}=\lambda P_1(t)-\mu P_2(t).$$

To obtain the instantaneous rate of probability change for States 1 and 2, we can take the limit as Δt approaches zero:

$$\dot{P}_1(t)=-\lambda P_1(t)+\mu P_2(t) \tag{9.12}$$

$$\dot{P}_2(t)=\lambda P_1(t)-\mu P_2(t). \tag{9.13}$$

9.3.1 Alternative Derivation

The path from State 1 to State 2 represents the probability of transitioning from State 1 to State 2. This directed path decreases the probability of remaining in State 1 and increases the probability of transitioning to State 2 over time. The path μ from State 2 to State 1 decreases the probability of remaining in State 2 and increases the probability of transitioning to State 1.

Due to the Markov property, once in State 1 again, the system is "good as new," that is, the previous history of transitioning from State 1 to State 2 to State 1 again does not affect the future behavior of the system. The behavior of the system is only dependent on the current state of the system and the transition probabilities of the system.

As an illustration to a real-life example, consider that we bought a brand new car. Let us define State 1 to be the operational state of the car, that is, the car is usable for day-to-day transportation requirements. Let us define State 2 as the state where the car has broken down and is under repair. In State 2, the car cannot be used for normal daily transportation. We can apply the Markov technique to the car during its useful life time if we make the assumption that after each repair, the car is as good as new, that is, the probability of failure of the car is a constant (constant failure rate) over the useful life time of the car. For example, if we assume that the useful life time is 10 years, the failure rate of the car in its first year is identical with the failure rate of the car in its second year and all the way up to its 10th year. Eventually, this assumption will no longer apply as the car becomes much older and the repairs become more frequent. That is why we need to carefully consider the time span during which the constant failure rate assumption is a good approximation of the system.

Now we would like to know the dynamic behavior of the system over time. Initially, the probability of being in State 1 at $t = 0$ is 1 and the probability of being in State 2 at $t = 0$ is 0. As time evolves, the probabilities of being in State 1 or State 2 change, but the sum of the probabilities of being either in State 1 or State 2 must always be 1. The directed line λ represents the flow of probability from State 1 to State 2. During the infinitesimal time increment dt, the flow of probability from State 1 to State 2 is λdt, thus decreasing the probability that if the system is in State 1, it remains in State 1 by an infinitesimal amount. The change in probability in State 1 is equal to the probability of being in State 1 times the flow of probability out of State 1, plus the probability of being in State 2 times the flow of probability into State 1 from State 2:

$$dP_1(t) = -P_1(t)\lambda dt + P_2(t)\mu dt,$$

where $P_1(t)$ is the probability of the system being in State 1 and $P_2(t)$ is the probability of the system being in State 2 at any time t.

The rate of probability change for State 1 is thus:

$$\frac{dP_1(t)}{dt} = -\lambda P_1(t) + \mu P_2(t). \tag{9.14}$$

Next, we consider State 2. The change in probability in State 2 is equal to the probability of being in State 2 multiplied by the flow of probability out of State 2 plus the probability of being in State 1 multiplied by the flow of probability into State 1 from State 2:

$$dP_2(t) = P_1(t)\lambda dt - P_2(t)\mu dt.$$

Rearranging:

$$\frac{dP_2(t)}{dt} = \lambda P_1(t) - \mu P_2(t). \tag{9.15}$$

Note that Equations (9.12) and (9.13) are the same as Equations (9.14) and (9.15).

Note also that the sum of the probabilities of the states must always equal 1, that is:

$$P_1(t) + P_2(t) = 1. \tag{9.16}$$

In general, the sum of the probabilities across all states of any n-state Markov system at any time t must be equal to 1:

$$\sum_{i=1}^{n} P_i(t) = 1, \tag{9.17}$$

where n = the number of states.

Equation (9.17) is called the *conservation of probability equation*. For the complete dynamic solution, we can either solve the two differential equations directly by applying the initial conditions or we can replace one of the equations by the conservation of probability equation and solve those two equations.

From Equations (9.14) and (9.15), and using the initial conditions $P_1(0) = 1$ and $P_2(0) = 0$, we can obtain the complete dynamic solution to $P_1(t)$ and $P_2(t)$. Step-by-step methods of obtaining these solutions are provided in Chapter 11.

9.3.2 State Matrix

Let us put the state equations in canonical matrix form:

$$\dot{P} = AP, \tag{9.18}$$

where:

$$\dot{P}(t) = \begin{bmatrix} \dot{P}_1(t) \\ \dot{P}_2(t) \end{bmatrix} \tag{9.19}$$

$$\mathbf{A} = \begin{bmatrix} -\lambda & \mu \\ \lambda & -\mu \end{bmatrix} \tag{9.20}$$

$$\mathbf{P} = \begin{bmatrix} P_1(t) \\ P_2(t) \end{bmatrix}. \tag{9.21}$$

This set of equations can be solved to obtain the dynamic behavior of **P**. This general concept can be extended to any set of n states.

When the system has n components, and each component has two states (functioning and failed), the system will have at most 2^n different states. For a complex system, the number of states may be overwhelming, and we may need to simplify the system model, and perhaps create several separate models. We explore these techniques in Chapter 22.

Compare the state transition rate matrix **A** with the DTMC probability matrix **P**. Notice for the matrix **A**, transitions from a state back to the same state are not shown. Since the beginning and ending states are the same, we can say no change has happened to the system. Thus, $p_{ii} = 0$ for every state. Instead, each rate β_{ii} represents the rate of transition out of State i (Eq. (9.9)).

9.4 STEPS TO CREATE A MARKOV RELIABILITY MODEL

Step 1: Create a model of the system that is to be analyzed.

Examine the system architecture and decompose it into components or subsystems that can be assigned individual failure rates and repair rates. An RBD that identifies the components and shows the relationships between them can then be created. For an RBD with n components, identify each component by a name and a unique numeric identifier (component 1 to component n). Since Markov analysis becomes unwieldy for systems with a large number of states (complicated systems or systems with many components), the analyst may want to simplify the model. For the simplified model, the analyst must ensure the major components (or aggregated components) and the reliability relationships between them are preserved and captured in a high-level reliability block diagram.

Step 2: Identify the system states.

The system to be analyzed will generally have several states, such as the fully operational states, degraded operational states, and one or more failure states. Define and number each state. Nonrelevant states should be removed. As a convention, number each state with a unique identifier from 1 to n. Also identify the fully functional state as State 1, and the completely failed state is State n.

Step 3: Construct the state transition diagram.

Show all states in the diagram and identify all possible transitions from the various states. States are represented in the diagram as small circles or bubbles with the number of the state inside the bubble. For each transition, identify the transition rate. Two types of transition rates are generally employed for reliability models: the failure rate of a component (λ) and the repair rate of a component (μ). From the reliability block diagram, the numbered components are used as reference. Thus, the failure rate of component 1 is λ_1, and if the component under analysis is maintained (i.e., repaired upon failure), it will also have an associated repair rate μ_1. The transitions from one state to another are shown as directed arcs. Each of these lines is labeled with the transition rate λ_j or μ_j.

Step 4: Create the state transition matrix and equations.

Once the state transition diagram is complete, the state transition matrix is written by inspecting the diagram. Arrange the transition rates per the state transition rate matrix, Equation (9.8). The matrix represents one part of the set of equations that can be solved to determine the probabilities of each state. As a check, make sure the sum of all entries in each *column* is equal to zero.

Step 5: Calculate dynamic or steady-state behavior.

The complete solution (probabilities dynamically change over time) or steady-state solution (asymptotic state probabilities) can be obtained by solving Equation (9.18). For a large matrix, a commercial or custom software program can quickly provide a solution.

Step 6: Calculate various items of interest from the state probabilities.

For example, you may want to calculate system component reliabilities and system availability as a function of time.

Step 7: Document results.

Identify specific actions to improve the system reliability (e.g., incorporate redundancy), or improve the transition rates (e.g., employ more reliable components, improve the reliability of existing components, or reduce the repair time).

9.5 ASYMPTOTIC BEHAVIOR (STEADY-STATE BEHAVIOR)

A set of linear, first-order differential equations, called the *Kolmogorov equations*, determine the probability distribution $\boldsymbol{P}(t) = P_1(t), P_2(t), \ldots, P_n(t)$ of the Markov process at time t, where $P_i(t)$ is the probability that the process (the system) is in state i at time t and n is the number of state. $\boldsymbol{P}(t)$, under specific conditions, will approach a limit $\boldsymbol{P}$ when $t \to \infty$. This limit is called the *steady-state* distribution of the system.

For the steady-state distribution, we will consider several system performance attributes and characteristics.

A Markov process starts out at time $t = 0$ in a certain state, say State 1, and stays in this state for duration of T_1. At time $t_1 = T_1$, the process transitions to another state where it stays for the duration of T_2. At time $t_2 = T_1 + T_2$, the process transitions to another state, and so on.

Consider a Markov process that enters State i at time 0, that is, $X(0) = i$. Let T_i be the duration of time the process spends in State i. We want to find the probability $P(T_i > t)$. Let us now assume that we observe that the process is still in State i at time s, that is, $T_i > s$, and that we are interested in finding the probability that it will remain in State i for t time units more, that is, $P(T_i > t + s|T_i > s)$. Since the process has the Markov property, the probability for the process to stay for t more time units is determined only by the current State i. The fact that the process has already been in State i for any amount of time is irrelevant. Thus

$$P(T_i > t + s|T_i > s) = P(T_i > t) \text{ for } s, t \geq 0. \tag{9.22}$$

The asymptotic probabilities are the steady-state probabilities for the Markov process.

A Markov process can be classified as reducible or irreducible. A Markov process is *irreducible* if every state is reachable from every other state. For an irreducible Markov process, the limits are:

$$\lim_{t \to \infty} P_j(t) = P_j \text{ for } j = 1, 2, \ldots, n;$$

for an n-state system, this limit always exist and are independent of the initial state of the process (at time $t = 0$). The process will converge to a steady-state behavior where the probability of being in state j will be a constant (independent of time).

If $P_j(t)$ tends to a constant value when $t \to \infty$, then the time derivative of the probabilities become 0:

$$\lim_{t \to \infty} P_j^{\bullet}(t) = 0 \text{ for } j = 1, 2, \ldots, n. \tag{9.23}$$

The steady-state probabilities $\boldsymbol{P}$ must therefore satisfy the matrix equation:

$$\begin{bmatrix} 0 \\ 0 \\ \vdots \\ 0 \end{bmatrix} = \begin{bmatrix} \beta_{11} & \beta_{21} & \cdots & \beta_{n1} \\ \beta_{12} & \beta_{22} & \cdots & \beta_{n2} \\ \vdots & \vdots & \ddots & \vdots \\ \beta_{1n} & \beta_{2n} & \cdots & \beta_{nn} \end{bmatrix} \cdot \begin{bmatrix} P_1 \\ P_2 \\ \vdots \\ P_n \end{bmatrix}. \tag{9.24}$$

P_j is the average, long-term proportion of time the system spends in State j. Observe from these equations, the initial state of the process has no influence on the steady-state probabilities.

We can write the earlier equation in compact form:

$$\mathbf{U} = \mathbf{AP}. \tag{9.25}$$

To solve this set of equations for $\mathbf{P}$, we take the inverse of matrix $\mathbf{A}$ and multiply it by $\mathbf{U}$:

$$\mathbf{P} = \mathbf{A}^{-1}\mathbf{U}. \tag{9.26}$$

As noted earlier, the sum of each column is 0; thus, the determinant of $\mathbf{A}$ is 0 and therefore the state equations in the state transition matrix $\mathbf{A}$ do not have a unique solution. The set of state equations are not independent. In general, for any $n \times n$ state transition matrix $\mathbf{A}$, only $n - 1$ of the equations will be independent.

However, we can find a unique solution if we add the necessary constraint that the sum of all the state probabilities must *always* equal 1, that is, the conservation of probability Equation (9.17) must hold. We choose one of the equations in the state matrix and replace it with the conservation of probability equation in order to solve the state equations.

To calculate the steady-state probabilities, $P_1, P_2, \ldots, P_n$, we use $n - 1$ of the n linear algebraic equations from the matrix Equation (9.24) plus Equation (9.17).

We can replace any of the equations, so let us replace the first equation in Equation (9.24):

$$\begin{bmatrix} 1 \\ 0 \\ \vdots \\ 0 \end{bmatrix} = \begin{bmatrix} 1 & 1 & \cdots & 1 \\ \beta_{12} & \beta_{22} & \cdots & \beta_{n2} \\ \vdots & \vdots & \ddots & \vdots \\ \beta_{1n} & \beta_{2n} & \cdots & \beta_{nn} \end{bmatrix} \cdot \begin{bmatrix} P_1 \\ P_2 \\ \vdots \\ P_n \end{bmatrix}. \tag{9.27}$$

With this updated matrix equation, Equation (9.26) can be solved to obtain the steady-state probabilities.

9.5.1 Characteristics of the Steady-State Markov Process

Several general relationships and parameters for the steady-state Markov process exist that we can apply to reliability theory.

The amount of time T_j a process spends in State j before making a transition into a different state (also known as the sojourn time) is dependent on the departure rate α_j. When the system is in State j, the system will remain in this state for a time equal to the reciprocal of the departure rate from that state. The mean duration of time spent in State j is thus:

$$T_j = 1/\alpha_j. \tag{9.28}$$

The frequency of transitions from State k into State j is equal to the probability of being in State k multiplied by the transition rate from State k to State j, that is, $P_k\beta_{kj}$. The total frequency of arrivals into State j, that is, f_{Aj} is therefore:

$$f_{Aj} = \sum_{k=1,k\neq j}^{r} P_k\beta_{kj} = f_{Dj} = P_j\alpha_j = f_j. \tag{9.29}$$

The frequency of departures from a State j (f_{Dj}) is the proportion of time spent in State j (P_j) times the transition rate α_j out of State j.

$$f_{Dj} = P_j \alpha_j \tag{9.30}$$

In the steady state, the frequency of arrivals to State j is equal to the frequency of departures from State j.

$$f_{Aj} = f_{Dj} = f_j. \tag{9.31}$$

The mean time between visits (MTBV) to State j is the reciprocal of the frequency of visits to State j.

$$\text{MTBV}_j = 1/f_j \tag{9.32}$$

The mean proportion of time the system spends in a state is the same as the probability of being in the state. Thus, the proportion of time, P_j, the system spends in State j is equal to the visit frequency to State j (f_j), multiplied by the mean duration of a visit to State j:

$$P_j = f_j T_j. \tag{9.33}$$

9.5.2 Frequency of System Failures

The frequency f_F of system failures is the steady-state frequency of transitions from a functioning state (G) to a failed state (F). That is, we sum the transition rates from each of the good states to each of the failed states multiplied by the probability of being in a good state:

$$f_F = \sum_{j \in G} \sum_{k \in F} P_j \beta_{jk}. \tag{9.34}$$

If only one failed state exists, Equation (9.34) reduces to:

$$f_F = \sum_{j \in G} P_j \beta_{jF} = \sum_{j \in G} P_j \alpha_F, \tag{9.35}$$

where α_F is the failure rate.

If only one failed and one good state exists, Equation (9.35) reduces to:

$$f_F = P_G \beta_{GF} = P_G \alpha_F, \tag{9.36}$$

where P_G is the probability of being in the good state, β_{GF} is the failure rate from the good state to the failed state, and α_F is the failure rate.

9.5.3 Mean Time between Failures

The mean time between system failures, $\mathrm{MTBF_S}$, is the mean time between consecutive transitions from a functioning state (G) into a failed state (F). The $\mathrm{MTBF_S}$ may be computed from the frequency of system failures by:

$$\mathrm{MTBF_s} = 1/f_F. \tag{9.37}$$

9.5.4 Mean Time to Repair

The frequency of system repairs is equal to the sum of the transition rates out of a failed state to a good state given that the system is in a failed state.

$$f_{FG} = \frac{\sum_{j\in F}\sum_{k\in G} P_j\beta_{jk}}{\sum_{j\in F} P_j}, \tag{9.38}$$

where f_{FG} is the frequency of transitions from $F \to G$:

$$f_{FG} = \left(f_F \sum_{j\in F} P_j \right). \tag{9.39}$$

If only one failed state exists, Equation (9.39) reduces to:

$$f_{FG} = \sum_{k\in G} \beta_{Fk}. \tag{9.40}$$

If only one failed state exists and only one possible transition path exists from that one failed state to a good state, Equation (9.40) reduces to:

$$f_{FG} = \beta_{FG}. \tag{9.41}$$

The mean duration of a system failure is defined as the mean time it takes for the system to be brought back into a functioning state (G).

The mean time to repair, $\mathrm{MTTR_S}$, is the reciprocal of f_{FG}:

$$\mathrm{MTTR_s} = 1/f_{FG}. \tag{9.42}$$

9.5.5 Mean Time to System Failure

The rate of transitioning from a good state to a failed state for a repairable system given that the process is in a good state is equal to the probability of the system being in a good state in which a transition to a failed state can occur multiplied by the departure rate of states that can transition from a good state to a failed state divided by the sum of the probabilities of being in a good state:

$$f_{GF} = \frac{\sum_{j \in G} \sum_{k \in F} P_j \beta_{jk}}{\sum_{j \in G} P_j}, \tag{9.43}$$

where f_{GF} is the frequency of transitions from $G \rightarrow F$.

If only one good state exists, Equation (9.43) above reduces to:

$$f_{GF} = \sum_{k \in F} \beta_{Gk}. \tag{9.44}$$

If only one good state and one state exists in which a transition to a failed state can occur, Equation (9.44) above reduces to:

$$f_{GF} = \beta_{GF}. \tag{9.45}$$

The reciprocal of the rate, f_{GF}, of transitions from a good state to a failed state is the MTTF:

$$\mathrm{MTTF}_s = 1/f_{GF}. \tag{9.46}$$

9.5.6 System Availability

One or more of the system states represent system functioning according to some specified criteria. Let G denote the subset of states in which the system is functioning, and let $F = X - G$ denote the states in which the system is failed, where X is the total number of states.

The long-term availability of the system is the mean proportion of time that the system is functioning, that is, a member of G. The average system availability A_S is thus defined as:

$$A_S = \sum_{j \in G} P_j. \tag{9.47}$$

The system unavailability $Q_S = (1 - A_S)$ is the mean proportion of time the system is in a failed state:

$$Q_S = \sum_{j \in F} P_j. \tag{9.48}$$

The unavailability, Q_S, is equivalently the frequency of system failures multiplied by the mean duration of a system failure. Thus:

$$Q_S = f_F \cdot \mathrm{MTTR}_s. \tag{9.49}$$

9.6 LIMITATIONS OF MARKOV MODELING

1. *Exponential Failure Rate and Repair Rate Assumption.* Many physical devices and software have time-dependent behavior, that is, the constant failure rate requirement for Markov modeling does not apply. However, the behavior of these failure rates can be approximated in the Markov model by breaking the time-dependent failure rates into several intervals, each of which have an average or constant failure rate. In many other cases, using the constant failure rate assumption is sufficient for the level of accuracy obtainable given the uncertainties with failure rate assumptions.
2. *State Space Explosion.* As the number of components to be modeled increases, the number of states increases exponentially. Although solving these models using commercial or other reliability software is not difficult, just capturing the state diagram itself can be quite challenging when the number of states exceeds 20 or 30. In addition, having a large number of states increases the chance of errors being introduced and reduces the comprehensibility of the state diagram. We typically address this by state reduction and state diagram partitioning techniques. These techniques will be explored in later chapters. Some of these reduction techniques will reduce the overall accuracy of the model, so we need to perform a tradeoff between tractability of the model and a reduction in the accuracy of the model.

9.7 MARKOV REWARD MODELS

The Markov Reward Model (MRM) provides a useful method for classifying the states of a Markov model based on their relative merit. The mapping of each state to this figure of merit is done by the reward function r. For each state i, r_i represents the reward obtained per unit time spent by the system in that state. The reward associated with a state may represent the operational reliability, or performance of the system while it is in that state. MRMs can be used to evaluate these and other metrics of interest for the modeled system.

The reward model can be extended to cover both performance and reliability of the system. For example, if a system has a two-component active–active configuration and the CPU utilization of each component is 40%, upon failure of one component, the system is still fully functional, but the failure rate of the remaining component may increase due to the increased load on this component, let us say of 80% CPU utilization. Furthermore, the system performance may become noticeably sluggish as delays in processing times increase and possibly overload conditions cause unacceptable delays or failures.

The graceful degradation of the system in terms of these parameters of interest can be systematically evaluated. For example, the loss of a component may reduce the level of service the system can provide. A different reward can then be assigned to this state to reflect this degraded but functional mode of operation. These dependencies are well-suited for Markov modeling since degraded modes of operation can be captured in a methodical way. Furthermore, each state can have a reward associated with it based on reliability, redundancy and/or performance. The combination of reliability and performance in these reward models is sometimes referred to in the literature as *performability*. The reader is referred to the literature for applying reward models to capture performability. We will limit our discussion to rewards associated with reliability and availability.

What measures of reliability can we obtain from a MRM? The most common types of measures are the following:

1. Expected value of the reward rate
2. Accumulated reward
3. Expected accumulated reward.

The reward rate, $X(t)$, is the instantaneous reward rate of the system.

Starting with the first measure, the expected value of the reward rate is defined as:

$$E[X(t)] = \sum_i r_i P_i(t). \tag{9.50}$$

In the simplest form, the reward rate of a state is a function of the state index. If we assign $r_i = 1$ for all system operational states and $r_i = 0$ for all failed states, then for a repairable system with no absorbing states, Equation (9.50) will give us the availability of the system. If we reverse the assignments, that is, the operational states have a reward of 0 and the failed states have a reward of 1, we obtain the unavailability of the system.

9.8 SUMMARY

In this chapter, we expanded on the concept of a continuous-time Markov processes previously introduced in Chapter 8.

Given the Markov properties and comparing with the DTMC, the state transition rate matrix was derived. With this canonical form of the continuous-time Markov process, we were able to derive the dynamic properties of each defined state and explore how they evolve over time. A two-state Markov model for a single component repairable reliability model was derived. We walked through the steps to create a Markov model and considered the characteristics of this model, including asymptotic behavior. We also provided a brief introduction to Markov reward models.

The continuous-time Markov process is one of the most powerful tools we have for modeling system behavior. With the appropriately defined model, we are able to extract many useful characteristics of the system under analysis. We will apply this powerful technique to a broad variety of problems in the latter half of this text. One of the drawbacks of this technique is state-space explosion. We will investigate this problem in the following two chapters and discuss ways in which we can mitigate this to make the problems more tractable.

CHAPTER 10

Markov Analysis: Nonrepairable Systems

10.1 INTRODUCTION

A nonrepairable system has a finite lifetime. Even with highly reliable components and redundancy, eventually, the system will fail if it operates for a sufficiently long period of time. Intuitively, this makes sense. The probability of a system failure will increase over time during system operation. A nonrepairable system remains failed after it fails once. In the next chapter, we will discuss repairable systems; if the system fails, it can be brought back to a partially or fully operational condition after a repair operation.

In this chapter, we will cover the following nonrepairable systems:

1. A system with one component with no repair
2. A system consisting of two parallel components with no repair—identical component failure rates
3. A two-component series system with no repair—identical component failure rates
4. A system consisting of two parallel components with partial repair—identical component failure rates, and finally
5. A system consisting of two parallel components with no repair—different component failure rates.

10.2 ONE COMPONENT, NO REPAIR

Let us start off with the simplest case: One-component system with no repair.

This is a Markov process with absorbing states. Recall that an absorbing state is a state that, once entered, cannot be left. For this example, the system is functioning

Designing High Availability Systems: Design for Six Sigma and Classical Reliability Techniques with Practical Real-Life Examples, First Edition. Zachary Taylor and Subramanyam Ranganathan.

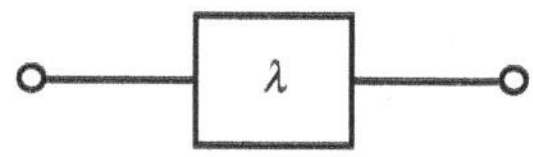

Figure 10.1 One-Component System RBD

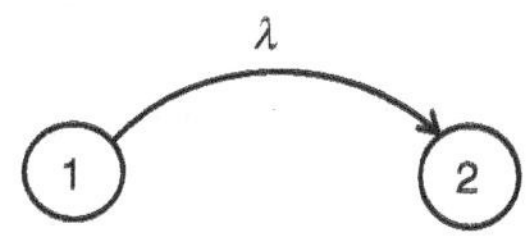

Figure 10.2 State Transition Diagram for One-Component System Without Repair

until it fails. Once failed, no repair has been defined so the system will remain in the failed state.

Following the prescribed steps in Chapter 8, for Step 1, we create an RBD for our one component system as shown in Figure 10.1.

Next, in Step 2, we define the states. This system will have just two states: the system is either functioning properly or it has failed completely. No degraded mode of operation is assumed for this simple example.

- *State 1: System fully functional.*
- *State 2: System is failed.*

In Step 3, we construct the state transition diagram. The two states are shown as numbered circles on the diagram. The system can only transition from the functional state to the failed state when the component fails. So we draw a directed arc from State 1 to State 2. The failure rate of the component is identified by λ and the arc is labeled with this failure rate (Fig. 10.2).

10.2.1 Dynamic Behavior

We next consider the dynamic behavior of the system, Step 4, in which we create the state transition matrix and equations:

$$\begin{bmatrix} \dot{P}_1 \\ \dot{P}_2 \end{bmatrix} = \begin{bmatrix} -\lambda & 0 \\ \lambda & 0 \end{bmatrix} \cdot \begin{bmatrix} P_1 \\ P_2 \end{bmatrix}. \tag{10.1}$$

We can choose to solve the set of equations in Equation (10.1) for the vector $\boldsymbol{P}(t)$, or we can remove one of the equations and replace it with the conservation of probability equation: $P_1 + P_2 = 1$ and solve that set of equations instead for the vector $\boldsymbol{P}(t)$. We replace the last equation in the matrix with the conservation of probability equation:

$$\begin{bmatrix} \dot{P}_1 \\ 1 \end{bmatrix} = \begin{bmatrix} -\lambda & 0 \\ 1 & 1 \end{bmatrix} \cdot \begin{bmatrix} P_1 \\ P_2 \end{bmatrix}. \tag{10.2}$$

The two equations from the matrix above are:

$$\dot{P}_1 = -\lambda P_1 \tag{10.3}$$

$$1 = P_1 + P_2. \tag{10.4}$$

We have two independent equations, and each equation provides new information about the unknown variables.

The solution of Equation (10.2) is of the form:

$$P_1(t) = e^{kt} + C_1.$$

The constant C_1 can be determined by evaluating the boundary conditions of the system. At time $t = 0$, the system is fully functional, so $P_1(0) = 1$, thus $C_1 = 0$:

$$P_1(t) = e^{kt} \tag{10.5}$$

By substituting this equation into Equation (10.3), we obtain:

$$\dot{P}_1 = ke^{kt} = -\lambda e^{kt}. \tag{10.6}$$

By inspection, we see that $k = -\lambda$.

Thus, the final equations become:

$$P_1(t) = e^{-\lambda t} \tag{10.7}$$

$$P_2(t) = 1 - e^{-\lambda t}. \tag{10.8}$$

10.2.1.1 Alternative Approach to Solve the Equation We can also solve these equations using Laplace transform techniques. Laplace transforms allow the solution of integral-differential equations using algebraic techniques.

Given that the system is in State 1 at $t = 0$, the Laplace transform of $\dot{P}_1$ is $sP^*(s) - 1$. The Laplace transform of $\dot{P}_2$ is $sP_2^*(s)$. The transition matrix in Equation (10.1) becomes:

$$\begin{bmatrix} sP_1^*(s) - 1 \\ sP_2(s) \end{bmatrix} = \begin{bmatrix} -\lambda & 0 \\ \lambda & 0 \end{bmatrix} \cdot \begin{bmatrix} P_1^*(s) \\ P_2^*(s) \end{bmatrix}, \tag{10.9}$$

which gives the equations:

$$sP_1^*(s) - 1 = -\lambda P_1^*(s)$$

$$sP_2^*(s) = \lambda P_1^*(s).$$

Solving for $P_1^*(s)$ and $P_2^*(s)$:

$$P_1^*(s) = \frac{1}{s + \lambda} \tag{10.10}$$

$$P_2^*(s) = \frac{\lambda}{s(s+\lambda)}. \tag{10.11}$$

Splitting Equation (10.11) into simpler transforms, we obtain:

$$P_2^*(s) = \frac{\lambda}{s(s+\lambda)} = \frac{A}{s} + \frac{B}{s+\lambda}.$$

The constants A and B are obtained by using the method of partial fraction expansion:

$$A = \lim_{s \to 0} \frac{\lambda}{s+\lambda} = 1$$

$$B = \lim_{s \to -\lambda} \frac{\lambda}{s} = -1.$$

$P_2^*(s)$ becomes:

$$P_2^*(s) = \frac{1}{s} - \frac{1}{s+\lambda}.$$

Taking the inverse Laplace transform to get the time solution:

$$P_1(t) = e^{-\lambda t}$$

$$P_2(t) = 1 - e^{-\lambda t}.$$

EXAMPLE 10.1

Given a one-component dynamic system with a failure rate $\lambda = 0.01$, what is the dynamic time response for States 1 and 2 over a mission time of 1000 hours?

Substituting the value for the failure rate λ into the above equations for P_1 and P_2, and graphing the results over a mission time of 1000 hours we obtain Figure 10.3. The system begins operation in State 1 with a probability of 1 and exponentially decreases until the probability is practically zero at around 600 hours. The probability of system failure corresponds to the probability of the system being in State 2 (the failed state). Note that the sum of the probabilities of being in either state must equal 1. We could also reverse the initial state of the system, with State 2 initially at 1 and State 2 initially at 0. However, by convention in this book, State 1 generally is the beginning state initialized to 1.

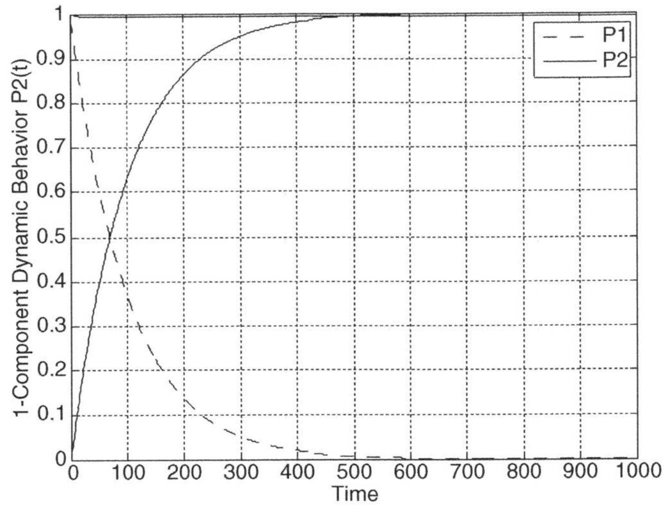

Figure 10.3 Single Component Two-State Dynamic Behavior

10.2.2 Reliability

From Chapter 7, we define reliability as the probability that a system or components will operate without a failure in a given period of time. From our example, the system is reliable if it is in State 1. Thus, the reliability equation for a single nonrepairable component is:

$$R(t) = P_1(t) = e^{-\lambda t}. \tag{10.12}$$

10.2.3 Availability

The system is functioning if it is in State 1, or equivalently if the system is ***not*** in the failed state (State 2). The probability that the system is functioning at time t, which corresponds to the availability of the system is:

$$A(t) = P_1(t) = e^{-\lambda t}. \tag{10.13}$$

When there is no repair, the availability is $P_1(t) = e^{-\lambda t}$ which corresponds to the reliability of the system, that is, for nonrepairable systems, the availability and reliability are the same.

The long-term availability is:

$$A_{lt} = \lim_{t \to \infty} A(t)$$

$$A_{lt} = 0.$$

For a nonrepairable system, the availability eventually goes to zero.

EXAMPLE 10.2

Determine the dynamic system behavior for a one-component system with an Mean Time between Failure (MTBF) of 100 hours and plot the system availability versus time.

First, calculate the failure rate. Since the failure rate is the reciprocal of the MTBF, $\lambda = 0.01$. The availability of the system is calculated using Equation (10.13):

$$A(t) = e^{-(0.01)t}.$$

The resulting dynamic availability is graphed in Figure 10.4.

For a system with an MTBF of 100 hours, the availability is close to 0 after only about 500 hours of operation.

Another interesting observation: Although the MTBF is 100 hours, there is only a 50% chance that the system will function beyond ~69 hours. This is a result of the nature of the constant failure rate model previously discussed. Architects should keep in mind that for any component that can be modeled with a constant failure rate assumption, 50% reliability is reached at 69% of the MTBF.

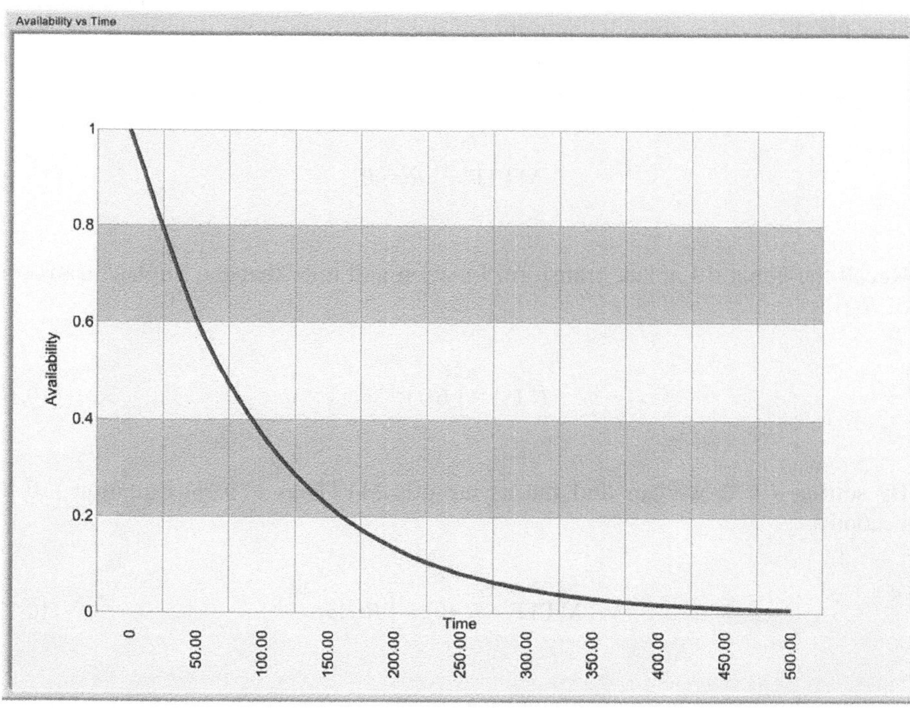

Figure 10.4 Single-Component Dynamic Availability

10.2.4 Markov Reward Model

More systematically, we can calculate the transient availability of the system using the Markov Reward Model (MRM) described in Section 9.7. We assign a reward of 1 when the system is in the operational states and a reward of 0 when the system is in the failure states. For this example, the system is operational (available) when in State 1 and is failed (unavailable) when in State 2. From Equation (9.50), the availability is the expected value of the reward rate:

$$A_s(t) = E[X(t)] = \sum_i r_i P_i(t) = 1 \cdot P_1 + 0 \cdot P_2 = e^{-\lambda t}. \tag{10.14}$$

This result is of course the same as Equation (10.13) as we would expect. The MRM in this case is straightforward, but in more complex cases, the MRM will help keep track of the rewards associated with each state using different weighted rewards. Furthermore, we may define weighted rewards that take into account performance, stress, overload conditions, partial capacity, and other degraded modes of operation that require fractional rewards.

10.2.5 Mean Time to Failure (MTTF)

Next, we would like to find the system MTBF. Strictly speaking, the MTBF does not exist for this system since it can only fail once. So, instead we determine the Mean Time to Failure or MTTF. The MTTF is equal to the mean value of the reliability of the single component. From Chapter 7, Equation (7.6), the MTTF is:

$$\text{MTTF} = \int_0^\infty R(t)dt. \tag{10.15}$$

Recall the general Laplace transform function and note that the Laplace transform of $R(t)$ is:

$$R^*(s) = \int_0^\infty R(t)e^{-st}dt. \tag{10.16}$$

By setting $s = 0$, we can find the asymptotic MTTF as $t \to \infty$. Equation (10.16) becomes:

$$\text{MTTF} = R^*(0) = \int_0^\infty R(t)dt. \tag{10.17}$$

Since the system is functioning as long as the system remains in State 1, the reliability of this system is equal to the probability of the system being in State 1.

$$R^*(s) = P_1^*(s)$$
$$R^*(s) = \frac{1}{s+\lambda}$$
$$R^*(0) = \frac{1}{\lambda} \tag{10.18}$$
$$\text{MTTF} = \frac{1}{\lambda}.$$

Of course, the MTTF is easily obtained by the definition of the system, that is, a one-component system with a failure rate of λ has an MTTF of $1/\lambda$. However, as the system becomes more complex, the MTTF will not be quite as obvious.

Since this example is an absorbing matrix, eventually the system will transition into the failed state and remain trapped in this state, that is, the system approaches a 100% certainty of becoming failed as $t \to \infty$.

10.2.6 Asymptotic Behavior (Steady-State Behavior)

If we start with the state transition matrix equations we previously created in Equation (10.1), we can obtain the steady-state behavior directly by setting the time derivatives of the probabilities to 0:

$$\begin{bmatrix} 0 \\ 0 \end{bmatrix} = \begin{bmatrix} -\lambda & 0 \\ \lambda & 0 \end{bmatrix} \cdot \begin{bmatrix} P_1 \\ P_2 \end{bmatrix}. \tag{10.19}$$

We can solve for P by taking the inverse of the transition matrix A and multiplying it by U:

$$\mathbf{P} = \mathbf{A}^{-1}\mathbf{U}, \tag{10.20}$$

where:

$$\mathbf{A} = \begin{bmatrix} -\lambda & 0 \\ \lambda & 0 \end{bmatrix}$$
$$\mathbf{P} = \begin{bmatrix} P_1 \\ P_2 \end{bmatrix}$$
$$\mathbf{U} = \begin{bmatrix} 0 \\ 0 \end{bmatrix}.$$

Before we can solve this system, note first that the above equations are not linearly independent. The equations of a linear system are independent if none of the equations can be derived algebraically from the others.

In general, the steady-state Markov state equations have the following properties:

1. They are linearly dependent. Any given equation is automatically satisfied if the other ones are satisfied (conservation of probability).

2. The solution is unique up to a constant factor.
3. The solution is uniquely determined by substituting the conservation of probability equation for one of the linear equations.

In the general case, let matrix $\mathbf{A} = [\boldsymbol{a}_1, \boldsymbol{a}_2, \ldots, \boldsymbol{a}_n]$, where $\boldsymbol{a}_i$ denotes the ith row vector of $\mathbf{A}$. This set of row vectors is linearly dependent if there exists multipliers $\beta_1, \beta_2, \ldots, \beta_n$ not all zero such that:

$$\beta_1 \begin{bmatrix} a_{11} \\ a_{12} \\ \vdots \\ a_{1n} \end{bmatrix} + \beta_2 \begin{bmatrix} a_{21} \\ a_{22} \\ \vdots \\ a_{2n} \end{bmatrix} + \cdots + \beta_n \begin{bmatrix} a_{n1} \\ a_{n2} \\ \vdots \\ a_{nn} \end{bmatrix} = \begin{bmatrix} 0 \\ 0 \\ \vdots \\ 0 \end{bmatrix}. \tag{10.21}$$

If Equation (10.21) is true only for all $\beta_i = 0$, then the set of equations is linearly independent.

For Equation (10.19):

$$\beta_1 \begin{bmatrix} a_{11} \\ a_{12} \end{bmatrix} + \beta_2 \begin{bmatrix} a_{21} \\ a_{22} \end{bmatrix} = \begin{bmatrix} 0 \\ 0 \end{bmatrix}$$

$$\beta_1 \begin{bmatrix} -\lambda \\ 0 \end{bmatrix} + \beta_2 \begin{bmatrix} \lambda \\ 0 \end{bmatrix} = \begin{bmatrix} 0 \\ 0 \end{bmatrix}$$

$$1 \begin{bmatrix} -\lambda \\ 0 \end{bmatrix} + 1 \begin{bmatrix} \lambda \\ 0 \end{bmatrix} = \begin{bmatrix} 0 \\ 0 \end{bmatrix}.$$

Since for $\beta_1 = 1$ and $\beta_2 = 1$, the equations evaluate to zero, the equations are not linearly independent. The two equations are the same equation scaled by a factor of -1.

Looking at it another way, the flow of probability into P_1 is the same as the flow of probability out of P_2 which is the same as the flow of probability into P_2.

Since the transition matrix is not full rank, that is, all of the equations in the matrix are not linearly independent, we can solve the set of equations by removing one of the equations and replacing it with the conservation of probability equation: $P_1 + P_2 = 1$. The matrix then becomes full rank (the determinant exists) and thus we can now solve the set of equations.

Replacing the first equation with the conservation of probability, the steady-state transition matrix becomes:

$$\begin{bmatrix} 1 \\ 0 \end{bmatrix} = \begin{bmatrix} 1 & 1 \\ \lambda & 0 \end{bmatrix} \cdot \begin{bmatrix} P_1 \\ P_2 \end{bmatrix}. \tag{10.22}$$

The state equations are:

$$1 = P_1 + P_2 \tag{10.23}$$

$$0 = \lambda P_1. \tag{10.24}$$

Solving Equations (10.23) and (10.24):

$$P_1 = 0 \tag{10.25}$$

$$P_2 = 1. \tag{10.26}$$

From the equations above, we interpret this as the probability that the system will be functioning, as $t \to \infty$ is 0. As Figure 10.3 illustrates, P_1 exponentially decreases from 1 and approaches 0 after about 600 hours.

Alternatively, we could start with Equation (10.22) identifying the matrices and solving for P using matrix algebra:

$$\mathbf{U} = \mathbf{A} \cdot \mathbf{P} \tag{10.27}$$

$$\mathbf{U} = \begin{bmatrix} 1 \\ 0 \end{bmatrix} \tag{10.28}$$

$$\mathbf{A} = \begin{bmatrix} 1 & 1 \\ \lambda & 0 \end{bmatrix} \tag{10.29}$$

$$\mathbf{P} = \begin{bmatrix} P_1 \\ P_2 \end{bmatrix}. \tag{10.30}$$

We can solve for P by taking the inverse of the transition matrix $\boldsymbol{A}$ and multiplying it by $\boldsymbol{U}$:

$$\mathbf{P} = \mathbf{A}^{-1}\mathbf{U} \tag{10.31}$$

$$\mathbf{P} = -\frac{1}{\lambda}\begin{bmatrix} 0 & -1 \\ -\lambda & 1 \end{bmatrix} \cdot \begin{bmatrix} 1 \\ 0 \end{bmatrix} \tag{10.32}$$

$$\begin{bmatrix} P_1 \\ P_2 \end{bmatrix} = \begin{bmatrix} 0 \\ 1 \end{bmatrix} \tag{10.33}$$

This solution matches Equations (10.25) and (10.26).

10.3 NONREPAIRABLE SYSTEMS: PARALLEL SYSTEM WITH NO REPAIR

In the previous example, we analyzed a one-component nonrepairable system. In this section, we will explore a two-component redundant system with no repair. The analysis follows the same steps as the previous section but with a different set of state equations.

The reliability block diagram for a two-component parallel system is shown in Figure 10.5.

For this example, let us assume both components have a constant failure rate λ.

The three possible states of the system are the following:

- *State 1: System functional and both components operating.*
- *State 2: System functional and one of the two components has failed.*
- *State 3: System failure; both components have failed.*

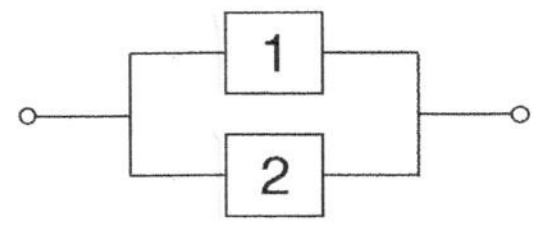

Figure 10.5 Two-Component RBD

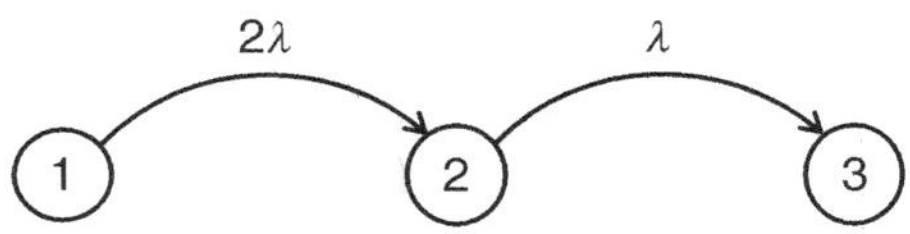

Figure 10.6 State Transition Diagram for Two-Component Parallel System: No Repair

From the definitions of these three states, the RBD, and assuming no repair on failed components, we can create the state transition diagram for the system, as shown in Figure 10.6.

Note that the transition rate from State 1 to State 2 is 2λ, since the probability of either one of the components failing is double the failure rate of a single component, that is, we have twice as many chances to fail since we have two components.

Notice that when drawing the state transition diagram, we consider a very short time interval, such that the transition diagram only records events of single transitions. We do not show a transition from State 1 to State 3, since we assume both components cannot fail at the exact same time or infinitesimally small interval. This constraint follows from the definition of the Poisson distribution.

Since the basis of the Markov process is a Poisson process, the probability of having two or more events in a short time Δt is negligible as $\Delta t \to 0$, and thus events of multiple transitions are not included in the state transition diagram.

We recall that this limitation did not apply in the discrete Markov model. Since the time intervals are discrete, we could consider a case in which multiple events occur in the same interval. Refer back to Chapter 8 for examples comparable with our current analysis and the approximations made.

The transitions from one state to another state occur instantaneously, and the time the transitions occur is determined by the exponential probability distribution with parameter λ, (β_{23}). This model captures the failure of both components by first forcing one of the components to enter State 2 before State 3 can be entered. Also note from this model that we do not identify which component fails first. Which component fails first is deemed unimportant and is not captured. We could, however, if desired, add a state to capture the failure of component 1 and another state to capture the failure of component 2. We will consider later in this chapter how to create a model that shows separate transitions for each component and why they are needed.

We also have not considered common mode failures. Common mode failures are single failures that can cause the system to fail even with redundancy in place, that is, the system can transition from State 1 to State 3 directly. We will not consider common mode failures in this example, but will discuss it in later chapters.

10.3.1 Dynamic Behavior

From the state transition diagram, Figure 10.6, the state transition matrix is:

$$\begin{bmatrix} \dot{P}_1 \\ \dot{P}_2 \\ \dot{P}_3 \end{bmatrix} = \begin{bmatrix} -2\lambda & 0 & 0 \\ 2\lambda & -\lambda & 0 \\ 0 & \lambda & 0 \end{bmatrix} \cdot \begin{bmatrix} P_1 \\ P_2 \\ P_3 \end{bmatrix}. \tag{10.34}$$

Although not required, we could choose to replace one of the equations with the conservation of probability equation. $P_1 + P_2 + P_3 = 1$.

We assume the system is fully operational at $t = 0$, that is, the system is in State 1. Taking the Laplace transform of the above matrix, we obtain:

$$\begin{bmatrix} sP_1^*(s)-1 \\ sP_2^*(s) \\ sP_3^*(s) \end{bmatrix} = \begin{bmatrix} -2\lambda & 0 & 0 \\ 2\lambda & -\lambda & 0 \\ 0 & \lambda & 0 \end{bmatrix} \cdot \begin{bmatrix} P_1^*(s) \\ P_2^*(s) \\ P_3^*(s) \end{bmatrix}. \tag{10.35}$$

The state equations are:

$$sP_1^*(s)-1=-2\lambda P_1^*(s) \tag{10.36}$$

$$sP_2^*(s)=2\lambda P_1^*(s)-\lambda P_2^*(s) \tag{10.37}$$

$$sP_3^*(s)=\lambda P_2^*(s). \tag{10.38}$$

Solving for $P_1^*(s)$, $P_2^*(s)$, and $P_3^*(s)$:

$$P_1^*(s)=\frac{1}{s+2\lambda} \tag{10.39}$$

$$P_2^*(s)=\frac{2\lambda}{(s+2\lambda)(s+\lambda)} \tag{10.40}$$

$$P_3^*(s)=\frac{2\lambda^2}{s(s+2\lambda)(s+\lambda)}. \tag{10.41}$$

Simplifying the transforms using the method of partial fraction expansion illustrated in Section 10.2.1:

$$P_1^*(s)=\frac{1}{s+2\lambda} \tag{10.42}$$

$$P_2^*(s)=\frac{-2}{s+2\lambda}+\frac{2}{s+\lambda} \tag{10.43}$$

$$P_3^*(s) = \frac{1}{s} + \frac{1}{s+2\lambda} - \frac{2}{s+\lambda}. \tag{10.44}$$

Taking the inverse transform, the time domain probability solutions are:

$$P_1(t) = e^{-2\lambda t} \tag{10.45}$$

$$P_2(t) = -2e^{-2\lambda t} + 2e^{-\lambda t} \tag{10.46}$$

$$P_3(t) = 1 + e^{-2\lambda t} - 2e^{-\lambda t}. \tag{10.47}$$

10.3.2 Reliability

The system is functioning if occupying either State 1 or State 2. The reliability of the system is thus:

$$R(t) = P_1(t) + P_2(t) = 2e^{-\lambda t} - e^{-2\lambda t}. \tag{10.48}$$

We can also calculate reliability directly from the RBD in Figure 10.5 to get the same answer.

$$R(t) = 1 - (1 - R_c(t))(1 - R_c(t))$$

$$R(t) = 1 - (1 - e^{-\lambda t})(1 - e^{-\lambda t}) = 2e^{-\lambda t} - e^{-2\lambda t}.$$

Observe that the reliability of the system is not in the form of e^{-kt}. Although each component has a constant failure rate, the system failure rate is not constant.

EXAMPLE 10.3

Consider a one-component and a two-component parallel system with identical per-component failure rates of $\lambda = 0.0005$ and plot the reliability of each system over a mission time of 10,000 hours.

Substituting the values for failure rates into the reliability Equations (10.12) and (10.48), we obtain the two reliability curves shown in Figure 10.7.

Note the reliability for a one-component is 36.78% at the MTTF time of 2000 hours. In other words, the system has a 63.22% chance of failing prior to the MTTF time. Another way to look at it is the probability of the one-component surviving at t = MTTF is 0.3678.

10.3.3 Availability

The system is functioning if it is either in State 1 or State 2, or equivalently if the system is not in the failed state (State 3). The availability of the system is the probability that the system is functioning at time t:

$$A(t) = P_1(t) + P_2(t) = 2e^{-\lambda t} - e^{-2\lambda t}. \tag{10.49}$$

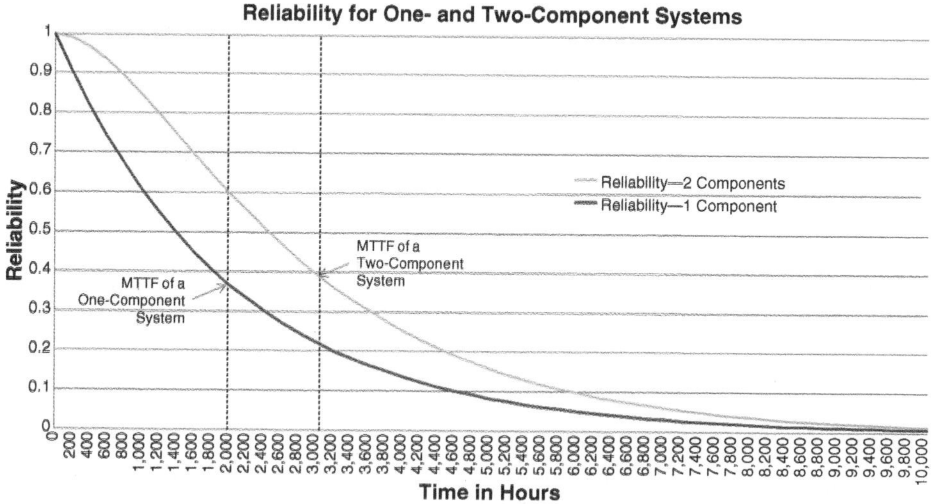

Figure 10.7 One-Component versus Two-Component Parallel System Comparison

As previously discussed, for nonrepairable systems, the system availability and reliability are the same.

Using the MRM technique, in this example, the system is operational (available) when in States 1 and 2, and is failed (unavailable) when in State 3. From Equation (9.50), the availability is:

$$A_s(t) = E[X(t)] = \sum_i r_i P_i(t) = 1 \cdot P_1 + 1 \cdot P_2 + 0 \cdot P_3 = 2e^{-\lambda t} - e^{-2\lambda t}. \tag{10.50}$$

This result is the same as Equation (10.49) as we would expect.

The long-term availability is:

$$A_{ss} = \lim_{t \to \infty} A(t)$$

$$A_{ss} = 0.$$

As in the first example, the availability eventually goes to zero and the availability and reliability are the same.

10.3.4 Mean Time to Failure (MTTF)

Since this example is an absorbing matrix, we can see that the system will eventually transition into the failed state and remain trapped in this state, that is, the system approaches a 100% certainty of becoming failed as $t \to \infty$.

We can obtain the MTTF of the system using the technique described in Section 10.2.5.

The reliability of this system is equal to the probability of the system being in either State 1 or State 2. Using Equation (10.42) and Equation (10.43), we obtain:

$$R^*(s) = P_1^*(s) + P_2^*(s) = \frac{1}{s+2\lambda} - \frac{2}{s+2\lambda} + \frac{2}{s+\lambda}$$
$$R^*(0) = P_1^*(0) + P_2^*(0) = \frac{-1}{2\lambda} + \frac{2}{\lambda} \tag{10.51}$$
$$R^*(0) = \frac{3}{2\lambda}.$$

The MTTF for a one-component system is simply the reciprocal of the component failure rate.

The MTTF for a two-component parallel system is 50% greater than the one-component system:

$$\text{MTTF}_s = 1.5 \times \text{MTTF}_c.$$

10.3.5 Asymptotic Behavior

If we have the complete time dependent state probability equations, Equations (10.45)–(10.47), we can obtain the steady-state probability distributions by allowing t to become very large. The initial transient behavior will be removed and what remains is the constant probability distribution for each state. For Equations (10.45)–(10.47), when t is very large, the exponential factor becomes very small and the equations reduce to:

$$P_1 = 0 \tag{10.52}$$

$$P_2 = 0 \tag{10.53}$$

$$P_3 = 1. \tag{10.54}$$

Alternatively, if we start with the state transition matrix equations we previously created in Equation (10.34), we can obtain the steady-state behavior directly.

Since the steady-state probabilities are constants, their time derivatives are zeros. Thus, the steady-state transition matrix becomes:

$$\begin{bmatrix} 0 \\ 0 \\ 0 \end{bmatrix} = \begin{bmatrix} -2\lambda & 0 & 0 \\ 2\lambda & -\lambda & 0 \\ 0 & \lambda & 0 \end{bmatrix} \cdot \begin{bmatrix} P_1 \\ P_2 \\ P_3 \end{bmatrix}. \tag{10.55}$$

We choose two of the above equations plus the conservation of probability equation $P_1 + P_2 + P_3 = 1$ to obtain the solutions for P_1, P_2, and P_3. Recall that for steady-state behavior, we must always replace one of the equations, since the equations are not linearly independent.

So let us replace the last equation with the conservation equation:

$$\begin{bmatrix} 0 \\ 0 \\ 1 \end{bmatrix} = \begin{bmatrix} -2\lambda & 0 & 0 \\ 2\lambda & -\lambda & 0 \\ 1 & 1 & 1 \end{bmatrix} \cdot \begin{bmatrix} P_1 \\ P_2 \\ P_3 \end{bmatrix}. \tag{10.56}$$

Solving for P_1, P_2, and P_3, we get:

$$P_1 = 0$$

$$P_2 = 0$$

$$P_3 = 1.$$

We can also solve the state equations directly using matrix algebra, noting:

$$\mathbf{U} = \begin{bmatrix} 0 \\ 0 \\ 1 \end{bmatrix} \tag{10.57}$$

$$\mathbf{A} = \begin{bmatrix} -2\lambda & 0 & 0 \\ 2\lambda & -\lambda & 0 \\ 1 & 1 & 1 \end{bmatrix} \tag{10.58}$$

$$\mathbf{P} = \begin{bmatrix} P_1 \\ P_2 \\ P_3 \end{bmatrix} \tag{10.59}$$

$$\mathbf{U} = \mathbf{AP}. \tag{10.60}$$

We can solve for **P** by taking the inverse of the transition matrix **A** and multiplying it by **U**:

$$\mathbf{P} = \mathbf{A}^{-1}\mathbf{U}. \tag{10.61}$$

EXAMPLE 10.4

We have a 2N redundant nonreparable system with perfect switching, and the MTBF of each of the components is 2000 hours.

What are the steady-state probabilities of the system being in each of the three states?

The failure rate of each component is the reciprocal of the MTBF:

$$\lambda = 0.0005.$$

The steady-state system probabilities are calculated using the procedure previously described. First, we calculate the state transition matrix:

$$\mathbf{A} = \begin{bmatrix} -0.001 & 0 & 0 \\ 0.001 & -0.0005 & 0 \\ 1 & 1 & 1 \end{bmatrix}. \tag{10.62}$$

The inverse of **A** is:

$$\mathbf{A}^{-1} = \begin{bmatrix} -1000 & 0 & 0 \\ -2000 & -2000 & 0 \\ 3000 & 2000 & 1 \end{bmatrix}. \tag{10.63}$$

Solving for the steady probability vector:

$$\mathbf{P} = \mathbf{A}^{-1}\mathbf{U} = \begin{bmatrix} -1000 & 0 & 0 \\ -2000 & -2000 & 0 \\ 3000 & 2000 & 1 \end{bmatrix} \cdot \begin{bmatrix} 0 \\ 0 \\ 1 \end{bmatrix} \tag{10.64}$$

$$\mathbf{P} = \begin{bmatrix} 0 \\ 0 \\ 1 \end{bmatrix}. \tag{10.65}$$

Actually, we could have chosen any value for λ. The end result will be the same. Although a system with a lower failure rate will take a longer time to fail, eventually all systems will fail, and the probability of being in State 3, the failed state, becomes 1.

10.4 SERIES SYSTEM WITH NO REPAIR: TWO IDENTICAL COMPONENTS

In the previous example, we analyzed two components in parallel with no repair. In this example, we will analyze a two-component system with components in series with no repair. Thus, the system will fail if either component fails.

The reliability block diagram is shown in Figure 10.8.

As in the previous example, let us assume both components have a constant failure rate λ.

The three states are:

- *State 1: System functional and both components operating.*
- *State 2: System failure, one of the two components has failed.*
- *State 3: System failure, both components have failed.*

From the definitions of these three states, the RBD, and assuming no repair on the failed components, the state transition diagram for the system can be created (Fig. 10.8). Note that the state transition diagram is *exactly* the same as in the previous example. Although the diagrams are the same, the definition of State 2 is different. In particular, State 2 is a failed state for this example, but is a functional state for the previous model.

10.4.1 Dynamic Behavior

From Figure 10.9, we write the state transition matrix directly:

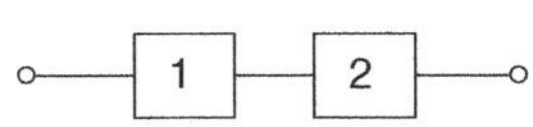

Figure 10.8 Reliability Block Diagram for a Two-Component Series System: No Repair

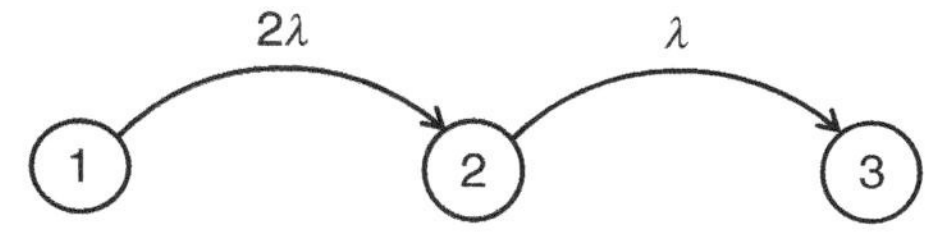

Figure 10.9 State Transition Diagram for Two-Component Series System: No Repair

$$\begin{bmatrix} \dot{P}_1 \\ \dot{P}_2 \\ \dot{P}_3 \end{bmatrix} = \begin{bmatrix} -2\lambda & 0 & 0 \\ 2\lambda & -\lambda & 0 \\ 0 & \lambda & 0 \end{bmatrix} \cdot \begin{bmatrix} P_1 \\ P_2 \\ P_3 \end{bmatrix}. \tag{10.66}$$

Taking the Laplace transform of the above matrix, we obtain:

$$\begin{bmatrix} sP_1^*(s)-1 \\ sP_2^*(s) \\ sP_3^*(s) \end{bmatrix} = \begin{bmatrix} -2\lambda & 0 & 0 \\ 2\lambda & -\lambda & 0 \\ 0 & \lambda & 0 \end{bmatrix} \cdot \begin{bmatrix} P_1^*(s) \\ P_2^*(s) \\ P_3^*(s) \end{bmatrix}. \tag{10.67}$$

The state equations are:

$$sP_1^*(s)-1=-2\lambda P_1^*(s) \tag{10.68}$$

$$sP_2^*(s)=2\lambda P_1^*(s)-\lambda P_2^*(s) \tag{10.69}$$

$$P_3^*(s)=\frac{\lambda P_2^*(s)}{s}. \tag{10.70}$$

Solving for $P_1^*(s)$, $P_2^*(s)$, and $P_3^*(s)$, we get:

$$P_1^*(s)=\frac{1}{s+2\lambda} \tag{10.71}$$

$$P_2^*(s)=\frac{2\lambda}{(s+2\lambda)(s+\lambda)} \tag{10.72}$$

$$P_3^*(s)=\frac{2\lambda^2}{s(s+2\lambda)(s+\lambda)}. \tag{10.73}$$

Note that these equations are exactly the same as Equations (10.39)–(10.41) from Section 10.3 for the two-component parallel system. How we interpret these equations reveals the differences between a two-component parallel system and a two-component series system.

Taking the inverse Laplace transform, the time domain probability solutions are:

$$P_1(t)=e^{-2\lambda t} \tag{10.74}$$

$$P_2(t) = -2e^{-2\lambda t} + 2e^{-\lambda t} \tag{10.75}$$

$$P_3(t) = 1 + e^{-2\lambda t} - 2e^{-\lambda t}. \tag{10.76}$$

10.4.2 Reliability

Our analysis for the series system diverges from the parallel system when we interpret the meaning of the three different states. For a series system, the system is only functioning in State 1. The reliability of the system is thus:

$$R(t) = P_1(t) = e^{-2\lambda t}. \tag{10.77}$$

We can also calculate reliability directly from the RBD in Figure 10.8 to get the same answer.

$$R(t) = R_c(t)R_c(t)$$

$$R(t) = e^{-\lambda t}e^{-\lambda t} = e^{-2\lambda t}.$$

Observe that the system failure rate remains constant for a series system:

$$\lambda_s = 2\lambda.$$

The failure rate doubles for a series system, that is, twice as many chances exist for the system to fail. Also recall the system failure rate is not constant for the parallel system we previously discussed.

10.4.3 Availability

Since the system is functioning only if it is in State 1, the availability of the system is the probability that the system is in State 1:

$$A(t) = P_1(t) = e^{-2\lambda t}. \tag{10.78}$$

Comparing this with the single-component system (Section 10.2.3), we see that the availability for the series case decreases at twice the rate of the single component system.

Using the MRM technique, in this example, the system is operational (available) when in State 1 and is failed (unavailable) when in States 2 and 3. From Equation (9.50), the availability is:

$$A_s(t) = E[X(t)] = \sum_i r_i P_i(t) = 1 \cdot P_1 + 0 \cdot P_2 + 0 \cdot P_3 = e^{-2\lambda t}. \tag{10.79}$$

This result is the same as Equation (10.78) as we would expect.

The long-term availability is:

$$A_{ss} = \lim_{t \to \infty} A(t)$$

$$A_{ss} = 0.$$

10.4.4 Mean Time to Failure (MTTF)

As discussed in the previous examples, the system approaches a 100% certainty of failure as $t \to \infty$.

We can obtain the MTTF of the system using the technique described in Section 10.2.5.

The reliability of this system is equal to the probability of the system being in State 1. Using Equation (10.71), we obtain:

$$R^*(s) = P_1^*(s) = \frac{1}{s+2\lambda} \tag{10.80}$$

$$\text{MTTF}_S = R^*(0) = P_1^*(0) = \frac{1}{2\lambda}. \tag{10.81}$$

We can see for two components in series, the MTTF is half the time for a single component Equation (10.18).

The equivalent system failure rate is twice the component failure rate:

$$\lambda_s = 2\lambda_c.$$

10.4.5 Asymptotic Behavior

If we have the complete time dependent state probability equations, as we obtained in Equations (10.74)–(10.76), we can obtain the steady-state probability distributions by allowing t to be very large. The initial transient behavior will be removed and what remains is the constant probability distribution for each state. For the Equations (10.74)–(10.76), when t is very large, the exponential factor becomes very small and the equations reduce to:

$$P_1 = 0 \tag{10.82}$$

$$P_2 = 0 \tag{10.83}$$

$$P_3 = 1 \tag{10.84}$$

Start with the state transition matrix equations (Eq. (10.66)). Replace one of the equations with the conservation of probability equation and substitute for the vector **P** above. We then obtain the steady-state behavior directly:

$$\begin{bmatrix} 0 \\ 0 \\ 1 \end{bmatrix} = \begin{bmatrix} -2\lambda & 0 & 0 \\ 2\lambda & -\lambda & 0 \\ 1 & 1 & 1 \end{bmatrix} \cdot \begin{bmatrix} P_1 \\ P_2 \\ P_3 \end{bmatrix}. \tag{10.85}$$

Solving for P_1, P_2, and P_3 we get:

$$P_1 = 0$$

$$P_2 = 0$$

$$P_3 = 1.$$

10.5 PARALLEL SYSTEM WITH PARTIAL REPAIR: IDENTICAL COMPONENTS

Earlier in this chapter, we analyzed two components in parallel with no repair. In this example, we will analyze the same two-component redundant system with the exception that if one of the components fails, the component can be repaired and restored to service. However, if the first failed component is not repaired, and subsequently the second component fails, the system is considered to be nonrepairable.

Again, we assume both components have a constant failure rate λ.

The three states we had in the previous model are the same for this model:

- *State 1: System functional and both components operating.*
- *State 2: System functional and one of the two components has failed.*
- *State 3: System failure; both components have failed.*

From the definitions of these three states, the RBD from Figure 10.5, and the repair assumptions just stated, the state transition diagram for the system is as shown in Figure 10.10.

Note that the transition rate from State 2 to State 1 is μ, which represents the repair rate of a single failed component, that is, restores the system to State 1.

10.5.1 Dynamic Behavior

From the state transition diagram, the state transition matrix is:

$$\begin{bmatrix} \dot{P}_1 \\ \dot{P}_2 \\ \dot{P}_3 \end{bmatrix} = \begin{bmatrix} -2\lambda & \mu & 0 \\ 2\lambda & -(\lambda+\mu) & 0 \\ 0 & \lambda & 0 \end{bmatrix} \cdot \begin{bmatrix} P_1 \\ P_2 \\ P_3 \end{bmatrix}. \tag{10.86}$$

Assuming the system is fully operational at $t = 0$, that is, the system is in State 1, the Laplace transform of the above matrix is:

$$\begin{bmatrix} sP_1^*(s)-1 \\ sP_2^*(s) \\ sP_3^*(s) \end{bmatrix} = \begin{bmatrix} -2\lambda & \mu & 0 \\ 2\lambda & -(\lambda+\mu) & 0 \\ 0 & \lambda & 0 \end{bmatrix} \cdot \begin{bmatrix} P_1^*(s) \\ P_2^*(s) \\ P_3^*(s) \end{bmatrix}. \tag{10.87}$$

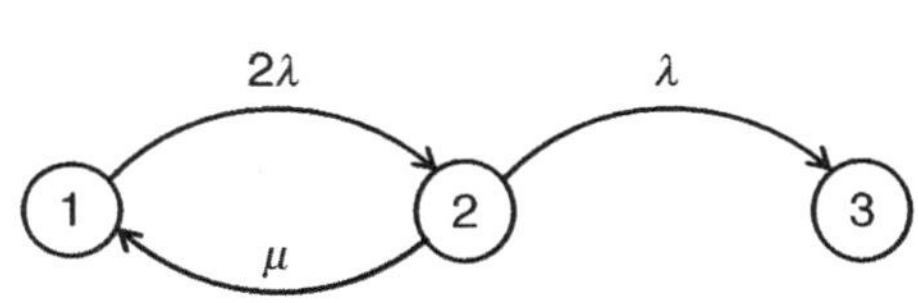

Figure 10.10 State Transition Diagram for Two-Component Parallel System: Partial Repair

The state equations are:

$$sP_1^*(s) - 1 = -2\lambda P_1^*(s) + \mu P_2^*(s) \tag{10.88}$$

$$sP_2^*(s) = 2\lambda P_1^*(s) - (\lambda + \mu) P_2^*(s) \tag{10.89}$$

$$P_3^*(s) = \frac{\lambda}{s} P_2^*(s). \tag{10.90}$$

Solving for $P_1^*(s)$, $P_2^*(s)$, and $P_3^*(s)$, we get:

$$P_1^*(s) = \frac{s + \lambda + \mu}{s^2 + s(3\lambda + \mu) + 2\lambda^2} \tag{10.91}$$

$$P_2^*(s) = \frac{2\lambda}{s^2 + s(3\lambda + \mu) + 2\lambda^2} \tag{10.92}$$

$$P_3^*(s) = \frac{2\lambda^2}{s(s^2 + s(3\lambda + \mu) + 2\lambda^2)}. \tag{10.93}$$

Next, we need to find the roots of $s^2 + s(3\lambda + \mu) + 2\lambda^2$:

$$r_1, r_2 = \frac{-3\lambda - \mu \pm \sqrt{\lambda^2 + 6\mu\lambda + \mu^2}}{2}.$$

This expression cannot be simplified further, but we can obtain the partial fraction expansion in terms of r_1, r_2:

$$P_1^*(s) = \frac{s + \lambda + \mu}{(s - r_1)(s - r_2)} \tag{10.94}$$

$$P_2^*(s) = \frac{2\lambda}{(s - r_1)(s - r_2)} \tag{10.95}$$

$$P_3^*(s) = \frac{2\lambda^2}{s(s - r_1)(s - r_2)}. \tag{10.96}$$

Splitting Equation (10.94) into simpler transforms, we obtain:

$$\frac{s + \lambda + \mu}{(s - r_1)(s - r_2)} = \frac{A}{s - r_1} + \frac{B}{s - r_2}.$$

The constants A and B are obtained by using the method of partial fraction expansion:

$$A = \lim_{s \to r_1} \frac{s + \lambda + \mu}{s - r_2} = \frac{\lambda + \mu + r_1}{r_1 - r_2}$$

$$B = \lim_{s \to r_2} \frac{s + \lambda + \mu}{s - r_1} = \frac{\lambda + \mu + r_2}{r_2 - r_1}.$$

Substituting the values for A and B, we get

$$P_1^*(s) = \frac{\lambda+\mu+r_1}{(r_1-r_2)(s-r_1)} + \frac{\lambda+\mu+r_2}{(r_2-r_1)(s-r_2)}. \tag{10.97}$$

Next, we split Equation (10.95) into simpler transforms:

$$\frac{2\lambda}{(s-r_1)(s-r_2)} = \frac{A}{s-r_1} + \frac{B}{s-r_2}$$

$$A = \lim_{s\to r_1} \frac{2\lambda}{s-r_2} = \frac{2\lambda}{r_1-r_2}$$

$$B = \lim_{s\to r_2} \frac{2\lambda}{s-r_1} = \frac{2\lambda}{r_2-r_1}.$$

Substituting the values for A and B, we get:

$$P_2^*(s) = \frac{2\lambda}{(r_1-r_2)(s-r_1)} + \frac{2\lambda}{(r_2-r_1)(s-r_2)}. \tag{10.98}$$

Finally, we split Equation (10.96) into simpler transforms:

$$\frac{2\lambda^2}{s(s-r_1)(s-r_2)} = \frac{A}{s} + \frac{B}{(s-r_1)} + \frac{C}{(s-r_2)}$$

$$A = \lim_{s\to 0} \frac{2\lambda^2}{(s-r_1)(s-r_2)} = \frac{2\lambda^2}{r_1 r_2}$$

$$B = \lim_{s\to r_1} \frac{2\lambda^2}{s(s-r_2)} = \frac{2\lambda^2}{r_1(r_1-r_2)}$$

$$C = \lim_{s\to r_2} \frac{2\lambda^2}{s(s-r_1)} = \frac{2\lambda^2}{r_2(r_2-r_1)}.$$

Substituting the values for A, B, and C, we get:

$$P_3^*(s) = \frac{2\lambda^2}{r_1 r_2 s} + \frac{2\lambda^2}{r_1(r_1-r_2)(s-r_1)} + \frac{2\lambda^2}{r_2(r_2-r_1)(s-r_2)}. \tag{10.99}$$

Taking the inverse transform, the time domain probability solutions are:

$$P_1(t) = \frac{\lambda+\mu+r_1}{(r_1-r_2)} e^{r_1 t} + \frac{\lambda+\mu+r_2}{(r_2-r_1)} e^{r_2 t} \tag{10.100}$$

$$P_2(t) = \frac{2\lambda}{(r_1-r_2)} e^{r_1 t} + \frac{2\lambda}{(r_2-r_1)} e^{r_2 t} \tag{10.101}$$

$$P_3(t) = \frac{2\lambda^2}{r_1 r_2} + \frac{2\lambda^2}{r_1(r_1-r_2)} e^{r_1 t} + \frac{2\lambda^2}{r_2(r_2-r_1)} e^{r_2 t}. \tag{10.102}$$

10.5.2 Reliability

The system is functioning if it is in State 1 or State 2. The reliability of the system is thus:

$$R(t) = P_1(t) + P_2(t) = \frac{3\lambda + \mu + r_1}{r_1 - r_2} e^{r_1 t} + \frac{3\lambda + \mu + r_2}{r_2 - r_1} e^{r_2 t}. \quad (10.103)$$

10.5.3 Availability

Since the system is functioning if it is in State 1 or State 2, the availability of the system is the probability that the system is in these states:

$$A(t) = P_1(t) + P_2(t) = \frac{3\lambda + \mu + r_1}{r_1 - r_2} e^{r_1 t} + \frac{3\lambda + \mu + r_2}{r_2 - r_1} e^{r_2 t}. \quad (10.104)$$

Using the MRM technique, in this example, the system is operational (available) when in States 1 and 2 and is failed (unavailable) when in State 3. From Equation (9.50), the availability is:

$$A_s(t) = E[X(t)] = \sum_i r_i P_i(t) = 1 \cdot P_1 + 1 \cdot P_2 + 0 \cdot P_3$$
$$A_s(t) = \frac{3\lambda + \mu + r_1}{r_1 - r_2} e^{r_1 t} + \frac{3\lambda + \mu + r_2}{r_2 - r_1} e^{r_2 t}. \quad (10.105)$$

This result is the same as Equation (10.103) as we would expect.

The long-term availability is:

$$A_{ss} = \lim_{t \to \infty} A(t)$$
$$A_{ss} = 0.$$

EXAMPLE 10.5

Let us now consider a two-component parallel system with the following values for the MTBF and MTTR of the components:

$$\text{MTBF}_1 = \text{MTBF}_2 = 100{,}000 \text{ hours}$$
$$\text{MTTR}_1 = \text{MTTR}_2 = 8 \text{ hours}.$$

(a) What is the dynamic availability of the system?
The failure rates and repair rates are the reciprocals of the MTBFs and MTTRs, respectively:

$$\lambda_1 = \lambda_1 = 1/100{,}000 \text{ hours} = 0.00001 \text{ failures/hour}$$
$$\mu_1 = \mu_2 = 1/8 \text{ hours} = 0.125 \text{ repairs/hour}.$$

(b) What is the availability of the system after 2,000,000 hours of operation? Substituting these values into Equation (10.104) and simplifying, we obtain:

$$A(t) = P_1(t) + P_2(t) = 1.000000013e^{-(1.59962E-9)t} - (1.27939E-8)e^{-0.1250299998t}. \quad (10.106)$$

Figure 10.11 shows dynamic availability initially at 1 and eventually approaching 0.

Since the component is highly reliable, the availability of the system remains high for a long period of time. The availability of this system after 2,000,000 hours is obtained using Equation (10.106): A = 0.9968. How can we test our assumption of the failure rate given the decades it would take to observe a failure? By itself, the MTBF value applied to a single component has little value. To make sense of these numbers, we need to consider a large number of these components. One common method hardware component manufacturers use to test this is to aggregate a large number of systems and run longevity tests for all systems in parallel. The time to first failure of one component is equivalent to an MTBF of the system multiplied by the number of components under test:

$$\text{MTBF}_{\text{est}} = (\text{Time to First Component Failure}) \times (\text{Number of Components}).$$

Thus if we have 1000 components under test, and the time to first component failure is 6 weeks, we will have an approximate MTBF point estimate of ~1.5 million hours.

We will explore life testing and reliability parameter estimation in greater detail in subsequent chapters.

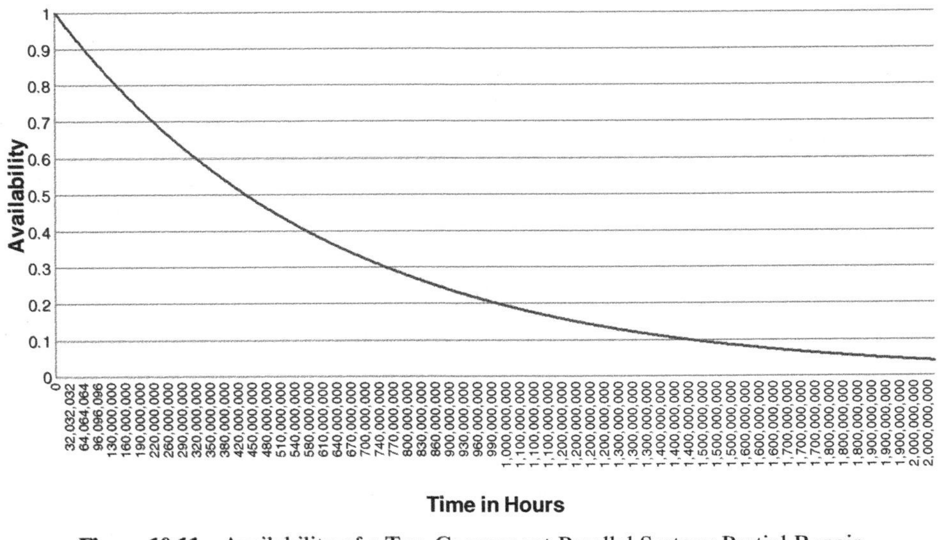

Figure 10.11 Availability of a Two-Component Parallel System: Partial Repair

10.5.4 Mean Time to Failure (MTTF)

Although for this system model, a repair transition exists, eventually the system approaches a 100% certainty of becoming failed as $t \to \infty$.

We are interested in the mean time it takes for the system to enter the failed state.

The reliability of the system is equal to the probability of the system being in either State 1 or State 2. Using Equations (10.91) and (10.92), we obtain:

$$R^*(s) = P_1^*(s) + P_2^*(s) = \frac{s+\lambda+\mu}{s^2+s(3\lambda+\mu)+2\lambda^2} + \frac{2\lambda}{s^2+s(3\lambda+\mu)+2\lambda^2}$$

$$R^*(0) = P_1^*(0) + P_2^*(0) = \frac{3\lambda+\mu}{2\lambda^2} \tag{10.107}$$

$$\text{MTTF}_S = \frac{3}{2\lambda} + \frac{\mu}{2\lambda^2}.$$

From Section 10.3.4, the MTTF_S of a two-component parallel system, without any repair (i.e., $\mu = 0$) is equal to $3/2\lambda$ (Eq. (10.51)). The addition of the component repair action increases the MTTF_S by $\mu/(2\lambda^2)$. If our repair time decreases, our MTTF increases. If our repair time is very small, the MTTF becomes very large.

10.5.5 Asymptotic Behavior

From the complete time-dependent state probability, Equations (10.100)–(10.102), the steady-state probability distributions are determined by making t very large. The exponential factor becomes very small and the equations reduce to:

$$P_1 = 0 \tag{10.108}$$

$$P_2 = 0 \tag{10.109}$$

$$P_3 = 1. \tag{10.110}$$

As can be observed from Figure 10.11, the availability approaches 0 as $t \to \infty$. This gives us the same solutions as Example 10.2, Equations (10.82)–(10.84).

Alternatively, starting with the state transition matrix, Equation (10.86), replacing the third equation by the conservation of probability equation, and setting the time derivative probabilities to zero, the matrix becomes:

$$\begin{bmatrix} 0 \\ 0 \\ 1 \end{bmatrix} = \begin{bmatrix} -2\lambda & \mu & 0 \\ 2\lambda & -(\lambda+\mu) & 0 \\ 1 & 1 & 1 \end{bmatrix} \cdot \begin{bmatrix} P_1 \\ P_2 \\ P_3 \end{bmatrix}. \tag{10.111}$$

The state equations are:

$$0 = -2\lambda P_1 + \mu P_2$$

$$0 = 2\lambda P_1 - (\lambda+\mu) P_2$$

$$1 = P_1 + P_2 + P_2.$$

Solving for P_1, P_2, and P_3 we get:

$$P_1 = 0$$

$$P_2 = 0$$

$$P_3 = 1.$$

Another method is to solve Equation (10.111) using matrix algebra. Note:

$$\mathbf{U} = \begin{bmatrix} 0 \\ 0 \\ 1 \end{bmatrix} \quad \textbf{(10.112)}$$

$$\mathbf{A} = \begin{bmatrix} -2\lambda & \mu & 0 \\ 2\lambda & -(\lambda+\mu) & 0 \\ 1 & 1 & 1 \end{bmatrix} \quad (10.113)$$

$$\mathbf{P} = \begin{bmatrix} P_1 \\ P_2 \\ P_3 \end{bmatrix} \quad (10.114)$$

$$\mathbf{U} = \mathbf{AP}. \quad (10.115)$$

We can solve for **P** by taking the inverse of the transition matrix **A** and multiplying by **U**:

$$\mathbf{P} = \mathbf{A}^{-1}\mathbf{U}. \quad (10.116)$$

EXAMPLE 10.6

For a two-component parallel system with $\lambda = 0.0005$ and $\mu = 0.04$, what are the steady-state probabilities?

Calculating the state transition matrix with these values:

$$\mathbf{A} = \begin{bmatrix} -0.001 & 0.04 & 0 \\ 0.001 & -0.0405 & 0 \\ 1 & 1 & 1 \end{bmatrix}. \quad (10.117)$$

The inverse of **A** is:

$$\mathbf{A}^{-1} = \begin{bmatrix} -81{,}000 & -80{,}000 & 0 \\ -2000 & -2000 & 0 \\ 83{,}000 & 82{,}000 & 1 \end{bmatrix} \quad (10.118)$$

$$\mathbf{P} = \mathbf{A}^{-1}\mathbf{U} = \begin{bmatrix} -81{,}000 & -80{,}000 & 0 \\ -2000 & -2000 & 0 \\ 83{,}000 & 82{,}000 & 1 \end{bmatrix} \cdot \begin{bmatrix} 0 \\ 0 \\ 1 \end{bmatrix} \tag{10.119}$$

$$\mathbf{P} = \begin{bmatrix} 0 \\ 0 \\ 1 \end{bmatrix}. \tag{10.120}$$

Regardless of the failure rate values chosen, in the end, the system will be in State 3, that is, the system will fail.

10.6 PARALLEL SYSTEM WITH NO REPAIR: NONIDENTICAL COMPONENTS

In Section 10.3, we analyzed two identical components in parallel with no repair. Now we will analyze a two-component redundant system with different failure rates and repair rates.

Let us assume component 1 has a constant failure rate λ_1 and component 2 has a constant failure rate λ_2.

Although we could have used the same model for the identical failure rates from the identical component example, we choose to assign unique states for the failure of each component since the failure rates are different.

- *State 1: System functional and both components operating.*
- *State 2: System functional and component 2 has failed.*
- *State 3: System functional and component 1 has failed.*
- *State 4: System failure; both components have failed.*

From the definitions of these states, the RBD from Figure 10.5, and assuming no repair on the failed component, we can create the state transition diagram for the system as shown in Figure 10.12.

10.6.1 Dynamic Behavior

From this diagram, we can write the state transition matrix:

$$\begin{bmatrix} \dot{P}_1 \\ \dot{P}_2 \\ \dot{P}_3 \\ \dot{P}_4 \end{bmatrix} = \begin{bmatrix} -(\lambda_1+\lambda_2) & 0 & 0 & 0 \\ \lambda_2 & -\lambda_1 & 0 & 0 \\ \lambda_1 & 0 & -\lambda_2 & 0 \\ 0 & \lambda_1 & \lambda_2 & 0 \end{bmatrix} \cdot \begin{bmatrix} P_1 \\ P_2 \\ P_3 \\ P_4 \end{bmatrix}. \tag{10.121}$$

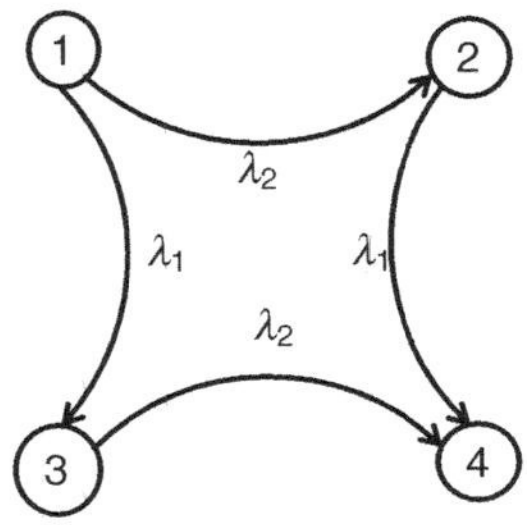

Figure 10.12 Transition Diagram for Two-Component Parallel System: No Repair

The system is initially in State 1 (probability of being in State 1 at $t = 0$ is 1). The Laplace transform of $\dot{P}_1(t)$ is therefore $P_1^*(s)-1$. The Laplace transforms of $P_2(t), P_3(t)$, and $P_4(t)$ are $P_2^*(s)$, $P_3^*(s)$, and $P_4^*(s)$, respectively (probability of being in States 2, 3, or 4 are all 0 at $t = 0$). The transition matrix becomes:

$$\begin{bmatrix} sP_1^*(s)-1 \\ sP_2^*(s) \\ sP_3^*(s) \\ sP_4^*(s) \end{bmatrix} = \begin{bmatrix} -(\lambda_1+\lambda_2) & 0 & 0 & 0 \\ \lambda_2 & -\lambda_1 & 0 & 0 \\ \lambda_1 & 0 & -\lambda_2 & 0 \\ 0 & \lambda_1 & \lambda_2 & 0 \end{bmatrix} \cdot \begin{bmatrix} P_1^*(s) \\ P_2^*(s) \\ P_3^*(s) \\ P_4^*(s) \end{bmatrix}. \tag{10.122}$$

The state equations are:

$$(1)\ sP_1^*(s)-1=-(\lambda_1+\lambda_2)P_1^*(s)$$
$$(2)\ sP_2^*(s)=\lambda_2P_1^*(s)-\lambda_1P_2^*(s)$$
$$(3)\ sP_3^*(s)=\lambda_1P_1^*(s)-\lambda_2P_3^*(s)$$
$$(4)\ sP_4^*(s)=\lambda_1P_2^*(s)+\lambda_2P_3^*(s).$$

Solving for $P_1^*(s)$, $P_2^*(s)$, $P_3^*(s)$, and $P_4^*(s)$:

$$\begin{aligned} P_1^*(s) &= \frac{1}{s+\lambda_1+\lambda_2} \\ P_2^*(s) &= \frac{\lambda_2}{(s+\lambda_1)(s+\lambda_1+\lambda_2)} \\ P_3^*(s) &= \frac{\lambda_1}{(s+\lambda_2)(s+\lambda_1+\lambda_2)} \\ P_4^*(s) &= \frac{\lambda_1\lambda_2}{s(s+\lambda_1+\lambda_2)(s+\lambda_1)}+\frac{\lambda_1\lambda_2}{s(s+\lambda_1+\lambda_2)(s+\lambda_2)}. \end{aligned} \tag{10.123}$$

Simplifying the transforms using partial fraction expansion we get:

$$\begin{aligned}
P_1^*(s) &= \frac{1}{s+\lambda_1+\lambda_2} \\
P_2^*(s) &= \frac{1}{s+\lambda_1} - \frac{1}{s+\lambda_1+\lambda_2} \\
P_3^*(s) &= \frac{1}{s+\lambda_2} - \frac{1}{s+\lambda_1+\lambda_2} \\
P_4^*(s) &= \frac{1}{s} - \frac{1}{s+\lambda_2} - \frac{1}{s+\lambda_1} + \frac{1}{s+\lambda_1+\lambda_2}.
\end{aligned} \tag{10.124}$$

Compare these equations with the identical component state Equations (10.42)–(10.44). If we combine States 2 and 3 into a single State 2 and make State 4 equal to State 3, the equations above are identical (assuming we also make $\lambda_1 = \lambda_2 = \lambda$).

Taking the inverse transform of the transform equations above, we obtain the time domain solutions:

$$P_1(t) = e^{-(\lambda_1+\lambda_2)t} \tag{10.125}$$

$$P_2(t) = e^{-\lambda_1 t} - e^{-(\lambda_1+\lambda_2)t} \tag{10.126}$$

$$P_3(t) = e^{-\lambda_2 t} - e^{-(\lambda_1+\lambda_2)t} \tag{10.127}$$

$$P_4(t) = 1 - e^{-\lambda_1 t} - e^{-\lambda_2 t} + e^{-(\lambda_1+\lambda_2)t}. \tag{10.128}$$

Obtaining these solutions by hand is quite tedious. We can solve these equations directly using the symbolic differential equation solver in Matlab. Select three of the four differential equations from the matrix (Eq. (10.121)), and then substitute the sum of probabilities for one of the probability equations. In this case, there is no need to take the Laplace transform of the equations. The set of differential equations we obtain is:

$$\dot{P}_4 = \lambda_1 P_2 + \lambda_2 P_3 \tag{10.129}$$

$$\dot{P}_3 = \lambda_1(1 - P_2 - P_3 - P_4) - \lambda_2 P_3 \tag{10.130}$$

$$\dot{P}_2 = \lambda_2(1 - P_2 - P_3 - P_4) - \lambda_1 P_2. \tag{10.131}$$

These three equations and the initial conditions are entered into Matlab's *dsolve* function:

```
>> [p1 p2 p3] = dsolve('Dp2=L2*(1-p2-p3-p4)-L1*p2', 'Dp3=L1*(1-
p2-p3-p4)-L2*p3', 'Dp4=L1*p2+L2*p3', 'p2(0)=0',  'p3(0)=0',
'p4(0)=0')

P2 = 1/exp(L1*t) - 1/exp(L1*t + L2*t)
P3 = 1/exp(L2*t) - 1/exp(L1*t + L2*t)
P4 = 1/exp(L1*t + L2*t) - 1/exp(L1*t) - 1/exp(L2*t) + 1
```

These equations match Equations (10.126)–(10.128). $P_1(t)$ is obtained in a straightforward manner by subtracting the sum of Equations (10.126)–(10.128) from 1, which gives us Equation (10.125).

If we make the failure rates for each component the same and are no longer interested in observing a unique state for each component, States 2 and 3 collapse into one state and the system Equation (10.124) become the same as Equations (10.39)–(10.41).

Since the failure rates are the same, we can combine States 2 and 3 from this example into a single state (State 2, Section 10.3). And State 4 then becomes State 3. By comparing Equations (10.126) and (10.127) to Equation (10.46), by observation, the summation of Equations (10.126) and (10.127) equals Equation (10.46), which is what we would expect for the combined States 2 and 3 from Section 10.3. Finally, Equation (10.128) is the same as Equation (10.47) when State 4 of Section 10.3 becomes State 3 in this example.

10.6.2 Availability and Reliability

The system is functioning if it is in State 1, State 2, or State 3. The probability that the system is functioning at time t, which corresponds to the availability of the system is thus:

$$\begin{aligned} A(t) &= R(t) = P_1(t) + P_2(t) + P_3(t) \\ A(t) &= e^{-(\lambda_1+\lambda_2)t} + e^{-\lambda_1 t} - e^{-(\lambda_1+\lambda_2)t} + e^{-\lambda_2 t} - e^{-(\lambda_1+\lambda_2)t} \\ A(t) &= e^{-\lambda_1 t} + e^{-\lambda_2 t} - e^{-(\lambda_1+\lambda_2)t}. \end{aligned} \tag{10.132}$$

EXAMPLE 10.7

For a two-component parallel system with $\lambda_1 = 0.005$ and $\lambda_2 = 0.007$, what is the availability over a mission time of 1500 hours? What is the availability of the system at 500 hours?

Plotting Equation (10.132) for these values from 1 to 1500 hours, we obtain the graph in Figure 10.13.

Using Equation (10.132), we can calculate the availability at $t = 500$ hours, giving a value of 0.1098.

We can also calculate reliability directly from the RBD in Figure 10.5 to get the same answer:

$$\begin{aligned} R(t) &= 1-(1-R_1(t))(1-R_2(t)) \\ R(t) &= 1-(1-e^{-\lambda_1 t})(1-e^{-\lambda_2 t}) \\ R(t) &= e^{-\lambda_1 t} + e^{-\lambda_2 t} - e^{-(\lambda_1+\lambda_2)t}. \end{aligned} \tag{10.133}$$

The long-term availability is:

$$A_{lt} = \lim_{t\to\infty} A(t)$$

$$A_{lt} = 0.$$

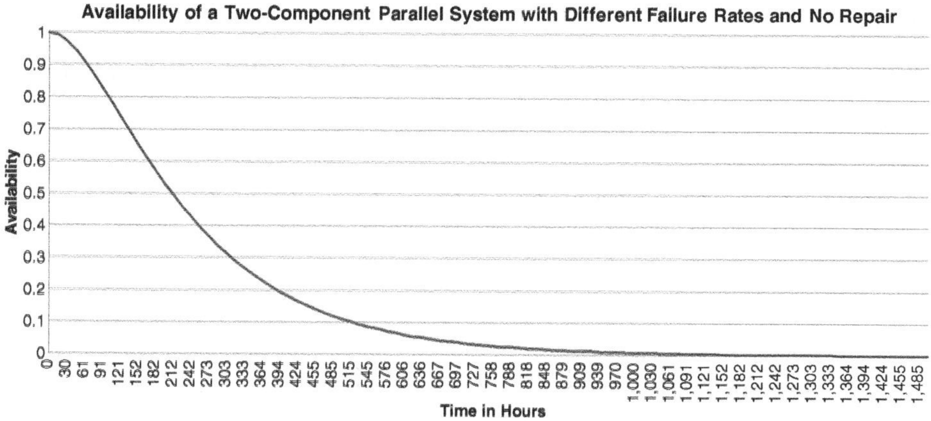

Figure 10.13 Availability of a Two-Component Parallel System with Different Failure Rates and No Repair

10.6.3 Mean Time to Failure (MTTF)

We can obtain the MTTF of the system using the technique described in the first example.

The reliability of this system is equal to the probability of the system being in State 1, State 2, or State 3. Using Equation (10.124), we obtain:

$$
\begin{aligned}
R^*(s) &= P_1^*(s) + P_2^*(s) + P_3^*(s) \\
R^*(s) &= \frac{1}{s+\lambda_1+\lambda_2} + \frac{1}{s+\lambda_1} - \frac{1}{s+\lambda_1+\lambda_2} + \frac{1}{s+\lambda_2} - \frac{1}{s+\lambda_1+\lambda_2} \\
R^*(s) &= \frac{-1}{s+\lambda_1+\lambda_2} + \frac{1}{s+\lambda_1} + \frac{1}{s+\lambda_2} \\
R^*(0) &= P_1^*(0) + P_2^*(0) + P_3^*(0) \\
R^*(0) &= \frac{-1}{\lambda_1+\lambda_2} + \frac{1}{\lambda_1} + \frac{1}{\lambda_2} \\
\text{MTTF} &= \frac{-1}{\lambda_1+\lambda_2} + \frac{1}{\lambda_1} + \frac{1}{\lambda_2}.
\end{aligned}
\qquad (10.134)
$$

Note that Equation (10.134) reduces to Equation (10.51) in Section 10.3 when both components are identical.

EXAMPLE 10.8

What is the MTTF for a two-component nonrepairable parallel system with component failure rates of $\lambda_1 = 0.005$ and $\lambda_2 = 0.007$?

From the equation above:

$$\text{MTTF} = \frac{-1}{0.005+0.007} + \frac{1}{0.005} + \frac{1}{0.007}$$

$$\text{MTTF} = 260 \text{ hours.}$$

10.6.4 System Failure Rate

We know the failure rate for each individual component is the constant λ_1 and λ_2. We can also calculate the equivalent system failure rate using the definition of hazard rate, Equation (7.38) repeated here for convenience:

$$\lambda(t) = \frac{f(t)}{R(t)} = \frac{-\frac{d}{dt}R(t)}{R(t)}.$$

Using the reliability equation for a two-component parallel system, Equation (10.133):

$$\lambda(t) = \frac{\lambda_1 e^{-\lambda_1 t} + \lambda_2 e^{-\lambda_2 t} - (\lambda_1 + \lambda_2) e^{-(\lambda_1+\lambda_2)t}}{e^{-\lambda_1 t} + e^{-\lambda_2 t} - e^{-(\lambda_1+\lambda_2)t}}. \tag{10.135}$$

Observe that although the failure rates of the individual components are constant, the failure rate of the two-component system depends on time and is thus not constant.

EXAMPLE 10.9

What is the conditional failure rate for a two-component nonrepairable parallel system over a period of 2000 hours with component failure rates $\lambda_1 = 0.005$ and $\lambda_2 = 0.007$? Plot the conditional failure rate for a mission time of 20,000 hours.

Using the same values for λ_1 and λ_2 in the previous example for calculation MTTF, we can plot the system failure rate over time (Fig. 10.14).

The failure rate starts at 0, ramps up to a peak value between 0.005 and 0.007, then asymptotically approaches a failure rate of 0.005 at around 2000 hours. Observe that the system asymptotically approaches the lowest failure rate of the two components. The speed at which this convergence happens is dependent on the failure rate of the lower failure rate component. Given sufficient time, the conditional failure rate approaches a constant value equivalent to the single component model with a failure rate equal to the lower failure rate of the two components.

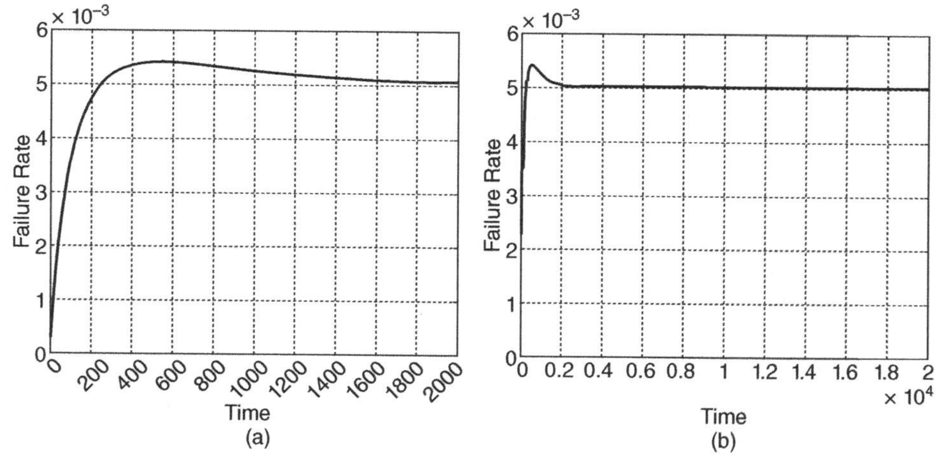

Figure 10.14 Two-Component Parallel System Failure Rate

Let us explore this some more by varying the failure rate of one of the components and plotting the resultant dynamic conditional failure rate over time. We keep one of the component failure rates, λ_1, at a fixed value of 0.005, and plot the values of the failure rate for λ_2 values of 0.007, 0.0005, and 0.005 (Fig. 10.15).

When λ_2 has a value of 0.0005 (one-tenth the rate of λ_1) the system quickly reaches the steady state failure rate of λ_1. Conversely, the slowest convergence is when the two failure rates are equal.

Another way to consider this behavior is to refer back to the original definition of reliability provided in Chapter 7, Equation (7.39):

$$R(t) = e^{-\lambda(t)t}. \tag{10.136}$$

If we substitute the failure rate equation we derived from Equation (10.135) into the equation above and plot it as a function of time, the curve will be the same as Figure 10.13. From these relationships, we can conclude the reliability of any system is an exponential function of the general failure rate $\lambda(t)$.

10.6.5 Asymptotic Behavior

For Equations (10.125)–(10.128), when t is very large, the exponential factor becomes very small and the equations reduce to:

$$P_1 = 0 \tag{10.137}$$

$$P_2 = 0 \tag{10.138}$$

$$P_3 = 0 \tag{10.139}$$

$$P_4 = 1. \tag{10.140}$$

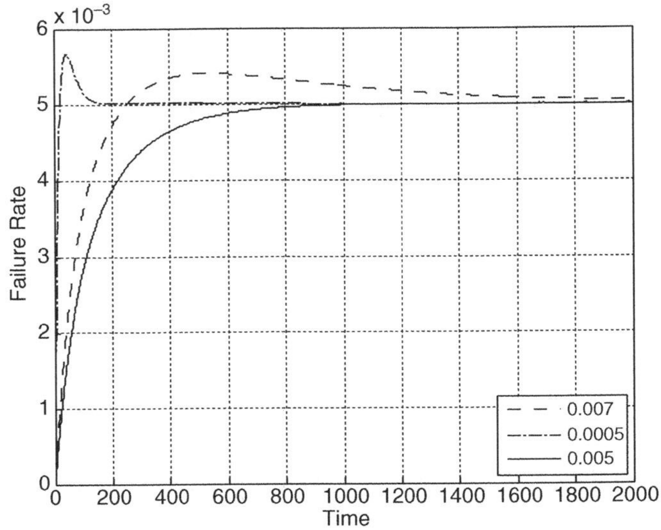

Figure 10.15 Two-Component Parallel System with Different Failure Rates

Replacing the first equation in the state transition matrix (Eq. (10.121)) with the conservation of probability equation, the steady-state transition matrix becomes:

$$\begin{bmatrix} 1 \\ 0 \\ 0 \\ 0 \end{bmatrix} = \begin{bmatrix} 1 & 1 & 1 & 1 \\ \lambda_2 & -\lambda_1 & 0 & 0 \\ \lambda_1 & 0 & -\lambda_2 & 0 \\ 0 & \lambda_1 & \lambda_2 & 0 \end{bmatrix} \cdot \begin{bmatrix} P_1 \\ P_2 \\ P_3 \\ P_4 \end{bmatrix}. \tag{10.141}$$

The state equations are:

$$1 = P_1 + P_2 + P_3 + P_4$$
$$0 = \lambda_2 P_1 - \lambda_1 P_2$$
$$0 = \lambda_1 P_1 - \lambda_2 P_3$$
$$0 = \lambda_1 P_2 + \lambda_2 P_3.$$

Solving for P_1, P_2, P_3, and P_4, we obtain:

$$P_1 = 0$$
$$P_2 = 0$$
$$P_3 = 0$$
$$P_4 = 1.$$

TABLE 10.1 Model Comparison for Nonrepairable Systems

	Dynamic Behavior		Asymptotic Behavior		
Availability Model	Reliability	Availability	MTTF	Reliability	Availability
One component with no repair	$e^{-\lambda t}$	$e^{-\lambda t}$	$\frac{1}{\lambda}$	0	0
Parallel system with no repair—identical components	$2e^{-\lambda t} - e^{-2\lambda t}$	$2e^{-\lambda t} - e^{-2\lambda t}$	$\frac{3}{2\lambda}$	0	0
Series system with no repair—identical components	$e^{-2\lambda t}$	$e^{-2\lambda t}$	$\frac{1}{2\lambda}$	0	0
Parallel system with partial repair—identical components	$\frac{3\lambda+\mu-r_1}{r_2-r_1}e^{r1t} + \frac{3\lambda+\mu-r_2}{r_1-r_2}e^{r2t}$	$\frac{3\lambda+\mu-r_1}{r_2-r_1}e^{r1t} + \frac{3\lambda+\mu-r_2}{r_1-r_2}e^{r2t}$	$\frac{3}{2\lambda}+\frac{\mu}{2\lambda^2}$	0	0
Parallel system with no repair—nonidentical components	$e^{-\lambda_1 t} + e^{-\lambda_1 t} - e^{-(\lambda_1+\lambda_1)t}$	$e^{-\lambda_1 t} + e^{-\lambda_2 t} - e^{-(\lambda_1+\lambda_2)t}$	$\frac{-1}{\lambda_1+\lambda_2}+\frac{1}{\lambda_1}+\frac{1}{\lambda_2}$	0	0

10.7 SUMMARY

This chapter introduced applications of Markov analysis to nonrepairable systems and illustrated both dynamic and steady state responses for nonrepairable and partially repairable systems with different configurations. Table 10.1 summarizes the availability models across different configurations.

CHAPTER 11

Markov Analysis: Repairable Systems

11.1 REPAIRABLE SYSTEMS

In the last chapter, we analyzed Markov models for nonrepairable systems, that is, systems in which the failed state once entered cannot be left (absorbing). In this chapter, we will examine repairable systems. For a repairable system, once the system fails, it can be brought back to full operational condition. During the system's expected lifetime or mission time, multiple repair actions can occur at different times based on the particular problem. The repair action can be any action that restores the system. Examples: manually replacing a failed processor card in the system, manually or automatically rebooting a hard drive, system task restart automatically upon failure detection, and replacing the entire system.

A note on Markov modeling and repair times: As a practical matter, repair times may not be exponentially distributed. However, the inaccuracies associated with this assumption are usually acceptable. The use of constant repair rates greatly simplifies the calculations and allows Markov modeling to be used. If this simplification or assumption is not valid, we can approximate the repair times with a piecewise model and employing additional states. If the alternative methods employed are unacceptable, the model fails.

The set of examples in this chapter illustrates the methodology for analyzing repairable (nonabsorbing) Markov processes with constant failure and repair rates.

We will cover the following examples:

1. One component with repair.
2. Two components in parallel with repair—identical failure rates and repair rates for the two components.
3. Two parallel components with repair—different component failure rates and repair rates.

Designing High Availability Systems: Design for Six Sigma and Classical Reliability Techniques with Practical Real-Life Examples, First Edition. Zachary Taylor and Subramanyam Ranganathan.

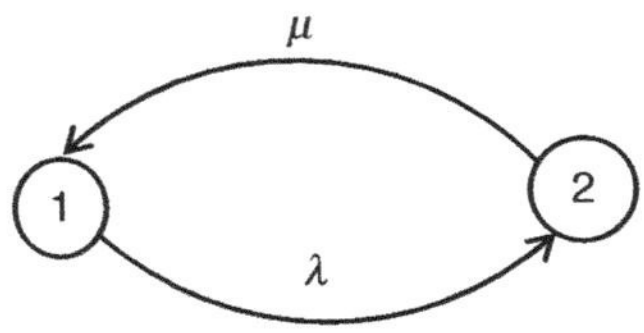

Figure 11.1 One Component with Repair State Transition Diagram

11.2 ONE COMPONENT WITH REPAIR

The reliability block diagram and states for this system looks the same as the system in Chapter 10, Section 10.2.

- *State 1: System is fully functional.*
- *State 2: System is failed.*

However, since we are now considering a repairable system, we will have a new transition from the failed state to the operating state at a defined constant repair rate μ.

The state transition diagram for this example is shown in Figure 11.1.

11.2.1 Dynamic Behavior

The system is initially in State 1 and remains in that state until a system failure occurs at which time the system transitions to State 2. The failure rate of the system is λ.

When the system fails, it remains in State 2 until repaired. The system is repaired with a repair rate μ. After the system is repaired, it becomes fully functional and thus transitions back to State 1. The system is then considered "good as new" and the same failure rate λ will apply to the system.

The component is assumed to be functioning at time $t = 0$, thus:

$$P_1(0) = 1, P_2(0) = 0.$$

The time spent in a state is called the *sojourn time*. The sojourn time in State 1 is therefore T_{12}, where T_{12} represents the time to the first transition from State 1 to State 2. The mean sojourn time in State 1 is $1/\lambda$.

When the system is in State 2, the next transition must be to State 1 (with rate μ). The memoryless property of the exponential distribution ensures that the system is as good as new when the system enters State 1.

The sojourn time in State 2, T_{21}, is exponentially distributed with rate μ, and the *mean downtime* (MDT) of the system is therefore $1/\mu$.

The state equations in matrix form are:

$$\dot{P_2}(t) = \mathbf{AP}(t) \tag{11.1}$$

$$\begin{bmatrix} \dot{P}_1(t) \\ \dot{P}_2(t) \end{bmatrix} = \begin{bmatrix} -\lambda & \mu \\ \lambda & -\mu \end{bmatrix} \cdot \begin{bmatrix} P_1(t) \\ P_2(t) \end{bmatrix}. \tag{11.2}$$

Let us replace the first equation from Equation (11.2) with the conservation of probability equation to obtain:

$$\begin{bmatrix} 1 \\ \dot{P}_2(t) \end{bmatrix} = \begin{bmatrix} 1 & 1 \\ \lambda & -\mu \end{bmatrix} \cdot \begin{bmatrix} P_1(t) \\ P_2(t) \end{bmatrix}. \tag{11.3}$$

From Equation (11.3), we obtain the following two linearly independent equations:

$$1 = P_1(t) + P_2(t) \tag{11.4}$$

$$\dot{P}_2(t) = \lambda P_1(t) - \mu P_2(t). \tag{11.5}$$

Now we have the information we need to solve for the probabilities $P_1(t)$ and $P_2(t)$. Solving for $P_2(t)$:

$$\dot{P}_2(t) = \lambda(1 - P_2(t)) - \mu P_2(t). \tag{11.6}$$

The solution of any general differential equation $\dot{y} = ky + a$ will be of the form:

$$C_2 e^{kt} + C_1.$$

From Equation (11.6), by inspection, we see that $k = -(\mu + \lambda)$. Thus:

$$P_2(t) = C_2 e^{-(\mu+\lambda)t} + C_1 \tag{11.7}$$

$$P_1(t) = 1 - C_2 e^{-(\mu+\lambda)t} - C_1. \tag{11.8}$$

Now, substitute Equation (11.7) into Equation (11.6) and solve for C_1:

$$-(\mu + \lambda) C_2 e^{-(\mu+\lambda)t} = \lambda(1 - C_2 e^{-(\mu+\lambda)t} - C_1) - \mu(C_2 e^{-(\mu+\lambda)t} + C_1)$$

$$0 = \lambda(1 - C_1) - \mu C_1$$

$$C_1 = \lambda/(\lambda + \mu).$$

And for $P_1(t)$: $1 - C_1 = 1 - \lambda/(\lambda + \mu) = \mu/(\lambda + \mu)$:

$$P_2(t) = C_2 e^{-(\mu+\lambda)t} + \frac{\lambda}{\lambda + u} \tag{11.9}$$

$$P_1(t) = 1 - C_2 e^{-(\mu+\lambda)t} - \frac{\lambda}{\lambda + u}. \tag{11.10}$$

A unique solution for C_2 cannot be found from the differential equations but can be found using the initial condition that at $t = 0$, the system is assumed to be fully functioning, thus $P_1(0) = 1$ and $P_2(0) = 0$.

Substituting the initial conditions at $t = 0$ into Equation (11.10):

$$1 = 1 - C_2 e^{-(\mu+\lambda)0} - \lambda/(\lambda+\mu)$$
$$1 = 1 - C_2 - \lambda/(\lambda+\mu)$$
$$\Rightarrow C_2 = -\lambda/(\lambda+\mu).$$

The complete solution is thus:

$$P_1(t) = \frac{\mu}{\lambda+\mu} + \frac{\lambda}{\lambda+\mu} e^{-(\mu+\lambda)t} \tag{11.11}$$

$$P_2(t) = \frac{\lambda}{\lambda+\mu} - \frac{\lambda}{\lambda+\mu} e^{-(\mu+\lambda)t}. \tag{11.12}$$

11.2.1.1 Laplace Transform Approach Let us solve for $P_1(t)$ and $P_2(t)$ using the Laplace transform technique. Since the system is fully operational at $t = 0$, and the probability is 0 that the system is in State 2 at $t = 0$, the Laplace transform of $\dot{P}_1(t)$ is $P_1^*(s) - 1$ and the Laplace transform of $\dot{P}_2(t)$ is $P_2^*(s)$.

The transition matrix (Eq. (11.3)) becomes:

$$\begin{bmatrix} sP_1^*(s) - 1 \\ sP_2^*(s) \end{bmatrix} = \begin{bmatrix} -\lambda & \mu \\ \lambda & -\mu \end{bmatrix} \cdot \begin{bmatrix} P_1^*(s) \\ P_2^*(s) \end{bmatrix}. \tag{11.13}$$

The state equations are:

$$sP_1^*(s) - 1 = -\lambda P_1^*(s) + \mu P_2^*(s) \tag{11.14}$$

$$sP_2^*(s) = \lambda P_1^*(s) - \mu P_2^*(s). \tag{11.15}$$

Solve for $P_2^*(s)$ in Equation (11.14) and substitute into Equation (11.15):

$$sP_1^*(s) - 1 = -\lambda P_1^*(s) + \mu\left(\frac{\lambda}{s+\mu} P_1^*(s)\right)$$

$$P_1^*(s) = \frac{s+\mu}{s(s+\lambda+\mu)}. \tag{11.16}$$

Splitting this into simpler transforms, we obtain:

$$P_1^*(s) = \frac{s+\mu}{s(s+\lambda+\mu)} = \frac{A}{s} + \frac{B}{s+\lambda+\mu}.$$

The constants A and B are obtained by using the method of partial fraction expansion:

$$A = \lim_{s \to 0} \frac{s+\mu}{s+\lambda+\mu} - \frac{sB}{s+\lambda+\mu}$$

$$A = \frac{\mu}{\lambda+\mu}$$

$$B = \lim_{s \to -\lambda-\mu} \frac{s+\mu}{s} - \frac{(s+\lambda+\mu)A}{s}$$

$$B = \frac{\lambda}{\lambda+\mu}.$$

Substituting the values for A and B:

$$P_1^*(s) = \frac{\lambda}{(\lambda+\mu)(s+\lambda+\mu)} + \frac{\mu}{s(\lambda+\mu)}. \tag{11.17}$$

We find $P_2^*(s)$ by substituting Equation (11.16) into Equation (11.15):

$$P_2^*(s) = \frac{(s+\mu)\lambda}{s(s+\lambda+\mu)(s+\mu)}$$

$$P_2^*(s) = \frac{\lambda}{s(s+\lambda+\mu)}.$$

Using partial fraction expansion to simplify:

$$P_2^*(s) = \frac{\lambda}{s(\lambda+\mu)} - \frac{\lambda}{(\lambda+\mu)(s+\lambda+\mu)}. \tag{11.18}$$

Now we can take the inverse Laplace transform to get the steady-state solution:

$$P_1(t) = \frac{\mu}{\lambda+\mu} + \frac{\lambda}{\lambda+\mu} e^{-(\lambda+\mu)t} \tag{11.19}$$

$$P_2(t) = \frac{\lambda}{\lambda+\mu} - \frac{\lambda}{\lambda+\mu} e^{-(\lambda+\mu)t}. \tag{11.20}$$

Compare with Equations (11.11) and (11.12) and note that the solutions are the same.

11.2.2 Availability

$P_1(t)$ denotes the probability that the system is functioning at time t, which corresponds to the availability of the system:

$$A(t) = \frac{\mu}{\lambda+\mu} + \frac{\lambda}{\lambda+\mu} e^{-(\lambda+\mu)t}. \tag{11.21}$$

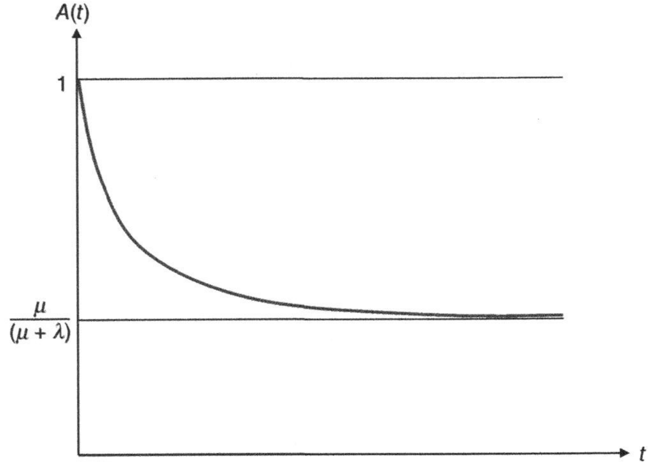

Figure 11.2 Dynamic and Steady-State Availability

The long-term availability is:

$$A_{ss} = \lim_{t \to \infty} A(t)$$
$$A_{ss} = \frac{\mu}{\lambda + \mu}. \tag{11.22}$$

The relationship between dynamic and steady-state availability is shown in Figure 11.2. Initially, the system is operating in State 1 (100% probability). The availability of the system then exponentially decreases and asymptotically approaches the steady-state nonzero value shown. In contrast with the systems we explored in Chapter 10, for a repairable system, the availability does not approach zero. The availability approaches some constant probability based on the model and the rate parameters.

From Figure 11.2, we observe that if the failure rate in comparison to the repair rate becomes smaller, the steady-state availability becomes larger, and if the repair rate becomes larger, that is the repair time is smaller, the availability approaches 1. For this repairable system, we have a couple of strategies when considering how to improve availability. We can either improve the reliability of the component or we can decrease the repair time, or some combination of both.

EXAMPLE 11.1

A one-component system has an MTBF of 100,000 hours and an MTTR of 1 hour. Calculate the dynamic availability at 1 hour and the steady-state availability.

Using Equations (11.21) and (11.22):

$$A(1)_{\text{dynamic}} = \frac{1}{0.00001 + 1} + \frac{0.00001}{0.00001 + 1} e^{-(0.00001+1)(1)}$$
$$A(t)_{\text{dynamic}} = 0.999994$$

$$A_{ss} = \frac{1}{0.00001 + 1}$$
$$A_{ss} = 0.99999.$$

The dynamic component of availability approaches zero very quickly and becomes insignificant in less than an hour. For that reason, we are normally justified in using the steady-state value of availability for practical purposes when the differences between the MTTR and MTBF values span several orders of magnitude.

11.2.3 Reliability

Recall that reliability is the probability of the system remaining in the operational state as a function of time. For a repairable system, it is the probability of remaining in an operational state, given that it is already in an operational state. Once the system transitions to a failed state, we consider this state to be an absorbing state, that is, once failed, the transitions from the failed state to an operational state do not contribute to system reliability. Equation (11.2) becomes:

$$\begin{bmatrix} P_1^{\bullet}(t) \\ P_2^{\bullet}(t) \end{bmatrix} = \begin{bmatrix} -\lambda & 0 \\ \lambda & 0 \end{bmatrix} \cdot \begin{bmatrix} P_1(t) \\ P_2(t) \end{bmatrix}. \tag{11.23}$$

Since State 2 is now the absorbing (failed) state, we only need to consider the operational state (State 1).

$$[P_1^{\bullet}(t)] = [-\lambda] \cdot [P_1(t)]. \tag{11.24}$$

The Laplace transform of Equation (11.24) is:

$$sP_1^*(s) - 1 = -\lambda P_1^*(s) \tag{11.25}$$
$$P_1^*(s) = \frac{1}{s + \lambda}.$$

Taking the inverse Laplace transform, we get:

$$R(t) = e^{-\lambda t}. \tag{11.26}$$

This is the same reliability we obtained for a nonrepairable single component in Chapter 10. The reliability of a component with repair is the same as the reliability of a component without repair. Remember, reliability provides the probability of a component working from the beginning of the mission time until t. Whether the system is repairable or not does not impact the time to first failure.

11.2.4 Mean Time to First Failure

We can obtain the *Mean Time to First Failure* (MTTFF) of the system using the technique described in Section 10.2.5.

The reliability of this system is equal to the probability of the system being in State 1 when State 2 is the failure (absorbing state). Using Equation (11.25), we obtain:

$$\begin{aligned} R^*(s) &= P_1^*(s) = \frac{1}{s+\lambda} \\ \text{MTTFF} &= R^*(0) = P_1^*(0) \\ \text{MTTFF} &= \frac{1}{\lambda}. \end{aligned} \tag{11.27}$$

Note that the reliability and MTTFF are the same for both the repairable and non-repairable single component systems. This follows directly from the definitions of reliability and MTTFF.

11.2.5 Asymptotic Solution

From the time-dependent state probability equations, Equations (11.19) and (11.20), we obtain the steady-state probability distributions by allowing t to be very large:

$$P_1 = \frac{\mu}{\lambda+\mu} \tag{11.28}$$

$$P_2 = \frac{\lambda}{\lambda+\mu}. \tag{11.29}$$

If we start with the state transition matrix equations we previously created (Eq. (11.13)), and replace the first equation with the conservation of probability equation, we can obtain the steady-state behavior directly:

$$\begin{bmatrix} 1 \\ 0 \end{bmatrix} = \begin{bmatrix} 1 & 1 \\ \lambda & -\mu \end{bmatrix} \cdot \begin{bmatrix} P_1 \\ P_2 \end{bmatrix}. \tag{11.30}$$

The state equations are:

$$1 = P_1 + P_2 \tag{11.31}$$

$$0 = \lambda P_1 - \mu P_2. \tag{11.32}$$

We will walk through the solution. Let's solve for P_1 in terms of P_2 in the first equation and substitute into the second equation to obtain:

$$0 = \lambda(1-P_2) - \mu P_2. \tag{11.33}$$

Solving for P_2, we get:

$$P_2 = \frac{\lambda}{\mu+\lambda}. \tag{11.34}$$

Substitute P_2 into Equation (11.31), we get P_1:

$$P_1 = \frac{\mu}{\mu+\lambda}. \tag{11.35}$$

Alternatively, we could start with Equation (11.30) above and solve by noting:

$$\mathbf{U} = \begin{bmatrix} 1 \\ 0 \end{bmatrix} \tag{11.36}$$

$$\mathbf{A} = \begin{bmatrix} 1 & 1 \\ \lambda & -\mu \end{bmatrix} \tag{11.37}$$

$$\mathbf{P} = \begin{bmatrix} P_1 \\ P_2 \end{bmatrix} \tag{11.38}$$

$$\mathbf{U} = \mathbf{AP}. \tag{11.39}$$

We can solve for **P** by taking the inverse of the transition matrix **A** and multiply by *U*:

$$\mathbf{P} = \mathbf{A}^{-1}\mathbf{U} \tag{11.40}$$

$$\mathbf{P} = \frac{-1}{\lambda+\mu}\begin{bmatrix} -\mu & -1 \\ -\lambda & 1 \end{bmatrix}\begin{bmatrix} 1 \\ 0 \end{bmatrix} \tag{11.41}$$

$$\begin{bmatrix} P_1 \\ P_2 \end{bmatrix} = \begin{bmatrix} \dfrac{\mu}{\lambda+\mu} \\ \dfrac{\lambda}{\lambda+\mu} \end{bmatrix}. \tag{11.42}$$

This solution matches Equations (11.34) and (11.35).

11.2.6 Mean Time to Failure

Assuming we are in an operational state, what is the mean time to failure?

From Equations (9.43) and (9.46), and noting we have one functional state and one failed state, the MTTF is:

$$\mathrm{MTTF}_s = 1/(P_1\beta_{12}/P_1)$$
$$\mathrm{MTTF}_s = \frac{1}{\lambda}. \tag{11.43}$$

Compare this with the MTTFF calculation and note they are the same and thus either the method described for MTTFF or the method described above can be used to get both values.

11.2.7 Mean Time to Repair

Assuming we are in a failed state, what is the Mean Time to Repair (MTTR) the system and restore it to an operational state?

From Equations (9.38) and (9.42), and noting we have just one functional state and one failed state, the $MTTR_s$ is:

$$MTTR_s = 1/\beta_{21}$$
$$MTTR_s = \frac{1}{\mu}. \tag{11.44}$$

If the system is nonrepairable, then $\mu = 0$ and $MTTR_s = \infty$ (i.e., never repaired) as we would expect.

11.2.8 Mean Time between Failures

The mean time between system failures, $MTBF_s$, is the mean time between consecutive transitions from a functioning state (G) to a failed state (F). The $MTBF_s$ can be calculated starting with Equation (9.34) to determine the failure frequency (note we have one functional state and one failed state):

$$f_F = P_1\beta_{12} = \left(\frac{\mu}{\mu+\lambda}\right)(\lambda) = \frac{\mu\lambda}{\mu+\lambda}. \tag{11.45}$$

The failure frequency is equal to the probability of being in State 1 multiplied by the transition rate out of State 1.

From Equation (9.37), the MTBF is the reciprocal of the failure frequency:

$$MTBF_S = \frac{1}{f_F}$$
$$MTBF_S = \frac{\mu+\lambda}{\mu\lambda} \tag{11.46}$$
$$MTBF_S = \frac{1}{\lambda} + \frac{1}{\mu}.$$

We can check the MTBF by using the relationship MTBF = MTTR + MTTF:

$$MTBF_S = MTTF_S + MTTR_S = \frac{1}{\lambda} + \frac{1}{\mu}. \tag{11.47}$$

If $\lambda \ll \mu$, then Equation (11.47) reduces to:

$$MTBF_S = \frac{1}{\lambda}. \tag{11.48}$$

11.2.9 System Availability

From Equation (9.47), system availability is the sum of the probabilities of all the good states. We have 1 good state, P_1:

$$A_S = P_1$$
$$A_S = \frac{\mu}{\lambda + \mu}. \tag{11.49}$$

More systematically, we can calculate the availability of the system using the Markov Reward Model (MRM) described in Section 9.7. We assign a reward of 1 when the system is in the operational states and a reward of 0 when the system is in the failure states. For this example, the system is operational (available) when in State 1 and is failed (unavailable) when in State 2. From Equation (9.50), the availability is the expected value of the reward rate:

$$A_s = E[X(t)] = \sum_i r_i P_i(t) = 1 \cdot P_1 + 0 \cdot P_2 = \frac{\mu}{\lambda + \mu}. \tag{11.50}$$

This result is of course the same as Equation (11.49) as we would expect.

11.2.10 System Unavailability

From Equation (9.48), system unavailability is the sum of the probabilities of all the bad states. We have one bad state, P_2:

$$Q_s = P_2$$
$$Q_S = \frac{\lambda}{\lambda + \mu}. \tag{11.51}$$

As a check, $Q_s = 1 - A_s$, which gives the same result.

In many cases, only the long-run (steady-state) probabilities are of interest, that is, the values of $P_j(t)$ when $t \to \infty$. In the previous example, the state probabilities $P_j(t)$ $(j = 1, 2)$ approached a steady-state P_j when $t \to \infty$. The same steady-state value would have been found irrespective of whether the system started in the operating state or in the failed state.

The mean sojourn time in State 1 is the mean time to failure, MTTF = $1/\lambda$, and the mean sojourn time in State 2 is the mean downtime, $MDT = 1/\mu$, also called the MTTR.

The limiting availability $P_1 = \lim_{t \to \infty} P_1(t)$ is from Equation (11.12):

$$P_1 = \frac{\mu}{\lambda + \mu}. \tag{11.52}$$

Since in the limiting case, μ = 1/MTTR and λ = 1/MTTF, Equation (11.52) can be rewritten in terms of MTTF and MTTR:

$$P_1 = \frac{\text{MTTF}}{\text{MTTF} + \text{MTTR}}. \tag{11.53}$$

EXAMPLE 11.2

Consider a one-component system with a failure rate $\lambda = 0.1$ and a repair rate $\mu = 0.01$. What are the dynamic system reliability, availability, and unavailability over a mission time of 50 hours? What is the MTTF of the system? Plot the results.

We solve by using Equation (11.21) for availability, Equation (11.26) for reliability, and Equation (11.27) for the MTTF value. Figure 11.3 illustrates the relationships between availability, reliability, MTTF, and asymptotic behavior.

11.3 PARALLEL SYSTEM WITH REPAIR: IDENTICAL COMPONENT FAILURE AND REPAIR RATES

Now we consider a system in which we have two components in parallel such that the system will remain fully operational if at least one of the two components is functioning, that is the system will fail only if both components have failed. This is similar to the nonrepairable system in Chapter 10, Section 10.3. However, in this case, we will assume that if either component fails or both components fail, the failed component(s) will be repaired with a given constant repair rate. We assume the system is a 2N redundant system with perfect switching. A discussion of common redundancy models and techniques is covered in Chapter 19. Our intent here is to introduce the basic model and explore these concepts in depth. The strategies and techniques learned here can readily be applied to much larger models.

The reliability block diagram is exactly the same as the RBD for the nonrepairable example in Section 10.3 (Fig. 10.5).

Let us assume both components have a constant failure rate λ and a constant repair rate μ.

The three possible states of the system are:

- *State 1: System functional and both components operating.*
- *State 2: System functional and one of the two components has failed.*
- *State 3: System failure; both components have failed.*

From the definitions of these three states and the RBD, we can create the state transition diagram for the system as shown in Figure 11.4.

Note that the transition rate from State 1 to State 2 is 2λ, since the probability of either one of the components failing is double the failure rate of a single component, that is, we have twice as many chances to fail with two components. Also, the repair rate from State 3 to State 2 is 2μ. We assume we have two independent repair crews each working on one of the failed components. We could also assume we only have one repair crew and only one component can be repaired at a time. If that is the case, then the transition rate from State 3 to State 2 becomes μ instead of 2μ.

11.3.1 Dynamic Behavior

From Figure 11.4, we can write the state transition matrix:

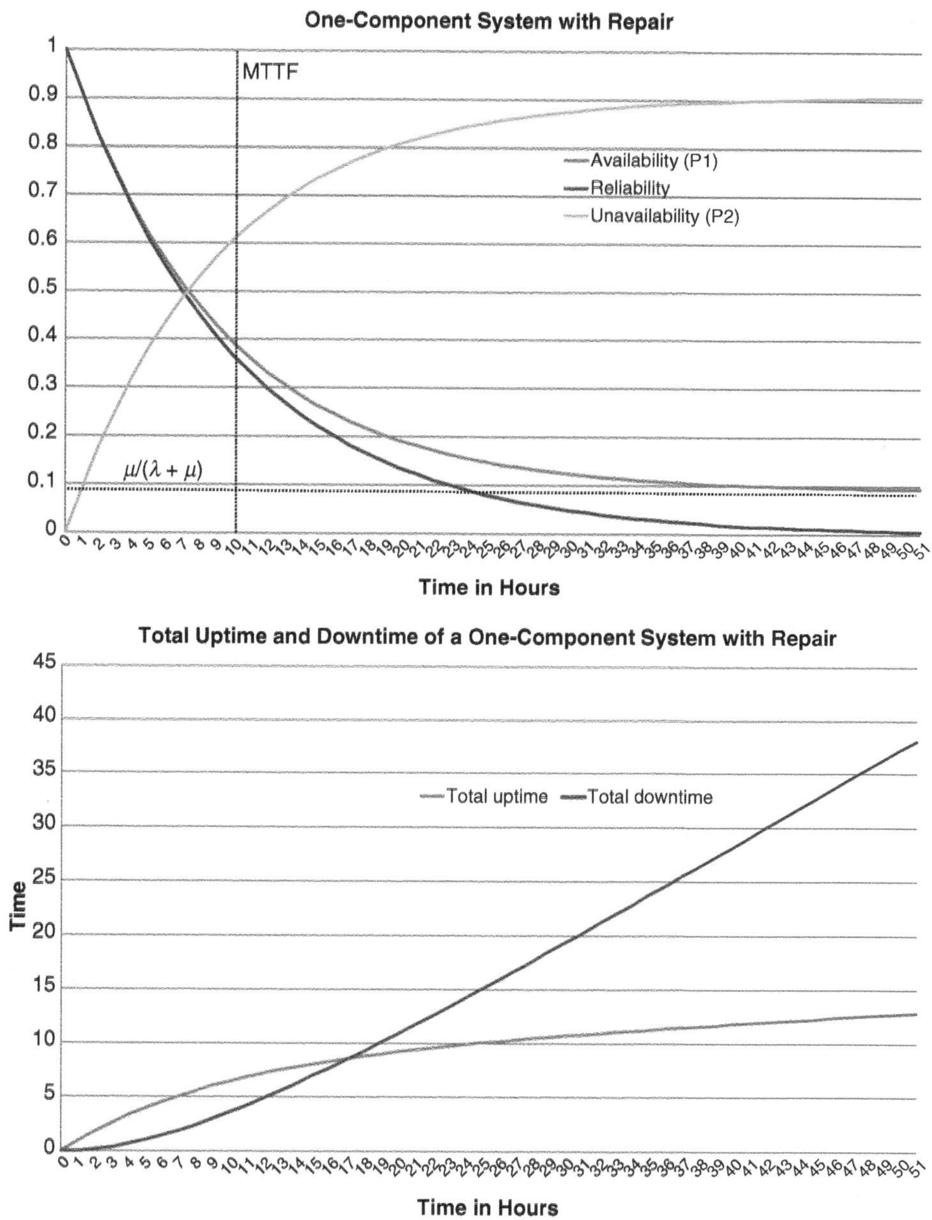

Figure 11.3 One-Component System with Repair

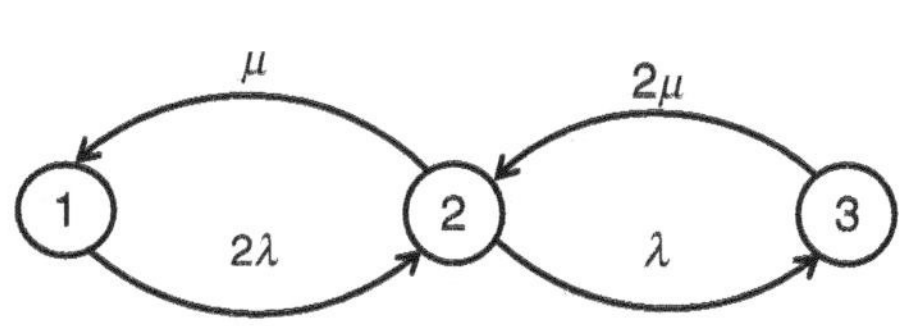

Figure 11.4 State Transition Diagram for Two-Component Parallel System

$$\begin{bmatrix} \dot{P}_1 \\ \dot{P}_2 \\ \dot{P}_3 \end{bmatrix} = \begin{bmatrix} -2\lambda & \mu & 0 \\ 2\lambda & -(\lambda+\mu) & 2\mu \\ 0 & \lambda & -2\mu \end{bmatrix} \cdot \begin{bmatrix} P_1 \\ P_2 \\ P_3 \end{bmatrix}. \tag{11.54}$$

For this example, we take one of the state equations above and replace it with the conservation of probability equation $P_1(t) + P_2(t) + P_3(t) = 1$.

$$\begin{bmatrix} \dot{P}_1 \\ \dot{P}_2 \\ 1 \end{bmatrix} = \begin{bmatrix} -2\lambda & \mu & 0 \\ 2\lambda & -(\lambda+\mu) & 2\mu \\ 1 & 1 & 1 \end{bmatrix} \cdot \begin{bmatrix} P_1 \\ P_2 \\ P_3 \end{bmatrix}. \tag{11.55}$$

We assume the system is fully operational at $t = 0$, that is, the system is in State 1. The Laplace transform of $\dot{P}_1(t)$ is therefore $P_1^*(s) - 1$. The Laplace transforms of $\dot{P}_2(t)$ and $\dot{P}_3(t)$ are $P_2^*(s)$, and $P_3^*(s)$, respectively. The state equation matrix becomes:

$$\begin{bmatrix} sP_1^*(s)-1 \\ sP_2^*(s) \\ 1/s \end{bmatrix} = \begin{bmatrix} -2\lambda & \mu & 0 \\ 2\lambda & -(\lambda+\mu) & 2\mu \\ 1 & 1 & 1 \end{bmatrix} \cdot \begin{bmatrix} P_1^*(s) \\ P_2^*(s) \\ P_3^*(s) \end{bmatrix}. \tag{11.56}$$

The corresponding state equations are:

$$sP_1^*(s) - 1 = -2\lambda P_1^*(s) + \mu P_2^*(s) \tag{11.57}$$

$$sP_2^*(s) = 2\lambda P_1^*(s) + -(\lambda+\mu)P_2^*(s) + 2\mu P_3^*(s) \tag{11.58}$$

$$1/s = P_1^*(s) + P_2^*(s) + P_3^*(s). \tag{11.59}$$

Let's solve Equation (11.59) in terms of $P_3(s)$ to obtain:

$$P_3^*(s) = 1/s - P_1^*(s) - P_2^*(s). \tag{11.60}$$

Next, let us choose the first two state equations, and substitute the probability equation above for P_3:

$$sP_1^*(s) - 1 = -2\lambda P_1^*(s) + \mu P_2^*(s)$$

$$sP_2^*(s) = 2\lambda P_1^*(s) + -(\lambda+\mu)P_2^*(s) + 2\mu(1/s - P_1^*(s) - P_2^*(s)).$$

By algebraic manipulation, we can rearrange these equations in the following form:

$$P_1^*(s) = \frac{\mu P_2^*(s) + 1}{s + 2\lambda} \tag{11.61}$$

$$P_1^*(s) = \frac{(s+3\mu+\lambda)P_2^*(s) - 2\mu/s}{2\lambda - 2\mu}. \tag{11.62}$$

Next, we set the right side of Equations (11.61) and (11.62) equal to each other and solve for $P_2^*(s)$.

$$P_2^*(s) = \frac{2\lambda(2\mu+s)}{s(s+\mu+\lambda)(s+2\mu+2\lambda)}. \tag{11.63}$$

Splitting Equation (11.63) into simpler transforms we obtain:

$$\frac{2\lambda(2\mu+s)}{s(s+\mu+\lambda)(s+2\mu+2\lambda)} = \frac{A}{s} + \frac{B}{s+\lambda+\mu} + \frac{C}{s+2\lambda+2\mu}.$$

The constants A through C are obtained by using the method of partial fraction expansion:

$$A = \lim_{s\to 0} \frac{2\lambda(2\mu+s)}{(s+2\lambda+2\mu)(s+\lambda+\mu)} = \frac{2\lambda\mu}{(\lambda+\mu)^2}$$

$$C = \lim_{s\to -2\lambda-2\mu} \frac{2\lambda(2\mu+s)}{s(s+\lambda+\mu)} = \frac{-2\lambda^2}{(\lambda+\mu)^2}$$

$$B = \lim_{s\to -\lambda-\mu} \frac{2\lambda(2\mu+s)}{s(s+2\lambda+2\mu)} = \frac{2\lambda(\lambda-\mu)}{(\lambda+\mu)^2}.$$

Substituting the values for $A - C$ and collecting terms, we get:

$$P_2^*(s) = \frac{2\lambda\mu}{s(\mu+\lambda)^2} + \frac{2(\lambda^2-\mu\lambda)}{(\mu+\lambda)^2(s+\mu+\lambda)} + \frac{-2\lambda^2}{(\mu+\lambda)^2(s+2\mu+2\lambda)}. \tag{11.64}$$

Next, we find $P_1^*(s)$ by substituting Equations (11.64) and (11.63) into Equation (11.61) and simplifying.

$$\begin{aligned} P_1^*(s) = {} & \frac{2\lambda\mu^2}{s(\lambda+\mu)^2(s+2\lambda)} + \frac{-2\lambda^2\mu}{(\lambda+\mu)^2(s+2\lambda)(s+2\mu+2\lambda)} \\ & + \frac{2\lambda\mu(\lambda-\mu)}{(\lambda+\mu)^2(s+\mu+\lambda)(s+2\lambda)} + \frac{1}{(s+2\lambda)}. \end{aligned} \tag{11.65}$$

Equation (11.65) is then split into simpler transforms using partial fraction expansion:

$$\frac{2\lambda\mu^2}{s(\lambda+\mu)^2(s+2\lambda)} = \frac{A}{s} + \frac{B}{s+2\lambda}$$

$$\frac{-2\mu\lambda^2}{(\lambda+\mu)^2(s+2\lambda+2\mu)(s+2\lambda)} = \frac{A}{s+2\mu+2\lambda} + \frac{B}{s+2\lambda}$$

$$\frac{2\mu\lambda(\lambda-\mu)}{(\lambda+\mu)^2(s+\lambda+\mu)(s+2\lambda)}=\frac{A}{s+\mu+\lambda}+\frac{B}{s+2\lambda}.$$

Solving for the constants and simplifying, we get

$$P_1^*(s)=\frac{2\lambda\mu}{(\mu+\lambda)^2(s+\mu+\lambda)}+\frac{\lambda^2}{(\mu+\lambda)^2(s+2\mu+2\lambda)}+\frac{\mu^2}{s(\mu+\lambda)^2}. \tag{11.66}$$

$P_3^*(s)$ is obtained by substituting Equations (11.64)–(11.66) into Equation (11.60) and simplifying:

$$P_3^*(s)=\frac{-2\lambda^2}{(\mu+\lambda)^2(s+\mu+\lambda)}+\frac{\lambda^2}{(\mu+\lambda)^2(s+2\mu+2\lambda)}+\frac{\lambda^2}{s(\mu+\lambda)^2}. \tag{11.67}$$

Now we can take the inverse Laplace transform to get the time domain equations:

$$P_1(t)=\frac{\mu^2}{(\mu+\lambda)^2}+\frac{2\lambda\mu}{(\mu+\lambda)^2}e^{-(\lambda+\mu)t}+\frac{\lambda^2}{(\mu+\lambda)^2}e^{-2(\lambda+\mu)t}. \tag{11.68}$$

$$P_2(t)=\frac{2\lambda\mu}{(\mu+\lambda)^2}+\frac{2(\lambda^2-\mu\lambda)}{(\mu+\lambda)^2}e^{-(\lambda+\mu)t}-\frac{2\lambda^2}{(\mu+\lambda)^2}e^{-2(\lambda+\mu)t} \tag{11.69}$$

$$P_3(t)=\frac{\lambda^2}{(\mu+\lambda)^2}-\frac{2\lambda^2}{(\mu+\lambda)^2}e^{-(\lambda+\mu)t}+\frac{\lambda^2}{(\mu+\lambda)^2}e^{-2(\lambda+\mu)t}. \tag{11.70}$$

11.3.2 Availability

The system is functioning if it is in State 1 or State 2. The probability that the system is functioning at time t, which corresponds to the availability of the system is:

$$A(t)=P_1(t)+P_2(t)$$
$$A(t)=\frac{\mu^2+2\lambda\mu}{(\mu+\lambda)^2}+\frac{2\lambda^2}{(\mu+\lambda)^2}e^{-(\mu+\lambda)t}-\frac{\lambda^2}{(\mu+\lambda)^2}e^{-2(\mu+\lambda)t}. \tag{11.71}$$

The long-term availability is:

$$A_{lt}=\lim_{t\to\infty}A(t)$$
$$A_{lt}=\frac{\mu^2+2\lambda\mu}{(\mu+\lambda)^2} \tag{11.72}$$
$$A_{lt}=\frac{\mu}{(\mu+\lambda)}+\frac{\lambda\mu}{(\mu+\lambda)^2}.$$

We see that the availability of a two-component system is increased by the amount $\lambda\mu/(\mu+\lambda)^2$ over the single-component repairable system availability (Eq. (11.22)). This is shown graphically in Figure 11.5.

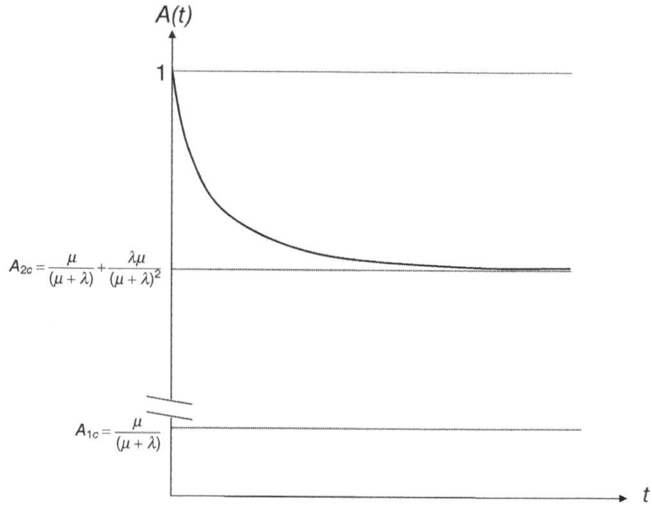

Figure 11.5 Two-Component and Single Component Availability

EXAMPLE 11.3

How much of an improvement in system availability occurs if we change our architecture from a single-component repairable system to a 2N redundant repairable system with identical components?

Choosing a MTBF of 100 hours and a MTTR of 10 hours, we can plot the availability for the one-component and two-component systems for a mission time of 100 hours (Fig. 11.6).

Availability increases significantly when redundancy is employed and perfect switching is assumed. More realistic scenarios will be investigated in Chapter 19.

11.3.3 Reliability

From Equation (11.54), State 3 is the single failure state and for purposes of reliability analysis, State 3 becomes the absorbing state, that is, to calculate reliability, transitions out of State 3 are not considered, and Equation (11.54) becomes:

$$\begin{bmatrix} \dot{P}_1 \\ \dot{P}_2 \\ \dot{P}_3 \end{bmatrix} = \begin{bmatrix} -2\lambda & \mu & 0 \\ 2\lambda & -(\lambda+\mu) & 0 \\ 0 & \lambda & 0 \end{bmatrix} \cdot \begin{bmatrix} P_1 \\ P_2 \\ P_3 \end{bmatrix}. \tag{11.73}$$

We can now truncate the state matrix to remove State 3, since only States 1 and 2 are of interest for the reliability calculation. The reduced state matrix becomes:

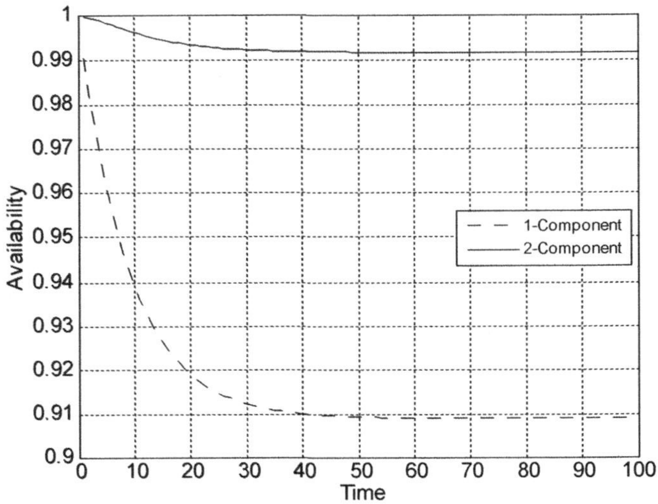

Figure 11.6 Two-Component Parallel System Availability

$$\begin{bmatrix} \dot{P}_1 \\ \dot{P}_2 \end{bmatrix} = \begin{bmatrix} -2\lambda & \mu \\ 2\lambda & -(\mu+\lambda) \end{bmatrix} \cdot \begin{bmatrix} P_1 \\ P_2 \end{bmatrix}. \tag{11.74}$$

Taking the Laplace transform and solving this equation, we get:

$$P_1^*(s) = \frac{s+\lambda+\mu}{s^2+s(3\lambda+\mu)+2\lambda^2} \tag{11.75}$$

$$P_2^*(s) = \frac{2\lambda}{s^2+s(3\lambda+\mu)+2\lambda^2}. \tag{11.76}$$

The roots of $s^2 + s(3\lambda + \mu) + 2\lambda^2$ are:

$$r_1, r_2 = \frac{-3\lambda-\mu\pm\sqrt{\lambda^2+6\mu\lambda+\mu^2}}{2} \tag{11.77}$$

$$P_1^*(s) = \frac{s+\lambda+\mu}{\left(s+\frac{3\lambda}{2}+\frac{\mu}{2}-\frac{1}{2}\sqrt{\lambda^2+6\lambda\mu+\mu^2}\right)\left(s+\frac{3\lambda}{2}+\frac{\mu}{2}+\frac{1}{2}\sqrt{\lambda^2+6\lambda\mu+\mu^2}\right)}$$

$$P_2^*(s) = \frac{2\lambda}{\left(s+\frac{3\lambda}{2}+\frac{\mu}{2}-\frac{1}{2}\sqrt{\lambda^2+6\lambda\mu+\mu^2}\right)\left(s+\frac{3\lambda}{2}+\frac{\mu}{2}+\frac{1}{2}\sqrt{\lambda^2+6\lambda\mu+\mu^2}\right)}. \tag{11.78}$$

Taking the inverse transform, the time domain probability solutions are:

$$P_1(t) = \frac{\lambda + \mu + r_1}{r_1 - r_2} e^{r_1 t} + \frac{\lambda + \mu + r_2}{r_2 - r_1} e^{r_2 t} \tag{11.79}$$

$$P_2(s) = \frac{2\lambda}{r_1 - r_2} e^{r_1 t} + \frac{2\lambda}{r_2 - r_1} e^{r_2 t}. \tag{11.80}$$

The reliability $R(t)$ is equal to the probability of being in State 1 or State 2:

$$R(t) = P_1(t) + P_2(t) = \frac{3\lambda + \mu + r_1}{r_1 - r_2} e^{r_1 t} + \frac{3\lambda + \mu + r_2}{r_2 - r_1} e^{r_2 t}. \tag{11.81}$$

EXAMPLE 11.4

What is the dynamic reliability for a 2N redundant system whose components have an MTBF of 100,000 hours and a repair rate of 8 hours?

The failure rate and repair rate are $\lambda = 0.00001$ and $\mu = 0.125$, respectively. Substituting the values for the failure rates and repair rates into the Equations (11.78) and (11.79), we get:

$$r_1 = -1.59962\text{E-}09$$

$$r_2 = -0.125029998.$$

Next, substitute the failure and repair rates into Equation (11.81):

$$R(t) = 1.00000016 e^{-(1.59962E-9)t} - 1.599616e - 07 e^{-0.12503t}. \tag{11.82}$$

11.3.4 Mean Time to First Failure

The reliability of this system is equal to the probability of the system being in either State 1 or State 2 given that State 3 is transformed into an absorbing state. Using Equations (11.75) and (11.76), we obtain:

$$R^*(s) = P_1^*(s) + P_2^*(s)$$

$$R^*(s) = \frac{s + \lambda + \mu}{s^2 + s(\mu + 3\lambda) + 2\lambda^2} + \frac{2\lambda}{s^2 + s(\mu + 3\lambda) + 2\lambda^2} \tag{11.83}$$

$$\text{MTTFF} = R^*(0) = P_1^*(0) + P_2^*(0)$$

$$\text{MTTFF} = \frac{3\lambda + \mu}{2\lambda^2}. \tag{11.84}$$

As a check, if we remove the repair action for both components, Equation (11.84) reduces to:

$$\text{MTTFF} = \frac{3}{2\lambda}. \tag{11.85}$$

This result agrees with the two-component no repair example, Equation (10.51). From Equation (11.84), we note that if the repair rate is much greater than the failure rate, the MTTFF can be approximated as:

$$\text{MTTFF} = \frac{\mu}{2\lambda^2}. \tag{11.86}$$

Thus, the MTTFF is significantly increased by the addition of redundancy with repair.

11.3.5 System Failure Rate

The equivalent system failure rate is calculated using the definition of the hazard rate, Equation (7.38):

$$\lambda(t) = \frac{-\frac{d}{dt}R(t)}{R(t)}$$
$$\lambda(t) = \frac{r_1 \frac{3\lambda+\mu+r_1}{r_1-r_2} e^{r_1 t} + r_2 \frac{3\lambda+\mu+r_2}{r_2-r_1} e^{r_2 t}}{\frac{3\lambda+\mu+r_1}{r_1-r_2} e^{r_1 t} + \frac{3\lambda+\mu+r_2}{r_2-r_1} e^{r_2 t}}. \tag{11.87}$$

EXAMPLE 11.5

A 2N redundant system has an MTBF of 200 hours and a MTTR of 2 hours. Determine the conditional failure rate as a function of time.

$\lambda = 0.005$ and $\mu = 0.5$. Next, determine r_1 and r_2 from Equation (11.77):

$$r_1 = \frac{-3(0.005) - 0.5 + \sqrt{(0.005)^2 + 6(0.5)(0.005) + (0.5)^2}}{2} = -9.71057E-05.$$
$$r_2 = \frac{-3(0.005) - 0.5 - \sqrt{(0.005)^2 + 6(0.5)(0.005) + (0.5)^2}}{2} = -0.514903.$$

We then calculate $\lambda(t)$ using Equation (11.87).

$$\lambda(t) = \frac{r_1 \frac{3\lambda+\mu+r_1}{r_1-r_2} e^{r_1 t} + r_2 \frac{3\lambda+\mu+r_2}{r_2-r_1} e^{r_2 t}}{\frac{3\lambda+\mu+r_1}{r_1-r_2} e^{r_1 t} + \frac{3\lambda+\mu+r_2}{r_2-r_1} e^{r_2 t}}$$

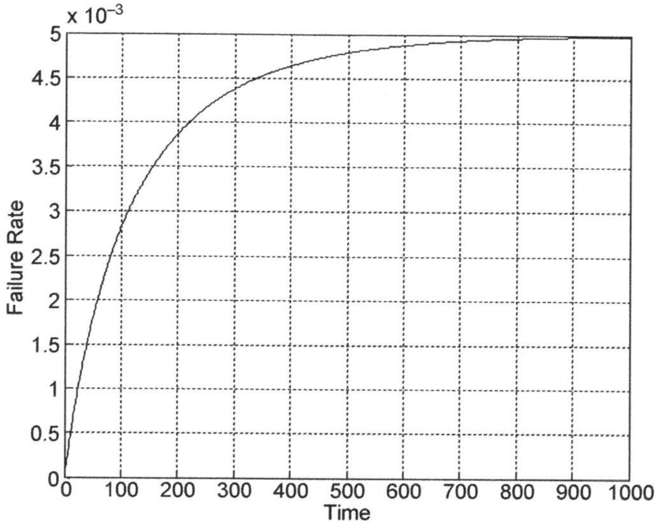

Figure 11.7 System Failure Rate

$$\lambda(t) = \frac{-(9.7124E-5)e^{-9.71057E-05t} + (9.7124E-5)e^{-0.514903t}}{1.00018863e^{-9.71057E-05t} - 0.00018863e^{-0.514903t}}.$$

Figure 11.7 depicts the failure rate over 1000 hours.

The system failure rate approaches the component failure rate of $\lambda = 0.005$ at around 1000 hours of operation.

11.3.6 Asymptotic Behavior

If we have the complete time dependent state probability equations, as we obtained in Equations (11.68)–(11.70), we do not need to solve the set of steady-state matrix equations. We can obtain the steady-state probability distributions by allowing t to be very large, that is, let $t \to \infty$, the equations reduce to:

$$P_1 = \frac{\mu^2}{(\mu+\lambda)^2} \tag{11.88}$$

$$P_2 = \frac{2\lambda\mu}{(\mu+\lambda)^2} \tag{11.89}$$

$$P_3 = \frac{\lambda^2}{(\mu+\lambda)^2}. \tag{11.90}$$

If we do not have the complete dynamic equations, we start with the state transition matrix equations we previously created (Eq. (11.55)), to obtain the steady-state behavior.

The state equations are:

$$0 = -2\lambda P_1 + \mu P_2 \tag{11.91}$$

$$0 = 2\lambda P_1 - (\lambda+\mu)P_2 + 2\mu P_3 \tag{11.92}$$

$$1 = P_1 + P_2 + P_3. \tag{11.93}$$

Let us solve for P_3 in terms of P_1 and P_2 in the third equation and substitute into the other two equations to obtain:

$$0 = -2\lambda P_1 + \mu P_2 \tag{11.94}$$

$$0 = 2\lambda P_1 - (\lambda+\mu)P_2 + 2\mu(1 - P_1 - P_2). \tag{11.95}$$

Solving Equations (11.94) and (11.95) for P_1, we get:

$$P_1 = \frac{\mu P_2}{2\lambda} \tag{11.96}$$

$$P_1 = \frac{(\lambda+3\mu)P_2 - 2\mu}{2(\lambda-\mu)}. \tag{11.97}$$

Now set Equations (11.96) and (11.97) equal to each other and solve for P_2:

$$P_2 = \frac{2\lambda\mu}{(\lambda+\mu)^2}. \tag{11.98}$$

Substitute P_2 into Equation (11.96), we get P_1:

$$P_1 = \frac{\mu^2}{(\lambda+\mu)^2}. \tag{11.99}$$

Finally, we obtain P_3 by substituting P_1 and P_2 into the probability equation and get:

$$P_3 = \frac{\lambda^2}{(\lambda+\mu)^2}. \tag{11.100}$$

11.3.7 Mean Time to Failure

Assuming we are in an operational state, what is the mean time to failure?

Using Equations (9.43) and (9.46), and noting we have two functional states and one failed state, the MTTF is:

$$\text{MTTF}_S = 1/f_{GF} = \frac{P_1 + P_2}{P_2\beta_{23}}$$

$$\text{MTTF}_S = \frac{2\lambda\mu + \mu^2}{(2\lambda\mu)\lambda}$$

$$\text{MTTF}_S = \frac{1}{\lambda} + \frac{\mu}{2\lambda^2}.$$

Comparing the MTTF of the single-component repairable system Equation (11.43), we see that the MTTF of a two-component parallel system increases by the amount $\mu/2\lambda^2$.

11.3.8 Mean Time to Repair

Assuming we are in a failed state, what is the mean time to repair the system and restore it to operational condition?

From Equations (9.38) and (9.42), and noting we have two functional states and one failed state, the MTTR_S is:

$$\begin{aligned}\text{MTTR}_S &= \frac{1}{\beta_{32}}\\ \text{MTTR}_S &= \frac{1}{2\mu}.\end{aligned} \tag{11.101}$$

If the system is nonrepairable, then $\mu = 0$ and $\text{MTTR}_S = \infty$ (i.e., never repaired) as we would expect. Note also the MTTR_S is half the repair rate MTTR of a single component since the system becomes functional once one of the components is repaired.

11.3.9 Mean Time between Failures

The mean time between system failures, MTBF_S, is the mean time between consecutive transitions from a functioning state into a failed state. The MTBF_S can be calculated from Equations (9.35)–(9.37), and noting we have two functional states and one failed state:

$$\begin{aligned}\text{MTBF}_S &= \frac{1}{P_2\beta_{23}} = \frac{(\lambda+\mu)^2}{(2\lambda\mu)\lambda}\\ \text{MTBF}_S &= \frac{1}{P_2\beta_{23}} = \frac{(\lambda+\mu)^2}{2\lambda^2\mu}.\end{aligned} \tag{11.102}$$

We can check the MTBF by using the relationship MTBF = MTTR + MTTF:

$$\begin{aligned}\text{MTBF}_S &= \text{MTTF}_S + MTTR_S\\ \text{MTBF}_S &= \frac{2\lambda\mu + \mu^2}{2\lambda^2\mu} + \frac{1}{2\mu}\\ \text{MTBF}_S &= \frac{(\lambda+\mu)^2}{2\lambda^2\mu}.\end{aligned} \tag{11.103}$$

If $\lambda \ll \mu$, then Equation (11.103) reduces to:

$$\mathrm{MTBF}_S = \frac{\mu}{2\lambda^2}. \tag{11.104}$$

Comparing this equation with the MTBF for a single-component repairable system, the MTBF increases by a factor of $\mu/(2\lambda)$.

11.3.10 System Availability

From Equation (9.47), the system availability is the sum of the probabilities of all the good states. We have two good states, P_1 and P_2:

$$\begin{aligned} A_S &= P_1 + P_2 \\ A_S &= \frac{2\lambda\mu + \mu^2}{(\lambda+\mu)^2}. \end{aligned} \tag{11.105}$$

For the MRM in this example, the system is operational (available) when in States 1 and 2 and is failed (unavailable) when in State 3. From Equation (9.50), the availability is:

$$A_S = E[X(t)] = \sum_i r_i P_i(t) = 1\cdot P_1 + 1\cdot P_2 + 0\cdot P_3 = \frac{2\lambda\mu + \mu^2}{(\lambda+\mu)^2}. \tag{11.106}$$

This result is the same as Equation (11.105) as we would expect.

11.3.11 System Unavailability

From Equation (9.48), system unavailability is the sum of the probabilities of all the bad states. We have one bad state, P_3:

$$\begin{aligned} Q_S &= P_3 \\ Q_S &= \frac{\lambda^2}{(\lambda+\mu)^2}. \end{aligned} \tag{11.107}$$

As a check, $Q_S = 1 - A_S$, which gives the same result.

EXAMPLE 11.6

For a 2N redundant system with an MTBF of 100 hours and a repair rate of 8 hours, determine the availability, reliability, MTTF, MTBF, and the system asymptotic behavior. Plot these values over a time frame of 100 and 800 hours.

$\lambda = 0.01$ and $\mu = 0.125$. Using the equations above, we can readily calculate the dynamic response of the system and other parameters. Figure 11.8 and Figure 11.9 illustrate the relationships between availability, reliability, MTTF, MTBF, and asymptotic behavior.

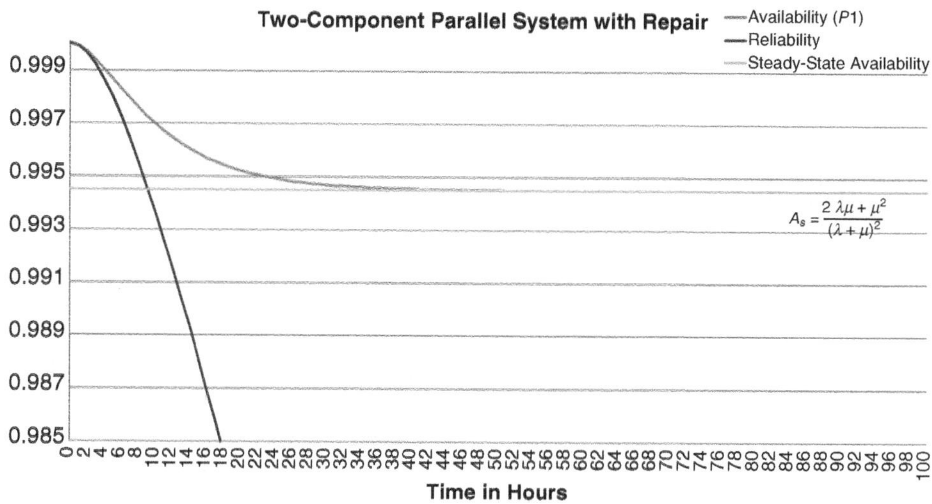

Figure 11.8 Two-Component Parallel System with Repair: 100 Hours

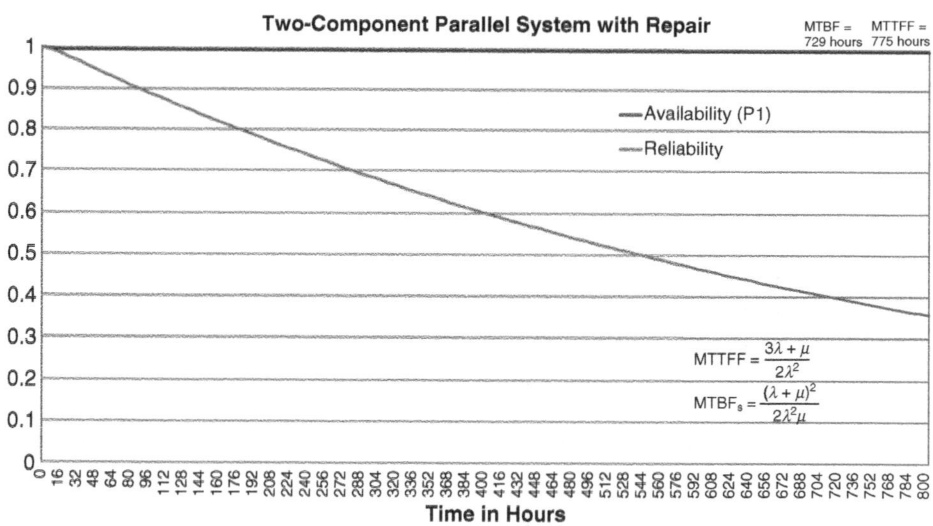

Figure 11.9 Two-Component Parallel System with Repair: 800 Hours

11.4 PARALLEL SYSTEM WITH REPAIR: DIFFERENT FAILURE AND REPAIR RATES

Referring back to the system from Section 11.3, we now consider different failure rates and repair rates for the two components in parallel, that is, each component has a different failure rate and repair rate. The reliability block diagram for such a system is shown in Figure 11.10.

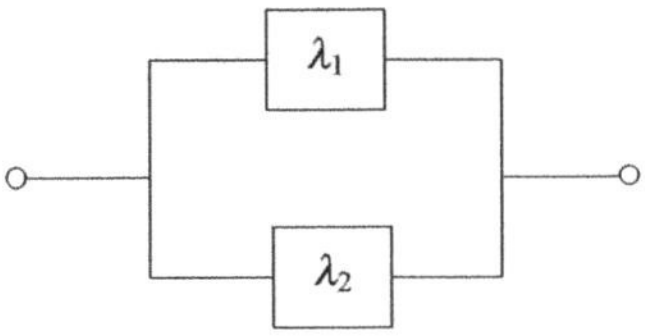

Figure 11.10 Reliability Block Diagram for a Two-Component Parallel System

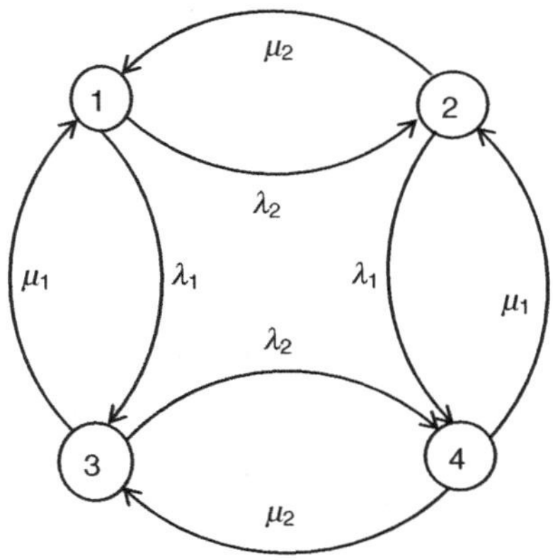

Figure 11.11 State Transition Diagram for Two-Component Parallel System

Component 1 has a constant failure rate λ_1 and a repair rate of μ_1. Likewise, component 2 has a constant failure rate of λ_2 and a repair rate of μ_2.

The four possible states of the system are:

- *State 1: System functional and both components operating.*
- *State 2: System functional and component 2 has failed.*
- *State 3: System functional and component 1 has failed.*
- *State 4: System failure; both components have failed.*

From the definitions of the four states, the state transition diagram for the system is depicted in Figure 11.11.

11.4.1 Dynamic Behavior

From the state transition diagram, write the state transition matrix and dynamic probability equations:

$$\begin{bmatrix} \dot{P}_1 \\ \dot{P}_2 \\ \dot{P}_3 \\ \dot{P}_4 \end{bmatrix} = \begin{bmatrix} -(\lambda_1+\lambda_2) & \mu_2 & \mu_1 & 0 \\ \lambda_2 & -(\lambda_1+\mu_2) & 0 & \mu_1 \\ \lambda_1 & 0 & -(\lambda_2+\mu_1) & \mu_2 \\ 0 & \lambda_1 & \lambda_2 & -(\mu_1+\mu_2) \end{bmatrix} \cdot \begin{bmatrix} P_1 \\ P_2 \\ P_3 \\ P_4 \end{bmatrix}. \tag{11.108}$$

Let us take one of the state equations above and replace it with the conservation of probability equation $P_1(t) + P_2(t) + P_3(t) + P_4(t) = 1$.

$$\begin{bmatrix} 1 \\ \dot{P}_2 \\ \dot{P}_3 \\ \dot{P}_4 \end{bmatrix} = \begin{bmatrix} 1 & 1 & 1 & 1 \\ \lambda_2 & -(\lambda_1+\mu_2) & 0 & \mu_1 \\ \lambda_1 & 0 & -(\lambda_2+\mu_1) & \mu_2 \\ 0 & \lambda_1 & \lambda_2 & -(\mu_1+\mu_2) \end{bmatrix} \cdot \begin{bmatrix} P_1 \\ P_2 \\ P_3 \\ P_4 \end{bmatrix}. \tag{11.109}$$

We assume the system is fully operational at $t = 0$, that is, the system is in State 1 (probability of being in State 1 at $t = 0$, is 1). The Laplace transform of $\dot{P}_1(t)$ is therefore $sP_1^*(s)-1$. The Laplace transforms of $\dot{P}_2(t)$, $\dot{P}_3(t)$, and $\dot{P}_4(t)$ are $sP_2^*(s)$, $sP_3^*(s)$, $sP_4^*(s)$, respectively (probability of being in States 2, 3, or 4 are all 0 at $t = 0$). The Laplace transform matrix becomes:

$$\begin{bmatrix} 1/s \\ sP_2^*(s) \\ sP_3^*(s) \\ sP_4^*(s) \end{bmatrix} = \begin{bmatrix} 1 & 1 & 1 & 1 \\ \lambda_2 & -(\lambda_1+\mu_2) & 0 & \mu_1 \\ \lambda_1 & 0 & -(\lambda_2+\mu_1) & \mu_2 \\ 0 & \lambda_1 & \lambda_2 & -(\mu_1+\mu_2) \end{bmatrix} \cdot \begin{bmatrix} P_1^*(s) \\ P_2^*(s) \\ P_3^*(s) \\ P_4^*(s) \end{bmatrix}.$$

The state equations are:

$$1/s = P_1^*(s) + P_2^*(s) + P_3^*(s) + P_4^*(s)$$

$$sP_2^*(s) = \lambda_2 P_1^*(s) - (\lambda_1+\mu_2)P_2^*(s) + \mu_1 P_4^*(s)$$

$$sP_3^*(s) = \lambda_1 P_1^*(s) - (\lambda_2+\mu_1)P_3^*(s) + \mu_2 P_4^*(s)$$

$$sP_4^*(s) = \lambda_1 P_2^*(s) + \lambda_2 P_3^*(s) - (\mu_1+\mu_2)P_4^*(s).$$

Solving these equations, we obtain:

$$P_1^*(s) = \frac{\mu_1\mu_2}{s(\lambda_1+\mu_1)(\lambda_2+\mu_2)} + \frac{\lambda_1\mu_2}{(s+\lambda_1+\mu_1)(\lambda_1+\mu_1)(\lambda_2+\mu_2)} + \frac{\lambda_2\mu_1}{(s+\lambda_2+\mu_2)(\lambda_1+\mu_1)(\lambda_2+\mu_2)} + \frac{\lambda_1\lambda_2}{(s+\lambda_1+\lambda_2+\mu_1+\mu_2)(\lambda_1+\mu_1)(\lambda_2+\mu_2)} \tag{11.110}$$

$$P_2^*(s) = \frac{\mu_1\lambda_2}{s(\lambda_1+\mu_1)(\lambda_2+\mu_2)} + \frac{\lambda_1\lambda_2}{(s+\lambda_1+\mu_1)(\lambda_1+\mu_1)(\lambda_2+\mu_2)} - \frac{\lambda_2\mu_1}{(s+\lambda_2+\mu_2)(\lambda_1+\mu_1)(\lambda_2+\mu_2)} - \frac{\lambda_1\lambda_2}{(s+\lambda_1+\lambda_2+\mu_1+\mu_2)(\lambda_1+\mu_1)(\lambda_2+\mu_2)} \tag{11.111}$$

$$P_3^*(s)=\frac{\lambda_1\mu_2}{s(\lambda_1+\mu_1)(\lambda_2+\mu_2)}-\frac{\lambda_1\mu_2}{(s+\lambda_1+\mu_1)(\lambda_1+\mu_1)(\lambda_2+\mu_2)}+\frac{\lambda_1\lambda_2}{(s+\lambda_2+\mu_2)(\lambda_1+\mu_1)(\lambda_2+\mu_2)}-\frac{\lambda_1\lambda_2}{(s+\lambda_1+\lambda_2+\mu_1+\mu_2)(\lambda_1+\mu_1)(\lambda_2+\mu_2)} \tag{11.112}$$

$$P_4^*(s)=\frac{\lambda_1\lambda_2}{s(\lambda_1+\mu_1)(\lambda_2+\mu_2)}-\frac{\lambda_1\lambda_2}{(s+\lambda_1+\mu_1)(\lambda_1+\mu_1)(\lambda_2+\mu_2)}-\frac{\lambda_1\lambda_2}{(s+\lambda_2+\mu_2)(\lambda_1+\mu_1)(\lambda_2+\mu_2)}+\frac{\lambda_1\lambda_2}{(s+\lambda_1+\lambda_2+\mu_1+\mu_2)(\lambda_1+\mu_1)(\lambda_2+\mu_2)}. \tag{11.113}$$

Now we can take the inverse Laplace transform to get the time domain equations:

$$P_1(t)=\frac{\mu_1\mu_2}{(\lambda_1+\mu_1)(\lambda_2+\mu_2)}+\frac{\lambda_1\mu_2}{(\lambda_1+\mu_1)(\lambda_2+\mu_2)}e^{-(\lambda_1+\mu_1)t}+\frac{\mu_1\lambda_2}{(\lambda_1+\mu_1)(\lambda_2+\mu_2)}e^{-(\lambda_2+\mu_2)t}+\frac{\lambda_1\lambda_2}{(\lambda_1+\mu_1)(\lambda_2+\mu_2)}e^{-(\lambda_1+\mu_1+\lambda_2+\mu_2)t} \tag{11.114}$$

$$P_2(t)=\frac{\mu_1\lambda_2}{(\lambda_1+\mu_1)(\lambda_2+\mu_2)}+\frac{\lambda_1\lambda_2}{(\lambda_1+\mu_1)(\lambda_2+\mu_2)}e^{-(\lambda_1+\mu_1)t}-\frac{\mu_1\lambda_2}{(\lambda_1+\mu_1)(\lambda_2+\mu_2)}e^{-(\lambda_2+\mu_2)t}-\frac{\lambda_1\lambda_2}{(\lambda_1+\mu_1)(\lambda_2+\mu_2)}e^{-(\lambda_1+\mu_1+\lambda_2+\mu_2)t} \tag{11.115}$$

$$P_3(t)=\frac{\mu_2\lambda_1}{(\lambda_1+\mu_1)(\lambda_2+\mu_2)}-\frac{\lambda_1\mu_2}{(\lambda_1+\mu_1)(\lambda_2+\mu_2)}e^{-(\lambda_1+\mu_1)t}+\frac{\lambda_1\lambda_2}{(\lambda_1+\mu_1)(\lambda_2+\mu_2)}e^{-(\lambda_2+\mu_2)t}-\frac{\lambda_1\lambda_2}{(\lambda_1+\mu_1)(\lambda_2+\mu_2)}e^{-(\lambda_1+\mu_1+\lambda_2+\mu_2)t} \tag{11.116}$$

$$P_4(t)=\frac{\lambda_2\lambda_1}{(\lambda_1+\mu_1)(\lambda_2+\mu_2)}-\frac{\lambda_1\lambda_2}{(\lambda_1+\mu_1)(\lambda_2+\mu_2)}e^{-(\lambda_1+\mu_1)t}-\frac{\lambda_1\lambda_2}{(\lambda_1+\mu_1)(\lambda_2+\mu_2)}e^{-(\lambda_2+\mu_2)t}+\frac{\lambda_1\lambda_2}{(\lambda_1+\mu_1)(\lambda_2+\mu_2)}e^{-(\lambda_1+\mu_1+\lambda_2+\mu_2)t}. \tag{11.117}$$

11.4.2 Comparison with Identical Parallel Component Example

Now that we have obtained the complete solution for a system composed of two different components with different failure rates and repair rates, let us compare this with the case in which both components are identical (repair rates and failure rates are the same).

In this case, Equations (11.114)–(11.117) simplify to:

$$P_1(t)=\frac{\mu^2}{(\lambda+\mu)^2}+\frac{2\lambda\mu}{(\lambda+\mu)^2}e^{-(\lambda+\mu)t}+\frac{\lambda^2}{(\lambda+\mu)^2}e^{-2(\lambda+\mu)t} \tag{11.118}$$

$$P_2(t)=\frac{\lambda\mu}{(\lambda+\mu)^2}+\frac{\lambda^2-\lambda\mu}{(\lambda+\mu)^2}e^{-(\lambda+\mu)t}-\frac{\lambda^2}{(\lambda+\mu)^2}e^{-2(\lambda+\mu)t} \quad (11.119)$$

$$P_3(t)=\frac{\lambda\mu}{(\lambda+\mu)^2}+\frac{\lambda^2-\lambda\mu}{(\lambda+\mu)^2}e^{-(\lambda+\mu)t}-\frac{\lambda^2}{(\lambda+\mu)^2}e^{-2(\lambda+\mu)t} \quad (11.120)$$

$$P_4(t)=\frac{\lambda^2}{(\lambda+\mu)^2}-\frac{2\lambda^2}{(\lambda+\mu)^2}e^{-(\lambda+\mu)t}+\frac{\lambda^2}{(\lambda+\mu)^2}e^{-2(\lambda+\mu)t}. \quad (11.121)$$

We compare Equation (11.118)–(11.121) with Equations (11.68)–(11.70), which represent the probability solution for a two-component parallel system with identical component failure rates and repair rates. Notice that Equations (11.68) and (11.118) are identical if we let $\mu_1 = \mu_2$ and $\lambda_1 = \lambda_2$ in Equation (11.114). Since the repair rates and failure rates are the same, we can combine States 2 and 3 from this example into a single state (State 2 in Section 11.3), and State 4 then becomes State 3. By comparing Equation (11.69) with Equations (11.119) and (11.120), by observation, the summation of Equations (11.119) and (11.120) equals Equation (11.69), which is what we would expect for the combined States 2 and 3 from this example. Finally, Equation (11.121) is the same as Equation (11.70), when State 4 of this example becomes State 3 in Section 11.3.

11.4.3 Obtaining Symbolic Solutions Using Matlab

Solving for Equations (11.114)–(11.117) by hand is quite tedious! As we showed in Section 11.3, we can solve these equations directly using the symbolic differential equation solver in Matlab. As we did previously, select 3 of the 4 equations from Equation (11.109) and substitute the conservation of probability equation for one of the probability variables. However, in this case, there is no need to take the Laplace transform of the equations. The set of differential equations we obtain is:

$$\dot{P}_1=-(\lambda_1+\lambda_2)P_1+\mu_2P_2+\mu_1P_3 \quad (11.122)$$

$$\dot{P}_2=(\lambda_2-\mu_1)P_1-(\lambda_1+\mu_1+\mu_2)P_2-\mu_1P_3+\mu_1 \quad (11.123)$$

$$\dot{P}_3=(\lambda_1-\mu_2)P_1-\mu_2P_2-(\lambda_2+\mu_1+\mu_2)P_3+\mu_2. \quad (11.124)$$

These three equations and the initial conditions are entered into Matlab's *dsolve* function:

```
[p1, p2, p3] =dsolve ('Dp1=-(L1+L2)*p1+M2*p2+M1*p3', 'Dp2=(L2-
M1)*p1 - (L1+M1+M2)*p2 - M1*p3 + M1', 'Dp3 = (L1-M2)*p1 - M2*p2 -
(L2+M1+M2)*p3 + M2', 'p1(0) = 1 p2(0) = 0 p3(0) = 0');
```

The result is:

```
p1 =

(M1*M2)/(L1*L2 + L1*M2 + L2*M1 + M1*M2) + (L1*M2)/(exp(L1*t +
M1*t)*(L1*L2 + L1*M2 + L2*M1 + M1*M2)) + (L2*M1)/(exp(L2*t +
```

```
M2*t)*(L1*L2 + L1*M2 + L2*M1 + M1*M2)) + (L1*L2)/(exp(L1*t + L2*t
+ M1*t + M2*t)*(L1*L2 + L1*M2 + L2*M1 + M1*M2))

p2 =

(L2*M1)/(L1*L2 + L1*M2 + L2*M1 + M1*M2) + (L1*L2)/(exp(L1*t +
M1*t)*(L1*L2 + L1*M2 + L2*M1 + M1*M2)) - (L2*M1)/(exp(L2*t +
M2*t)*(L1*L2 + L1*M2 + L2*M1 + M1*M2)) - (L1*L2)/(exp(L1*t + L2*t
+ M1*t + M2*t)*(L1*L2 + L1*M2 + L2*M1 + M1*M2))

p3 =

(L1*M2)/(L1*L2 + L1*M2 + L2*M1 + M1*M2) + (L1*L2)/(exp(L2*t +
M2*t)*(L1*L2 + L1*M2 + L2*M1 + M1*M2)) - (L1*M2)/(exp(L1*t +
M1*t)*(L1*L2 + L1*M2 + L2*M1 + M1*M2)) - (L1*L2)/(exp(L1*t + L2*t
+ M1*t + M2*t)*(L1*L2 + L1*M2 + L2*M1 + M1*M2))
```

These equations match Equations (11.114)–(11.117). p4 is obtained in a straightforward manner by subtracting the sum of Equations (11.114)–(11.116) from the value 1 giving us Equation (11.117).

EXAMPLE 11.7

Let us now consider an example in which we are given the following MTBFs and MTTRs for a two-component parallel system:

$$MTBF_1 = 100{,}000 \text{ hours}$$

$$MTBF_2 = 50{,}000 \text{ hours}$$

$$MTTR_1 = 8 \text{ hours}$$

$$MTTR_2 = 16 \text{ hours.}$$

Determine the dynamic behavior of the system and plot the four different states over a period of 100 hours.

The failure rates and repair rates are the reciprocals of the MTBFs and MTTRs, respectively.

$$\lambda_1 = 1 \text{ failure/100,000 hours} = 0.00001 \text{ failures/hour}$$

$$\lambda_2 = 1 \text{ failure/50,000 hours} = 0.00002 \text{ failures/hour}$$

$$\mu_1 = 1 \text{ repair/8 hours} = 0.125 \text{ repairs/hour}$$

$$\mu_2 = 1 \text{ repair/16 hours} = 0.0625 \text{ repairs/hour.}$$

Substituting these values into Equation (11.114)–(11.117) and simplifying, we obtain:

$$P_1(t) = 0.999600124 + 0.000319872e^{-0.0652t} + (7.9968\text{E-}5)e^{-0.12501t} + (2.55898\text{E-}8)e^{-0.18753t} \quad (11.125)$$

$$P_2(t) = 0.000319872 + 0.999600134e^{-0.12501t} - 0.000319872e^{-0.06252t} - 2.55898\text{E-}8e^{-0.18753t} \quad (11.126)$$

$$P_3(t) = 7.9968\text{E-}5 + 2.55898\text{E-}8e^{-0.0652t} - (7.9968\text{e-}5)e^{-0.12501t} - (2.55898\text{E-}8)e^{-0.18753t} \quad (11.127)$$

$$P_4(t) = 2.55898\text{E-}8 + (2.55898\text{E-}8)e^{-0.18753t} - (2.55898\text{E-}8)e^{-0.12501t} - (2.55898\text{E-}5)e^{-0.0652t}. \quad (11.128)$$

These probabilities plotted over a period of 100 hours using Matlab (markovDynamicModel.m) are shown in Figure 11.12, Figure 11.13, Figure 11.14, and Figure 11.15.

From the diagrams, we note the following:

1. Each of the four probabilities asymptotically approaches a constant value.
2. The probability of being in State 1 starts at 1 for $t = 0$ (due to our initial condition assumptions) and quickly approaches a steady-state value of 0.9996.
3. The probabilities of being in the other four states rises from a value of 0 for $t = 0$ and approaches a steady-state value significantly smaller than 1.
4. The asymptotic value approached is equal to the constant term in each of the probability equations.
5. After about 100 hours, we can assume the system behaves with constant probabilities for each state.

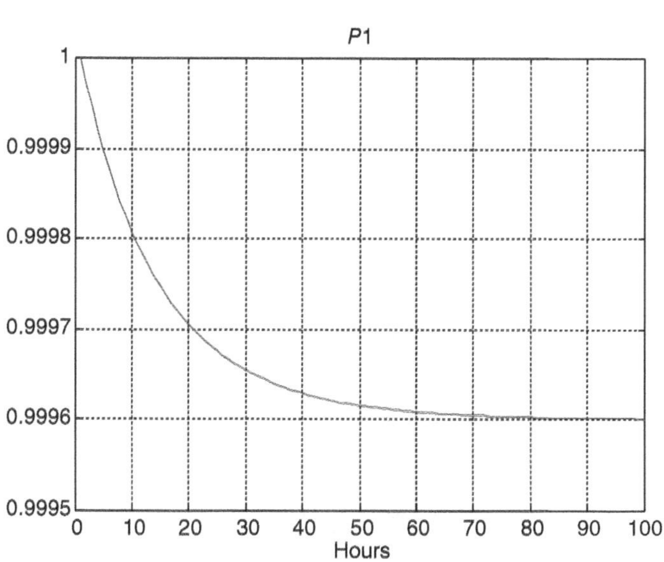

Figure 11.12 $P_1(t)$

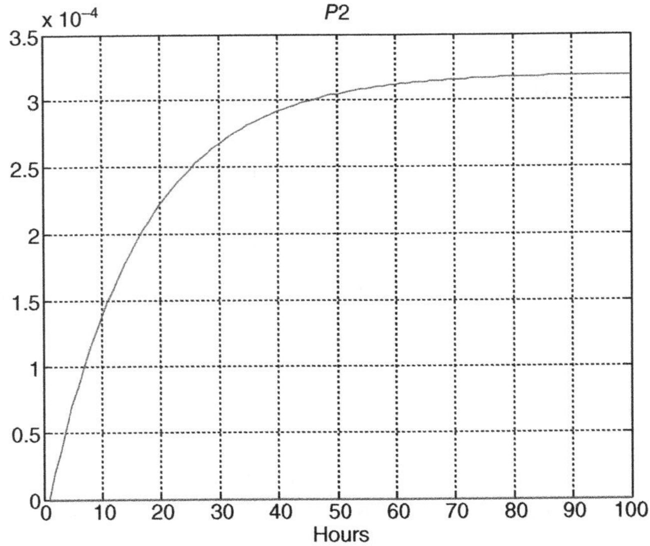

Figure 11.13 $P_2(t)$

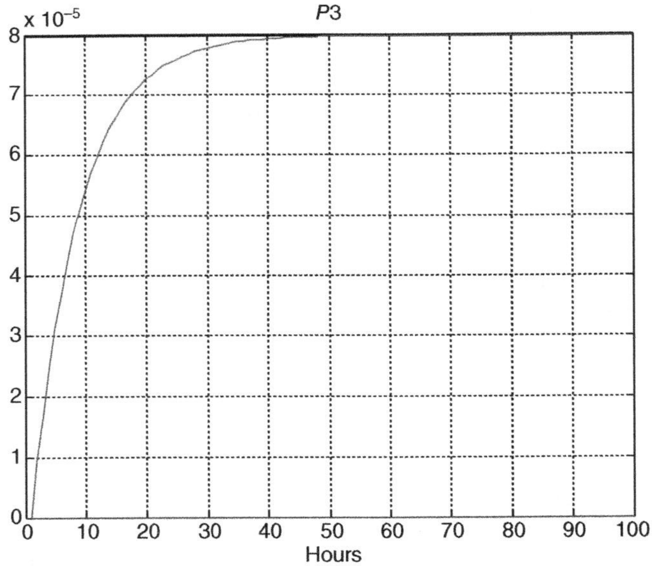

Figure 11.14 $P_3(t)$

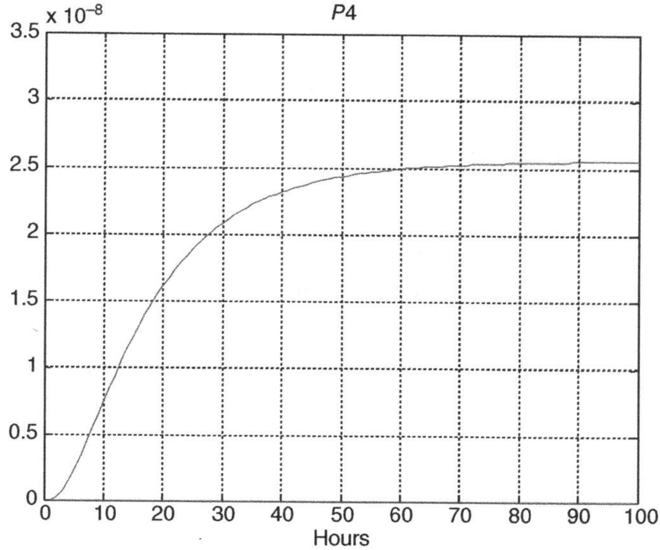

Figure 11.15 $P_4(t)$

11.4.4 Availability

The system is functioning if it is in State 1, 2, or 3, or equivalently if the system is *not* in the failed state (State 4). The probability that the system is functioning at time t (the availability of the system) is:

$$A(t) = 1 - P_4(t) = 1 - \frac{\lambda_1\lambda_2}{(\lambda_1+\mu_1)(\lambda_2+\mu_2)} + \frac{\lambda_1\lambda_2}{(\lambda_1+\mu_1)(\lambda_2+\mu_2)}e^{-(\lambda_1+\mu_1)t} + \frac{\lambda_1\lambda_2}{(\lambda_1+\mu_1)(\lambda_2+\mu_2)}e^{-(\lambda_2+\mu_2)t} - \frac{\lambda_1\lambda_2}{(\lambda_1+\mu_1)(\lambda_2+\mu_2)}e^{-(\lambda_1+\mu_1+\lambda_2+\mu_2)t}. \tag{11.129}$$

The long-term availability is:

$$A_{lt} = \lim_{t\to\infty} A(t)$$

$$A_{lt} = 1 - \frac{\lambda_1\lambda_2}{(\lambda_1+\mu_1)(\lambda_2+\mu_2)}.$$

11.4.5 Reliability

Since State 4 is the system failure state, we consider State 4 in Equation (11.108) to be an absorbing state when calculating reliability, that is, transitions out of State 4 become 0:

$$\begin{bmatrix} \dot{P}_1 \\ \dot{P}_2 \\ \dot{P}_3 \\ \dot{P}_4 \end{bmatrix} = \begin{bmatrix} -(\lambda_1+\lambda_2) & \mu_2 & \mu_1 & 0 \\ \lambda_2 & -(\lambda_1+\mu_2) & 0 & 0 \\ \lambda_1 & 0 & -(\lambda_2+\mu_1) & 0 \\ 0 & \lambda_1 & \lambda_2 & 0 \end{bmatrix} \cdot \begin{bmatrix} P_1 \\ P_2 \\ P_3 \\ P_4 \end{bmatrix}. \tag{11.130}$$

The state matrix can be truncated to remove State 4, since only States 1, 2, and 3 are used to calculate reliability. The reduced state matrix becomes:

$$\begin{bmatrix} \dot{P}_1 \\ \dot{P}_2 \\ \dot{P}_3 \end{bmatrix} = \begin{bmatrix} -(\lambda_1+\lambda_2) & \mu_2 & \mu_1 \\ \lambda_2 & -(\lambda_1+\mu_2) & 0 \\ \lambda_1 & 0 & -(\lambda_2+\mu_1) \end{bmatrix} \cdot \begin{bmatrix} P_1 \\ P_2 \\ P_3 \end{bmatrix}. \tag{11.131}$$

The Laplace transform of these equations is:

$$\begin{aligned} P_1^*(s) &= \frac{(s+\lambda_1+\mu_2)(s+\lambda_2+\mu_1)}{s^3+(2\lambda_1+2\lambda_2+\mu_1+\mu_2)s^2+(3\lambda_1\lambda_2+\lambda_1^2+\lambda_2^2+\lambda_1\mu_1 + \lambda_2\mu_2+\mu_1\mu_2+\lambda_2\mu_1+\lambda_1\mu_2)s+\lambda_1\lambda_2(\lambda_1+\lambda_2+\mu_1+\mu_2)} \\ P_2^*(s) &= \frac{\lambda_2(s+\lambda_2+\mu_1)}{s^3+(2\lambda_1+2\lambda_2+\mu_1+\mu_2)s^2+(3\lambda_1\lambda_2+\lambda_1^2+\lambda_2^2+\lambda_1\mu_1 + \lambda_2\mu_2+\mu_1\mu_2+\lambda_2\mu_1+\lambda_1\mu_2)s+\lambda_1\lambda_2(\lambda_1+\lambda_2+\mu_1+\mu_2)} \\ P_3^*(s) &= \frac{\lambda_1(s+\lambda_1+\mu_2)}{s^3+(2\lambda_1+2\lambda_2+\mu_1+\mu_2)s^2+(3\lambda_1\lambda_2+\lambda_1^2+\lambda_2^2+\lambda_1\mu_1 + \lambda_2\mu_2+\mu_1\mu_2+\lambda_2\mu_1+\lambda_1\mu_2)s+\lambda_1\lambda_2(\lambda_1+\lambda_2+\mu_1+\mu_2)}. \end{aligned} \tag{11.132}$$

The explicit solution to these equations in terms of the symbolic values λ_1, λ_2, μ_1, and μ_2 becomes unwieldy. So let us substitute the values for the failure rates and repair rates from the previous example into the equations above and solve:

$$P_1(t) = (0.9996)e^{-(4.7973\text{E-}09)t} + (0.00031969)e^{-0.06253t} + 0.000080019e^{-0.12503t}$$

$$P_2(t) = 0.00031982e^{-(4.7973\text{E-}09)t} - 0.0003198e^{-0.06253t} + (2.5594\text{E-}8)e^{-0.12503t}$$

$$P_3(t) = 0.000079955e^{-(4.7973\text{E-}09)t} + (5.1175\text{E-}6)e^{-0.06253t} - 0.000080006e^{-0.12503t}$$

$$R(t) = P_1(t) + P_2(t) + P_3(t)$$

$$R(t) = 0.99999e^{-(4.7973\text{E-}09)t} + 0.0000050075e^{-0.06253t} + 0.0001650581e^{-0.12503t}.$$

If we assume both the failure rates and the repair rates are identical, then the Laplace transform equations above simplify to:

$$P_1^*(s) = \frac{(s+\lambda+\mu)^2}{s^3+(4\lambda+2\mu)s^2+(5\lambda^2+4\lambda\mu+\mu^2)s+\lambda^2(2\lambda+2\mu)}$$
$$P_2^*(s) = \frac{\lambda(s+\lambda+\mu)}{s^3+(4\lambda+2\mu)s^2+(5\lambda^2+4\lambda\mu+\mu^2)s+\lambda^2(2\lambda+2\mu)}$$
$$P_3^*(s) = \frac{\lambda(s+\lambda+\mu)}{s^3+(4\lambda+2\mu)s^2+(5\lambda^2+4\lambda\mu+\mu^2)s+\lambda^2(2\lambda+2\mu)}.$$

We need the roots of the equations above to solve for the inverse Laplace transform. The above equations written with the solved roots are:

$$P_1^*(s) = \frac{(s+\lambda+\mu)}{\left(s+3\lambda/2+\mu/2-\frac{1}{2}\sqrt{\lambda^2+6\lambda\mu+\mu^2}\right)\left(s+3\lambda/2+\mu/2+\frac{1}{2}\sqrt{\lambda^2+6\lambda\mu+\mu^2}\right)}$$
$$P_2^*(s) = \frac{\lambda}{\left(s+3\lambda/2+\mu/2-\frac{1}{2}\sqrt{\lambda^2+6\lambda\mu+\mu^2}\right)\left(s+3\lambda/2+\mu/2+\frac{1}{2}\sqrt{\lambda^2+6\lambda\mu+\mu^2}\right)}$$
$$P_3^*(s) = \frac{\lambda}{\left(s+3\lambda/2+\mu/2-\frac{1}{2}\sqrt{\lambda^2+6\lambda\mu+\mu^2}\right)\left(s+3\lambda/2+\mu/2+\frac{1}{2}\sqrt{\lambda^2+6\lambda\mu+\mu^2}\right)}.$$

Let us choose $\lambda = 0.00001$ and $\mu = 0.125$. Substituting the values for the failure rates and repair rates into the equations above, we get:

$$P_1^*(s) = \frac{(s+0.1251)}{(s+1.599616E\text{-}9)(s+0.12503)}$$
$$P_2^*(s) = \frac{0.00001}{(s+1.599616E\text{-}9)(s+0.12503)}$$
$$P_3^*(s) = \frac{0.00001}{(s+1.599616E\text{-}9)(s+0.12503)}.$$

Taking the inverse Laplace transform of the above equations, the complete dynamic solution for $P_1(t)$, $P_2(t)$, and $P_3(t)$ is:

$$P_1(t) = 1.0006e^{-(1.5996\text{E-}9)t} - (5.5988\text{E-}4)e^{-0.12503t}$$
$$P_2(t) = 0.000079981e^{-(1.5996\text{E-}9)t} - 0.000079981e^{-0.12503t}$$
$$P_3(t) = 0.000079981e^{-(1.5996\text{E-}9)t} - 0.000079981e^{-0.12503t}.$$

The reliability is:

$$R(t) = P_1(t) + P_2(t) + P_3(t)$$
$$R(t) = 1.0007599e^{-(1.5996\text{E-}9)t} - (7.19842\text{E-}4)e^{-0.12503t}.$$

11.4.6 Mean Time to First Failure

The reliability of this system is equal to the probability of the system being in States 1, 2, or 3 given that State 4 is transformed into an absorbing state. Using Equation (11.137), we get:

$$R^*(s) = P_1^*(s) + P_2^*(s) + P_3^*(s)$$

$$R^*(s) = \frac{(s+\lambda_1+\mu_2)(s+\lambda_2+\mu_1)+\lambda_2(s+\lambda_2+\mu_1)+\lambda_1(s+\lambda_2+\mu_1)}{s^3+(2\lambda_1+2\lambda_2+\mu_1+\mu_2)s^2+(3\lambda_1\lambda_2+\lambda_1^2+\lambda_2^2+\lambda_1\mu_1 + \lambda_2\mu_2+\mu_1\mu_2+\lambda_2\mu_2+\lambda_1\mu_2)s+\lambda_1\lambda_2(\lambda_1+\lambda_2+\mu_1+\mu_2)} \qquad (11.133)$$

$$R^*(0) = P_1^*(0) + P_2^*(0) + P_3^*(0)$$

$$\text{MTTFF} = \frac{(\lambda_1+\mu_2)(\lambda_2+\mu_1)+\lambda_2(\lambda_2+\mu_1)+\lambda_1(\lambda_2+\mu_1)}{\lambda_1\lambda_2(\lambda_1+\lambda_2+\mu_1+\mu_2)}.$$

If the failure rates and repair rates are equal, we get:

$$R^*(0) = \frac{(\lambda+\mu)^2+2\lambda(\lambda+\mu)}{\lambda^2(2\lambda+2\mu)}$$

$$R^*(0) = \frac{3\lambda+\mu}{2\lambda^2}.$$

As a check, compare with Equation (11.84), the parallel system with repair—identical components example, and note that the MTTFFs are the same.

11.4.7 Asymptotic Behavior

If we have the complete time dependent state probability equations, as we obtained in Equations (11.114)–(11.117), we can obtain the steady-state probability distributions by allowing t to be very large. The initial transient behavior will be removed, and what remains is the constant probability distribution for each state. For Equations (11.114)–(11.117), when t is very large, the exponential factor becomes very small and the equations reduce to:

$$P_1(t) = \frac{\mu_1\mu_2}{(\mu_1+\lambda_1)(\mu_2+\lambda_2)} \qquad (11.134)$$

$$P_2(t) = \frac{\lambda_2\mu_1}{(\mu_1+\lambda_1)(\mu_2+\lambda_2)} \qquad (11.135)$$

$$P_3(t) = \frac{\lambda_1\mu_2}{(\mu_1+\lambda_1)(\mu_2+\lambda_2)} \qquad (11.136)$$

$$P_4(t) = \frac{\lambda_1\lambda_2}{(\mu_1+\lambda_1)(\mu_2+\lambda_2)}. \qquad (11.137)$$

Alternatively, if we are not interested in the complete solution but rather the long-term behavior of the system over time, we do not need to go through the laborious exercise of obtaining the complete solution. We start with the state transition matrix equations we previously created (Eq. (11.109)).

To obtain the steady-state behavior, we set the change in probability (derivatives) for each state to 0. The matrix then becomes:

$$\begin{bmatrix} 1 \\ 0 \\ 0 \\ 0 \end{bmatrix} = \begin{bmatrix} 1 & 1 & 1 & 1 \\ \lambda_2 & -(\lambda_1+\mu_2) & 0 & \mu_1 \\ \lambda_1 & 0 & -(\lambda_2+\mu_1) & \mu_2 \\ 0 & \lambda_1 & \lambda_2 & -(\mu_1+\mu_2) \end{bmatrix} \cdot \begin{bmatrix} P_1 \\ P_2 \\ P_3 \\ P_4 \end{bmatrix}. \tag{11.138}$$

The state equations are:

$$1 = P_1 + P_2 + P_3 + P_4 \tag{11.139}$$

$$0 = \lambda_2 P_1 - (\lambda_1 + \mu_2) P_2 + \mu_1 P_4 \tag{11.140}$$

$$0 = \lambda_1 P_1 - (\lambda_2 + \mu_1) P_3 + \mu_2 P_4 \tag{11.141}$$

$$0 = \lambda_1 P_2 + \lambda_2 P_3 - (\mu_1 + \mu_2) P_4. \tag{11.142}$$

Let us solve for P_1 in terms of the other probabilities in the first equation and substitute into the other three equations to obtain:

$$0 = \lambda_2 (1 - P_2 - P_3 - P_4) - (\lambda_1 + \mu_2) P_2 + \mu_1 P_4 \tag{11.143}$$

$$0 = \lambda_1 (1 - P_2 - P_3 - P_4) - (\lambda_2 + \mu_1) P_3 + \mu_2 P_4 \tag{11.144}$$

$$0 = \lambda_1 P_2 + \lambda_2 P_3 - (\mu_1 + \mu_2) P_4. \tag{11.145}$$

Rearranging terms for each probability, we get:

$$0 = -(\lambda_1 + \mu_2 + \lambda_2) P_2 - \lambda_2 P_3 + (\mu_1 - \lambda_2) P_4 + \lambda_2 \tag{11.146}$$

$$0 = -\lambda_1 P_2 - (\lambda_1 + \lambda_2 + \mu_1) P_3 + (\mu_2 - \lambda_1) P_4 + \lambda_1 \tag{11.147}$$

$$0 = \lambda_1 P_2 + \lambda_2 P_3 - (\mu_1 + \mu_2) P_4. \tag{11.148}$$

Now solve Equation (11.148) for P_2:

$$P_2 = (\mu_1 + \mu_2)/\lambda_1 P_4 + -\lambda_2 P_3/\lambda_1 \tag{11.149}$$

In Equations (11.146) and (11.148), substitute Equation (11.149) for P_2:

$$0 = (\mu_2 - \lambda_1) P_4 - (\lambda_1 + \lambda_2 + \mu_1) P_3 - \lambda_1 ((\mu_1 + \mu_2)/\lambda_1 P_4 - \lambda_2 P_3/\lambda_1) + \lambda_1 \tag{11.150}$$

$$0 = (\mu_1 - \lambda_2) P_4 - \lambda_2 P_3 - (\lambda_1 + \mu_2 + \lambda_2)((\mu_1 + \mu_2)/\lambda_1 P_4 - \lambda_2 P_3/\lambda_1) + \lambda_2. \tag{11.151}$$

Rearranging terms to group P_4 and P_1 expressions and multiplying Equation (11.151) by λ_1:

$$0 = -(\mu_1 + \lambda_1)P_4 - (\lambda_1 + \mu_1)P_3 + \lambda_1 \tag{11.152}$$

$$0 = [\lambda_1(\mu_1 - \lambda_2) - (\lambda_1 + \mu_2 + \lambda_2)(\mu_1 + \mu_2)]P_4 + [(\lambda_1 + \mu_2 + \lambda_2)\lambda_2 - \lambda_2\lambda_1]P_3 + \lambda_1\lambda_2. \tag{11.153}$$

Multiply out terms for P_4 and P_3 and rearrange:

$$0 = -(\mu_2 + \lambda_2)(\lambda_1 + \mu_1 + \mu_2)P_4 + (\mu_2 + \lambda_2)\lambda_2 P_3 + \lambda_1\lambda_2. \tag{11.154}$$

Now solve Equation (11.154) for P_3:

$$P_3 = \frac{-\lambda_1\lambda_2}{\lambda_2(\lambda_2 + \mu_2)} + \frac{(\mu_2 + \lambda_2)(\lambda_1 + \mu_1 + \mu_2)}{\lambda_2(\lambda_2 + \mu_2)} P_4 \tag{11.155}$$

$$P_3 = \frac{\mu_1 + \mu_2 + \lambda_1}{\lambda_2} P_4 - \frac{\lambda_1}{\mu_2 + \lambda_2}. \tag{11.156}$$

Next, solve Equation (11.152) for P_3:

$$P_3 = \frac{-(\mu_1 + \lambda_1)}{\mu_1 + \lambda_1} P_4 + \frac{\lambda_1}{\mu_1 + \lambda_1} \tag{11.157}$$

$$P_3 = -P_4 + \frac{\lambda_1}{\mu_1 + \lambda_1}. \tag{11.158}$$

Finally, set Equations (11.156) and (11.158) equal to each other and solve for P_4:

$$P_4 = \frac{\lambda_1\lambda_2}{(\mu_1 + \lambda_1)(\mu_2 + \lambda_2)}.$$

Now the other solutions for P_1, P_2, and P_3 can be readily obtained, and the resulting equations are:

$$P_1 = \frac{\mu_1\mu_2}{(\mu_1 + \lambda_1)(\mu_2 + \lambda_2)} \tag{11.159}$$

$$P_2 = \frac{\mu_1\lambda_2}{(\mu_1 + \lambda_1)(\mu_2 + \lambda_2)} \tag{11.160}$$

$$P_3 = \frac{\lambda_1\mu_2}{(\mu_1 + \lambda_1)(\mu_2 + \lambda_2)} \tag{11.161}$$

$$P_4 = \frac{\lambda_1\lambda_2}{(\mu_1 + \lambda_1)(\mu_2 + \lambda_2)}. \tag{11.162}$$

Referring back to Equations (11.114)–(11.117), we see these equations are the same when the dynamic terms are removed. When t becomes very large, these transient terms become very small and can be ignored, leaving us with Equations (11.159)–(11.162) above.

Alternatively, we could start with Equation (11.138) and solve by noting:

$$\mathbf{U} = \begin{bmatrix} 1 \\ 0 \\ 0 \\ 0 \end{bmatrix} \tag{11.163}$$

$$\mathbf{A} = \begin{bmatrix} 1 & 1 & 1 & 1 \\ \lambda_2 & -(\lambda_1 + \mu_2) & 0 & \mu_1 \\ \lambda_1 & 0 & -(\lambda_2 + \mu_1) & \mu_2 \\ 0 & \lambda_1 & \lambda_2 & -(\mu_1 + \mu_2) \end{bmatrix} \tag{11.164}$$

$$\mathbf{P} = \begin{bmatrix} P_1 \\ P_2 \\ P_3 \\ P_4 \end{bmatrix} \tag{11.165}$$

$$\mathbf{U} = \mathbf{AP}. \tag{11.166}$$

EXAMPLE 11.8

Calculate the steady-state availability, unavailability, and Five 9s availability for a 2N redundant system with the MTBF and MTTR values from the previous example.

First, calculate the probability vector by substituting the values for λ_1, μ_1, λ_2, and μ_2 into Equations (11.159)–(11.162):

$$P_1 = 0.999600124 \tag{11.167}$$

$$P_2 = 0.000319872 \tag{11.168}$$

$$P_3 = 7.9968\text{E-}5 \tag{11.169}$$

$$P_4 = 2.55898\text{E-}8. \tag{11.170}$$

Referring back to Equations (11.125)–(11.128), the equations above are identical when the transient terms are removed.

Substituting the values for λ_1, μ_1, λ_2, and μ_2 into the state transition matrix, we obtain:

$$\mathbf{A} = \begin{bmatrix} 1 & 1 & 1 & 1 \\ 0.00002 & -0.06251 & 0 & 0.125 \\ 0.00001 & 0 & -0.12502 & 0.0625 \\ 0 & 0.00001 & 0.00002 & -0.1875 \end{bmatrix}. \tag{11.171}$$

Taking the inverse of this matrix using Matlab:

$$\mathbf{A}^{-1} = \begin{bmatrix} 0.9996 & 15.9940 & 7.9985 & 18.6601 \\ 0.0003 & -15.9940 & 0.0009 & -10.6607 \\ 0.0001 & 0.0009 & -7.9985 & -2.6652 \\ 0.0000 & -0.0009 & -0.0009 & -5.3342 \end{bmatrix}. \tag{11.172}$$

We can solve for **P** by taking the inverse of the transition matrix **A** and multiply by **U**:

$$\mathbf{P} = \mathbf{A}^{-1}\mathbf{U} \tag{11.173}$$

$$\begin{bmatrix} P_1 \\ P_2 \\ P_3 \\ P_4 \end{bmatrix} = \begin{bmatrix} 0.9996 \\ 0.0003 \\ 7.9968E-05 \\ 2.55898E-08 \end{bmatrix}. \tag{11.174}$$

This result agrees with Equations (11.167)–(11.170).

The system is available if the system is in States 1, 2, or 3:

$$\begin{aligned} A_S &= P_1 + P_2 + P_3 \\ A_S &= 0.9996 + 0.0003 + 7.99968\text{E}{-05} \\ A_S &= 0.99999997 \end{aligned} \tag{11.175}$$

The Five 9s value is calculated from Equation (7.13):

$$\begin{aligned} \text{Five 9s} &= -\log(10, 1 - 0.99999997) \\ \text{Five 9s} &= 7.592. \end{aligned} \tag{11.176}$$

The unavailability is:

$$\begin{aligned} Q_S &= P_4 \\ Q_s &= 2.55898\text{E}{-08}. \end{aligned} \tag{11.177}$$

11.4.8 Mean Time to Failure

Assuming the system is in an operational state, what is the mean time to failure?

Using Equations (9.43) and (9.46) and noting we have three functional states and one failed state, the MTTF is:

$$\begin{aligned} \text{MTTF}_S &= \frac{P_1 + P_2 + P_3}{P_2\beta_{24} + P_3\beta_{34}} \\ \text{MTTF}_S &= \frac{\mu_1\mu_2 + \lambda_2\mu_1 + \mu_2\lambda_1}{(\mu_1 + \mu_2)\lambda_1\lambda_2}. \end{aligned} \tag{11.178}$$

Substituting the failure rates and repair rates into the above equation, we obtain:

$$\text{MTTF}_S = 208{,}416{,}672 \text{ hours.}$$

11.4.9 Mean Time to Repair

If the system is in a failed state, what is the mean time to repair the system and restore it to operational condition?

From Equations (9.38) and (9.42), and noting we have three functional states and one failed state, the MTTR_S is:

$$\begin{aligned}
\text{MTTR}_S &= \frac{1}{\beta_{43}+\beta_{42}} \\
\text{MTTR}_S &= \frac{1}{\alpha_4} \\
\text{MTTR}_S &= \frac{1}{\mu_1+\mu_2}.
\end{aligned} \tag{11.179}$$

If the system is nonrepairable, then $\mu = 0$ and $\text{MTTR}_S = \infty$ (i.e., never repaired), as we would expect.

11.4.10 Mean Time between Failures

The mean time between system failures, MTBF_S, is the mean time between consecutive transitions from a functioning state (G) into a failed state (F). The MTBFs can be calculated from Equations (9.37) and (9.35), and noting we have three functional states and one failed state:

$$\begin{aligned}
\text{MTBF}_S &= 1/(P_2\beta_{24}+P_3\beta_{34}) \\
\text{MTBF}_S &= \frac{(\lambda_2+\mu_2)(\lambda_1+\mu_1)}{(\mu_1+\mu_2)\lambda_1\lambda_2}.
\end{aligned} \tag{11.180}$$

We can check the MTBF by using the relationship MTBF = MTTR + MTTF:

$$\begin{aligned}
\text{MTBF}_S &= \text{MTTF}_S + \text{MTTR}_S \\
\text{MTBF}_S &= \frac{\mu_1\mu_2+\lambda_2\mu_1+\mu_2\lambda_1}{(\mu_1+\mu_2)\lambda_1\lambda_2} + \frac{1}{\mu_1+\mu_2} \\
\text{MTBF}_S &= \frac{(\lambda_2+\mu_2)(\lambda_1+\mu_1)}{(\mu_1+\mu_2)\lambda_1\lambda_2}.
\end{aligned} \tag{11.181}$$

Alternatively, we can calculate the MTBF by using Equations (9.35)–(9.37):

$$\begin{aligned}
\text{MTBF}_S &= 1/(P_4\alpha_4) \\
\text{MTBF}_S &= \frac{(\lambda_2+\mu_2)(\lambda_1+\mu_1)}{(\mu_1+\mu_2)\lambda_1\lambda_2}.
\end{aligned} \tag{11.182}$$

If $\lambda_1 \ll \mu_1$ and $\lambda_2 \ll \mu_2$, then Equation (11.182) reduces to:

$$\text{MTBF}_S = \frac{\mu_1\mu_2}{(\mu_1+\mu_2)\lambda_1\lambda_2}. \quad (11.183)$$

If $\lambda_1 \ll \mu_1$ and $\lambda_2 \ll \mu_2$ and the repair rates and failure rates of both components are the same, then Equation (11.182) reduces to:

$$\text{MTBF}_S = \frac{\mu}{\lambda^2}. \quad (11.184)$$

11.4.11 System Availability

From Equation (9.47), system availability is the sum of the probabilities of all the good states. We have three good states: P_1, P_2, and P_3:

$$\begin{aligned}
A_S &= P_1 + P_2 + P_3 \\
A_S &= \frac{\mu_1\mu_2}{(\lambda_1+\mu_1)(\lambda_2+\mu_2)} + \frac{\mu_1\lambda_2}{(\lambda_1+\mu_1)(\lambda_2+\mu_2)} + \frac{\mu_2\lambda_1}{(\lambda_1+\mu_1)(\lambda_2+\mu_2)} \\
A_S &= \frac{\mu_1\mu_2+\mu_1\lambda_2+\mu_2\lambda_1}{(\lambda_1+\mu_1)(\lambda_2+\mu_2)}.
\end{aligned} \quad (11.185)$$

If $\lambda_1 \ll \mu_1$ and $\lambda_2 \ll \mu_2$:

$$A_S \approx \frac{\mu_1\mu_2 - \lambda_2\lambda_1}{\mu_1\mu_2}. \quad (11.186)$$

EXAMPLE 11.9

A 2N redundant system has an active card with an MTBF of 100,000 hours and an MTTR of 8 hours. The standby card takes over the functionality of the active card when the standby card has an MTBF of 50,000 hours and an MTTR of 16 hours. What is the long-term availability of the system?

The failure and repair rates of the two cards are:

$$\begin{aligned}
\lambda_1 &= 1/\text{MTBF}_1 = 0.00001 \\
\mu_1 &= 1/\text{MTTR}_1 = 0.125 \\
\lambda_2 &= 1/\text{MTBF}_2 = 0.00002 \\
\mu_2 &= 1/\text{MTTR}_2 = 0.0625.
\end{aligned}$$

From Equation (11.185):

$$A_S = \frac{(0.125)(0.0625)+(0.125)(0.00002)+(0.0625)(0.00001)}{(0.00001+0.125)(0.00002+0.0625)}$$
$$A_S = 0.999999974.$$

We obtain a very high availability for this simple system.

Now we calculate the availability using the approximate Equation (11.186):

$$A_S = \frac{(0.125)(0.0625)-(0.00002)(0.00001)}{(0.125)(0.0625)}$$
$$A_S = 0.999999974,$$

which provides the same answer.

11.4.12 Markov Reward Model

For the MRM in this example, the system is operational (available) when in States 1, 2, and 3 and is failed (unavailable) when in State 4. From Equation (9.50), the availability is:

$$A_S = E[X(t)] = \sum_i r_i P_i(t) = 1\cdot P_1 + 1\cdot P_2 + 1\cdot P_3 + 0\cdot P_4$$
$$A_S = \frac{\mu_1\mu_2 + \mu_1\lambda_2 + \mu_2\lambda_1}{(\lambda_1+\mu_1)(\lambda_2+\mu_2)}. \qquad (11.187)$$

11.4.13 System Unavailability

From Equation (9.46), system unavailability is the sum of the probabilities of all the bad states. We have one bad state, P_4:

$$Q_S = P_4$$
$$Q_S = \frac{\lambda_1\lambda_2}{(\lambda_1+\mu_1)(\lambda_2+\mu_2)}. \qquad (11.188)$$

As a check, $Q_S = 1 - A_S$, which gives the same result.

11.4.14 Calculating State Probabilities Using Component Availability

Let us next consider the probability of being in any of the four states.

The probability of being in State 1 is equal to the probability of component 1 functioning and the probability of component 2 functioning. The availability of a component in the steady-state condition is the probability that the component is functioning. Recall that the definition of steady-state availability is:

$$A = \text{MTTF}/(\text{MTTF} + \text{MTTR}),$$

and unavailability, Q, is 1 minus the availability:

$$Q = 1 - A = 1 - \text{MTTF}/(\text{MTTF} + \text{MTTR})$$

$$Q = \text{MTTR}/(\text{MTTF} + \text{MTTR}).$$

MTTF is the reciprocal of the failure rate λ and MTTR is the reciprocal of the repair rate μ:

$$\text{MTTF} = 1/\lambda$$

$$\text{MTTR} = 1/\mu.$$

Availability and unavailability for a component in terms of λ and μ for that component is therefore:

$$A = \frac{\mu}{\lambda + \mu}$$

$$Q = \frac{\lambda}{\lambda + \mu}.$$

Thus, the probability of being in State 1 is the availability of component 1 multiplied by the availability of component 2:

$$P_1 = A_1 A_2 = \frac{\mu_1}{\lambda_1 + \mu_1} \frac{\mu_2}{\lambda_2 + \mu_2}$$
$$P_1 = A_1 A_2 = \frac{\mu_1 \mu_2}{(\lambda_1 + \mu_1)(\lambda_2 + \mu_2)}. \qquad (11.189)$$

Now, based on the definitions for the other two states, we have:

$$P_2 = Q_1 A_2 = \frac{\mu_1 \lambda_2}{(\lambda_1 + \mu_1)(\lambda_2 + \mu_2)} \qquad (11.190)$$

$$P_3 = Q_1 A_2 = \frac{\mu_2 \lambda_1}{(\lambda_1 + \mu_1)(\lambda_2 + \mu_2)} \qquad (11.191)$$

$$P_4 = Q_1 Q_2 = \frac{\lambda_1 \lambda_2}{(\lambda_1 + \mu_1)(\lambda_2 + \mu_2)}. \qquad (11.192)$$

Compare these equations with Equation (11.159) through Equation (11.162). You will note they are the same (which we would expect).

11.4.15 Duration and Frequency of State Visits

The mean duration of visits to each state from Figure 11.11 and using Equation (9.28):

$$T_4 = 1/(\mu_1 + \mu_2) \quad (11.193)$$

$$T_2 = 1/(\lambda_1 + \mu_2) \quad (11.194)$$

$$T_3 = 1/(\mu_1 + \lambda_2) \quad (11.195)$$

$$T_1 = 1/(\lambda_1 + \lambda_2). \quad (11.196)$$

Note that Equation (11.193) represents the average steady-state duration of system failures since State 4 is the system failed state.

Employing Equation (9.30), the frequency of visiting each state is:

$$f_4 = P_4(\mu_1 + \mu_2) \quad (11.197)$$

$$f_2 = P_2(\lambda_1 + \mu_2) \quad (11.198)$$

$$f_3 = P_3(\mu_1 + \lambda_2) \quad (11.199)$$

$$f_1 = P_1(\lambda_1 + \lambda_2). \quad (11.200)$$

Equation (11.197) is the frequency of system failures since State 4 is the system failed state.

EXAMPLE 11.10

Consider a two-component system with $\lambda_1 = 0.001, \lambda_2 = 0.002$ and $\mu_1 = 0.125, \mu_2 = 0.0625$.

(a) What are the dynamic system reliability, availability, and unavailability over a time frame of 50 and 6000 hours?

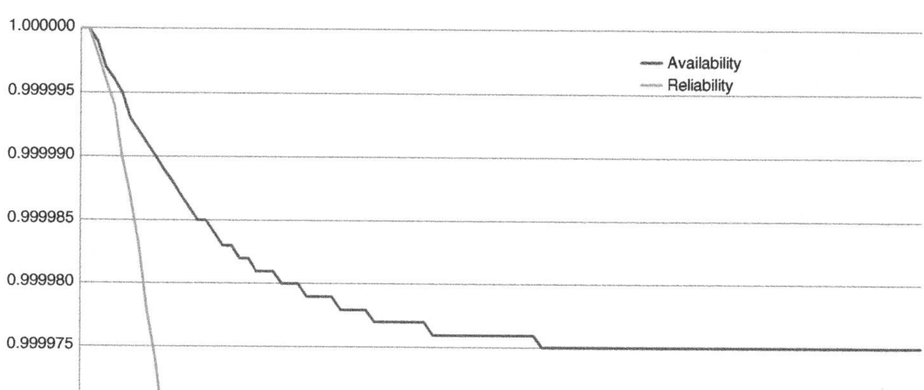

Figure 11.16 Two-Component Parallel System with Repair: Different Failure and Repair Rates (0–50 hours)

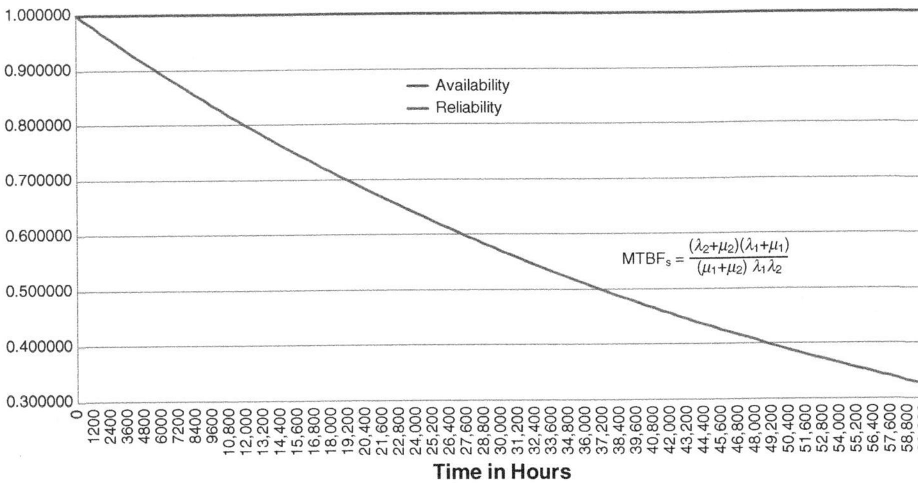

Figure 11.17 Two-Component Parallel System with Repair: Different Failure and Repair Rates (0–60,000 hours)

TABLE 11.1 Dynamic Model Comparisons for Repairable Systems

	Dynamic Behavior	
Availability Model	Reliability	Availability
One component with repair	$e^{-\lambda t}$	$\frac{\mu}{\lambda+\mu}+\frac{\lambda}{\lambda+\mu}e^{-(\lambda+\mu)t}$
Parallel system with repair—identical component failure and repair rates	$\frac{3\lambda+\mu-r_1}{r_2-r_1}e^{r_1 t}+\frac{3\lambda+\mu-r_2}{r_1-r_2}e^{r_2 t}$	$1-\frac{\lambda^2}{(\mu+\lambda)^2}+\frac{2\lambda^2}{(\mu+\lambda)^2}e^{-(\mu+\lambda)t}-\frac{\lambda^2}{(\mu+\lambda)^2}e^{-2(\mu+\lambda)t}$
Parallel system with repair—different Component failure and repair rates	$0.99999e^{-(4.7973\text{E-}09)t}+0.0000050075e^{-0.06253t}+0.0001650581e^{-0.12503t}$ (This is the numeric solution example)	$1-\frac{\lambda_1\lambda_2}{(\lambda_1+\mu_1)(\lambda_2+\mu_2)}-\frac{\lambda_1\lambda_2}{(\lambda_1+\mu_1)(\lambda_2+\mu_2)}e^{-(\lambda_1+\mu_1)t}-\frac{\lambda_1\lambda_2}{(\lambda_1+\mu_1)(\lambda_2+\mu_2)}e^{-(\lambda_2+\mu_2)t}+\frac{\lambda_1\lambda_2}{(\lambda_1+\mu_1)(\lambda_2+\mu_2)}e^{-(\lambda_1+\mu_1+\lambda_2+\mu_2)t}$

TABLE 11.2 Static Model Comparisons for Repairable Systems

	Asymptotic Behavior			
Availability Model	MTTF	MTBF	MTTR	Availability
One component with repair	$\frac{1}{\lambda}$	$\frac{1}{\lambda}+\frac{1}{\mu}$	$\frac{1}{\mu}$	$\frac{\mu}{\lambda+\mu}$
Parallel system with repair—identical component failure and repair rates	$\frac{1}{\lambda}+\frac{\mu}{2\lambda^2}$	$\frac{(\lambda+\mu)^2}{2\lambda^2\mu}$	$\frac{1}{2\mu}$	$\frac{2\lambda\mu+\mu^2}{(\lambda+\mu)^2}$
Parallel system with repair—different component failure and repair rates	$\frac{\mu_1\mu_2+\lambda_2\mu_1+\mu_2\lambda_1}{(\mu_1+\mu_2)\lambda_1\lambda_2}$	$\frac{(\lambda_2+\mu_2)(\lambda_1+\mu_1)}{(\mu_1+\mu_2)\lambda_1\lambda_2}$	$\frac{1}{\mu_1+\mu_2}$	$\frac{\mu_1\mu_2+\mu_1\lambda_2+\mu_2\lambda_1}{(\lambda_1+\mu_1)(\lambda_2+\mu_2)}$

(b) What are the MTTFF and MTBF of the system?

(c) Plot the results.

We solve by using Equation (11.21) for availability, Equation (11.133) for MTTFF, and Equation (11.181) for the MTBF value. Dynamic reliability can be obtained by creating the state diagram and plotting the calculation results using Relex.

Figure 11.16 and Figure 11.17 illustrate the relationships between availability, reliability, MTTF, and asymptotic behavior.

11.5 SUMMARY

This chapter introduced applications of Markov analysis to repairable systems and illustrated both dynamic and steady-state responses for repairable systems with different configurations. Table 11.1 and Table 11.2 summarize the reliability characteristics for the different configurations we considered.

CHAPTER 12

Analyzing Confidence Levels

12.1 INTRODUCTION

The *chi-square distribution* (or χ^2 *distribution*) is one of the most widely used probability distributions in inferential statistics. Karl Pearson, one of the key contributors to modern statistics, developed the chi-square (χ^2) statistical technique in 1900.

We will show that a random variable has a chi-square distribution if the random variable is the sum of the squares of a set of statistically independent normal random variables. We apply this knowledge to "Goodness-of-Fit" tests, which helps us evaluate how likely our assumption is of the underlying probability distribution based on test information we have available.

We will also explore an important relationship between Poisson distributions and chi-square distributions that will allow us to describe the confidence level of reliability parameters—in particular, MTBF values based on available test data.

The chi-square distributions are a family of distributions, with each distribution identified by a parameter known as the number of degrees of freedom (df). Figure 12.1 shows the *pdf* of chi-square distributions for various degrees of freedom.

The chi-square distribution is a special case of the gamma distribution. We will introduce the gamma distribution and show the relationship to the chi-square distribution.

12.2 *pdf* OF A SQUARED NORMAL RANDOM VARIABLE

If $X_1, \ldots, X_k$ are k independent, normally distributed random variables with mean 0 and variance 1, then the random variable

$$Q = \sum_{i=1}^{k} \chi_i^2 \tag{12.1}$$

is distributed according to the chi-square distribution with k degrees of freedom.

Designing High Availability Systems: Design for Six Sigma and Classical Reliability Techniques with Practical Real-Life Examples, First Edition. Zachary Taylor and Subramanyam Ranganathan.

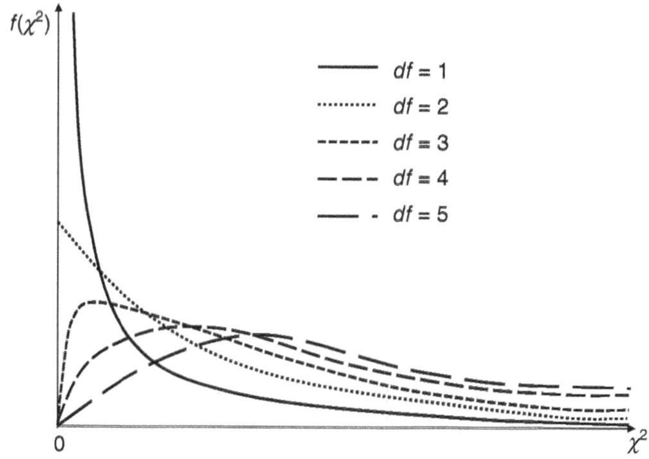

Figure 12.1 Chi-Square Probability Distribution

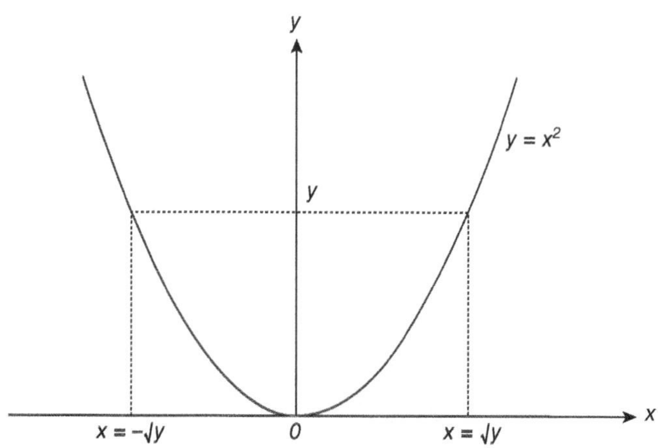

Figure 12.2 The Square of a Random Variable

To prove Equation (12.1), let us start first with finding the *pdf* of the square of a random variable for any general distribution. Let X be a random variable with *pdf* $f_x(x)$. Let $Y = X^2$. We then find the *pdf* of Y. The event $A = (Y \le y)$ is equivalent to the event $B = \left(-\sqrt{y} \le X \le \sqrt{y}\right)$ in Figure 12.2.

Since y is the square of x, it can never be negative. Thus, the probability $y \le 0$ is always zero and the cumulative distribution function of Y is:

$$F_Y(y) = P(Y \le y) = 0, \quad \text{for } y \le 0.$$

If $y \le 0$, the *pdf* of Y is:

$$f_Y(y) = 0, \quad \text{for } y \le 0.$$

If $y > 0$, then

$$F_Y(y) = P(Y \le y) = P\left(-\sqrt{y} \le X \le \sqrt{y}\right) = F_X\left(\sqrt{y}\right) - F_X\left(-\sqrt{y}\right). \tag{12.2}$$

If $y > 0$, the *pdf* of Y is:

$$\begin{aligned} f_Y(y) &= \frac{d}{dy}F_Y(y) = \frac{d}{dy}F_X\left(\sqrt{y}\right) - \frac{d}{dy}F_X\left(-\sqrt{y}\right) \\ f_Y(y) &= \frac{1}{2\sqrt{y}}\left(f_X\left(\sqrt{y}\right) - f_X\left(-\sqrt{y}\right)\right), \quad \text{for } y>0. \end{aligned} \tag{12.3}$$

Next, let us find the *pdf* of Y if X is a standard normally distributed random variable.

The *pdf* of a normally distributed random variable (from Eq. (5.33) repeated here):

$$f_X(x) = \frac{1}{\sqrt{2\pi\sigma^2}} e^{\frac{-(x-\mu)^2}{2\sigma^2}}. \tag{12.4}$$

The *pdf* of a normally distributed random variable with a mean of 0 and standard deviation of 1 is called the standard normal distribution:

$$f_X(x) = \frac{1}{\sqrt{2\pi}} e^{\frac{-x^2}{2}}. \tag{12.5}$$

Substituting $f_X(x)$ into Equation (12.3), we obtain:

$$f_Y(y) = \frac{1}{\sqrt{y}} f_X\left(\sqrt{y}\right) = \frac{1}{\sqrt{2\pi y}} e^{\frac{-y}{2}}, \quad y > 0 \tag{12.6}$$

$$f_Y(y) = 0, \quad y \le 0. \tag{12.7}$$

Next, let us create a general sum of these squared random variables and find the *pdf* of the resulting random variable.

Let $X_1, \ldots, X_n$ be n independent standard normal random variables. The sum of these squared random variables is:

$$Y = X_1^2 + \cdots + X_n^2 = \sum_{i=1}^{n} X_i^2. \tag{12.8}$$

We now show the relationship between Equation (12.8) and the gamma function. A gamma random variable has two parameters (α, λ), where α is called the shape parameter and λ is the rate parameter. The *pdf* of a gamma random variable x is defined as:

$$f_X(x) = \frac{\lambda e^{-\lambda x}(\lambda x)^{\alpha-1}}{\Gamma(\alpha)}, \quad x > 0. \tag{12.9}$$

where the gamma function is defined as:

$$\Gamma(\alpha) = \int_0^\infty e^{-x} x^{\alpha-1} dx, \quad \alpha > 0. \tag{12.10}$$

Note that a gamma function has the following property:

$$\Gamma(1/2) = \sqrt{\pi}. \tag{12.11}$$

For brevity, we would not derive the above equation. Refer to mathematical texts for this common relationship.

We can rewrite Equation (12.6) to incorporate Equation (12.11) for a single term in Equation (12.8):

$$\begin{aligned} f_{Y_i}(y) &= \frac{1}{\sqrt{2\pi y}} e^{\frac{-y}{2}} = \frac{1}{2\sqrt{\pi}} e^{-\frac{y}{2}} (y/2)^{1/2-1} \\ f_{Y_i}(y) &= \frac{1}{2\Gamma(1/2)} e^{\frac{-y}{2}} (y/2)^{1/2-1}. \end{aligned} \tag{12.12}$$

We recognize the above as the *pdf* of a gamma random variable with parameters ($\alpha = ½$, $\lambda = ½$) (Eq. (12.9)). This equation is also a special form of the gamma distribution and is referred to as the chi-square (χ^2) density function with n degrees of freedom, where in this case, $n = 1$ (Eq. (5.29)). We conclude that the square of a standard normal variable is a chi-square random variable with one degree of freedom.

Our next step is to show that the sum of two independent gamma random variables is also a gamma random variable. But before we continue, we need to consider the general case of determining the *pdf* of the sum of any two random variables.

12.3 *pdf* OF THE SUM OF TWO RANDOM VARIABLES

We will now calculate the random variable $Z = X + Y$.

Given two random variables X and Y, a new random variable Z can be formed as a function of X and Y:

$$Z = g(X, Y)$$

$$F_Z(z) = P(Z \le z) = P(g(x, y) \le z).$$

Referring to Figure 12.3, in the xy plane, $x + y \le z$ is the shaded area to the left of the line $x + y \le z$ (z is a fixed value).

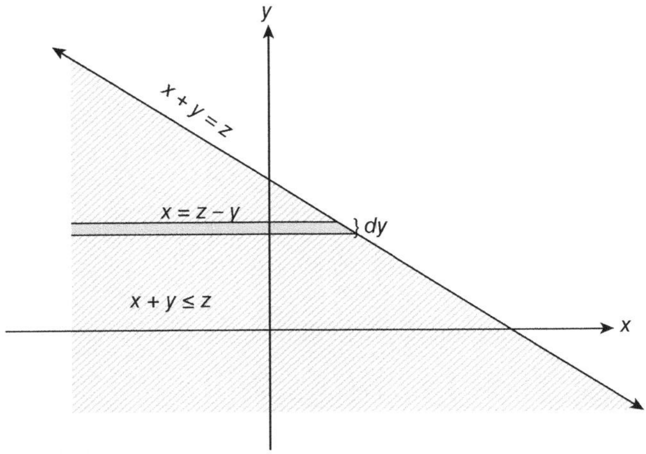

Figure 12.3 Evaluating the Integrals for $F(z)$

For the sum of two random variables X and Y, we have:

$$Z = g(X, Y) = X + Y$$

$$F_Z(z) = P(Z \le z) = P(X + Y \le z) = \int_{y=-\infty}^{\infty} \int_{x=-\infty}^{z-x} f_{xy}(x, y)dxdy. \tag{12.13}$$

This area is determined by first integrating from $x = -\infty$ to $x = z - y$ to obtain the horizontal strip shown (inner integral). We then take this horizontal strip and integrate vertically from $y = -\infty$ to $+\infty$ to obtain the entire shaded area (outer integral).

The *pdf* of Equation (12.13) is:

$$f_Z(z) = \frac{d}{dz} F_Z(z) = \int_{y=-\infty}^{\infty} \left[\frac{d}{dz} \int_{x=-\infty}^{z-y} f_{xy}(x, y)dx \right] dy \tag{12.14}$$

$$f_Z(z) = \int_{-\infty}^{\infty} f_{xy}(x, z - y)dy. \tag{12.15}$$

The above equation is also known as the *convolution* of $f_X(z)$ and $f_Y(z)$.

If the two random variables X and Y are independent, then the *pdf* of Z is:

$$f_Z(z) = f_{xy}(x, y) = f_X(x) f_Y(y). \tag{12.16}$$

Inserting Equation (12.16) into Equation (12.15), we obtain:

$$f_Z(z) = \int_{-\infty}^{\infty} f_X(x) f_Y(z - x)dx. \tag{12.17}$$

This integral sums the various ways that Z could equal any particular value z. Suppose $Y = t$; then Z has the value z *iff* $X = z - t$ and $Y = t$, for any value of t. We integrate the product of the densities of X and Y with respect to these quantities, over all possible values of t.

If $f_X(x) = 0$ for $x < 0$ and $f_Y(y) = 0$ for $y < 0$, then our new limits of integration for Equation (12.13) become:

$$F_Z(z) = \int_{y=0}^{z} \int_{z=0}^{z-y} f_{xy}(x, y)dxdy \tag{12.18}$$

$$f_Z(z) = \frac{d}{dz} F_Z(z) = \int_0^z \left[\frac{d}{dz} \int_0^{z-y} f_{xy}(x, y)dx \right] dy \tag{12.19}$$

$$f_Z(z) = \int_0^z f_X(x) f_Y(z-x)dx, z > 0. \tag{12.20}$$

12.4 *pdf* OF THE SUM OF TWO GAMMA RANDOM VARIABLES

Let X be a gamma random variable with parameters (α, λ) and Y be a gamma random variable with parameters (β, λ).

From Equation (12.9):

$$f_X(x) = \frac{\lambda e^{-\lambda x}(\lambda x)^{\alpha-1}}{\Gamma(\alpha)}, \quad x > 0 \tag{12.21}$$

$$f_Y(y) = \frac{\lambda e^{-\lambda y}(\lambda y)^{\beta-1}}{\Gamma(\beta)}, \quad y > 0. \tag{12.22}$$

Substituting Equations (12.21) and (12.22) into Equation (12.20), we obtain:

$$\begin{aligned} f_Z(z) &= \frac{1}{\Gamma(\alpha)\Gamma(\beta)} \int_0^z \lambda e^{-\lambda x}(\lambda x)^{\alpha-1} \lambda e^{-\lambda(z-x)}[\lambda(z-x)]^{\beta-1} dx \\ f_Z(z) &= \frac{\lambda^{\alpha+\beta}}{\Gamma(\alpha)\Gamma(\beta)} e^{-\lambda z} \int_0^z x^{\alpha-1}(z-x)^{\beta-1} dx. \end{aligned} \tag{12.23}$$

By substituting $w = x/z$ we have:

$$\begin{aligned} f_Z(z) &= \frac{\lambda^{\alpha+\beta}}{\Gamma(\alpha)\Gamma(\beta)} e^{-\lambda z} z^{\alpha+\beta-1} \int_0^1 w^{\alpha-1}(1-w)^{\beta-1} dw \\ f_Z(z) &= k e^{-\lambda z} z^{\alpha+\beta-1}. \end{aligned} \tag{12.24}$$

where k is a constant that does not depend on z. To find k, we note the following:

$$f_Z(z) = \int_{-\infty}^{\infty} f(x)dx = 1. \tag{12.25}$$

k is found by substituting Equation (12.24) into Equation (12.25):

$$\int_{-\infty}^{\infty} f_Z(z)dz = k\int_{0}^{\infty} e^{-\lambda z} z^{\alpha+\beta-1}dz. \tag{12.26}$$

Substitute v for λz:

$$\int_{-\infty}^{\infty} f_Z(z)dz = \frac{k}{\lambda^{\alpha+\beta}}\int_{0}^{\infty} e^{-v} v^{\alpha+\beta-1}dv. \tag{12.27}$$

The integral part of the above equation is a form of the gamma function. Substituting the definition of the gamma function (Eq. (12.10)), we obtain:

$$\frac{k}{\lambda^{\alpha+\beta}}\Gamma(\alpha+\beta) = 1. \tag{12.28}$$

Thus:

$$k = \frac{\lambda^{\alpha+\beta}}{\Gamma(\alpha+\beta)}. \tag{12.29}$$

And:

$$f_Z(z) = \frac{\lambda^{\alpha+\beta}}{\Gamma(\alpha+\beta)} e^{-\lambda z} z^{\alpha+\beta-1} \tag{12.30}$$

$$f_Z(z) = \frac{\lambda}{\Gamma(\alpha+\beta)} e^{-\lambda z} (\lambda z)^{\alpha+\beta-1}. \tag{12.31}$$

Thus Z is a gamma random variable with parameters $(\alpha+\beta, \lambda)$.

We conclude that the sum of two gamma random variables is a gamma random variable.

12.5 *pdf* OF THE SUM OF *n* GAMMA RANDOM VARIABLES

Our next step is to calculate the sum of n independent gamma random variables with parameters (α_i, λ), $i = 1, \ldots, n$. Let:

$$Y = X_1 + \cdots + X_n = \sum_{i=1}^{n} X_i. \tag{12.32}$$

We want to show that Y is also a gamma random variable with parameters:

$$\left(\sum_{i=1}^{n}\alpha_i, \lambda\right)$$

By induction, it suffices to prove it for two random variables: X of parameters (α, λ) and Y of parameters (β, λ).

Using induction, let us first assume that the sum for $n = k$ is a gamma random variable:

$$Z = X_1 + \cdots + X_k = \sum_{i=1}^{k} X_i, \tag{12.33}$$

with parameters

$$(\beta, \lambda) = \left(\sum_{i=1}^{n}\alpha_i, \lambda\right). \tag{12.34}$$

Next let:

$$W = Z + X_{k+1} = \sum_{i=1}^{k+1} X_i. \tag{12.35}$$

Since we have shown from Equation (12.31) that the sum of two gamma random variables with parameters (α, λ) and (β, λ) is a gamma random variable with parameters $(\alpha+\beta, \lambda)$, then W must be a gamma random variable with parameters $(\beta + \alpha_{k+1}, \lambda)$. And from Equation (12.33), also has the parameters:

$$\left(\sum_{i=1}^{k+1}\alpha_i, \lambda\right), \tag{12.36}$$

for the sum of gamma random variables from $n = 1$ to $n = \text{k} + 1$. Thus, the sum of gamma random variables for *any* number of gamma random variables for $n \geq 2$ must be true.

Another way to prove this is with the simpler case of considering:

$$Y = X_1 + X_2, \tag{12.37}$$

where X_1 and X_2 are gamma random variables with parameters (α_1, λ) and (α_2, λ), respectively. Based on Equation (12.31), we know that Y is a gamma random variable with parameters $(\alpha_1 + \alpha_2, \lambda)$. Now, let us add to the sum a third gamma random variable:

$$W = Y + X_3 \tag{12.38}$$

$$W = X_1 + X_2 + X_3. \tag{12.39}$$

Since the sum of two independent gamma random variables is a gamma random variable, from Equation (12.38), W must also be a gamma random variable. But Equation (12.39) must also be true, so W is a gamma random variable with parameters:

$$\left(\sum_{i=1}^{3} \alpha_i, \lambda\right). \tag{12.40}$$

We can continue this logic for any number of gamma random variables. Thus, the sum of any number of gamma random variables, n, must be a gamma random variable with parameters:

$$\left(\sum_{i=1}^{n} \alpha_i, \lambda\right). \tag{12.41}$$

And the *pdf* of a gamma random variable Y,which is a sum of n gamma random variables each with parameters (α, λ), is:

$$f_Y(y) = \frac{\lambda e^{-\lambda y}(\lambda y)^{n\alpha-1}}{\Gamma(n\alpha)}. \tag{12.42}$$

We previously determined that the *pdf* of the square of a standard normal random variable is a gamma random variable with parameters ($\alpha = ½$, $\lambda = ½$) (Eq. (12.12)).

$$f_{Y_i}(y) = \frac{\frac{1}{2}e^{-\lambda/2}(y/2)^{\frac{1}{2}-1}}{\Gamma(1/2)}. \tag{12.43}$$

A chi-square random variable with n degrees of freedom is:

$$\begin{aligned} F_Y(y) &= \frac{\frac{1}{2}e^{-y/2}(y/2)^{n/2-1}}{\Gamma(n/2)} \\ F_Y(y) &= \frac{\frac{1}{2}e^{-y/2}(y/2)^{n/2-1}}{2^{n/2}\Gamma(1/2)}, \quad \text{for } y > 0. \end{aligned} \tag{12.44}$$

When n is an even integer, $\Gamma(n/2) = [n/2-1]!$, whereas when n is odd, $\Gamma(n/2)$ can be obtained from $\Gamma(\alpha) = (\alpha - 1)\Gamma(\alpha - 1)$ and $\Gamma(1/2) = \sqrt{\pi}$.

Equation (12.44) (chi-square distribution with n degrees of freedom) is the same as the Equation (12.42) (gamma distribution with parameters ($n/2$, ½)).

Thus, from Equations (12.43) and (12.42), we can conclude that the sum of the squares of n independent standard normal random variables is a chi-square random variable with n degrees of freedom.

Now that we have hopefully convinced ourselves that the relationship defined in Equation (12.1) is true, how do we apply it? One important application is the Goodness of Fit (GoF) test.

12.6 GOODNESS-OF-FIT TEST USING CHI-SQUARE

When analyzing failure data, we make an assumption regarding the underlying probability distribution of the failures. The assumed distribution affects all of our subsequent calculations, and if the real distribution is different than our assumed distribution, our calculations and the conclusions we draw from these data may be wrong. Various techniques are available that provide a quantifiable degree of confidence that the data we are analyzing is a sample set that follows a known distribution.

One common test that can be used is a GoF test. GoF tests are generally divided into two categories:

1. *Distance Test.* Based on the cumulative distribution function,
2. *Area Test.* Based on the probability density function.

We will examine the well-known chi-square GoF test (area test).

Given a set of data, we first make an assumption regarding the *pdf* of the population from which the sampled data came from. We could test for a variety of probability distributions, such as the normal distribution, Weibull distribution, exponential distribution, uniform distribution, and so on. We will limit our discussion to testing against the exponential distribution assumption.

Next, we calculate or estimate the distribution statistics (i.e., mean, variance, etc.) either from the data or from our engineering judgment based on previous projects or experience.

We test the assumed (hypothesized) distribution using the data set. The hypothesis is rejected when the hypothesis is not supported by the data within the confidence levels required.

Intuitively, if we have a set of data that we believe follows a certain distribution curve; we overlay the data on the idealized theoretical curve and measure how far apart the real data are from the idealized distribution. One way to measure how far apart they are is to divide the idealized *pdf* into a set of bins or subintervals based on the available data and allocate the data into the appropriate bins. If the assumed distribution is correct, its *pdf* should closely match the data range. We thus select convenient values in the data range that divide it into several subintervals. Next, we compute the expected number that should be in the same subintervals based on the *pdf* of the assumed distribution. To determine how closely the data agree, we can take the square of the difference between the expected e_i and actual values o_i in each subinterval, and then divide the result by the expected number. We then sum up this value for each of the k subintervals to get the total:

$$\chi^2 = \sum_{i=1}^{k} \frac{(e_i - o_i)^2}{e_i} \tag{12.45}$$

As we previously explored in the last section, the sum of the squares of n normal random variables results in a chi-square distribution with n degrees of freedom (the chi-square distribution applies only to the square of normally distributed random variables). It turns out that the square of the difference of the expected number of

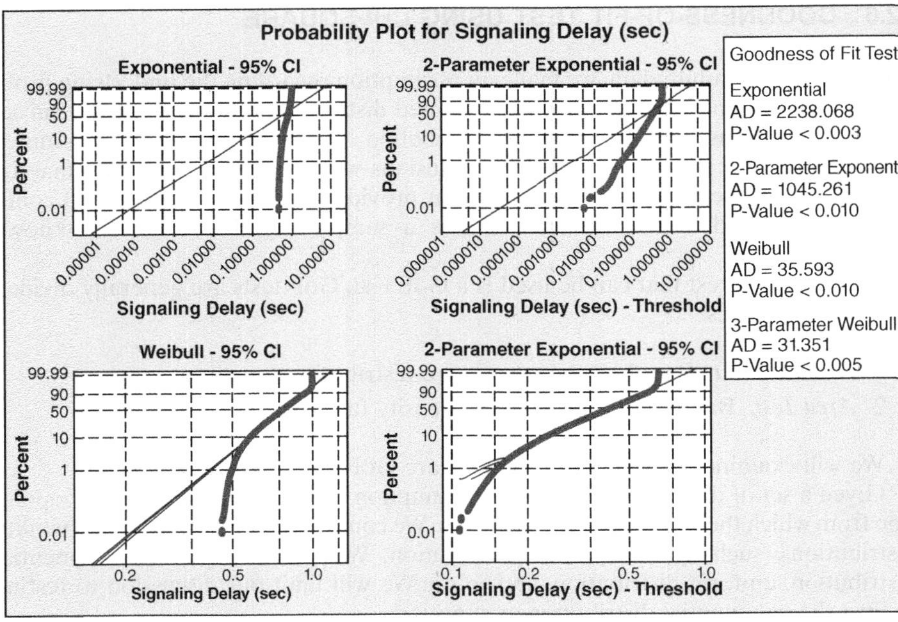

Figure 12.4 Anderson–Darling Test for Goodness of Fit

data items and the observed data items follow a normal distribution for any underlying distribution curve.

As a constraint, the chi-square test requires at least five observed values in every subinterval. If some of the bin counts are less than five, some bins may need to be combined. If we have insufficient data, such that we cannot place at least five data points in every subinterval and we are not able to create five or more subintervals, we cannot apply the chi-square GoF test and other techniques will need to be explored. Other tests are available, such as the Anderson–Darling test included in the Minitab statistics software (Fig. 12.4), to test a set of data points for goodness of fit with different probability distributions. We will not consider these other methods within this text. The interested reader can consult the references for more information.

The chi-square statistic is one of the most popular statistics because it is easy to calculate and interpret. The chi-square tests determine how much the observed frequencies (counts) differ from the frequencies that we would expect by chance. The chi-square statistic is calculated based on the sum of the contributions from each of the individual cells (data points). Every cell in a table contributes something to the overall chi-square statistic. If a given cell differs significantly from the expected frequency, then the contribution of that cell to the overall chi-square is large. If a cell is close to the expected frequency for that cell, then the contribution of that cell is small. The smaller the chi-square value, the better the data represent the assumed distribution.

Assuming we have met the minimum data requirements, the GoF test procedure is the following:

1. Divide the data into n subintervals (histogram).
2. Count the number of data points in each subinterval.
3. Superimpose the *pdf* on the assumed distribution.
4. Compare the empirical histogram with the theoretical *pdf*.
5. Select a desired confidence level and determine if our data agree with the assumed distribution.

12.6.1 Degrees of Freedom

The chi-square distribution is defined as the sum of the squares of n normal independent random variables. The number of degrees of freedom for a chi-square random variable in this case is n. If the n random variables are not independent, that is, a dependency exists among the data, then we must reduce the degrees of freedom by the constraints we impose on the data that affects the independence of the data. We can say that the number of *degrees of freedom* (DF) is the number of observed data minus the number of parameters computed from the data and used in our calculations.

Increasing the sample size provides more information about the population, and thus increases the degrees of freedom present in your data. Adding parameters to your model that are dependent on the sampled data reduces the degrees of freedom available to estimate the goodness of fit.

For example, in the goodness of fit test, we have the following summation:

$$(O_1 - np_1)^2/(np_1) + \cdots + (O_k - np_k)^2/(np_k).$$

This places the constraint that the sum of the observed counts of data in all subintervals be equal to the total number of sample points, k, that is, we have to calculate one parameter from the data: k. This constraint reduces the number of degrees of freedom from k independent freely-varying random variables $O_1 \ldots O_k$ to $k - 1$ independent random variables, since we have the constraint that the sum of $O_1 + \cdots + O_k = k$ creates a dependent relationship. For example, we could say that all of the observations from 1 to $k - 1$ are independent. However, the last observation is not independent, that is, $O_k = k - O_1 + \cdots + O_{k-1}$. We have constrained one of the observations and are left with $k - 1$ independent variables.

Next, consider how our other data estimates are determined and how these estimates may affect the degrees of freedom. Generally, we estimate the mean from the data, which results in another reduction of DF by 1. We can see how the mean estimate affects the DF by noting for n sampled observations, $x_1, x_2, \ldots x_n$, from a population with unknown mean and variance, we have the following relationships:

$$\mu = \sum_{j=1}^{n} \frac{x_j}{n}, \tag{12.46}$$

and

$$x_n - \mu = -\sum_{j=1}^{n-1}(x_j - \mu). \tag{12.47}$$

From these two equations, we see that x_n is not independent, thus, we have $n - 1$ degrees of freedom when we estimate the mean from the data.

In summary, DF is the number of observations minus the number of necessary relations among these observations. Since we estimated the mean from the data, and we constrained the sum of the interval counts, the resulting chi-square has DF = $k - 2$ (sum of counts) − 2 (interval counts + estimate of mean). If the number of bins we select is 5 ($k = 5$), then our resultant DF is 3 (DF = 5 − 2 = 3 > 0). Note the resultant DF must be greater than 0.

EXAMPLE 12.1

Let us assume we have deployed a Base Station Controller system to various customer sites at staggered intervals. A hardware failure occurs when a board is unable to perform its required function and must be returned to the manufacturer for repair and failure analysis. On-site recovery of the board is attempted first, usually be resetting the board. If the reset fails and other troubleshooting techniques fail to resolve the problem, then the board is declared failed by the customer and returned to the vendor. We can readily obtain hardware failure statistics for each board type by tracking the following information:

1. Number of systems deployed at each site
2. Number of boards of a particular type for each system
3. Date of deployment of each system
4. Date of board failure
5. Date of board replacement.

Table 12.1 contains the actual raw data for deployments and returns for a particular project within the telecom industry. The data are sorted by MTTF from lowest to highest values. Table 12.2 contains the number of payload boards deployed at each site and the total number of days the system has been operational since deployment. For this example, the date of return of a failed board is roughly the same as the date of board failure and the date of board replacement.

The analysis covers the time period from the earliest deployment date, say, from June 1 through the end of the analysis period of September 30, the following year.

From these data, we would like to determine if the failure rate distribution follows an exponential *pdf*. More precisely, we would like to determine the percent degree of confidence that the board failure rate follows an exponential *pdf*.

First, we create statistics from the data based on the assumption that the data are exponential.

TABLE 12.1 Payload Board Failure Data

	Payload Board				Payload Board		
Site	Deploy Date	Return Date	MTTF (Hours)	Site	Deploy Date	Return Date	MTTF (Hours)
4	11/19/2011	11/1/2011	72	20	5/8/2012	1/1/2012	2688
4	11/19/2011	11/1/2011	72	1	4/14/2012	12/1/2011	2856
4	11/19/2011	11/1/2011	72	9	3/25/2012	11/1/2011	3096
4	11/19/2011	11/1/2011	72	11	5/1/2012	12/1/2011	3264
7	12/19/2011	12/1/2011	72	11	5/5/2012	12/1/2011	3360
18	12/19/2011	12/1/2011	72	16	3/30/2012	10/1/2011	3960
18	12/19/2011	12/1/2011	72	16	3/30/2012	10/1/2011	3960
16	10/21/2011	10/1/2011	120	16	3/30/2012	10/1/2011	3960
11	12/26/2011	12/1/2011	240	11	6/8/2012	12/1/2011	4176
11	12/26/2011	12/1/2011	240	11	6/8/2012	12/1/2011	4176
20	1/27/2012	1/1/2012	264	11	6/8/2012	12/1/2011	4176
20	1/27/2012	1/1/2012	264	11	6/8/2012	12/1/2011	4176
21	1/13/2012	12/1/2011	672	2	6/11/2012	12/1/2011	4248
1	1/19/2012	12/1/2011	816	5	7/2/2012	12/1/2011	4752
3	1/19/2012	12/1/2011	816	7	7/7/2012	12/1/2011	4872
11	1/19/2012	12/1/2011	816	12	6/9/2012	11/1/2011	4920
11	1/19/2012	12/1/2011	816	20	8/25/2012	1/1/2012	5304
11	1/19/2012	12/1/2011	816	12	7/1/2012	11/1/2011	5448
11	1/19/2012	12/1/2011	816	12	7/1/2012	11/1/2011	5448
11	1/27/2012	12/1/2011	1008	15	8/10/2012	12/1/2011	5688
21	1/27/2012	12/1/2011	1008	16	6/12/2012	10/1/2011	5736
11	1/29/2012	12/1/2011	1056	11	8/19/2012	12/1/2011	5904
17	8/22/2011	6/1/2011	1608	3	8/28/2012	12/1/2011	6120
12	1/27/2012	11/1/2011	1728	17	3/13/2012	6/1/2011	6480
12	2/5/2012	11/1/2011	1944	11	9/21/2012	12/1/2011	6696
12	2/5/2012	11/1/2011	1944	12	8/25/2012	11/1/2011	6768
14	1/5/2012	10/1/2011	1944	14	8/28/2012	10/1/2011	7584
2	3/10/2012	12/1/2011	2016	16	9/1/2012	10/1/2011	7680
11	3/13/2012	12/1/2011	2088	14	9/2/2012	10/1/2011	7704
20	4/16/2012	1/1/2012	2160	23	5/13/2012	6/1/2011	7944
19	1/19/2012	10/1/2011	2280	17	5/27/2012	6/1/2011	8280
20	4/23/2012	1/1/2012	2328	17	6/2/2012	6/1/2011	8424
20	4/23/2012	1/1/2012	2328	8	8/10/2012	6/1/2011	10080
21	3/30/2012	12/1/2011	2496				

The total Payload Board hours of operation for all systems is calculated by summing the number of Payload Boards deployed at each site multiplied by the number of days of operation multiplied by 24 hours.

Total Payload Board hours: 35,645,760
Number of Payload Board Failures: 67

TABLE 12.2 Payload Boards per Site

Site	Payload Board	Days	Site	Payload Board	Days	Site	Payload Board	Days	Site	Payload Board	Days
1	224	300	7	128	300	13	448	300	19	176	360
2	128	300	8	176	480	14	128	360	20	80	270
3	272	300	9	224	330	15	80	300	21	128	300
4	176	330	10	224	300	16	176	360	22	496	390
5	320	300	11	128	300	17	224	480	23	80	480
6	176	270	12	76	330	18	128	300			

Average MTBF: 532,026 hours
Failure rate (λ): 1.880×10^{-6}

It is important to note that this calculated MTBF and failure rate information is our best guess of the point MTBF applied to any board currently in service and includes boards that have failed and are yet to fail. Thus, if our assumption of a constant failure rate is accurate, we can expect the average MTBF of a board to approach this value.

For the purpose of determining GoF, we only need to focus on the boards that have failed, since obviously we do not know the time to failure of boards that have not yet failed or boards that have been replaced and are yet to fail. Thus, we will use the data for the failed boards to determine how closely the distribution of the board failure rate matches the idealized exponential distribution using the assumption that all boards have failed at least once. We ignore boards that have not failed. If the distribution of failures for boards that have already failed (and have been replaced) is exponential, then we can extrapolate this information and say that the exponential distribution applies to all boards in the pool.

Using the data in Table 12.1, we calculate the following statistics:

Data Points (Payload Board Failures): 67
Mean MTBF of failed boards: 3210 hours
Median: 2496 hours
Std Dev.: 2640 hours
Min: 72 hours
Max: 10,080 hours
Quartile 1: 816 hours
Quartile 3: 5112 hours

Per the GoF test procedure, the first step is to divide these data into n bins. We need at least five bins. How do we determine the number of bins? We can apply Sturges' rule to determine the number of bins when representing data by a histogram. Applying Sturges' rule (a rule for determining how to choose the number of bins when representing data in a histogram):

$$k = 1 + 3.222\log_{10}(N),$$

where N = number of data points.

The number of bins is:

$$k = 1 + 3.222\log_{10}(67)$$

$$k = 6.8836 \approx 7.$$

So we will create seven bins or subintervals using the data as a guide. Let us choose the following sub-intervals:

Bin No.	MTBF Interval
1	0–250
2	>250–1000
3	>1000–2000
4	>2000–3000
5	>3000–5000
6	>5000–7000
7	>7000

Refer to Table 12.1 for the raw data.

For end points, we now select 250, 1000, 2000, 3000, 5000, and 7000, which in turn define seven subintervals (Fig. 12.5).

Table 12.3 shows the compiled information needed for the chi-square test. The first column identifies the bin number. We have seven bins identified.

Column 2 (End Point): The maximum value of the MTBF for this bin.

Column 3 (Cumulative Probability) is the cumulative probability for values ranging from zero to the maximum MTBF for this bin based on the theoretical *cdf* we are comparing against.

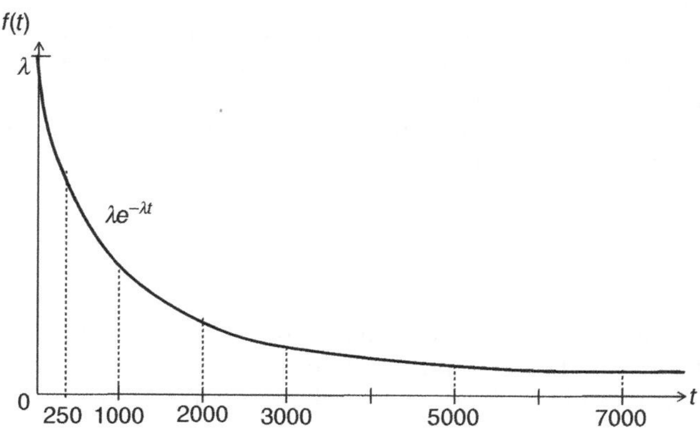

Figure 12.5 Subintervals for Goodness of Fit

TABLE 12.3 Chi-Square Data

1	2	3	4	5	6	7
Bin No.	End Point	Cumulative Probability	Cell Probability	Expected Number	Observed Number	(e-o)^2/e
1	250	0.074928	0.074928	5.020181	10	4.939781
2	1000	0.267678	0.19275	12.91425	9	1.186394
3	2000	0.463705	0.196027	13.13378	8	2.00671
4	3000	0.607259	0.143555	9.618154	9	0.039728
5	5000	0.789375	0.182116	12.20175	14	0.265018
6	7000	0.887043	0.097668	6.543744	10	1.825516
7	∞	1	0.112957	7.568133	7	0.042649
Totals			1	67	67	10.3058

For an exponential distribution, the *cdf* is:

$$cdf = 1 - e^{-\lambda t}. \quad (12.48)$$

For example, for the first end point, 250 hours:

$$P = 1 - e^{-250/3210} = 0.074928.$$

This indicates that for a theoretical exponential *cdf*, we expect 7.49% of all values to be between 0 and the end point value for this bin.

Column 4 (cell probability) is the probability that a particular MTBF value will fall into this bin. This can be calculated by taking the difference between the value in the cumulative probability column and the previous cumulative probability value.

Column 5 (expected) is the number of data values expected in this subinterval given that n total sample values were observed. It is calculated by multiplying the subinterval probability (column 4) by the number of data points.

Column 6 (observed) is simply a count of the number of observed data values that fall into this subinterval.

Finally, column 7 represents the chi-square calculation for that subinterval. The chi-square calculation is the square of the difference of the expected number and the observed number divided by the expected number.

When we sum up the values in the last column, we obtain the result of the chi-square GoF test statistic for the data set:

$$\chi^2 = \sum_{i=1}^{k} \frac{(e_i - o_i)^2}{e_i} = 10.3058 < \chi^2_{0.05,5} = 11.0705.$$

χ^2 is a well-known statistical factor that can be obtained from a chi-square table or statistical software. For example, to obtain the chi-square value for a 0.95 confidence and 5 degrees of freedom, use the following formula in Excel:

```
CHIINV(0.05,5) = 11.070409
```

Here, the chi-square statistic calculated value (10.3058) is smaller than the chi-square theoretical value (11.0705) for $DF = 7 - 2 = 5$ and $1 - \alpha = 0.95$. This indicates that the data set we have agrees with the theoretical exponential distribution with a confidence level greater than 95%. We assume the null hypothesis, H_0: the distribution of the population from which this data set is sampled is exponential. Strictly speaking, we can conclude that we cannot reject the null hypothesis.

12.6.2 Chi-Square GoF Test Procedure Summary

The steps required for the chi-square GoF test are:

1. Identify the assumed underlying distribution.
2. Estimate or calculate the distribution statistics from the data.
3. Determine the subintervals.
4. Obtain the probability of occurrence of an event within each of the subintervals for the theoretical distribution.
5. Choose the confidence level, for example, >95%.
6. Calculate the chi-square value.
7. Calculate the Test Statistic result.
8. If chi-square value is greater than the GoF test calculation, the data fit the assumed distribution for the confidence level chosen.

12.7 CONFIDENCE LEVELS

The failure rate λ of a component is calculated via analysis, testing, or field data. The best estimate of the failure rate based on field data or testing is determined by dividing the number of failures that occur by the number of components deployed (or under test). The greater the number of components and the longer the time these components are tested, the more accurate the estimate of λ becomes and the more confident we are that our estimate is close to the true failure rate of the component. But how do we determine if our estimate is close enough? As a practical matter, the amount of data available to us is limited, so how can we incorporate this information to improve our estimates?

The widely used chi-square statistic $X^2_{(1-CL,2r+2)}$ provides us with a quantifiable confidence that the true failure rate lies between calculated upper and lower bounds. The chi-square test uses the chi-square distribution to calculate failure rates for a given confidence level based on the number of failures observed. For example, to determine the upper bound of a system operating for T hours with r observed failures using the chi-square test, the following equation is used:

$$\lambda = \frac{\chi^2_{(1-CL,2r+2)}}{2T}, \tag{12.49}$$

where CL is the confidence level, r and $2r + 2$ is the degrees of freedom for the chi-square distribution.

How is Equation (12.49) related to confidence levels for constant failure rates? To do this, we first need to review the binomial cumulative distribution function previously introduced in Chapter 4, Equation (4.34):

$$P(\text{failures} \le r) = \sum_{x=0}^{r} \frac{n!}{x!(n-x)!} p^x (1-p)^{n-x}. \tag{12.50}$$

Equation (12.50) gives us the probability that r or fewer events will occur after n trials with failure probability p. The probability that more than r events occur after n trials $P(\textit{failures} > r)$ is then $1 - P(Y \le r)$:

$$P(\text{failures} > r) = \sum_{x=r+1}^{n} \frac{n!}{x!(n-x)!} p^x (1-p)^{n-x}. \tag{12.51}$$

Our confidence level that we have r or fewer failures is:

$$1 - CL = P(\text{failures} \le r) = \sum_{x=0}^{r} \frac{n!}{x!(n-x)!} p^x (1-p)^{n-x}. \tag{12.52}$$

If n is very large and p is very small, that is, $n \to \infty$, $p \to 0$, then the product np is a constant, and the Poisson approximation, Equation (4.37), can be used for each term in Equation (12.52):

$$\frac{n!}{x!(n-x)!} p^x (1-p)^{n-x} \approx \frac{(np)^x}{x!} e^{-np}.$$

Refer to Section 4.8 for a proof of this approximation.

For constant failure rate components, the probability of failure p is given by:

$$p = 1 - e^{-\lambda t}. \tag{12.53}$$

When the quantity λt is small, Equation (12.53) becomes:

$$p = \lambda t. \tag{12.54}$$

Proof:

$$e^{-\lambda t} = 1 - \lambda t + \frac{(\lambda t)^2}{2!} - \frac{(\lambda t)^3}{3!} + \frac{(\lambda t)^4}{4!} + \cdots$$

$$1 - e^{-\lambda t} = \lambda t - \frac{(\lambda t)^2}{2!} + \frac{(\lambda t)^3}{3!} - \frac{(\lambda t)^4}{4!} + \cdots$$

$$1 - e^{-\lambda t} = \lambda t\left(1 - \frac{\lambda t}{2!} + \frac{(\lambda t)^2}{3!} - \frac{(\lambda t)^4}{4!} + \cdots\right)$$

$$\lim_{\lambda t \to 0}\left(1 - \frac{\lambda t}{2!} + \frac{(\lambda t)^2}{3!} - \frac{(\lambda t)^3}{4!} + \cdots\right) = 1$$

$$\lim_{\lambda t \to 0}(1 - e^{-\lambda t}) = \lambda t.$$

Inserting this approximation into Equation (12.52) and substituting λt for p, we obtain:

$$1 - CL = \sum_{x=0}^{r} \frac{(n\lambda t)^x}{x!} e^{-n\lambda t} \tag{12.55}$$

$$1 - CL = e^{-n\lambda t}\left(1 + n\lambda t + \cdots + \frac{(n\lambda t)^{r-1}}{(r-1)!} + \frac{(n\lambda t)^r}{r!}\right). \tag{12.56}$$

The first term gives the probability of zero failures; the second term gives the probability of one failure, and so on up to r failures.

Each term of Equation (12.55) is the discrete Poisson distribution. The summation of all terms in Equation (12.55) for $x = 0$ to $x = \infty$ is the Poisson *cdf* and equals 1.

Our problem now is to solve for λt for a given CL. We cannot obtain a closed-form solution for λt for Equation (12.56). We can however obtain a solution using the chi-square distribution.

A well-known relationship exists between the cumulative Poisson distribution and the cumulative chi-square distribution:

$$P(\text{Poisson}(\lambda t) \le r) = P(\chi^2_{1-CL,2(r+1)} > 2\lambda t), \tag{12.57}$$

where $2(r + 1)$ is the degrees of freedom.

12.7.1 Proof 1

The left side of the Equation (12.57) is:

$$P(\text{Poisson}(\lambda t) \le r) = e^{-\lambda t}\left(1 + \lambda t + \cdots + \frac{(\lambda t)^{x-1}}{(x-1)!} + \frac{(\lambda t)^x}{x!}\right).$$

The right side of the equation is per definition the gamma cumulative distribution function. The left side of Equation (12.57) is set equal to the chi-square distribution (Eq. (5.32)):

$$P(\chi^2_{1-CL,2(r+1)} > 2\lambda t) = \frac{\int_{\lambda t}^{\infty} t^{n/2-1} e^{-t/2} dt}{2^{n/2}\Gamma(n/2)}. \tag{12.58}$$

Now, let us evaluate Equation (12.58) for failures r from 0 to n. For $r = 0$, the degrees of freedom are two:

$$P(\chi^2_{1-CL,2(0+1)} > 2\lambda t) = \frac{\int_{\lambda t}^{\infty} t^{2/2-1} e^{-t/2} dt}{2^2\Gamma(1)} = \frac{\int_{2\lambda t}^{\infty} e^{-t/2} dt}{2(0)!} = -e^{-t/2}\Big|_{\lambda t}^{\infty} = e^{-\lambda t}.$$

For $r = 1$:

$$P(\chi^2_{1-CL,2(1+1)} > 2\lambda t) = \frac{\int_{2\lambda t}^{\infty} te^{-t/2}dt}{2^2\Gamma(2)} = \frac{\int_{2\lambda t}^{\infty} te^{-t/2}dt}{4}.$$

Using integration by parts:

$$\frac{\int_{2\lambda t}^{\infty} te^{-t/2}dt}{4} = \frac{2te^{-t/2}}{4}\bigg|_{2\lambda t}^{\infty} + \frac{\int_{2\lambda t}^{\infty} e^{-t/2}dt}{2}$$

$$P(\chi^2_{1-CL,2(1+1)} > 2\lambda t) = \lambda te^{-\lambda t} + e^{-\lambda t}$$

$$P(\chi^2_{1-CL,2(1+1)} > 2\lambda t) = e^{-\lambda t}(1 + \lambda t).$$

Repeating the process, for any value of r:

$$P\left(\chi^2_{1-CL,2(r+1)} > 2\lambda t\right) = e^{-\lambda t}\left(1 + \lambda t + \frac{(\lambda t)^2}{2!} + \cdots + \frac{(\lambda t)^r}{r!}\right). \tag{12.59}$$

We note that the chi-square distribution above is the same as the Poisson cumulative distribution shown in Equation (4.38). Thus, Equation (12.57) is proven. α is defined as $1 - CL$. Thus:

$$\lambda_{\text{MAX}} = \frac{\chi^2_{(\alpha,2r+2)}}{2T} \tag{12.60}$$

provides the maximum failure rate λ for a given confidence level CL, $2r + 2$ degrees of freedom and test time T.

Equivalently:

$$\text{MTBF}_{\text{MIN}} = \frac{2T}{\chi^2_{(\alpha,2r+2)}}. \tag{12.61}$$

Equivalent forms of Equations (12.60) and (12.61) are:

$$\lambda_{\text{MAX}} = \lambda_{est}\frac{\chi^2_{(\alpha,2r+2)}}{2r} \tag{12.62}$$

$$\text{MTBF}_{\text{MIN}} = \frac{2r\text{MTBF}_{\text{est}}}{\chi^2_{(\alpha,2r+2)}}, \tag{12.63}$$

where MTBF_{est} is the point MTBF estimate calculated by:

$$\text{MTBF}_{\text{est}} = \frac{\text{(Test duration)(Number of components under test)}}{\text{Number of failures}}. \tag{12.64}$$

12.7.2 Proof 2

Let $X_1, \ldots, X_n$ be independent and identically distributed exponential variables with mean $1/\lambda$. Show that the sum $X_1 + \cdots + X_n$ has a gamma distribution with parameters $(n, 1/\lambda)$.

$$G(n, 1/\lambda) = \frac{\lambda e^{-\lambda t}(\lambda t)^{n-1}}{\Gamma(n)}$$

$$f(t) = \frac{\lambda e^{-\lambda t}(\lambda t)^{n-1}}{(n-1)!}.$$

We want to prove $G(n,1/\lambda) = f(t)$.

To show this, we first start with the cumulative gamma distributions (Eq. (5.27) repeated here for convenience):

$$P(\text{gamma}(\alpha, \beta) \le x) = \frac{\int_0^x t^{\alpha-1} e^{-t/\beta} dt}{\beta^\alpha \Gamma(\alpha)}. \tag{12.65}$$

By definition, the chi-square cumulative distribution is equal to the gamma cumulative distribution with the shape parameter α set to $n/2$, and the scale parameter β set to 2. The chi-square cumulative distribution obtained from Equation (5.32) is:

$$P[\chi^2_{(n)} > \lambda] = \frac{\int_\lambda^\infty t^{n/2-1} e^{-t/2} dt}{2^{n/2}\Gamma(n/2)} \tag{12.66}$$

where n is the degrees of freedom.

$$G(n-1, 1/\lambda) = f(t) = \frac{\lambda e^{-\lambda t}(\lambda t)^{n-2}}{(n-2)!}$$

Next, we add the random variable X_n to this sum:

$$G(n, 1/\lambda) = f_{x1+\cdots+xn}(t).$$

Using the convolution integral to obtain the *pdf* of the two densities, we have:

$$G(n, 1/\lambda) = \int_0^\infty f_{x_n}(t-s)(f_{x1+\cdots x_{n-1}}(s))ds$$

$$G(n, 1/\lambda) = \int_0^t [\lambda e^{-\lambda(t-s)}][\lambda e^{-\lambda s}(\lambda s)^{n-2}/(n-2)!]ds$$

$$G(n, 1/\lambda) = \frac{\lambda^2 e^{-\lambda t}}{(n-2)!}\int_0^t (\lambda s)^{n-2} ds$$

$$G(n, 1/\lambda) = \frac{\lambda e^{-\lambda t}\lambda^{n-1}}{(n-2)!}\int_0^t s^{n-2} ds$$

$$G(n,1/\lambda)=\frac{\lambda e^{-\lambda t}(\lambda t)^{n-1}}{(n-1)!},$$

which proves the relationship.

EXAMPLE 12.2

We test 100 components for 1000 hours, and observe 20 failures. We want to know with 95% confidence, the maximum failure rate.

First, let us calculate the MTBF point estimate of the components (Equation (12.64)):

$$\text{MTBF}_{\text{est}}=\frac{\text{Total Test Time}}{\text{No. of Failures}}=\frac{1000\times100}{20}=5000\text{ hours.}$$

Employing the chi-square Equation (12.62), the maximum failure rate is:

$$\lambda_{\text{MAX}}=\frac{\chi^2_{(0.05,2(20)+2)}}{2\times20\times5000}=\frac{58.12404}{20{,}000}=0.00029062.$$

This corresponds to a minimum MTBF we can expect with 95% confidence:

$$\text{MTBF}_{\text{MIN}}=\frac{1}{0.00029062}=3{,}441\text{ hours.}$$

12.7.3 Test Strategies

Various types of testing strategies are employed. Let us consider two industry standard types.

Type I: Time censored test. The test ends at a predetermined time. T_{test} is fixed. At the end of the test, we count the number of failures for the system or components under test, calculate the point MTBF and then calculate the lower bound of the MTBF with the desired confidence level. Equation (12.65) gives us the lower bound (worst case) for the MTBF for the desired confidence level. This is shown graphically in Figure 12.6a with a confidence level of 90%. The area to the right of the chi-square value x represents the probability that the true MTBF value is less than or equal to the calculated MTBF_{MIN} value with a 90% level of confidence.

If we want to determine the maximum MTBF value with the same confidence level, we use the following equation:

$$\text{MTBF}_{\text{MAX}}=\frac{2T}{\chi^2_{(1-\alpha,2r)}}.\tag{12.67}$$

Or equivalently:

$$\text{MTBF}_{\text{MAX}}=\frac{2r\text{MTBF}_{\text{est}}}{\chi^2_{(1-\alpha,2r)}}.\tag{12.68}$$

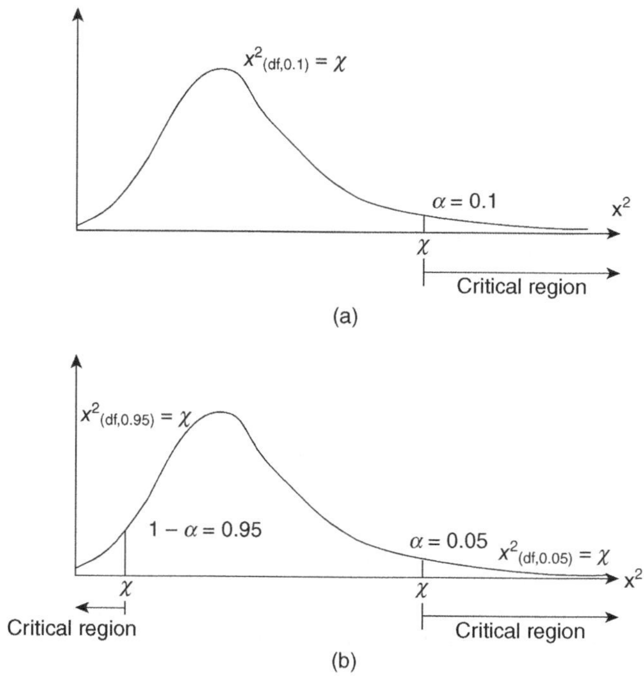

Figure 12.6 Chi-Square Upper/Lower Bounds Statistic

Note that the degrees of freedom in this case have changed from $2(r + 1)$ to $2r$. We discount the number of failures observed by 1 to obtain the best case estimate for MTBF.

This type of testing is Type II, failure censored tests. The test time is unknown beforehand and the test ends after a predetermined number of failures have occurred (at least one). For example, we test n devices in parallel, and terminate the test on the first failure.

We can perform either a one-sided or two-sided test. For a one-sided test, we are interested only on the worst case value of MTBF or failure rate. But we can also perform a two-sided test to determine our best case value and worst case value for MTBF. For the two-sided test, we allocate half of the confidence level to one tail and the other half of the confidence level to the other tail of the chi-square distribution. Thus, for a two-sided Type II test, we have the following:

$$\mathrm{MTBF_{MIN}} = \frac{2T}{\chi^2_{(\alpha/2,2r)}} \tag{12.69}$$

$$\mathrm{MTBF_{MAX}} = \frac{2T}{\chi^2_{(1-\alpha/2,2r)}} \tag{12.70}$$

$$\frac{2T}{\chi^2_{(\alpha/2,2r)}} \leq \mathrm{MTBF_{actual}} \leq \frac{2T}{\chi^2_{(1-\alpha/2,2r)}}. \tag{12.71}$$

Or equivalently:

$$\frac{2r\text{MTBF}_{\text{est}}}{\chi^2_{(\alpha,2r+2)}} \leq \text{MTBF}_{\text{actual}} \leq \frac{2r\text{MTBF}_{\text{est}}}{\chi^2_{(\alpha,2r+2)}}. \quad (12.72)$$

The upper bound is equivalent to a Poisson distribution with $r - 1$ failures. By decrementing the number of failures by 1, we obtain the best case MTBF estimate. These upper and lower bounds confidence intervals with a confidence level of 90% are shown in Figure 12.6b. The area between the two extreme critical regions represents the probability that the true MTBF value is greater than or equal to the calculated MTBF_{MIN} value and less than or equal to the calculated MTBF_{MAX} value with a 90% level of confidence.

We can readily compute the chi-square values for any confidence level and degrees of freedom using the Excel statistic formula, CHIINV previously mentioned.

EXAMPLE 12.3

A test is executed for a predefined duration with n components and one failure is detected. From this, an MTBF value of 1000 hours is calculated. Assume either the test duration is increased and/or the number of components under test is increased, resulting in additional failures. We want to determine the one-sided MTBF maximum and minimum estimates with a confidence level of 90% based on the number of failures.

Starting with $r = 1$, the MTBF_{MAX} and MTBF_{MIN} are calculated using Equations (12.69) and (12.70).

$$\text{MTBF}_{\text{MIN}} = \frac{2(1)(1000)}{\chi^2_{(0.01,2(1+1))}} = 257\text{ hours}$$

$$\text{MTBF}_{\text{MAX}} = \frac{2(1)(1000)}{\chi^2_{(1-0.01,2(1))}} = 9491\text{ hours.}$$

This procedure is repeated for each additional failure. We can plot the results of the $MBTF_{MIN}$ and $MTBF_{MAX}$ values as a function of the number of failures as shown in Figure 12.7.

From the figure, if only 1 failure occurs, the MTBF ranges are quite large. However, as additional failures are observed, the variation between maximum MTBF and minimum MTBF values decrease and approach the MTBF point estimate with 90% confidence. The additional failures provide more evidence that the MTBF maximum and minimum values are closer to the MTBF point estimate of 1000.

12.8 SUMMARY

To increase the predictive usefulness of reliability models, we need good estimates of reliability parameters, such as MTBF and MTTR. A key source of information that can be applied to verify or improve our reliability models is test data. These test data

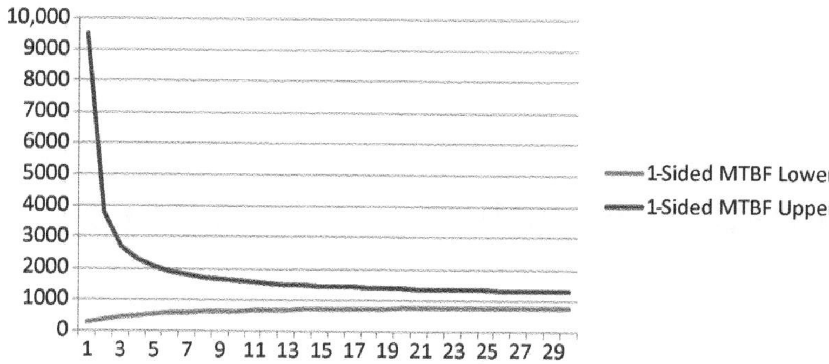

Figure 12.7 One-Sided MTBF Values versus Number of Failures

may be obtained from projects with similar characteristics as our model or be based directly on test data or field data from the system we are modeling.

GoF tests provide us with a measure of how well our assumed probability distributions for our model match the actual test data. If a large disparity is identified, then our model may need to be amended to improve its accuracy of modeling the actual system behavior.

Confidence levels, on the other hand, tell us the range of values for reliability estimates, such as MTBF, that fall within a certain confidence level, for example, minimum and maximum values that fall within 95% confidence levels. The more data we have, the more narrow the range between the maximum and minimal values. And the tighter the confidence level, the wider the range will be.

The confidence level provides us with feedback on how likely our model will be in predicting future reliability based on data we have—including the case where no failure data exist but the system(s) has been operational for a given period of time.

In Chapter 18, we will apply these techniques to a case study.

CHAPTER 13

Estimating Reliability Parameters

13.1 INTRODUCTION

At the beginning of a high availability project, a set of baseline reliability and availability goals and requirements is established. Based on the high availability design, choice of hardware, new software to be created, software to be ported, and other factors—including engineering judgment, the reliability engineer assesses the reliability of the constituent components and the initial availability of the system.

Once the engineer compiles the baseline information, the initial availability model is created. The next task is to reassess this availability model as additional information, such as failure data, updated Mean Time between Failures (MTBF) data, and other relevant data, become available. We should also note that the model should be revised after the system is built based on soak test results, extended life testing, field test results, and so on. Regardless of the amount of data initially available, some estimate of reliability is required to determine if the top-level architecture satisfies our reliability requirements. The initial reliability estimates may only reflect the engineer's "*best guess*." Later, we can incorporate previous best guess (*a priori*) data with new data to create new and hopefully more accurate reliability estimates (*a posteriori*).

What is the best estimate of the system failure rate? We have several choices:

(a) We could discard our initial estimate and update our predicted system failure rate to match the observed failure rate for the first 6 months.

(b) We could also take the mean of our initial prediction and the observed failure rate, giving equal weight to our initial prediction.

(c) We could update our prediction by incorporating the previous estimate with our newer data using Bayes' estimation techniques.

How do we come up with the initial reliability probability distributions? Our two main parameters of interest are the Mean Time between Failures (MTBF) and Mean

Designing High Availability Systems: Design for Six Sigma and Classical Reliability Techniques with Practical Real-Life Examples, First Edition. Zachary Taylor and Subramanyam Ranganathan.

Time to Repair (MTTR) for each component in the system we model. If we can make the simplifying assumption that the component failure rates follow the exponential probability distribution, a wealth of techniques are available as described in this book. We start with the MTBF point estimates for each of the components and its associated MTTR. The MTTR may be easier to determine if reset times are available and repair policies are known beforehand. If we are modeling a restartable software component for example, then we can measure software restart time in the lab under nominal and stress test conditions and obtain a relatively accurate MTTR. For a nonrestartable software component or a restartable software component that cannot be successfully restarted, we may then need to consider a hardware reset of the board or platform on which the software component resides. The hardware reset time of the board likewise can be measured in the lab.

The MTBF of an off-the-shelf unit is normally provided by the vendor. If it is not published, then the reliability engineer should request the MTBF estimate from the vendor. This MTBF may include hardware, firmware, OS software, or other vendor-provided platform software. It is important that the assumptions regarding the MTBF and reliability data provided by the vendor are clearly specified such that proper allocation of software and firmware MTBF can be made.

Determining software MTBF is much more challenging—especially since the probability distribution of software failures can be significantly different from that of hardware. Many software reliability engineering models have been proposed in the literature and a variety has been employed. We will examine a few of these models and discuss simplifying assumptions and general rules of thumb on their use.

Historical Data. The best estimate of MTBF is obtained from historical data. Ideally, if we have deployed a large number of systems in the field over several years and the system configuration and operating conditions are similar to future deployment, the system MTBF estimate should be based on the historical data. Carefully collecting all reliability data, outages, board return rates, software problems, and so on is very important for understanding the current system behavior and being able to predict future reliability. It also serves as a database of reliability information for future projects. In many cases, these data are required contractually.

For complex systems, software is periodically modified or upgraded to address software bugs or to introduce new functionality. The historical software failure rate data may no longer provide the best estimate for future failure rates. We will consider approaches for obtaining initial MTBF estimates in this case.

Expert Opinion. Another approach for estimating MTBF is to rely on expert opinion. This subjective approach may seem more of a wild guess, but certain techniques can be employed to reduce the uncertainty of the estimates.

Simulations. We can run simulations such as Monte Carlo simulations to quantify how much of an impact a range of MTBF estimates will have on the overall system availability. This will help us direct resources to those areas that have the most impact to system availability, with the goal of either increasing the accuracy of our estimates or mitigating the impact of these failures on the system.

For redundant systems, assumptions are made regarding failover performance to the redundant resources in the event of an unplanned hardware or software failure (probability of successful failover).

For nonredundant systems, unplanned outages include software faults that typically manifest themselves in the form of a software task restart or hardware reset. When software faults occur, assumptions must be made about the probability of a successful restart or hardware reset and the time it takes to recover when the reset is not successful. These assumptions are accounted for in the software Markov or fault tree models. Planned downtime is affected by the probability of successful reboots and the restart times when a software upgrade or configuration change occurs.

13.2 BAYES'S ESTIMATION

Two types of paradigms can be used for system reliability valuation—the classical paradigm and Bayesian paradigm. For the classical paradigm, the MTBF is one fixed value. There is no probability associated with it; failure data from test and/or deployment allow us to improve the estimates of the "true" MTBF value. One major drawback is that the actual MTBF value cannot be known until after all devices deployed in the field have completed their operational life and the failure information has been collected and analyzed. To be useful, our model must be able to predict to some degree of accuracy what this true value may be at the beginning of the project.

For the Bayesian paradigm, the MTBF is considered to be a random variable with a probability distribution. Before running our test, we have some knowledge of what the MTBF probability distribution looks like, based on engineering judgment, and so on. When we are testing the system, we observe failure data that follow a particular failure distribution. When more data are accumulated, the probability distribution tends to have a much smaller variance and thus gets closer to the mean value of the MTBF. The Bayesian approach is a subjective interpretation of probability since the initial distribution and MTBF are unknown. The analyst must make some assumptions regarding the probability distribution. So with the Bayesian approach, we consider all of the discrete values and their associated probabilities provided by the analyst. The basic transformation is:

$$\text{Posterior Estimate} = \frac{(\text{Likelihood})(\text{Prior belief})}{\text{Evidence}}.$$

We take our previous estimate on the nature of the MTBF, and given the new evidence, we determine the likelihood that the new evidence is indicative of the previous estimate and create a new estimate. Suppose we have made our best guess for the MTBF of a device based on similarity to previous projects, combining existing component MTBFs, industry data, and our own engineering judgment. How accurate is this initial estimate? Can we improve on our estimate? If we obtain more information about the MTBF, such as field data or lab testing, do we throw away our initial estimate and replace it with the new data? If we collect only a few new data points, the new data may be statistically insufficient to be used to calculate new MTBF with a high degree of confidence.

One approach might be to incorporate the new data to augment or improve upon our previous estimate; that is, do not discard our previous estimate but combine our new data with our previous estimate to create a new estimate. Our original knowledge of the system should be increased by incorporating any new data we obtain with the existing data.

For estimating system failure rates, we can mathematically express the above relationship in terms of previously assumed failure rate distributions and new data.

From Bayes's theorem, Equation (7.51),

$$P(\lambda_i \mid D) = \frac{P(D \mid \lambda_i)P(\lambda_i)}{P(D)}. \tag{13.1}$$

$P(D)$ is the average data likelihood across all possible models.

$P(\lambda_i)$ is our prior belief regarding the probability of the *i*th possible value of the failure rate.

$P(D|\lambda_i)$ is the likelihood of these data occurring given that the *i*th possible value of the failure rate is correct.

$P(\lambda_i|D)$ is the posterior probability of the *i*th possible value of the failure rate, given the new data.

Bayes's estimate takes each prior estimation of the probability and creates an improved estimate. One of the constraints of this technique is that once the prior probability distribution is established, new data points cannot be added and existing data points cannot be removed.

Bayes's theorem mathematically indicates a decrease in uncertainty as a result of an increase in knowledge. One of the sources of this knowledge are test data. Bayes's theorem has proven useful in estimating rare events, such as large MTBFs, when little or no product reliability data are available.

EXAMPLE 13.1

We have developed an embedded system that controls air conditioners. We do not know the MTBF of the system with a high degree of certainty. However, we have a reasonable estimate of the probability distribution of the MTBF. We estimate that after 1 year of continuous operation, the probability of one or more failures in *any* system is as follows:

1 failure = 0.15
2 failures = 0.25
3 failures = 0.45
4 failures = 0.15.

Let λ_i represent the failure rate of the system where *i* is the number of failures of that system in 1 year. The probability of the failure rate of any λ_i is $P(\lambda_i)$. We can now express the probabilities of the number of failures as:

$$P(\lambda_i) = P(i \text{ failures/year})$$
$$P(\lambda_1) = P(1 \text{ failures/year}) = 0.15$$
$$P(\lambda_2) = P(2 \text{ failures/year}) = 0.25$$
$$P(\lambda_3) = P(3 \text{ failures/year}) = 0.45$$
$$P(\lambda_4) = P(3 \text{ failures/year}) = 0.15.$$

After 1 year of operation of a system, we observe that the system failed twice. Given these new data, what is the revised MTBF distribution?

Solution

We would like to *modify* our initial MTBF failure distribution by incorporating these new data.

Using Bayes's theorem, our prior estimate of the failure rate is proportional to the prior probability λ_i multiplied by the likelihood that the data would be observed given that the true failure rate is λ_i.

$$P(\lambda_i \mid D=2) \propto P(D=2 \mid \lambda_i)P(\lambda_i)$$
$$P(\lambda_i \mid D=2) = \frac{P(D=2 \mid \lambda_i)P(\lambda_i)}{\sum_j P(D=2 \mid \lambda_j)P(\lambda_j)}. \tag{13.2}$$

As we have previously discussed, the Poisson distribution represents the number of failures in a time interval, assuming that the failure rate is a constant. The probability that *exactly* two failures will occur after 1 year (8760 hours) is:

$$P(D=2 \mid \lambda_i) = \frac{(8760\lambda_i)^2}{2!}e^{-8760\lambda_i}. \tag{13.3}$$

Now we can calculate the updated estimate of the failure rate distribution. $\lambda_1 = 1/8760 = 0.000114$, $\lambda_2 = 2/8760 = 0.000228$, $\lambda_3 = 3/8760 = 0.000342$, and $\lambda_4 = 4/8760 = 0.000457$:

$$P(D=2 \mid \lambda_1) = 0.1839$$
$$P(D=2 \mid \lambda_2) = 0.2707$$
$$P(D=2 \mid \lambda_3) = 0.2240$$
$$P(D=2 \mid \lambda_4) = 0.1465$$
$$P(D=2 \mid \lambda_1)P(\lambda_1) = 0.0276$$
$$P(D=2 \mid \lambda_2)P(\lambda_2) = 0.0677$$
$$P(D=2 \mid \lambda_3)P(\lambda_3) = 0.1008$$

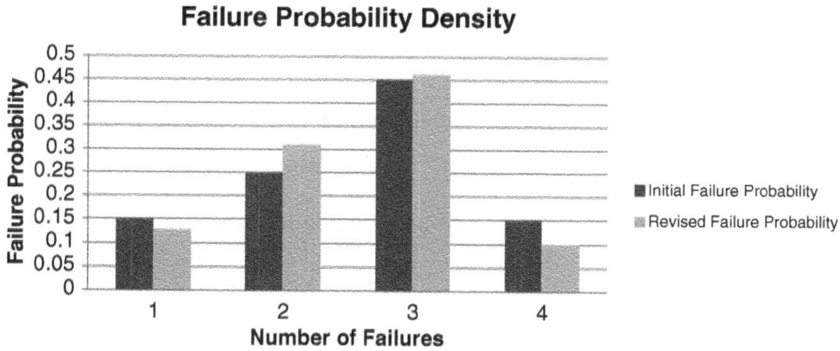

Figure 13.1 Failure Probability Density

$$P(D=2|\lambda_4)P(\lambda_4)=0.0220.$$

The proportionality constant $K = 1/(0.0276 + 0.0677 + 0.1008 + 0.0220) = 4.586$

$$P(\lambda_1|D=2)=P(D=2|\lambda_1)P(\lambda_1)\,K=0.1265$$
$$P(\lambda_2|D=2)=P(D=2|\lambda_2)P(\lambda_2)\,K=0.3103$$
$$P(\lambda_3|D=2)=P(D=2|\lambda_3)P(\lambda_3)\,K=0.4623$$
$$P(\lambda_4\mid D=2)=P(D=2|\lambda_4)P(\lambda_4)\,K=0.1008.$$

The $P(\lambda_1)$ changes from 0.15 to 0.1265, $P(\lambda_2)$ changes from 0.25 to 0.3103, $P(\lambda_3)$ changes from 0.45 to 0.4623, and $P(\lambda_4)$ changes from 0.15 to 0.1008 (Fig. 13.1).

The probability of two or three failures in a year has increased, and the probability that the system will experience one or four failures in a year has decreased. We can conclude from this example from our revised estimate that it is *more likely* that the actual number of failures in a year is either two or three, compared with our previous estimate. Thus, our revised estimate has given us more confidence about the number of failures.

We can also use Bayes's Equation (13.1) to update existing *a priori* failure rate assessments with new information.

EXAMPLE 13.2

Let us consider a system in which the initial predicted failure rate was 2×10^{-5} failures/hours. After deployment of 20 systems to various sites, five failures were observed. With this new failure information, what is the revised failure rate estimate?

Solution

The observed failure rate is:

$$\lambda_o = (5\text{ failures})/((20\text{ systems})(6\text{ months})(720\text{ hours/month}))$$

$$\lambda_o = 5.8 \times 10^{-5}\text{ failures/hour.}$$

The assumed initial failure rate (λ_i) and observed failure rate (λ_o) are:

$$\lambda_i = 2.0 \times 10^{-5}$$

$$\lambda_o = 5.8 \times 10^{-5}.$$

Let us update our prediction using Bayes's estimation. If we assume that the true failure rate is either our initial estimate or the observed failure rate with both equally likely, then

$$P(\lambda_i) = 0.5$$

$$P(\lambda_o) = 0.5.$$

Given that λ is the true failure rate, what is the probability that N failures would occur during the 6-month time interval? Assuming a constant failure rate and instantaneous repair, we can model the events as a Poisson process (see Section 6.3 for a description of a Poisson process):

$$P(N|\lambda) = (20\lambda t)^N e^{-20\lambda t}/N!$$

The probability of observing five failures over a 6-month period, given a failure rate of 5.8×10^{-5} is:

$$P(N|\lambda_o) = [(20)(5.8 \times 10^{-5})(720)(6)]^5 exp[-(20)(5.8 \times 10^{-5})(720)(6)]/5!$$

$$P(N|\lambda_o) = 0.1755.$$

And the probability of observing five failures over a 6-month period, given a failure rate of 2.0×10^{-5} is:

$$P(N|\lambda_i) = [(20)(2.0 \times 10^{-5})(720)(6)]^5 \exp[-(20)(2.0 \times 10^{-5})(720)(6)]/5!$$

$$P(N|\lambda_i) = 0.02281.$$

From Equation (13.1), the probability the failure rate is equal to λ_o given that N failures occurred during the observation time period is:

$$P(\lambda_o|N) = P(\lambda_o)P(N|\lambda_o)/[P(\lambda_i)P(N|\lambda_i) + P(\lambda_o)P(N|\lambda_o)].$$

Substituting the values previously obtained, we get:

$$P(\lambda_o|N) = (0.5)(0.1755)/[(0.5)(0.1755) + (0.5)(0.02281)]$$

$$P(\lambda_o|N) = 0.885.$$

Similarly, the probability the failure rate is equal to λ_i given that N failures occurred during the observation time period is:

$$P(\lambda_i|N) = (0.5)(0.02281)/[(0.5)(0.1755) + (0.5)(0.02281)]$$

$$P(\lambda_i|N) = 0.115.$$

The best updated estimate of λ is thus:

$$\lambda_{\text{new}} = P(\lambda_i|N)\lambda_i + P(\lambda_o|N)\lambda_o$$

$$\lambda_{\text{new}} = (0.115)(2.0 \times 10^{-5}) + (0.885)(5.8 \times 10^{-5})$$

$$\lambda_{\text{new}} = 5.363 \times 10^{-5}.$$

13.3 EXAMPLE OF ESTIMATING HARDWARE MTBF

As a practical example, during the development phase of a new Base Station Controller system, we evaluate processing blades from several vendors. The initial MTBF for an ATCA processing blade was estimated to be 550,000 based on three similar cards:

(a) Emerson ATCA Blade, MTBF = 1,200,000 hours
(b) Compact PCI Card, MTBF = 1,400,000 hours
(c) Kontron AT8400 Card, MTBF = 150,000 hours.

We perform a similar evaluation for other hardware components that compose the system. The estimation comparison worksheet is shown in Table 13.1. The second column contains the estimates we used for the reliability model.

Armed with these initial estimates, we can create an updated estimate worksheet that incorporates new information as it becomes available. A snapshot of a project worksheet is shown in Table 13.2. Both Bayesian techniques (described in this chapter), as well as chi-square estimates (covered in Chapter 12) are incorporated. The worksheet is dynamic, that is once new field data or other information become available, we can automatically recalculate the worksheet using the embedded formulas. The revised MTBFs and confidence levels are then fed into our reliability model to give us a hopefully more accurate representation of the current system reliability behavior, as well as be able to more accurately predict future system reliability performance.

13.4 ESTIMATING SOFTWARE MTBF

For software, MTBF estimates depend on whether we are reusing existing software that has already been working in the field or developing new software. If we build on software from a previous project, adding additional functionality on top of this, we

TABLE 13.1 MTBF Comparison Worksheet

		Comparable Industry Survey			
Device Name	MTBF	Device Name	MTBF	Delta	Comments
PAM	1500000	AMC 9210	1000000		Greater than >1000000
		AMC131	263505		
		AMC131	224205		
		AMC121	653948		
		AM4011	223951		Kontron
		AMC10G-604	1622712	8.18%	10 Gigabit CX4 Ethernet and 1 Gigabit Ethernet
Carrier	550000	S110	750000	36.36%	
			1200000		Emerson ATCA Blade
			1400000		Compact PCI
		AT8400	150000		Kontron
SAM	1300000				
SSC	280000	ATCA Switch	122173		
			103986		
		ZX5000	210000	–25.00%	
		AT8902	150000		
EMHBLADE	310000		1400000		
Upper FTM	160000		60000	–62.50%	Dual Tray
PEM	2624672		1362480		PIM300X
			400000		
			2000000	–23.80%	Power supply

can extract software failure rate information from existing projects and extrapolate it to the current project. We can also identify failure rates from industry information for common commercial software. It may also be possible to obtain some reliability information from third-party suppliers.

13.5 REVISING INITIAL MTBF ESTIMATES AND TRADE-OFFS

This generally happens after the system gets deployed in the field and we obtain field data that reflect the system performance.

Another practical example comes from the telecommunications industry. One hundred twenty-one processing blades were deployed in the field in various markets within North America. The operational time for each card in every system since initial deployment was tracked, and the sum total of all operational months calculated. In this case, we had 14,966 months of in-service time for the 121 cards (Table 13.3) or 10,775,520 hours of service. During this observation period, 39 cards (almost 33%) were returned due to failure. Based on these data, the point MTBF point estimate is calculated: total number of in-service hours divided by the total number of failures = 276,295 hours. With this information, the highest MTBF and lowest MTBF for

TABLE 13.2 Component Reliability Estimates

				MTBF Confidence Bounds							
				60%		80%		90%		95%	
WAB Hardware Failures		Number of Failures	Point MTBF	Lower MTBF	Upper MTBF	Lower MTBF	Upper MTBF	Lower MTBF	Upper MTBF	Lower MTBF	Upper MTBF
MTTF Estimate (hours)	275,000	0	275,000	170,867	NA	119,431	NA	91,797	NA	74,548	NA
MTTR (hours)	8.00	1	72	24	323	19	683	15	1,404	13	2,844
		2	36	17	87	14	135	11	203	10	297
		3	24	39	141	32	196	28	264	25	349

MTBF Estimate (Bayes's)	Failure Rate Est (Bayes's)	MTBF Estimate (Bayes's Classic)					
		P(N|λe)	P(N|λo)	P(λe|N)	P(λo|N)	λ	MTBF
N/A							
137,536	7.2708E-06	2.6175E-04	3.6788E-01	0.0007	0.9993	0.013879	72.05122
91,691	1.0906E-05	3.4265E-08	2.7067E-01	0.0000	1.0000	0.027778	36
68,768	1.4542E-05	2.9904E-12	2.2404E-01	0.0000	1.0000	0.041667	24

				MTBF Confidence Bounds							
				60%		80%		90%		95%	
WAB Software Failures		Number of Failures	Point MTBF	Lower MTBF	Upper MTBF	Lower MTBF	Upper MTBF	Lower MTBF	Upper MTBF	Lower MTBF	Upper MTBF
MTTF Estimate (hours)	35,000	0	35,000	21,747	NA	15,200	NA	11,683	NA	9,488	NA
MTTR (hours)	0.10	1	72	24	323	19	683	15	1,404	13	2,844
		2	36	17	87	14	135	11	203	10	297
		3	24	39	141	32	196	28	264	25	349

MTBF Estimate (Bayes's)	Failure Rate Est (Bayes's)	MTBF Estimate (Bayes's Classic)					
		P(N|λe)	P(N|λo)	P(λe|N)	P(λo|N)	λ	MTBF
N/A							
17,536	5.7026E-05	2.0529E-03	3.6788E-01	0.0055	0.9945	0.013812	72.40168
11,691	8.5538E-05	2.1116E-06	2.7067E-01	0.0000	1.0000	0.027778	36.00028
8,768	1.1405E-04	1.4479E-09	2.2404E-01	0.0000	1.0000	0.041667	24

TABLE 13.3 MTBF Analysis Example

Scope of analysis: Analysis covers blade hardware failures from June 2011 to September 2012
Assumptions:

1 Number of deployed systems, and number of payload boards is defined in the availability prediction deployment spreadsheet
2 Payload boards have a constant failure rate
3 The MTBF confidence intervals for a payload board is obtained by multiplying the MTBF point estimate by the chi-square confidence intervals:

Total number of cages deployed	121
Number of payload boards/cage	10
Total payload boards—months of deployment	14966
Total payload boards hours	10,775,520
Total payload boards board failures	39
Original MTBF estimate prior to deployments	200,000
MTBF point estimate	276,295

	Chi-Square Factors							
	60%		80%		90%		95%	
Num Failures	Lower MTBF	Upper MTBF	Lower MTBF	Upper MTBF	Lower MTBF	Upper MTBF	Lower MTBF	Upper MTBF
39	0.86278	1.158275976	0.8076356	1.24834911	0.7656105	1.32983381	0.73151128	1.4062764

	MTBF Confidence Bounds							
	60%		80%		90%		95%	
MTBF (Point Estimate)	Lower MTBF	Upper MTBF	Lower MTBF	Upper MTBF	Lower MTBF	Upper MTBF	Lower MTBF	Upper MTBF
276,295	**238,382**	**320,026**	**223,146**	**344,913**	**211,535**	**367,427**	**202,113**	**388,548**

the expected range determine the accuracy of the initial MTBF numbers, assuming that we have a constant failure rate.

We also used a chi-square calculation to obtain the MTBF numbers for different confidence levels, that is, 60, 80, 90, and 95% (Table 13.3). The higher the confidence level, the wider is the MTBF range.

At 60% confidence, the lower MTBF = 239,000 hours and the upper MTBF = 320,000. If we look at the 95% confidence level, the MTBF range is 200,000–390,000 hours. So more confidence implies a higher range of values. We can use these confidence levels to analyze "what-if" scenarios to determine how big an impact a lower MTBF will have on the system if this lower value is closer to the true MTBF of the card. These MTBF estimates can then be plugged into the fault tree analysis to estimate the impact to overall system availability. So based on the chi-square calculation and the Bayesian updates, we can reaffirm our point estimate from the field.

Postmortem of the boards indicated a higher than expected failure frequency of a pull-up resistor on the card. An analysis was conducted and determined if all the cards in the system were replaced with new versions of the cards that incorporated the resistor fix, the MTBF would increase from 275,000 to 325,000 hours (18% increase). Had this fix resulted in the doubling of the MTBF, then it would make sense to replace all of the cards in the field. In this scenario, however, since the MTBF increased only by 18%, the system availability was still well-within the required margins. Thus, it was concluded that the previous versions of the boards would be replaced only when necessary or if a major upgrade was scheduled. The analysis led us to significant cost avoidance.

13.6 SUMMARY

For our model to be as accurate as possible, we required "good" estimates of reliability parameters. Without any data related to the system we are designing, expert opinions of reliability engineers and analysts are used. These experts will usually base their estimates on reliability data similar projects, systems, or components. Vendors may provide some reliability data as well. If we have test data from the system being built, or better yet, data from deployed systems, we use that information to update our estimates. Monte Carlo simulations are also helpful for our estimation.

A powerful technique is the use of Bayes's theorem, which provides a method for combining previous estimates with new data to create a new estimate that combines both.

CHAPTER 14

Six Sigma Tools for Predictive Engineering

14.1 INTRODUCTION

In this chapter, we discuss several Design for Six Sigma (DFSS) tools and methods that are used for high availability system analysis and design. This is not targeted to be a comprehensive illustration of Six Sigma for Product Development but is meant to provide a high level overview of several of the most useful techniques, along with some illustrative examples for the practitioner. Subsequent chapters use some of the techniques described here in greater detail for practical real-life applications. This chapter also provides a context of where these tools fit across the product development life cycle and how they are useful in mitigating or eliminating design risks.

DFSS is a relatively new methodology with a primary objective of "Doing it right the first time." DFSS is product oriented, not process oriented, unlike other Six Sigma methodologies, but nevertheless DFSS should be integrated into a product development framework so that it becomes repeatable as well as optimized to yield sustained success for any organization. DFSS sharpens the development process by offering qualitative- and quantitative-driven engineering improvement techniques. DFSS is the application of specific tools and methods that aid in designing products and systems to achieve Six Sigma levels of operating performance. DFSS is used to identify and manage Critical Parameters for enhanced risk management, design robustness, customer requirements fulfillment, and product launch with Six Sigma quality. DFSS supports predictive engineering where the capability of the system can be well understood even as we design the product. Hence, it offers *leading indicators*, unlike the conventional *lagging indicators*, which provide after the fact view of quality. It helps identify and manage risks throughout the development life cycle from product concept through deployment.

At the heart of DFSS is critical parameter management, which is a technique to identify those vital elements within a system that drive product differentiation, customer satisfaction, and engineering excellence. A critical parameter can often be thought of as a high priority Key Performance Indicator (KPI) that has maximum

Designing High Availability Systems: Design for Six Sigma and Classical Reliability Techniques with Practical Real-Life Examples, First Edition. Zachary Taylor and Subramanyam Ranganathan.

impact on the successful delivery of the system. The Critical Parameter Scorecard identifies and track these parameters and risks transparently at a program level. It also helps manage inherent variability of a system's performance to ensure Six Sigma quality and reliability, specifically with respect to meet its customer requirements.

Skeptics often ask if DFSS is added overhead cost. In fact, an organization that begins using DFSS can realize significant benefits and overall development cost reduction quite quickly. DFSS is an enabler of better design and creativity; be it an Agile or Conventional development methodology, DFSS adds significant value and builds momentum right from start. Just as a surgeon benefits through advanced tools, such as microscopic navigation, DFSS offers a palette of techniques that engineers can leverage to improve their effectiveness at designing systems that reliably meet customer requirements. To get started, identify what is key to your organization, select a small set of the highest leveraged techniques initially and then continue to expand the scope of application for even greater benefits as your capabilities mature. Once engrained into the organization's DNA, the company can realize multifaceted benefits. Though some experts classify Six Sigma for Product Development into Marketing for Six Sigma (MFSS), Technology Design for Six Sigma (TDFSS), and so on, we have considered DFSS to encompass both of these flavors and to traverse the complete life cycle of any product.

14.2 GATHERING VOICE OF CUSTOMER (VOC)

Obtaining and understanding VOC is critical for any product to be successful. There are several techniques that can be employed to capture VOC. Though this is an ongoing activity for a healthy business, it typically needs to be done with great rigor at the very beginning of the life cycle. Some of the techniques, such as interviews, Focus Groups, Ethnography, and Surveys, are introduced here.

Interviews. Having direct interviews with customers helps us understand what their real needs are, their strategic roadmaps, what market risks they perceive, and how we as a company, can enable them to mitigate those risks to be successful. The key is to align our strategic plans and product roadmaps with their roadmaps so that there is synergy and capability to fulfill their requirements. Thus, face-to-face interviews are critical to effectively achieve this understanding. They also help build strong working relationships and closer partnerships with customers. A company can be successful only if its customers are successful with its products and solutions; this is a key mindset driven by DFSS.

Interviews can be conducted at various levels. Talking to senior executives helps understand their vision and business strategy. Talking to engineers and operators who have used our system can provide valuable insights for product improvements and how our products compare with respect to the competition. Engineers can also offer a graphic perspective of the key issues needing resolution and even new solutions that can be implemented as product enhancements. For example, in a face-to-face interview, we may learn that a card could need several resets over a week due to unstable software; or there could be issues with interoperability with a third-party product causing additional downtime, or there could be inputs related to how a user interface can be simplified for better operations and control, and so on.

Talking to different levels across different functions is a very valuable approach to gather richer insights into how we can design the best product for the customer. Though some of their remarks could be brutally honest and critical, they are good in the long run because they will help us develop better products and eventually serve our customers better. We can also bring about drastically positive changes in our products and services that result in increased customer delight and loyalty, as well as competitive differentiation.

Focus Groups. This is another method to solicit customer requirements in a group setting. The group could have a mix of different experts who can provide innovative perspectives. For example, for a telecom product, we could have people from RF planning, network operations, lab support, and so on if we are gathering inputs related to performance and ease of use. For a medical device, we could have doctors, nurses, specialists, and some consumers who have used a similar device. Focus groups typically work well to get concurrence of the top few "pain points" and priorities.

Ethnography. Ethnography is an extremely effective way of gathering insights and is used more often in consumer/commodity businesses. Ethnography is a technique where we spend time with our customers or consumers and understand what they go through in their day-to-day aspects of using our products and services. Examples include observing and studying how a field engineer operates a system, or a patient who uses a medical device every few days, and so on. We have to invest the time and effort to be there and understand the functions they do and grasp their pain points. The key is to discover insights and develop breakthrough solutions that can eliminate their pain points. These improvements could also improve their productivity substantially and thus increase their loyalty towards our products and services. So especially for this exercise, a combination of experience, empathy, and creativity will be a huge plus. Imagine putting a medical device on a child or a senior citizen struggling to do it herself. Think about these powerful scenarios that can drive breakthrough innovations and make a huge positive impact. Ethnography is a valuable technique, as it sometimes gives us the chance to observe product improvement opportunities first hand where otherwise they may never be discovered.

Surveys. Surveys are very common and less expensive. Unlike interviews, which ask open ended questions, surveys typically ask specific pointed questions to get a deeper understanding of the area that is being probed. For example, a customer satisfaction survey could contain specific questions on outages and their impact.

Surveys may sometimes not be as effective as other techniques. Given the low response rates and depending on the way in which the answer scales are crafted the true response of the target population may be diluted. People might tend to put things in the middle that might not really offer constructive input into what should really be improved. This is not to dismiss surveys; they are effective, provided they are crafted well, and targeted at the right set of people. They complement other VOC gathering techniques and are more often used to solicit feedback and understand the relative priority of different known requirements rather than to obtain new requirements.

A combination of these VOC techniques can yield significant actionable inputs, as well as provide clear insights on what customers really need. We can consider this as an *outside-in* perspective to product development.

From an availability perspective, customers typically expect Five 9s performance (99.999% availability for intended use). This may not be explicitly expressed, but may have to be elicited through a formal communication. For a brand-new product where there is incremental development in redundancies and fault management, there may be room for negotiating less than Five 9s availability for the initial field trial; but subsequently, a Five 9s availability will have to be achieved. Meanwhile, in other critical systems, such as in medical and aviation domains, Five 9s performance may not be good enough.

Gathering VOC on availability may also include specific requirements related to repair policies; for example, how frequently they repair, how frequently they manage site visits, will repair be driven in-house or will it be outsourced to third parties, and so on. There could also be expectations related to keeping the return rates low, which will have direct impact on the reliability of the components.

In later chapters, we talk about how we design high availability systems making sure we have redundancy techniques and high availability components. We present detailed models and analysis techniques. However, once the system is built, that is not the end of the story. Hence, we need to keep eliciting and even imagine some evolving creative use cases during the life of the product. For example, what if a data-only link is used to overlay voice, but we did not explicitly design the system for that? What about interworking with nonstandard signaling systems in developing countries in remote areas?

The way in which the system is maintained in the field has equal if not more impact on the system availability. Thus, we need to consider the entire life cycle that includes how the customer plans to deploy the system, use the system, maintain the system, repair the system, and so on. The assumptions associated with that must be negotiated between the system provider and system user in order for the availability requirements given by the customer to be met. In some cases, the system provider may be responsible for the maintenance of the product, and on the other extreme, the customer will take complete responsibility of the product without vendor involvement.

14.3 PROCESSING VOICE OF CUSTOMER

After gathering customer requirements, a Kawakito Jira (KJ) or affinity analysis can be performed. Herein, we try to analyze, aggregate, and categorize information that has been gathered. Examples of categories include screen size of a television, web browsing speed of a mobile phone, subscriber capacity of a Base Station Controller, cost of ownership of a hi-tech system, and so on. We should bubble up a list of distinct requirements as a summary and go through a prioritization process with the customer(s). This will not only help confirm the understanding of customer requirements, but also help focus resources on the right areas for development. After VOC

TABLE 14.1 Example of VOC Prioritization

Voice of the Customer	Customer 1	Customer 2
Low deployment cost	8	10
Low total cost of ownership	10	10
Advanced radio capabilities	10	7
High throughput	8	8
5 nines availability	10	10
Ease of operability and maintainability	9	9
Video on demand	9	8
Interoperability with legacy systems	9	9
Interoperability with other vendor networks	9	9
Interoperability with new technologies	7	8

is analyzed and prioritized, it immensely helps system engineers to create technical requirements that align with the VOC.

Table 14.1 is an example of VOC prioritization for different customers. The table uses a 10-point scale (10 being highest priority) to prioritize the VOC. This is especially helpful if requirements are still trickling down from various sources. Another method to prioritize VOC is to use a 100-point prioritization process. In this method, a list of key stakeholders who were directly involved in VOC gathering are given 100 points each, and asked to distribute these 100 points among the list of VOCs. The idea is similar to having $100 to spend, and how much each key stakeholder is willing to allocate to each requirement. The overall priorities are then calculated. In either case, it will be highly beneficial to confirm these priorities with the customer to make sure that the program is well aligned with their requirements.

In Table 14.1, we have kept these VOCs at a high level without specifying their targets. However, we need to discuss these specifics with the customer as well as with engineers to make these VOCs solid.

14.4 KANO ANALYSIS

Kano analysis is used to understand how a customer could potentially feel regarding presence or absence of a given product feature or attribute. In addition, it helps understand how the customer would feel depending on how well a given feature or attribute is provided (Fig. 14.1).

Curve 1 represents features that are generally taken for granted. However, if these features are either absent or delivered with poor quality, the level of satisfaction can rapidly deteriorate to the extent that it could even result in loss of business. A simple example is delivering a high quality TV without a remote. Hence, attributes corresponding to Curve 1 are also referred to as hygiene factors or “must-haves.” If these are not present or if their quality is not acceptable, the customer can become extremely dissatisfied. On the flip side, if these factors are done well, they are not going to delight the customer because these are taken for granted as basic or essential.

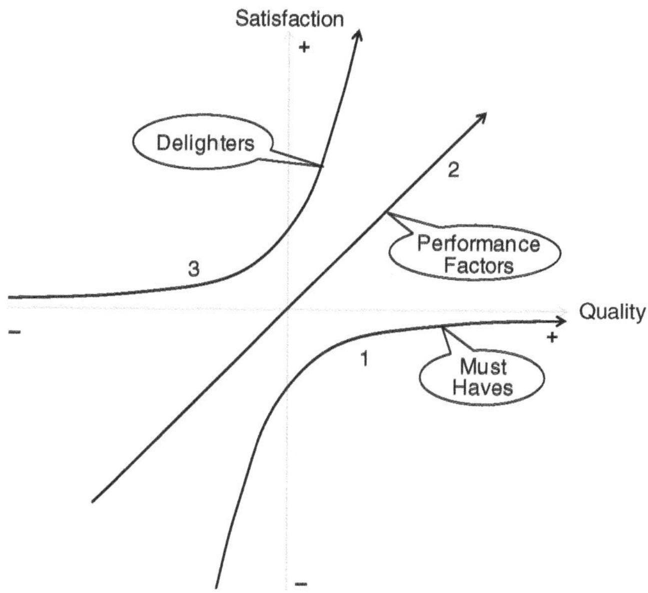

Figure 14.1 Kano Analysis

Curve 2 describes reaction to what are called "performance factors." Examples include gas mileage in a car, increased system capacity in a Base Station Controller, accuracy and speed of response of a medical device, a large television screen. Thus, the higher the performance, the greater is the customer satisfaction. Such attributes are also referred to as "linear satisfiers."

Curve 3 corresponds to attributes that can bring about customer delight. For example, a simple, easy-to-use menu in a user interface, providing the senior citizen market with a large display phone, and providing a one-touch button to accomplish a critical task, could be delighters.

A good strategy is to have a feature basket that has a balanced combination of must-haves, performance factors, and delighters. This analysis will also help realize if a feature could be treated as "indifferent." Thus, Kano analysis can help system engineers and architects to channel their creativity in alignment with customer expectations. If there is knowledge that historically, user interface design has been below expectations, the architects could focus their creative energies to convert this into a delighter. A good strategy is to have a "feature basket" that has a balanced combination of must-haves, performance factors, and delighters.

With respect to high availability requirements, we could say that a Five 9s availability is typically a must-have. If we design a system that falls short of this expectation, the dissatisfaction could rapidly go up. On the other hand, if we are able to exceed the Five 9s availability requirement, say with zero field outages over an extended period of time, which could even become a delighter. System initialization time could be considered as a performance factor. The faster a system initializes and comes into service, the higher will be the customer satisfaction. It must also be noted

TABLE 14.2 Example of Kano Analysis

Feature	Customer 1	Customer 2
Access attempt improvement	Performance factor	Performance factor
Increased capacity	Must-have	Performance factor
Redundant cage support	Must-have	Must-have
Carrier card redundancy and failover	Must-have	Must-have
System performance enhancements	Performance factor	Performance factor
Transport layer enhancements	Delighter	Delighter

that in a given feature set, different customers could have different perspectives; thus, a must-have for one customer could be a performance factor for another customer (Table 14.2), and so on. These factors can also transition from a higher to a lower curve with passage of time and as expectations increase.

To summarize, Kano analysis is a simple but very useful analysis technique. A good mix of must haves, performance factors, and delighters can help set the stage for further success in designing and developing the best product.

14.5 ANALYSIS OF TECHNICAL RISKS

Now that the feature set has been identified and prioritized, a technical risk analysis is performed to focus DFSS tools and methods on the right feature set (Table 14.3). High risk typically stems from a feature being relatively new to the engineers or technically challenging to implement or having market differentiation, or a combination of these flavors. High risk and high value could also stem from contractual commitments, such as revenue realization or penalties. Typically, DFSS is applied to high risk high value features so that they get special emphasis on design and quality. For example, if there are 50 features in a given product or service, it is not necessary that all of them go through predictive engineering techniques. A subset of those 50 features that are short-listed as high risk and high value will stand to benefit through these techniques.

High availability typically belongs to this category. Thus, it becomes imperative to use predictive modeling and simulation techniques to proactively manage the associated risks and deliver it right the first time. An example set of features impacting availability is shown in Table 14.3.

14.6 QUALITY FUNCTION DEPLOYMENT (QFD) OR HOUSE OF QUALITY

A QFD associates a prioritized set of customer requirements with a set of technical requirements and quantifies each technical requirement with a priority. Typically, the top few ranked technical requirements are identified as “Critical Parameters” for the program. QFD serves as a key input to systems engineers to design and develop

TABLE 14.3 Technical Risk Analysis

Feature	Risk
Access attempt improvement	Medium
Increased geographic coverage	High
Increased capacity	High
Redundant cage support	High
Carrier card redundancy and failover	High
System performance enhancements	Medium
Transport layer enhancements	Low

the product with high customer alignment. Creating, populating, and analyzing a QFD should be a collective effort between Engineering and Product Management at the very least.

As shown in Figure 14.2, the Customer Requirements and their priorities are in the rows, and the technical requirements are in the columns. The technical requirements are also referred to as technical parameters in case they are measurable. The direction of goodness just above the technical requirements specifies which direction a particular requirement should trend. For example, the direction of goodness for Session Capacity is + (positive) because the higher the capacity, the better it is for the product. The direction of goodness for Command Response Time on the contrary is a – (negative) because smaller the response lag time, the better it is for the product being designed.

The correlation matrix (roof of the house) indicates how each technical requirement is correlated with the rest of the technical requirements. For example, Failover Time and Availability are negatively correlated because the higher the Failover Time, the lower the Availability will be. The correlation matrix provides engineers an idea of the design tradeoffs.

Each technical parameter should typically have a target nominal value, that is, performance expectation during normal operating conditions. Each technical parameter should also include an Upper Specification Limit (USL) or a Lower Specification Limit (LSL) or both. For example, Failover Time can have a USL of 30 seconds; that is, it should not exceed 30 seconds at any cost.

The columns corresponding to customer perception are together referred to as the competitive matrix. This matrix illustrates how better or worse the product being developed is, when compared with competitor products. This will provide both engineers and product managers an idea of the degree of differentiation. As a side note, competitive intelligence is often difficult to obtain and thus this information is usually sparsely populated in a QFD.

The heart of a QDF is the interrelationship matrix indicated by dotted lines. This is typically filled out in a team setting where system engineers and architects assess how each technical parameter fulfills the VOC list. The entries in the matrix are High, Medium, Low, or None, corresponding to 9, 3, 1, and 0, respectively. The scales are purposely designed to be nonlinear so that the ones that are most critical bubble up to the top. An entry of High (9) in a cell indicates that the technical parameter has a high degree of fulfillment or impact on that VOC. For example, the technical requirement "Failover Detection %" will have a high impact on the customer requirement

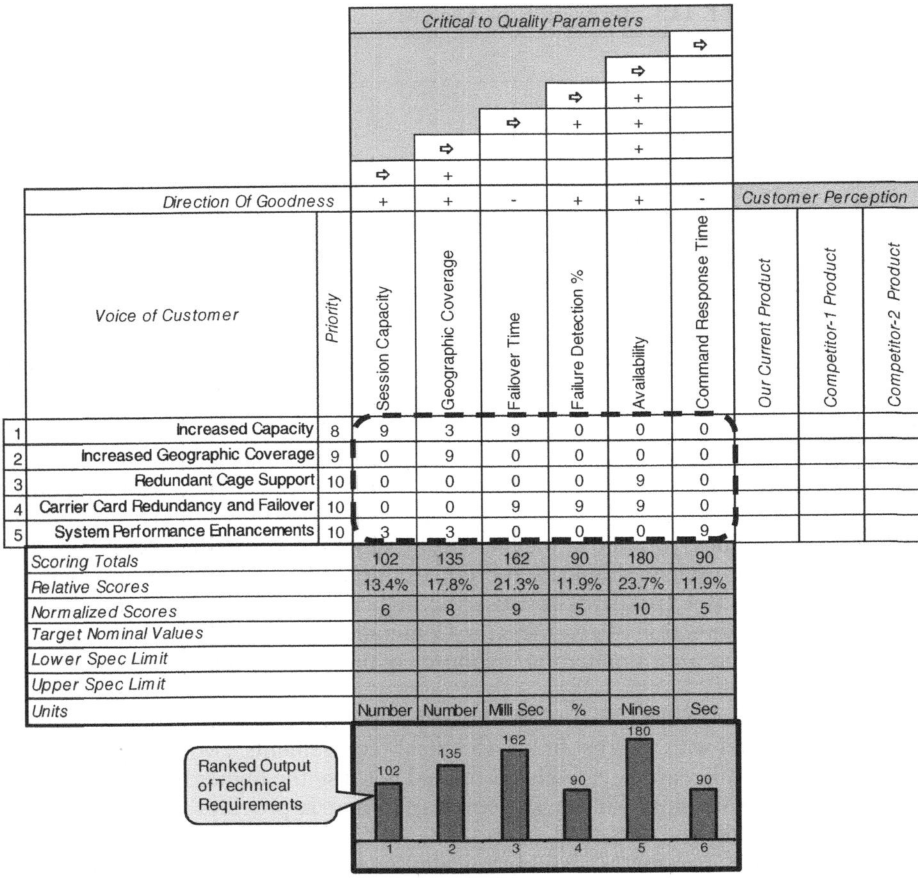

Figure 14.2 Example of a QFD

"Carrier Card Redundancy and Failover." After each cell in the interrelationship matrix is filled out, the sum of the product of each cell with the customer requirement priority is computed resulting in the scoring totals that are graphed.

Thus, the final outcome of the QFD is the ranked output of the technical parameters. This will provide the core team a view of what is really critical for the product tying back to the customer needs. Typically, a subset of the highest ranked technical parameters is selected and formally referred to as *Critical Parameters* (as previously discussed). This is the main output of the QFD. Remember that alignment and traceability with VOC is the key to this process.

A QFD can be considered to be a one-stop-shop that brings together customer requirements, their priorities, technical requirements, performance targets, correlations, tradeoffs, and competitive information in a crisp graphical representation. Thus, this is a key tool for any new product development to bridge VOC with technical product attributes. It can also be used during subsequent product revisions or even with new product lines. A QFD is not just for hi-tech products; a shampoo, detergent, medicine, or any other product will benefit from a QFD analysis.

14.7 PROGRAM LEVEL TRANSPARENCY OF CRITICAL PARAMETERS

We know that a key outcome of the QFD is a set of Critical Parameters along with their targets. For example, ease-of-use of a User Interface can also be a Critical Parameter; this is more abstract to define and quantify but could be critical for a product's success.

Whether it is a conventional framework or an Agile framework, Critical Parameters must be identified early in the development cycle even before system requirements and architecture are defined. The idea is that all the team members, including Engineers, Product Managers, and Program Managers are aware of and highly focused on the critical parameters. This fosters a culture of "Critical Parameter Transparency" in a program, which is really the key to the success of any product development. What we mean here is that the Product Managers, Program Managers, Engineering Managers, System Architects, Development Engineers, System, and Product Test Teams know what the Critical Parameters and their targets are. The entire team is aware of and proactively works on these parameters, mitigating the associated risks and at the same time building the product capabilities. Tracking the progress of these critical parameters through proper governance reviews is crucial so that the associated high risks are proactively controlled and mitigated.

This awareness at different levels, including third parties (which will be addressed later in the chapter) enables the team to focus on critical aspects of the design and manage them effectively—increasing transparency enables better risk mitigation and control. Thus, it is very important to establish transparency of Critical Parameters at the program level. Once established, DFSS techniques can be used to manage the critical parameters throughout the product development life cycle.

14.8 MAPPING DFSS TECHNIQUES TO CRITICAL PARAMETERS

Once Critical Parameters are identified, the Black Belt or system engineer typically maps each critical parameter to a Six Sigma tool or method for further analysis and risk mitigation. Some of the critical parameters can lend themselves to Critical Parameter Management (CPM), Design Failure Modes and Effects Analysis (DFMEA), Fault Tree Analysis (FTA), a combination of CPM and DFMEA, or prototyping, and so on. Depending on the parameter and the experience of the Black Belt, we can actually associate the Six Sigma tools and methods with these parameters. An example mapping is provided in Table 14.4.

14.9 CRITICAL PARAMETER MANAGEMENT (CPM)

CPM refers to the process of breaking down a Critical Parameter into its atomic and controllable lower level parameters in a hierarchical manner, and then interconnecting

TABLE 14.4 Example of Mapping DFSS Techniques to Critical Parameters

Critical Parameter	DFSS Technique
Availability	DFMEA, FTA, Markov
Failover time	CPM
Geographic coverage	CPM, DFMEA
Session capacity	CPM

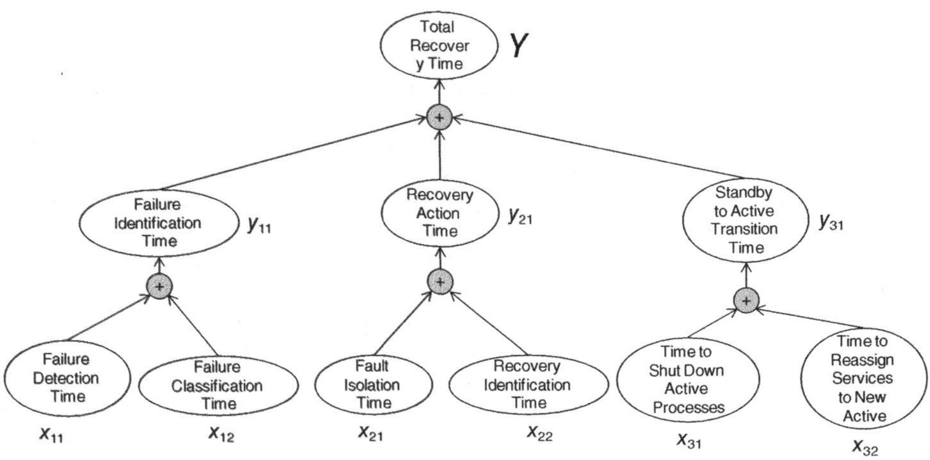

Figure 14.3 Example CPM Tree

the atomic-level parameters through mathematical relationships back to the top-level Critical Parameter. It can be summarized as a hierarchy of requirements *flowing down* and the corresponding mathematical relationships *flowing up*. If we represent a Critical Parameter as "Y," and the atomic lower level parameters as "x_i," CPM flow down and flow up will result in a mathematical equation $Y = f(x_1, x_2, \ldots x_i)$. The beauty of this process is that we can influence the outcome of Y by controlling the x_i's. The intermediate decompositions between Y and x's can be referred to as y's. In essence, we carve out the performance of Y by controlling the variation with the x's.

Figure 14.3 provides an example that is used during availability analysis. In this case, the Total Recovery Time of a card can be considered to be a simple addition of the various atomic-level times assuming that these activities occur in sequence:

$$Y = y_{11} + y_{21} + y_{31}$$

$$Y = x_{11} + x_{12} + x_{21} + x_{22} + x_{31} + x_{32}.$$

Once a mathematical relationship is established between Y and x, Monte Carlo simulations can be used to inject variations into x to predict and control the performance of Y.

14.10 FIRST PRINCIPLES MODELING

As we know, a model is a mathematical equation that represents any system behavior. Creating or deriving the equation that characterizes this behavior can be done directly from known mathematical and physical principles. This is referred to as first principles modeling. Figure 14.3 is an example of First Principles Modeling because in this case, we know that the times just add up. Another example is calculating the output voltage and current from a circuit with the use of both physical and mathematical principles, such as basic laws of circuit theory. There are many situations where the behavior is not straightforward and cannot be derived solely from first principles; in those cases, Design of Experiments along with regression techniques need to be employed.

14.11 DESIGN OF EXPERIMENTS (DOE)

There are many situations where the model or mathematical equation cannot be derived fully using first principles. In order to understand and thus control the relationship between the inputs and outputs, we need to discover the mathematical relationship between them. DOE is a technique that helps understand the system behavior and create a mathematical equation based on the results of the experiment. DOE goes hand-in-hand with another statistical technique called regression, which in simple terms is finding the best fit among a given set of data points and hence mathematically associating the output with the inputs. There could be simple or multiple regressions involved depending on the number of outputs that are in the experiment, as well as linear or nonlinear regression depending on the outcome. DOE helps set up the input and output array along with the number of runs required for a given experiment. Different DOE techniques are described in greater detail in many sources of literature. Regression helps process the input data and output results to generate the mathematical equation between them. Once we have established a good regression model, we can use Monte Carlo simulations to inject variations into the independent variables and look at the behavior of the dependent variables. DOE and regression are techniques for predictive engineering of complex parameters.

Minitab is a commonly used tool that provides capabilities for DOE and Regression analysis. A case study that uses these techniques is described in Chapter 21.

14.12 DESIGN FAILURE MODES AND EFFECTS ANALYSIS (DFMEA)

Design Failure Modes and Effects Analysis (DFMEA) is an analysis technique by which failure modes are anticipated, brainstormed, and the associated technical risks are quantified and prioritized. The outcome of an effective DFMEA is a more robust and defect-free product for our customers. Since defects are anticipated and fixed early in the development cycle, this tool drives what is called "*left-shifting*" of defects. Subsequent chapters delve more deeply into this technique and also include a case study.

14.13 FAULT TREE ANALYSIS

Fault tree analysis is another technique which is very helpful in availability modeling and prediction. It can also be used for analyzing and prioritizing the highest reliability risks for mitigation.

Many times, a customer may specify a certain level of availability as part of their requirements. Fault tree analysis helps identify and break down a system architecture into potential points of failures in a top-down hierarchical tree structure. These failures are characterized by probabilities and are also related through logic gates. If a failure happens at one node, several logic gates will fire resulting in a top-level impact or degradation in service. Though fault tree analysis provides only a single point estimate of the unavailability, it can be innovatively combined with Monte Carlo simulations to provide a probability distribution of the top-level failure. Subsequent chapters provide more insights, and a case study of this technique in designing high availability systems.

14.14 PUGH MATRIX

A Pugh Matrix is used for concept selection and optimization. During a new product or feature design, it is quite possible that different ideas are evaluated and typically the one that is most feasible (in terms of meeting customer requirements) is chosen. A Pugh Matrix helps in this process. Table 14.5 is an example of a Pugh Matrix where different concepts (ideas) are evaluated against a set of attributes as well the baseline (reference). The baseline reference is often chosen to be the previous version of the product or system. Whether a concept fulfills an attribute or not is represented as a + (better), – (worse), or 0 (on par) with respect to the baseline reference. These are then arithmetically summed up in each column to get the column scores. The concept with the highest score is generally considered as the superior concept and can be chosen for the product design. Other variations, such as weighted Pugh matrices, exist, and the reader is referred to available literature for more information.

TABLE 14.5 Example of a Pugh Matrix

High Availability Attributes	Reference	Concept 1	Concept 2	Concept 3
Low downtime	0	0	–	+
Fast repair time	–	+	0	+
Automatic recovery	+	+	+	+
System adaptability	–	+	+	+
Redundancy configurations	–	0	+	–
Over provisioning	0	0	+	–
Overload protection	0	0	–	–
Sum of pluses	1	3	4	4
Sum of minuses	–3	0	–2	–3
Total score	–2	3	2	1

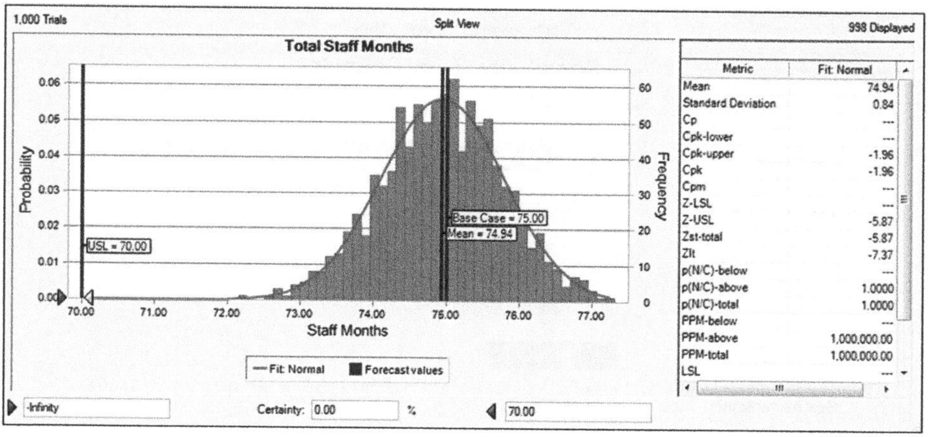

Figure 14.4 Example of Monte Carlo Simulation

14.15 MONTE CARLO SIMULATION

Monte Carlo simulation is a powerful analytical method that can be used to simulate system performance and predict system capability. Once a system is mathematically characterized, Monte Carlo simulations can be used to inject variations to the inputs to observe the behavior and resilience of the output with respect to input variations. It can be used at various stages of the design, development, and test to look at capability growth or deterioration and how the actual performance fares with respect to the predicted design performance and customer requirements. Sensitivity analysis is an approach within Monte Carlo simulation that provides a pareto of how much each input contributes to the variation of the output, and thus identifies which inputs to optimize in which order so that the desired output performance and design margin can be achieved most efficiently.

Monte Carlo simulations can also be used in project management to generate and optimize schedule planning (Fig. 14.4 and Fig. 14.5). Other powerful uses of Monte Carlo simulations can include more accurate sales and demand forecasting, as well as higher confidence prediction of a product's overall business case. Monte Carlo simulation is described in detail in Chapter 17.

14.16 COMMERCIAL DFSS TOOLS

This section provides a brief introduction to several commercially available DFSS tools that have been used in our examples and case studies. All of these are excellent tools that can be used by system engineers and Black Belts as part of predictive analysis and design. Associating a tool with DFSS by no means precludes their capabilities for other applications.

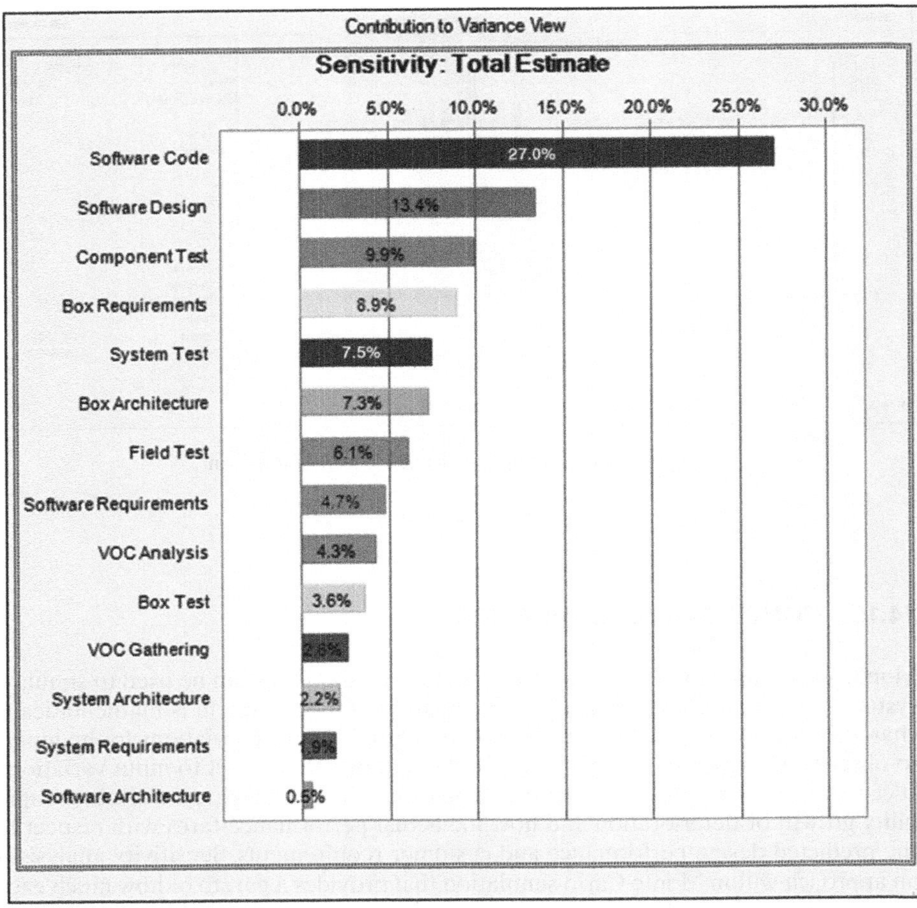

Figure 14.5 Sensitivity Analysis

14.16.1 Cognition Cockpit®

Cognition provides a comprehensive suite of tools to support end-to-end Six Sigma Product development. Capabilities include creating, characterizing, and maintaining VOCs, identifying and defining Critical Parameters, establishing traceability between Critical Parameters and VOCs, generating QFDs, Performing Critical Parameter Management Flow Down and Flow Up, running Monte Carlo Simulations, providing Capability Analysis, and summarizing outcomes through Critical Parameter Scorecards. It is an excellent tool for end-to-end DFSS-based product development in a distributed development environment. An example of application of the Cognition Cockpit tool is provided in Section 14.17.

14.16.2 WindChill (Relex)

Relex is a sophisticated tool suite that offers a variety of techniques for reliability analysis and prediction. In our examples, we have used Relex extensively to create and simulate Fault Trees, RBDs, and Markov Models.

14.16.3 Crystal Ball (Oracle)

Crystal Ball is a very useful tool to perform Monte Carlo simulations, correlation analysis, sensitivity analysis, and a variety of advanced decision-making functions. It is available as an Excel add-on, which makes the tool very convenient to use. Several Monte Carlo simulations in this book have been generated using Crystal Ball.

14.16.4 Minitab

Minitab is an extremely helpful tool that is used for a wide variety of statistical data analysis. Some of the capabilities of Minitab include conducting statistical tests, setting up Design of Experiments, performing Regression Analysis, identifying the best fit probability distribution, and Capability Analysis. Minitab is a versatile house of statistical tools for analysis and design. We have also used Minitab-based examples in this book.

14.16.5 Matlab

Matlab is a flexible engineering analysis tool that allows us to model and simulate a wide variety of random processes and systems to visualize system behavior. Since the tool is script-driven, the engineer can write innovative programs as part of modeling, simulations, and visualizing output results. Matlab models and outputs are used in several chapters of this book.

14.17 MATHEMATICAL PREDICTION OF SYSTEM CAPABILITY INSTEAD OF "GUT FEEL"

This section provides an example of using DFSS techniques to design and predict the capability of a Base Station system to support a certain level of data session capacity. It contrasts how a design that relies on just "gut-feel" can result in a system failing to meet a customer requirement, and emphasizes the importance of using data-driven mathematical analysis to ensure expected system performance. This is an example of using a selective set of techniques from the DFSS palette and includes CPM flow down and flow up, Monte Carlo simulations, Sensitivity, and Capability Analyses.

In the telecom world, system capacity is critical for revenue generation. System capacity controls the ability to support mobile device subscriber volume. From an availability perspective, loss of capacity can result in a partial or complete outage. Thus, design of system capacity with an acceptable margin becomes critical. Margin also plays a significant role in overload control. If the system is overloaded, the system may no longer function properly and that in turn impacts the availability of the system. Hence, it is very important to fully understand the architecture and performance

Memory Allocation	Controller Blade	Payload Blade-1	Payload Blade-2	Ownership	Original Design
Platform RAM Disk	x	x	x	Company	256 MB
Platform Software Run Time	x	x	x		
Platform Software Shared	x	x	x		
Platform Temp File System	x	x	x		
Call Processing RAM Disk	x	x	x	Third Party	1500 MB
CallProcessing Software Run Time	x	x			
Call Processing System, Neighbor	x	x	x		
Performance Management	x	x			
Call Processing Temp File System	x				
Controller Session Memory	x				
Payload-1 Session Memory		x			
Payload-2 Session Memory			x		
Reserve	x	x	x	Reserve	356MB
				Cpk	0.32
				Mean Session Capacity	1.6M

Figure 14.6 Session Capacity Memory Budgeting

constraints of the system, design it well, and prevent capacity failures. One technique to ensure robustness is to provide overload protection when the system attempts to exceed its designed capacity limits.

If a system fails to provide the specified capacity for a certain period of time, the system has failed and is no longer available. The system can be considered either as degraded or unavailable. Thus, sufficient due diligence needs to be exercised rather than relying on just gut instinct.

In this story, the original design for session capacity was based on "gut feel." It was assumed that this design would aggressively exceed the customer requirement by meeting a much higher internal target. An analysis of the original design was performed using a systematic CPM approach. Since capacity is a function of memory, the flow down started decomposing Session Capacity into memory allocations for different functions and included RAM, runtime, and hard disk distributed across different cards (Fig. 14.6).

Figure 14.7 represents the CPM flow down and flow up that was created using Cognition Cockpit tool. As explained in Section 14.9, Session Capacity was the Y and was broken down into smaller components—y's and x's.

As part of the budgeting process, memory blocks had to be assigned to different engineering teams and even to third parties based on the development plan. A CPM flow up was created to associate Session Capacity with its constituent functional memory blocks and the mathematical equation was derived using first principles modeling. A Monte Carlo simulation was run by injecting variations to the individual memory blocks to obtain the Session Capacity probability distribution.

The realization: in contrast with what was assumed in the original design, the CPM results showed that it was not possible to meet not only the more aggressive internal requirement, but also the customer requirement. There was also lack of sufficient design margin with respect to the lower specification limit for session capacity. The C_{pk} (a performance metric with respect to an upper or lower specification limit)

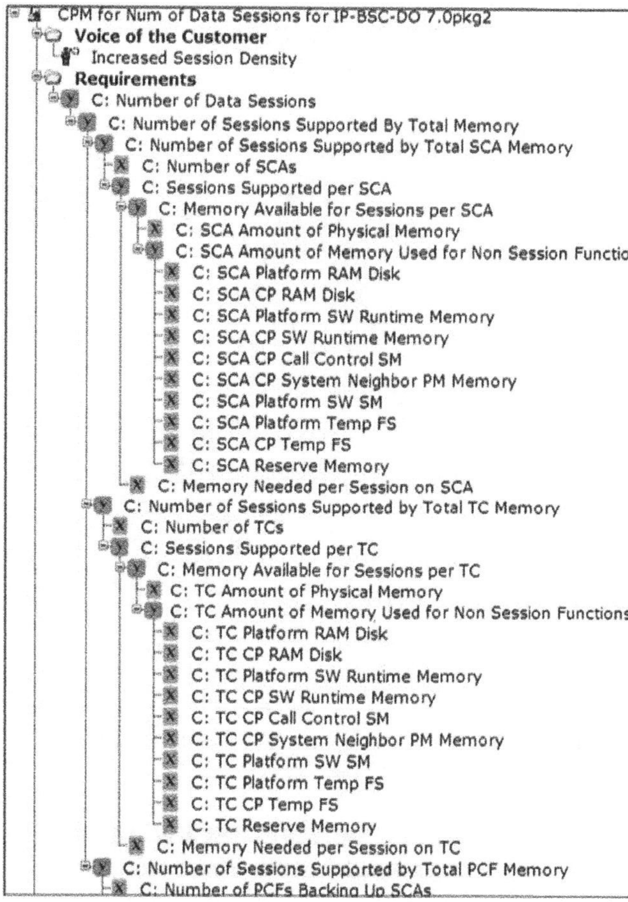

Figure 14.7 Session Capacity CPM Tree Modeled Using Cognition Cockpit

estimate of this design was only 0.32. A Minitab output of Session Capacity distribution is given in Figure 14.8.

The original design fell short of the customer requirement by 16.9%. Thus, the system engineers had to redo the memory budgets and supply revised requirements to the development teams. This was the fundamental take away from this exercise.

A byproduct of the Monte Carlo simulation is the sensitivity analysis (Fig. 14.9). This analysis provides the x's that are the highest contributors to the variation of session capacity, Y. This information helped in directing rework efforts on the right memory blocks.

A Monte Carlo simulation after several design iterations is given in Figure 14.10. The revised design was able to yield a 17% increase in session capacity and a

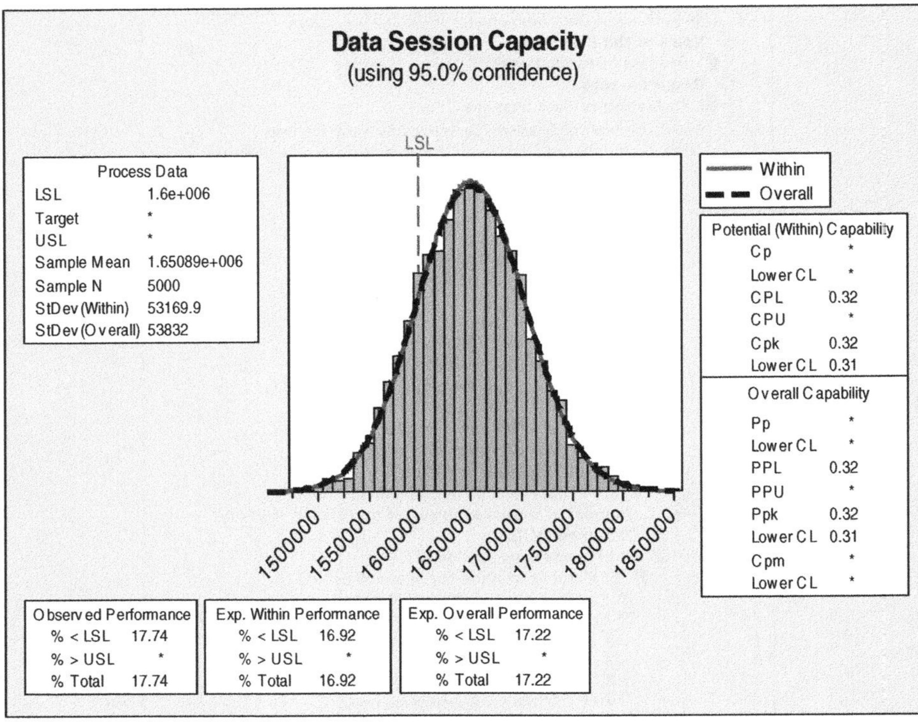

Figure 14.8 Session Capacity Probability Distribution

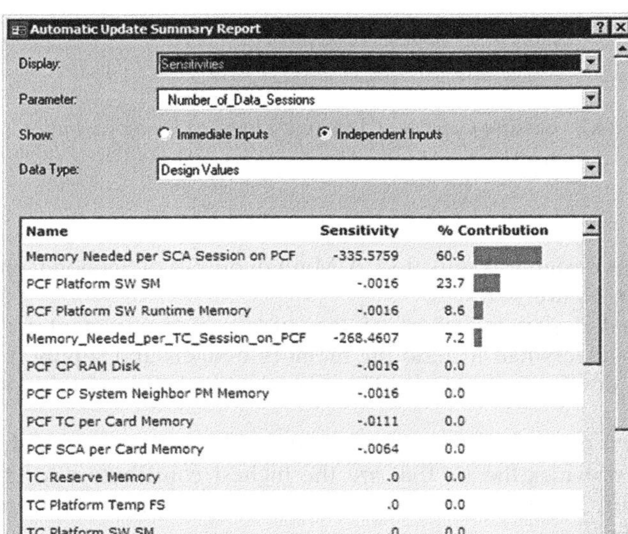

Figure 14.9 Session Capacity Sensitivity Analysis

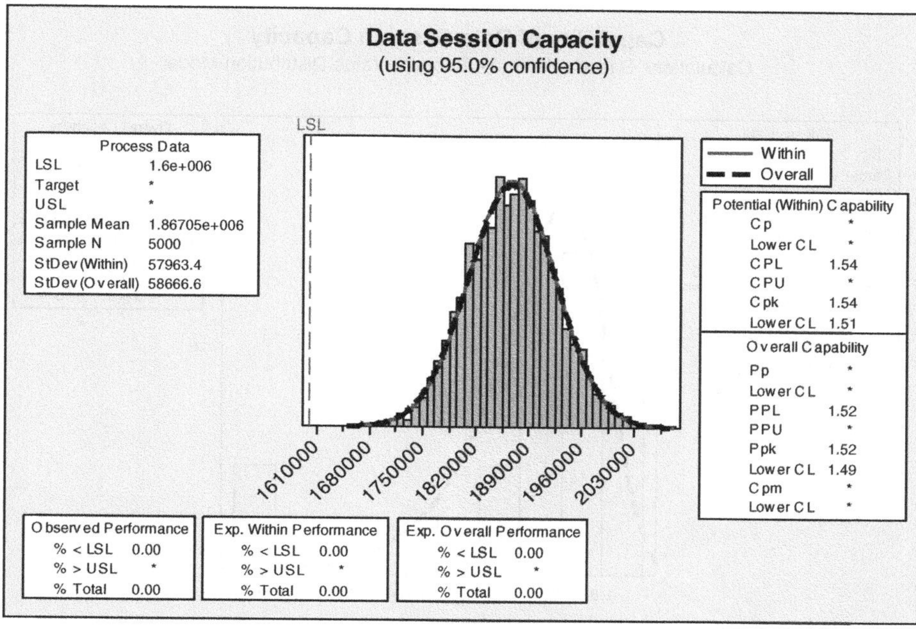

Figure 14.10 Revised Session Capacity Prediction

significant right shift of the mean. This corresponded to 220,000 additional sessions from the *same* hardware. The design margin significantly increased with a new C_{pk} of 1.54, which corresponds to Six Sigma capability. Note the dramatic transformation that was achieved leveraging DFSS techniques.

Note that when the system was tested, the mean performance corresponded to almost exactly what was predicted for the revised design (Fig. 14.11).

14.18 VISUALIZING SYSTEM BEHAVIOR EARLY IN THE LIFE CYCLE

Simulations offer a powerful way to model as well as visualize the system behavior far ahead of testing. As an example, one of the design challenges of a Base Station Controller product was to recommend an optimal set of configuration parameters to the customer. This is usually possible only after the system is deployed and tuned in the field. The model below provides an example of visualizing downlink RF messages under different configuration settings, which enabled recommendation of optimal settings to the customer (Fig. 14.12).

14.19 CRITICAL PARAMETER SCORECARD

A Critical Parameter Scorecard is an efficient mechanism to document and track Critical Parameter performance across key milestones during the life cycle. Thus, it

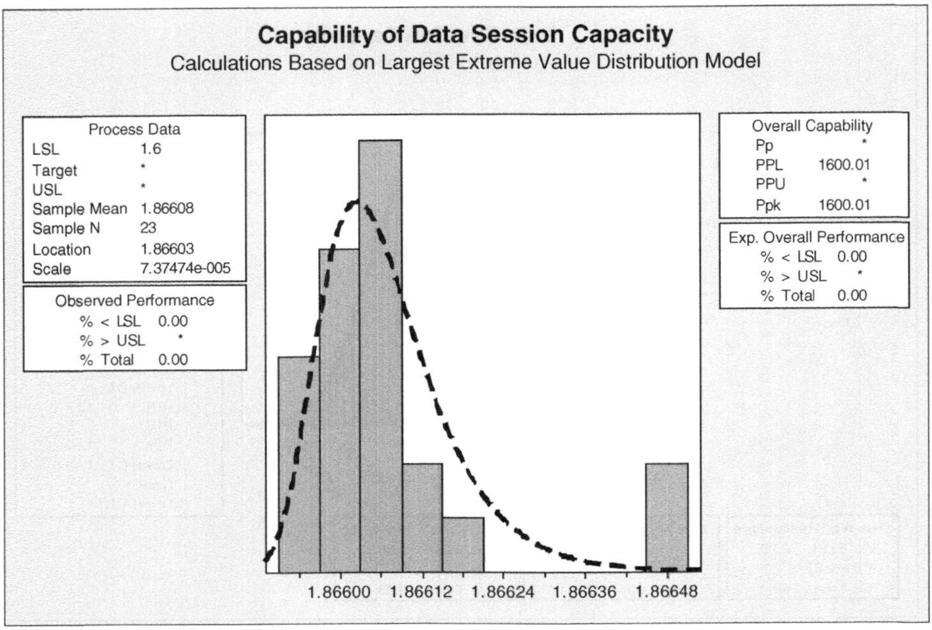

Figure 14.11 Actual Session Capacity

is a very efficient and effective tool to proactively monitor and control risks across a program. This is also key to promoting critical parameter transparency, leading quality indicators, as well as monitoring capability growth. A simple Critical Parameter Scorecard is shown in Table 14.6.

14.20 APPLYING DFSS IN THIRD-PARTY INTENSIVE PROGRAMS

In the context of globally distributed development and system integration, third-party hardware and software suppliers are an integral part of many programs. This creates program-level risks and dependencies with third parties, and thus the success of the overall program, including on-quality, on-time, on-cost, and on-plan, depends highly on the successful management of these suppliers. While there are various business models of engagement with third parties, a common engagement model is distributed hardware and software development. Herein, a company provides requirements and overall system architecture to third-party suppliers and then acts as a system level integrator of all the third-party deliverables. Obtaining high-quality defect-free deliverables is often a challenge, and the companies typically expend intensive effort during this process, resulting in additional costs, not meeting planned schedules, and reduced overall quality. From an availability perspective, it becomes all the more important as numerous defects during and after integration can leave a dent on customer satisfaction as well as operating costs. DFSS can be leveraged to successfully manage some of these challenges with third party intensive programs.

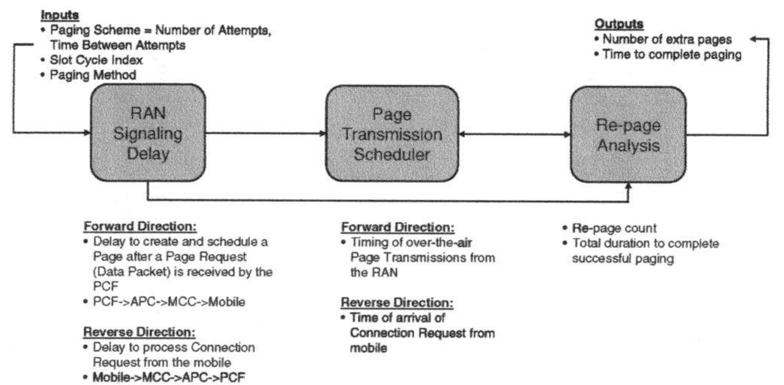

STAGE	STEP	delay (ms)	type	pdf	mean	sd	low	high
Paging Trigger PDSN & RAN	Data Packet Arrival at PDSN	0.0	trigger	uniform			1	213.35
	PCF Determines Start of Major Paging	0.00	s/w delay	normal	2	0.1		
	A14 Page Req from SCA arrives at Ctl APC	0.0	transport delay	triangle	2		1	10
	Ctl APC Determines Start of Minor Paging	0.0	s/w delay	normal	2	0.1		
AT Paging Response	Page arrives at AT (air prop delay)	0.0			0.5		0.1	1
	AT decides to respond	0.0	s/w delay	normal	2	0.1		
	Access Probe with ConnReq arrives at MCC-DO	0	access probes, poor RF	triangle for now	80		55	715
MCC-DO Paging Response	Px Connection Request from MCC-DO arrives at APC	0.0	transport delay	triangle		17.5	5	30
Inter BSC Paging Response	A15 delay distribution "normalized"	0	transport delay	triangle	15		10	35
	Distance based paging enabled	0		fixed				
	Probability of inter BSC messaging	0	scenario enable	custom			0	1
Major Paging Response	A9 Setup A8 msg from APC arrives at SCA	0.0	transport delay	triangle	2		1	10

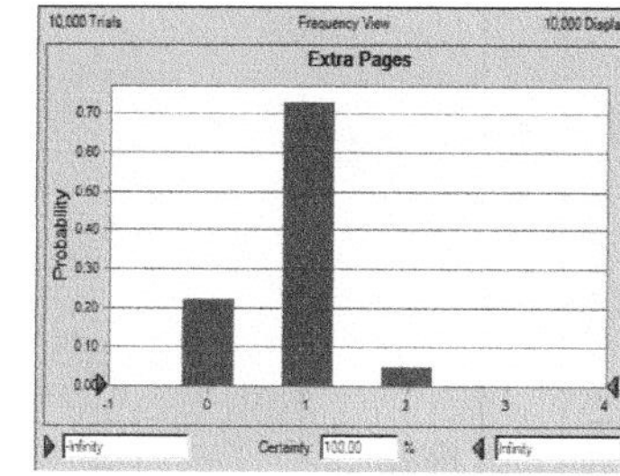

Q-Chat Mobile
SCI = 5
Repage Interval = 313ms

Result: 1 extra Page 80% of the time

Q-Chat Mobile
SCI = 6
Repage Interval = 526ms

Result: 1 extra Page 47% of the time

Q-Chat Mobile
SCI = 9
Repage Interval = 5220ms

Result: 1 extra Page 4% of the time

- Created a model for paging retries and used field data as inputs
- Performed Monte Carlo Simulation & measured extra pages for different applications, SCI values, and re-page intervals
- Simulation helped visualize how the feature would work which otherwise wouldn't have been possible without testing
- Helped better describe the feature design and functionality

Figure 14.12 Visualizing Feature Performance

TABLE 14.6 Critical Parameter Scorecard (a Non-Sophisticated Example)

Critical Parameter	Predicted and Actual Performance			
	Concept	Design	Optimize	Verify
Capacity (number of sessions)	100,000	80,000	110,000	99,000
Availability (nines)	99.999	99.567	99.988	99.991
Failover time (seconds)	20	15	12	16

Working with suppliers poses several challenges such as:

1. *Supplier Selection.* Lack of sufficient due diligence during third party selection can cause poor strategic fit and mismatch in expectations.
2. *Contracts.* Poorly crafted contracts can result in renegotiations during the program, which can inject higher risk levels that could even derail the project.
3. *Delivery Schedule and Quality.* Project management complexities can result in critical path risks impacting delivery schedules and product quality. Aggressive and proactive management of schedule and quality is a sine qua non for success.
4. *Requirements and System Integration.* In many situations, especially with new product development, there will be significant churn of requirements and feature content cascading to the suppliers, having enormous negative effects.

DFSS can be leveraged to manage these challenges and reduce the associated risks. It is not necessary that a supplier be knowledgeable about Six Sigma techniques. The company that is gathering VOC and creating requirements can identify critical parameters, perform DFSS analysis, and then supply these processed requirements to the third parties. In our example of capacity modeling, the memory budgets, including tolerable variations were supplied to third parties. This makes the requirements a lot more unambiguous and the engineers can be a lot more confident in their expectations. If identifying Critical Parameters is performed early in the life cycle, they can also be specified in contracts, and various revenue and penalty clauses can be embedded. This will also help third parties focus on the most critical aspects of their deliverables.

14.21 SUMMARY

In this chapter, we introduced a variety of DFSS tools and methods and talked about their practical applications and advantages. Here are some of the key takeaways:

- Application of DFSS tools should not be considered as an overhead cost. Used in the right context, these tools reduce cost and are extremely beneficial to successful product development and launch.
- There is generally lots of resistance within organizations to invest the extra time and effort up front to ensure quality and typically schedule pressure takes pre-

cedence over quality. This often results in taking unproductive shortcuts in the front end activities that ultimately lead to poor product quality released to customers and lots of time and effort wasted in firefighting defects. This directly increases cost of poor quality, including MoL (Maintenance on Line) costs. In addition, precious engineering time and resources get pulled away from new feature creation activities. The engineering leverage also becomes significantly diluted.

- It is critical that we apply DFSS at the start of the program from VOC gathering and formulating product features. Early engagement of feature architects on the critical parameters helps left shift defects significantly and can dramatically reduce overall costs.
- Sharing critical parameters with third parties enforces our expectations in their deliverables and helps unambiguous definition of requirements and thus increases the quality of the deliverables.
- Cross-functional team discussions enable better information flow from Product Management into Engineering on understanding customer requirements and clarifying priorities.

Integrating and optimizing DFSS into an organization's DNA can enable a creative, efficient, and excellent high performance team to deliver breakthrough products and services.

CHAPTER 15

Design Failure Modes and Effects Analysis

15.1 INTRODUCTION

Reliability problems are expressed as failures, and failures cost money. The later in the development cycle the problem is detected, the more costly the problem becomes. Even with development methodologies like Agile and its variants, an escaped customer-reported defect or failure is usually the most costly to fix. Failure events drive organizations to take action. Too often the actual funding for reliability improvements comes from the cost of unreliability. At times, recognizing people's efforts and ability for firefighting in the organization becomes very important, thus promoting a culture of firefighting rather than doing it right the first time.

What actions can we take to minimize problems in the field? One very effective tactic is to employ DFMEA.

15.2 WHAT IS DESIGN FAILURE MODES AND EFFECTS ANALYSIS (DFMEA)?

DFMEA (often shortened to FMEA) is an effective DFSS technique that is used to identify *potential* failure scenarios that could occur during system operations and analyze how these failures affect the overall system behavior. The results of the failure analysis are used to create more robust designs that mitigate, reduce, or eliminate the impact of these potential failures. DFMEA is perhaps one of the most important and most significantly underutilized DFSS tools and techniques.

DFMEA provides a systematic framework that allows the design team to document what they know and suspect about a product's failure modes prior to completing the design. This information is used to optimize the design and mitigate or eliminate the causes of failure.

DFMEA is used to explore ways that a product design might fail during real-world operation, document those scenarios, assign risk values to these failure scenarios,

Designing High Availability Systems: Design for Six Sigma and Classical Reliability Techniques with Practical Real-Life Examples, First Edition. Zachary Taylor and Subramanyam Ranganathan.

prioritize these risks, and eliminate or reduce the impact of the higher risk failure scenarios to ensure maximal improvement in the product *prior* to test or deployment. The DFMEA technique is intended to detect design deficiencies or errors that have been embedded into a design or inherent in the operating environment and provide recommendations for corrective action to minimize unacceptable risks. By providing a rigorous, systematic, and proven methodology, DFMEA increases the probability that the majority of potential failure modes and their effects have been considered and mitigated during the design/development process.

DFMEA provides a framework to identify, classify, and rank the various failures based upon certain criteria, details of which are described in this chapter. This technique is a general approach to prioritizing any set of risks or issues by quantifying what might otherwise be very subjective opinions. DFMEA provides additional rigor and structure to failure analysis and is a well-known and validated process used for designing high availability systems:

- Requires that all known or suspected potential failures be considered.
- Relies on the collective expertise of the technical leaders representing all areas impacted by the functionality.
- Stimulates open communication of potential failures and their outcomes.
- Results in actions that improve product robustness and reliability.
- Alignment of product capabilities with customer requirements.
- Bringing an outside-in perspective by involving test and field support.

15.3 DEFINITIONS

The American Society for Quality (http://www.asq.org) provides the following FMEA definitions:

FMEA. A systematized group of activities to recognize and evaluate the potential failure of a product or process and its effects, identify actions that could eliminate or reduce the occurrence of the potential failure, and document the process.

Failure Modes. The ways (modes) in which something might fail—especially ones that affect the customer

Effects Analysis. A study of the consequences of those failures. Also documents current knowledge and actions about risks and failures for continuous improvement.

15.4 BUSINESS CASE FOR DFMEA

The objective of DFMEA is to reduce or eliminate the risk of potential failures *before* they occur.

Figure 15.1 compares DFMEA with traditional firefighting. Once an escaped defect causes a failure in the field, the customer will first notice the device or system fails to operate correctly—Step 1. The immediate problem must be diagnosed using available information from the effect of the failure and possibly the observation of the failure mode itself—Steps 2 and 3. Once the operator has come to some conclusion on the

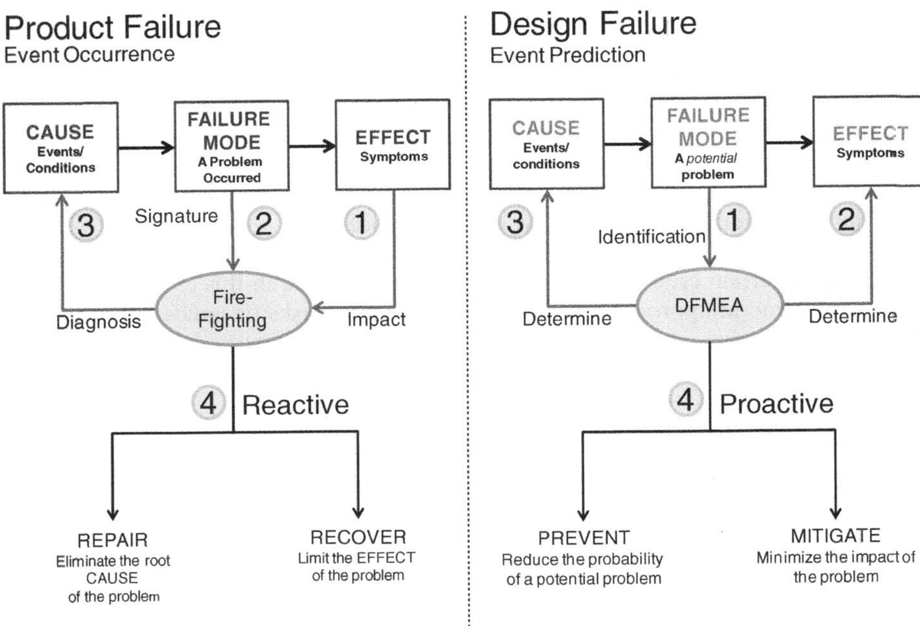

Figure 15.1 Using DFMEA to Prevent Firefighting

nature of the problem, the next step is to recover from the problem and/or repair the problem (Step 4). For example, if the failure is on a processing card, the operator may attempt to reboot that card, and failing that, may replace the faulty card.

Once the immediate problem has been addressed in some fashion, the device or system may be returned to the manufacturer or customer support may be engaged. The vendor will perform additional diagnosis and testing to make a determination on the root cause of the problem. Unfortunately, not all root causes can be easily determined. The exact root cause may not be found in spite of extensive analysis (e.g., RCA) and testing to reproduce the problem. If the root cause is found, an effective solution must then be created, the system repaired, and/or new hardware or software provided to the customer. If the root cause is not found, the firefighting team can introduce some other fixes that will limit the effect of the problem if it occurs again in the future.

DFMEA, on the other hand, addresses problems *before* they occur. Potential problems are identified by brainstorming and analysis. The effect of these potential failure modes is identified. Then the reason these failure modes occur is identified. Armed with this information, the DFMEA team makes proactive recommendations to prevent or minimize the impact of these potential failures that could occur in the deployed product. Thus, DFMEA reduces the firefighting mode of operation by reducing or eliminating as many problems as possible early in the product life cycle.

The DFMEA identifies potential failure modes of a system, product, or process, evaluates the impact on the system behavior, and proposes appropriate countermeasures to suppress these effects.

15.5 WHY CONDUCT DFMEA?

The most important reason for conducting a DFMEA is to improve the product or feature. To receive maximal benefits of a DFMEA program, the importance of always thinking of ways a product can fail and how we can improve our products early in the development cycle must be engrained in the organization's culture. If not, the DFMEA program will not succeed, and will be relegated to the process improvement trash bin.

The key goal and benefit of DFMEA is to identify and remove potential failures at the design phase—instead of the test or deployment phases. It is "proactive" since it is typically conducted before the product or system is implemented. It is thus preventive in nature.

One of the most important and well-established tools in the software development toolkit is peer reviews and software inspections. These techniques have been proven over the last several decades as a key method to find and remove defects in software requirements, architecture, design, and code, that is, the best way to find problems is to get the experts together and look for them. Similarly, DFMEAs offer a proactive way to remove defects at the requirements, architecture, and design stages of development, and in a more structured and comprehensive way. The DFMEA team looks for potential problems and failures that reveal defects or omissions in the design. Reviews and inspections are not a replacement for DFMEA. DFMEA focuses specifically on the failure scenarios and how to effectively eliminate or mitigate them.

15.6 WHEN TO PERFORM DFMEA

DFMEAs should be considered when a new process, product, or component is to be developed, whenever a revision to the product occurs, when the operating environment or operating conditions significantly change, and whenever a significant number of problems with the process, product or component have occurred.

Ideally, the DFMEA is launched at the earliest stages of concept development, and can also be used to help analyze competing designs and generate new, more robust concepts. One should start a DFMEA as soon as *sufficient* information is available. Do not wait for all of the information to be available, or the design to be completed, or the code to be completed, and so on. If the team waits for complete information, the DFMEA may never begin. On any development project, complete data are never available. With a set of customer requirements and preliminary concept architecture, the DFMEA can commence.

A DFMEA should be launched whenever:

- New systems or designs are initiated.
- Existing systems or products are modified.

- New features are added to the existing product.
- New architectures or improvements are being analyzed.
- Operating environments or operating conditions significantly change.

Though DFMEAs are launched during the requirements or design phases, they should continue through test and deployment. Problems found in test or out in the field are used to fine-tune the DFMEA, and serve as a document for "lessons learned" to be applied to the next product or product iteration.

After the DFMEA begins, it becomes an integrated part of the project life cycle from cradle to grave with important contributions throughout the project phases. It is a true dynamic tool of improvement regardless of the beginning phase. Each time a new feature or modification is made to the product, each time a failure is found in the field or lab, the DFMEA should be revisited, and new scenarios considered.

15.7 APPLICABILITY OF DFMEA

DFMEAs can be applied to the following scenarios:

- Development of robust system requirements that minimize the likelihood of introducing unexpected failures.
- Evaluation of customer requirements to ensure that those do not give rise to potential failures.
- Identification of certain design characteristics that contribute to failures, and minimize or eliminate those effects.
- Development of methods to design and test systems to ensure that the failures have been eliminated.
- Tracking and managing potential risks in the design. This helps avoid the same failures in future projects.
- Identifying software problems during software development before they occur.
- Ensuring that hazardous failures are prevented, that is, will not injure the customer or seriously impact the system.
- Enhancing test plans. Failure scenarios identified in the DFMEA can be used directly to drive test plans.
- Designing Built-In Tests (BIT) or Fault Management that can automatically detect and isolate system failures when they occur and then automatically recover from these failures with minimal or no impact to the system.

15.8 DFMEA TEMPLATE

Figure 15.2 is an example DFMEA template. Though there are variations across different DFMEA templates, we will describe some of the primary fields that are typically used in any DFMEA. This template can be customized to provide additional columns or details as required in your specific project.

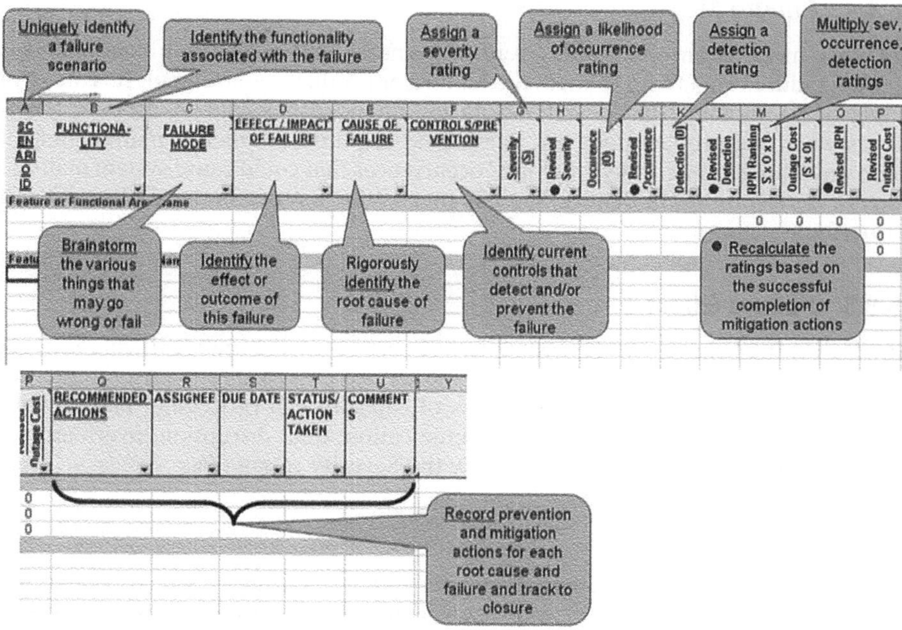

Figure 15.2 Example of a DFMEA Template

The fields in the example DFMEA template are defined in the next sections.

15.8.1 Scenario ID

Each unique failure scenario is assigned a tracking ID for reference.

15.8.2 Functional Area

This field identifies an aggregated functional area. For example, it could be Telecom Call Processing, Alarm Management, or Flight Control that categorizes a set of scenarios applicable to the function. The identification of functionality or customer features facilitates the identification of applicable scenarios. Typically, we would expect multiple scenarios for each functional area.

15.8.3 HW/SW/UI/System

Each failure scenario is mapped to a category, such as hardware, software, user interface, system, and environment. Firmware could be aggregated with either hardware or software. System could mean a subsystem or a system consisting of multiple hardware and software units.

15.8.4 Failure Mode

A failure mode is an observable event or scenario that results in a component or product failure or a failure to meet the feature or product requirement. It is an error, defect or an anomaly that deviates from the intended behavior. A product failure occurs when the product does not function as required or when it malfunctions in some way. These are typically "real-life" occurrences that could cause system malfunctions or deterioration in quality. See Chapter 19 for a more in-depth discussion of errors, faults, and failures.

15.8.5 Failure Effect

An effect is the behavior of the system in response to a failure mode. It characterizes the consequences of the failure mode on the service provided to the user. Examples of failure effects: loss of redundancy, loss of capacity, call processing failure, voice quality degradation, loss of control/communication, data disruption, overload, and service delays. Each effect is quantified by the Severity rating (S).

15.8.6 Cause of Failure

The cause of failure is the specific defect in the requirement, usage scenario, operating environment, design, process, or software which is the root cause of the failure, that is, the fundamental reason for the potential failure. Examples include: software transient failure, hardware failure, memory leak, unimplemented requirement, and unexpected interactions. The cause of failure is "why" the failure happened (or could happen) in the first place. Contrast this with Root Cause Analysis (RCA), where a postmortem is performed after a failure has occurred to determine the cause. In DFMEA, the *potential* failure is analyzed to determine the defect that could cause the failure.

15.8.7 Severity

Severity is a measure of the effect of a failure mode. How severe is the system or feature impacted when this failure mode occurs? These numbers help the analyst to prioritize the failure modes and their effects. Severity indicates the relative effect of this particular failure mode on the system and is quantified on a scale of 1–10, with 1 being the least severe impact and 10 being the most severe impact.

15.8.8 Revised Severity

This entry reflects the severity of a failure scenario after the failure has been completely or partially addressed by revising the system design.

15.8.9 Occurrence

The occurrence reflects the probability that the failure will occur or the frequency of the failure mode. This is normalized on a scale of 1–10 in relationship to all the other failure probabilities: 1 being the least likely and 10 being the most likely probability of occurrence.

15.8.10 Revised Occurrence

This entry reflects the probability that the failure mode will occur after that failure has been completely or partially addressed by revising the system design. Typically, this is targeted to be lower than the initial Occurrence rating.

15.8.11 Detection

Detection is a measure of the likelihood of detecting a failure when the system is operational. How difficult is it to detect the problem? Is the failure detected immediately upon occurrence or detected some time after the failure event impacts the system? This is normalized on a scale of 1–10 with 1 being the highest likelihood of detection and 10 being the least.

15.8.12 Revised Detection

This entry reflects the likelihood of detection of that particular failure mode after it has been completely or partially addressed by revising the system design. Typically, this is targeted to be lower than the initial Detection rating.

15.8.13 Severity, Occurrence, and Detection Scales

Each of the numerical rankings 1–10 for severity, occurrence, and detection need to have a precise definition that applies to the project. For example, all participants must be able to understand and differentiate between a severity of 3 versus a severity of 4. Although the rankings are by nature somewhat subjective, it is important to have a relative understanding between the different rankings of each one of these categories. A well-thought-out set of definitions should precede any numerical assignments and Risk Priority Number (RPN) rankings.

We recommend that the entire organization or project have the same scale. One should not have to change the scale or definitions of the column based upon individual functional areas or features associated with the same product or set of products being developed by the organization, because in effect, you need to tie all these DFMEAs together to get the big picture and ensure that there are no significant gaps in analysis coverage. For example, we have a transport problem but how does that relate to the overall system behavior and how does it compare with other DFMEAs? Although DFMEAs in different functional areas can and as a practical matter should be conducted independently, the DFMEA of one function should be relative to other DFMEAs so that a high RPN on a transport DFMEA and a high RPN on a redundancy DFMEA can be compared and ranked to ensure resources are focused on the most important problems.

An example set of definitions is shown in Table 15.1.

The generic ranking scale should be customized to your organization's products or projects. Instead of struggling with some abstract or generalized scale each time a

TABLE 15.1 Severity, Occurrence, and Detection Numerical Definitions Example

Severity	Occurrence	Detection
10—Hazardous effect	10—Almost inevitable	10—Almost impossible
9—Very serious effect	9—Very high	9—Remote
8—Serious effect	8—High	8—Very slight
7—Major effect	7—Moderately high	7—Slight
6—Significant effect	6—Moderate	6—Low
5—Moderate effect	5—Occasional	5—Medium
4—Minor effect	4—Few	4—Moderately high
3—Slight effect	3—Low	3—High
2—Very slight effect	2—Very low	2—Very high
1—None	1—Unlikely	1—Almost certain

failure scenario is considered, this additional detail and familiarity will allow more accuracy and reduce analysis time when assigning Severity, Occurrence, and Detection scores.

We may want to add a set of rankings that the analyst can select based on the nature of the DFMEA. For example, if the feature is Five 9s, high availability, the scores would reflect reliability, down time, initialization time, and partial outages. If the feature is call processing centric, the scales may reflect call attributes such as maximum users served, latency time, call drop rates, and so on.

The scale may also be different based on the complexity of the design and the third-party vendors, and hardware selected. A balance is needed however. Too many scales are difficult to create and manage and may not be consistent. Too few scales and the mismatch between the particular DFMEA and the general scales become larger introducing assignment and interpretation errors. An example severity scale is shown in Table 15.2.

15.8.14 Control Type

This specifies existing mechanisms in the system that will indicate the failure mode to the operator. The control mode can be either detection or prevention. Detection implies that when the failure mode occurs, an indication is given to the user so that the failure can be resolved prior to the failure escalating into a higher level of severity. Prevention is a control mechanism built into the system to prevent the failure mode from occurring in the first place.

15.8.15 Controls/Prevention

Controls and prevention are methods employed by the architecture, design, or software to detect or prevent the failure. What alarms or events including sympathetic events occur when the failure mode is detected? How does the system mitigate the problem and recover from the failure mode (message retries, failover, restart, error

TABLE 15.2 Severity Scale Illustration

Severity Rating	Reliability Description	Availability Description
1	No impact	No impact
2	Occasional degradation in voice/data quality and not perceptible to user but calls are maintained.	Out of service (OOS) ≤ 1 second
3	Occasional degradation in voice/data quality and perceptible to user, but calls are maintained.	OOS ≤ 5 seconds
4	Constant degradation in voice/data quality and perceptible to user, but calls are maintained.	OOS ≤ 30 seconds
5	Constant degradation in voice/data quality and perceptible to user, but calls are occasionally dropped	OOS ≤ 1 minute/soft reset required
6	Dropped calls between 1% and 5%	OOS ≤ 2 minutes/failover required
7	Dropped calls between 5% and 10%	OOS ≤ 10 minutes, global reset required
8	Dropped calls > 10%	OOS ≤ 30 minutes. hard reset required
9	Calls almost always drops during the call	OOS ≤ 1 hour/system replacement required
10	Calls always drops/never complete call setup.	OOS > 1 hours—major repair/ replacement required

correction, etc.)? What are the detection and recovery times? A combination of both detection and prevention controls could apply to the same scenario.

15.8.16 RPN

RPN, referred to as Risk Priority Number, is calculated as the product of Severity, Occurrence, and Detection ratings. RPN = (Severity) × (Occurrence) × (Detection) [range of 1–1000]. This is a quantitative representation of the overall risk associated with that particular failure mode. The higher the RPN score, the higher is the risk. The RPN ranking is important since it is used to determine the relative risk between one failure scenario and another. The RPN is used to prioritize the failure modes that require additional scrutiny or corrective action to reduce or eliminate the risk. This score provides guidance on the relative order for which failure scenarios should be addressed. Additional factors may need to be considered. For example, a high severity failure mode should be addressed irrespective of the RPN score.

Organizations can define a particular threshold of the RPN earlier, which mitigation actions will be mandated. For example, an RPN threshold > 100 is a candidate for further analysis and risk mitigation. It may not be necessary to address RPNs of lower values because of the minimal impact of the failure modes on the system.

15.8.17 Recommended Actions

Recommended actions are specific actions recommended by the DFMEA team to be taken to eliminate or mitigate the risks in the product or feature (which reduces the RPN). These should be tracked with the same project management discipline and rigor as any other project issue or risk.

15.8.18 Revised RPN

Once the failure or risk has been reduced by the recommended actions for requirements, architecture or design improvements, we recalculate the RPN based on the revised Severity, Occurrence, and Detection estimates. We then compare the revised RPN with the Original RPN to assess the magnitude of risk reduction. Typically, the revised RPN is much less than the original RPN—conveying that the risk has been well mitigated. The revised RPN may often need to be tracked and recalculated during the project as incremental actions are completed.

If the change is not significant, we may want to consider other scenarios that have high impact improvement recommendations.

15.9 DFMEA LIFE CYCLE

The DFMEA is started after the initial design concept and continues through implementation of recommended actions. The modified design, test results, field failures, and so on are reviewed in the context of the DFMEAs previously performed and the results of the new recommendations. Thus, a DFMEA does not end after the initial set of recommendations. The DFMEA spreadsheet becomes a living document. The iterative cycle is shown in Figure 15.3.

We start a DFMEA with a statement of the problem domain. What is the scope of the DFMEA? What particular component, subsystem, system, product, feature, or

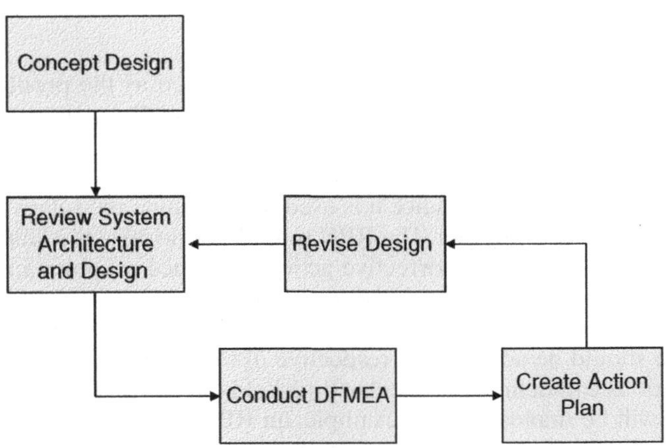

Figure 15.3 DFMEA Improvement Cycle

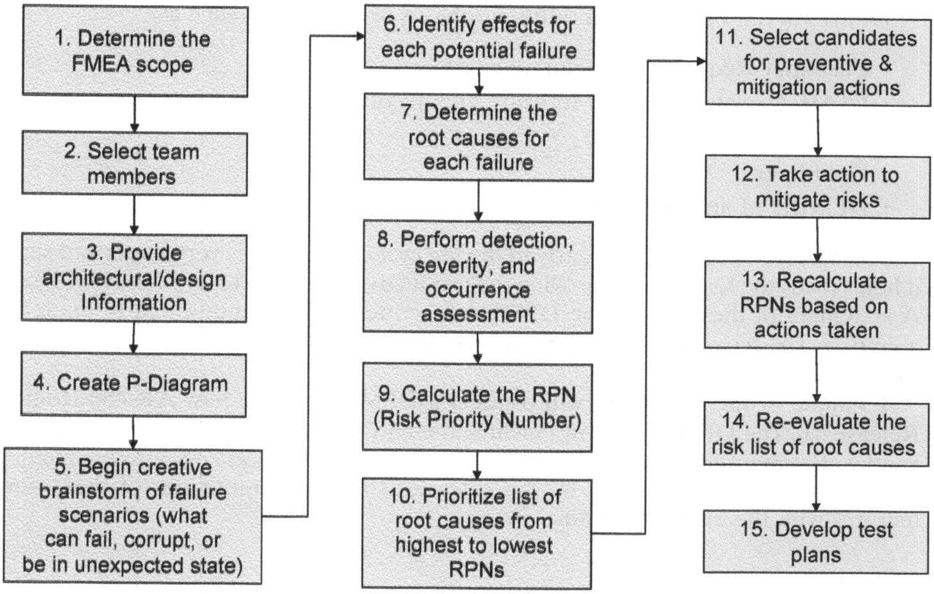

Figure 15.4 DFMEA Steps

functionality are we evaluating? Once the scope has been defined, and the analysis domain is understood, our focus is on the potential failures that could result from this new architecture or design. What are the ways the system or feature can potentially fail? DFMEAs can be applied to a variety of problem domains, such as: a large complex system (a jet fighter), a new software application (new database server), a communication system (cellular base station), a medical device (pace maker), or a new product (lawn mower).

DFMEA is a step-by-step methodical approach to identifying all possible failures in a product or service and implement improvements to the product or service. Each step in the DFMEA is important, and by following these steps, we maximize the number of potential failures identified, and recommend actions that have the most impact to system improvement (Fig. 15.4).

15.9.1 Determine the DFMEA Scope

What does the DFMEA cover? The scope should be well defined. This will minimize questions such as "What are we trying to accomplish here?" and "Which part of the project are we talking about?" The DFMEA may be limited to an enhancement to the existing product or a broader DFMEA covering the new hardware platform being developed, for example. How far should we drill down? Should separate detailed DFMEAs be held for the individual software components? What are the boundaries between DFMEAs? At which stage in the development process is it being introduced?

DFMEAs can be applied at different phases in a life cycle, such as:

- Customer Requirement failure analysis
- DFMEA of the proposed software architecture
- Software design DFMEA
- Field failure DFMEA
- DFMEA of an enhancement to an existing function.

Ideally, DFMEA is applied early in the life cycle. The scope is very important since it helps focus the team and improving the productivity and efficiency of the DFMEAs. A statement of the scope of the DFMEA is written and communicated to the team members. The scope is provided by the DFMEA lead prior to the start of the brainstorming session.

Example Scope: Our team will conduct a DFMEA on the new AR-400 Router, Release 3.0. The DFMEA will not include the hardware or firmware provided by the vendor. The main focus will be on the new overload protection and message latency reduction features introduced by this release, as well as interactions of these new features with the existing software.

Issues with Scoping an FMEA

For very large and complex systems in which analysis is needed, we should consider breaking up the DFMEA into separate distinct DFMEAs, focusing on different aspects of the system. For example, transport is a large area that may need a separate DFMEA. The user interface, especially the GUI interactions, is always a source of investigation and may also be a separate DFMEA. Another example is redundancy management, which may warrant a separate intense analysis. And lastly, if you have a specific enhancement of a product, then a DFMEA that limits its scope to that particular enhancement should suffice.

If the DFMEA becomes too long or has too many entries in it, it becomes very difficult to manage and logistically difficult to gather the appropriate experts and have them available for the sessions. When you focus on specific functional areas or capabilities then you can identify a limited group of appropriate experts required for that particular session.

15.9.2 Select Team Members

The most important key to the success of the DFMEA is the selection and participation of the product experts. A cross-functional team with diverse knowledge of the product, feature, requirements, and test should be assembled.

Finding problems in system behavior that has yet to be built is a difficult task (and maybe more so than the normal functional capabilities that the system is required to deliver to the end user), and thus the expert will know more about particular failure modes and issues or constraints in the system that can be impacted. That relates to our discussion on breaking the DFMEA sessions into smaller pieces so that the DFMEA sessions include the appropriate experts and tailor the meeting times to the schedule of this more restricted group of participants.

15.9.3 Provide Architectural/Design Information

Prior to the DFMEA, the architecture of the system or feature is provided and described to the team members (although it may still be in a preliminary state). What is the purpose of the new function or product from the perspective of the customer? Identify each component, interface, and/or function within the scope of the DFMEA. Typically, a block diagram or other diagram is provided prior to the beginning of the DFMEA session. This overview is intended to accomplish the following:

- Provide all team members the scope of the DFMEA.
- Provide team members with a baseline understanding of the product or feature and terminology for which the DFMEA is being conducted.
- Identify the functionality, components, and interfaces that are to be analyzed as part of the DFMEA.

15.9.4 Create P-Diagram

The Parameter Diagram (P-Diagram) is a tool used to help refine the scope of a DFMEA and identify failure modes. A P-Diagram consists of input, output, control, and noise factors. Inputs also known as signal factors are processed by the function being analyzed at hand to generate the outputs. The outputs can be of two types: intended functions or behavior, and unintended functions or failures.

Control Factors are typically elements within the system architecture and design that can be used to tune the desired output response. Noise Factors are parameters that are outside of the control of the designer/engineer, such as environmental factors, customer usage, functional interfaces, and so on. These Noise Factors can negatively impact the design, introducing failure modes into the system.

As an example, the following P-Diagram illustrates the scoping of one of the DFMEAs conducted on a telecommunications product. The unintended results captured in the P-Diagram were transferred to the DFMEA template as failure modes for further analysis on the root causes and effects (Fig. 15.5).

The unintended results from the P-Diagram will be used as input Failure Modes in the DFMEA spreadsheet.

15.9.5 Begin Creative Brainstorm of Failure Scenarios

A brainstorming session is set up with the knowledgeable participants (software engineers, hardware engineers, designers, architects, test engineers, customer support or field support engineers, product managers, and other key participants). This is the most important phase of the DFMEA in which the team identifies as many possible ways the product or function could fail to meet the customer requirements or feature. For each function or component, brainstorm with the team ways in which the product might fail to meet its intended function. Remember, successful DFMEAs are highly dependent on the subject matter experts who can identify a large set of potential failure modes and bring insight into the possible problems. The synergy of the team can result in many more scenarios than could be generated by an individual working

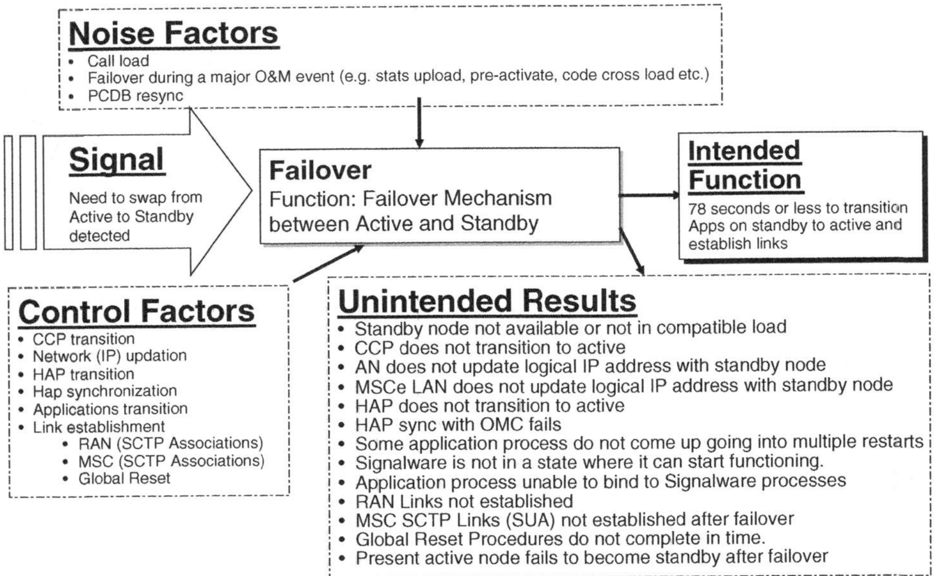

Figure 15.5 P-Diagram Example

separately. Consider what could happen to the product during unusual operating or stress conditions. How could operator actions result in undesirable behavior? Also note that each function/component may have multiple failure modes. At this stage of the DFMEA, we are concerned with identifying and capturing every conceivable failure mode. The DFMEA lead captures all possible failure mode ideas from the team, including those that may initially be perceived as trivial or would occur only very rarely or would occur in very unusual circumstances.

Brainstorming is a technique used in DFMEA for generating a large number of ideas within a short period of time. Encourage as many ideas as possible; allow a thought from one person to inspire an idea from another person. Encourage each team member to "throw ideas into the ring." A round-robin technique may be used to encourage each team member to express her ideas, which in turn helps stimulate ideas from others. Ideally, each participant should be prepared by coming to the brainstorming session with a list of ideas to start with.

The intent during this phase of the DFMEA is to be thorough and not overlook or disregard scenarios. All ideas from all team members should be accepted and added to the list without analyzing or judging them.

Using a rapid-fire approach, the team can easily capture dozens of ideas in a single session.

To help generate ideas either before or during the brainstorming session, a checklist can be used. Figure 15.6 is an example checklist that provides guidance to the team on specific functionality or general architectural issues that can facilitate the generation of quite a few failure modes that can then be discussed with the team.

			Failure Modes			
	Generic Design Considerations (Not Exhaustive)	Functionality	No Function	Partial Function	Intermittent Function	Unintended Function
User Interface	Data Dictionary Mapping					
	Read - Latency	Read latency < 2 seconds	Read Failure	< 2 seconds under low load conditions	Read fails every half an hour	None identified
	Read - Concurrent Reads					
	Read - Auto Referesh					
	Write - Effective Time					
	Write - Number of Concurrent Writers					
	Cross functional interactions					
	GUI Screen Relationships					
	Feature Specific					
	Ease of Use	Adjustable column width for Alarm viewer	Columns cant be adjusted	Columns adjusted but goes back to default width in the next access	Works on PC but not Unix	One user settings override another user preference
External Interface	Data Dictionary Mapping					
	Response Latency					
	Message Processing Times					
	Concurrency of Processing (Eg. NBI)					
	Concurrency of Data Access					
	Feature Specific					
	Interoperability					
	Version compatibility					
	Synchronization/checkpointing of data					
HW Platforms	External Interfaces					
	Expansion					
	Memory Usage					
	Performance					
	Component Failures					
	Failover Times					
	Redundancy					
	Feature Specific					
SW Platforms	Process Failures					
	Init Times					
	Downloads					
	Software Upgrade / Downgrade					
	Software Migration / Installation					
	Performance					
	Feature Specific					
	File System Management					
	Database Performance					
	Database Integrity					
	Overload Protection					
	Fault recovery escalation policy					

FAILURE MODES

Figure 15.6 Structured Aid for Brainstorming

15.9.6 Determine Failure Effects and Root Causes

Once we have identified a set of failure modes, we can then focus our attention on identifying the potential effects of each failure mode. What happens to the system when this failure occurs? How does the customer experience and perceive this problem? Review the product design or system architecture to identify potential causes of each failure mode. Look at failures captured in previous versions of the product or similar products. How difficult was it to debug these problems? Focus on the unique capabilities of the new functionality. What new complexities are being introduced? How does this new functionality interact with other existing functions? These questions are asked and analysis is done at the DFMEA brainstorming session with the list of potential failure modes already identified. The DFMEA team also reviews each failure mode and asks the question, if this failure occurs, what would happen? What is the impact to the system? What is the impact to the customer? What could cause this failure? A failure may have only one effect or it could have several effects. All of the potential effects, even seemingly inconsequential effects, should be captured.

Root cause analysis is a key part of DFMEA. Consider the sources of failures. These could be software bugs, component failure, user errors, external events, overload, communication errors, and so on. Cast a wide net to help identify "unexpected" sources of errors.

A fishbone diagram is another tool used to help identify the root cause within the scope of the DFMEA as illustrated in Figure 15.7.

5 Whys is an additional technique employed to obtain root causes. This is a technique to help connect the failure mode to the root cause. We start with the failure mode and ask why it would happen? Once you have an answer, and the root cause is not yet identified, we ask another question and repeat this process until we have a satisfactory answer.

For example, while cutting the grass on your lawn, the lawnmower stops. Some friends come over to help. Someone asks the question: Why did the mower stop?

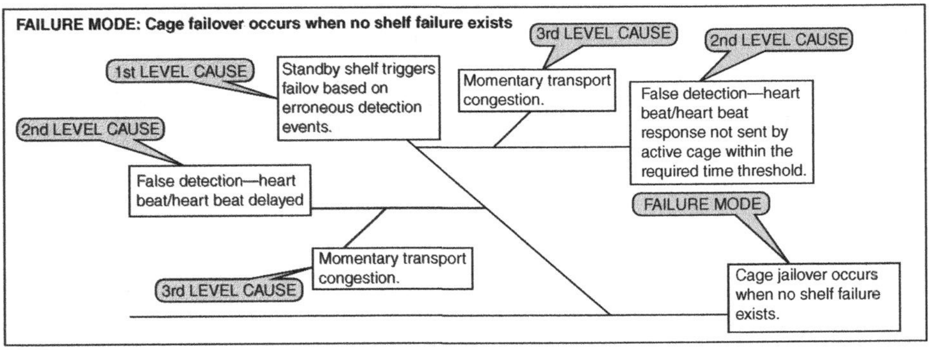

Figure 15.7 Identifying Root Causes Example

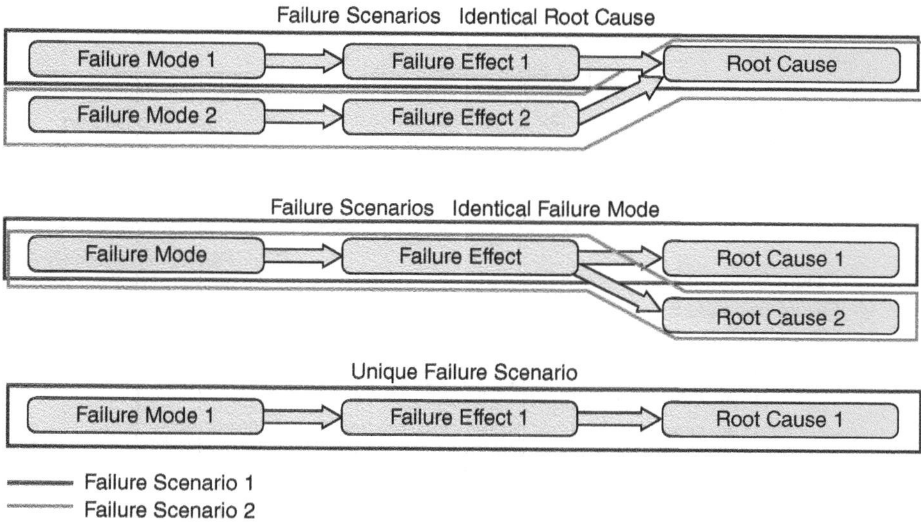

Figure 15.8 Failure Scenarios

Answer: The engine died. He then asks: Why did the engine die? Someone answers that perhaps there is no gas in the tank. You check the tank and see it has gas. Someone else offers: Perhaps the filter is dirty. You check the filter and sure enough, the filter is dirty. The next question then becomes: Why is the filter dirty? Answer: I haven't changed the filter in over a year. Why haven't you changed the filter in over a year? You respond that it had not occurred to you until this moment. You then change the filter and the lawnmower starts up properly. We solved the problem! We also have a lesson learned!

15.9.7 Review Failure Scenarios

A failure scenario is a unique combination of a failure mode, failure effect, and a root cause. For example, a single root cause event can trigger multiple failure modes and effects (Fig. 15.8). Or we can have multiple root causes that result in identical failure modes and effects as shown in the example captured in Figure 15.9. Each unique failure scenario should have a separate entry in the DFMEA spreadsheet.

15.9.8 Perform Severity, Occurrence, and Detection Assessment

After all ideas have been offered and captured during the brainstorming sessions, and the team determines they have exhausted the failure scenario possibilities, the next step is to review these ideas. Clarifications of ideas are captured, questions addressed, and scenarios are reworded as appropriate. Similar ideas may be combined and ideas that do not fit within the scope of the DFMEA can be eliminated.

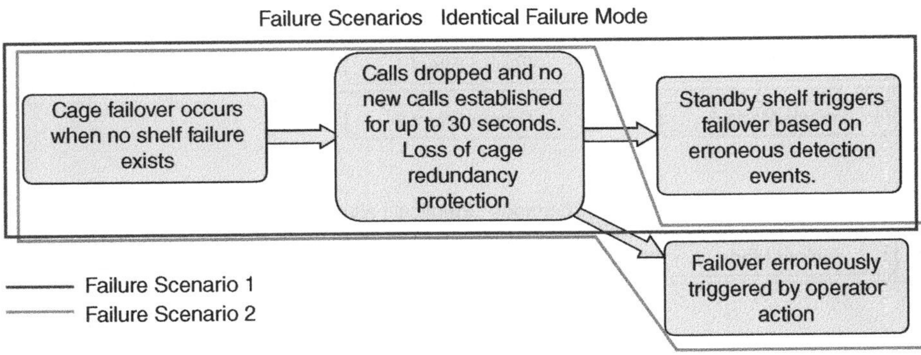

FAILOVER MODE/ FAILURE SCENARIO	EFFECT/IMPACT OF FAILURE	CAUSE OF FAILURE
Cage failover occure when no shelf failure exists	Calls dropped and no new calls established for up to 30 seconds. Loss of cage redundancy protection.	standby shelf triggers failover based on erroneous detection events.
Cage failover occurs when no shelf failure exists	Calls dropped and no new calls established for up to 30 seconds. Loss of cage redundancy protection.	Failover erroneously triggered by operator action

Figure 15.9 Failure Scenarios Example

The scrubbed list of failure scenarios is numbered and the assignment of scores commences. Each issue is scored based on the frequency of failure occurrence, the severity of the failure, and the detectability of the failure.

Assign Severity Rankings. The feature or design team ranks the severity of the failure mode, using a 1 to 10 scale, where a "1" indicates that there is no discernible effect and a "10" indicates the worst case impact of a failure on the system. The severity ranking is driven by the failure effect. This ranking is important since it is used to help quantify risk between one failure mode and the others. The DFMEA lead assigns the final ranking after inputs/concerns from the team have been considered, erring on the side of caution when disparities arise. The DFMEA champion can provide additional information and guidance on the rankings, but the DFMEA team makes the final decision.

Assign Occurrence Rankings. Assign the frequency of occurrence of the failure mode using a scale from "1" (remote likelihood of failure), to "10," (persistent or continuous failures). Once the potential causes have been identified for all of the failure modes, the team uses this information to help estimate the frequency of the failure mode.

Assign Detection/Prevention Rankings. For each failure cause, evaluate the ability of the design to detect and/or prevent the failure mode. Prevention mechanisms prevent the failure mode from occurring, but may not prevent the failure from manifesting itself in all scenarios. Detection mechanisms detect the failure, but do not prevent the failure from manifesting itself. The detectability is also ranked on a scale from "1" (almost always prevented) to "10" (never detected).

15.9.9 Calculate the RPN

For each scenario, the Risk Priority Number (RPN) is calculated by multiplying the severity, occurrence, and detection rankings together. The RPN score is used in relation to other RPN scores to determine the priority of actions to improve system robustness and functionality. The RPN number only indicates one risk is more significant or less significant than another—it is not intended to quantify how much the risk differs. It is also used to compare with the revised RPN after the recommended actions have been adopted.

15.9.10 Prioritize the Scenarios

An important objective of the DFMEA is to direct available resources to addressing the most critical issues that will have the most significant impact in improving the robustness and reliability of the product. The prioritization is based on the RPN score. The failure scenarios are ranked from the highest RPN score to the lowest. In accordance with the 80/20 rule, as a rule of thumb, the top 20% of the failure modes account for 80% of the risks. The initial RPNs and revised RPNs are often ranked in a Pareto diagram to provide a clearer representation of the risks and how the recommended actions impact those risks (Fig. 15.10). Potential risks are inherent in any new or revised product or feature. The RPN Pareto helps quantify risks and design decisions. Targets for improvements can be evaluated using the completed DFMEA.

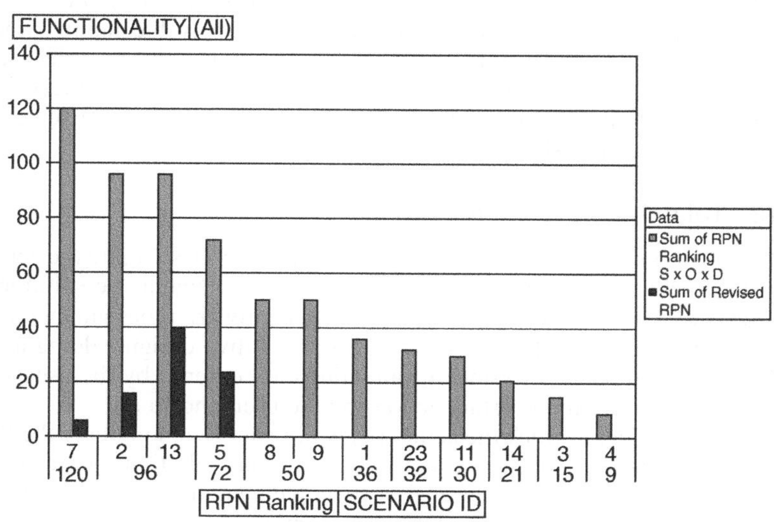

Figure 15.10 RPN Pareto Chart

The failure modes that have the highest RPN are given the highest priority for corrective action. For each failure mode, one or more of the following actions should be taken in priority order:

- Eliminate the failure mode if possible.
- Minimize the severity of the failure.
- Reduce the occurrence of the failure mode.
- Improve the failure detection.

15.9.11 Select Candidates for Preventive & Mitigation Actions

When the RPNs are calculated for all of the failure mode entries, the next action by the design team depends on the stage of development. In the concept development stage, the RPNs may be used to compare various designs and select the best among them, or can be used as an input to sensitivity analysis when combining features of the best concepts. Later in the development process, the RPNs should be used by the developers as a guide to focus development effort. After product deployment, design changes could be used across products and/or releases for further improvement activities.

Usually, it is not necessary to address all failure scenarios for risk reduction. Those with very low RPN scores, for example, need not be addressed, unless the severity of the failure is very significant, perhaps classified as a safety issue. It is best to add a justification for not addressing the failure scenario in the DFMEA spreadsheet or documentation.

What Is the RPN Threshold?

What is the maximum RPN value requiring corrective action? This depends on your organization and project. In some cases, an RPN value of 100 or greater may be the threshold to take action. In other cases, perhaps 80 is the threshold. However, in either case, the DFMEA has already captured all of the conceivable failure modes and ranked them for all to see. The DFMEA begins to give us benefits as we start addressing one or more failures in priority order. These are failures that otherwise would have escaped and only discovered later during product testing or worse yet, discovered by a customer after shipping.

15.9.12 Take Action to Mitigate Risks

Take action to eliminate or reduce the high-risk failure modes. Create the action plan: determine specific actions that need to be done to implement the solution, assign action items and due dates. The action could be a software redesign, increasing the robustness of the current design, selecting an alternative design, adding additional error handling to detect the problem, validating data entered by the user, updating the user manual, preventing certain actions by the user, and so on.

Assign an identifier to track the action items. For simple action items, the DFMEA is sufficient. For larger and more complex action items, additional project management tools may be required. The goal is to implement the improvements recommended by the DFMEA team. The DFMEA team ensures an action plan exists and is being followed.

15.9.13 Recalculate the RPNs Based on Actions Taken

The relative benefits of adopting the recommended actions are reflected by the revised RPN scores. This revised RPN score is compared with the original RPN score to determine if the recommended action provides a significant impact in reducing the risk. It must be ensured that the RPN score has been reduced to an acceptable level.

If we increase the detectability of the failure, we lower the detection score. Early detection may limit the damage the failure causes (which in turn affects the severity score). We are in effect reducing the potential time the failure manifests itself by improving the time to correctly communicate the failure to the user. The Detection score can be reduced by implementing or enhancing the prevention or detection mechanisms. Reducing the severity improves the product by limiting the impact of the failure. For example, if a particular failure mode causes us to lose communication with a server, we might add redundancy such that a single failure will not result in total loss of communication. A containment strategy to limit the impact of the severity or to reduce the downtime of the system results in a reduction of the severity score. A reduction in the Severity score usually requires a design change.

The best improvement is to reduce (or better yet eliminate) the likelihood of the failure occurring. If the failure mode cannot be entirely eliminated, reduce the probability of occurrence. That is, if the failure rarely occurs, the system reliability increases, thus improving customer satisfaction. The Occurrence Ranking can be reduced by removing or controlling the potential causes of failure.

15.9.14 Update Documentation

The DFMEA document captures all of the important information about the DFMEA sessions. The DFMEA document captures a priority system for design improvements and provides an open issue format for recommending and tracking risk reduction actions and future reference to aid in analyzing field concerns.

The DFMEA documents serve as an excellent communication tool. Copies of all DFMEAs should be kept in a central online location so they are easily accessible by team members and other project members outside of the particular DFMEA team for review and reference. These DFMEAs are also useful to support audits or internal process and product reviews.

15.9.15 Reevaluate the Risk List of Root Causes

After the RPNs are recalculated, it may be necessary to reevaluate these scores to see if any further action is necessary. This step ensures that we do not leave out any crucial actions that might result in reduced robustness of the design.

15.9.16 Develop Test Plans

Typically, DFMEAs result in high risk failure scenarios that the design team implement to prevent product defects. In order to ensure that these design changes are indeed implemented, it is important that test cases mapping to these high risk failure scenarios are created, documented, and incorporated as part of the test strategy and test plans. This will ensure that lapses with the development teams are identified as part of testing.

DFMEA example: The number of transactions supported by the system was designed to be 1000 per minute. The failure scenario identified: If we see more than 1000 transactions per minute, the system will no longer function properly. The mitigation action: limit the number of transactions to 1000 and drop any new transaction that exceeds this rate. Although the DFMEA resulted in overload protection being implemented, we cannot be certain that the code is implemented correctly. Thus, testing is mandatory to ensure that this scenario matches the design. A software bug could be introduced outside the scope of the DFMEA; however, the test will verify the implementation conforms to the design that was recommended by the DFMEA. The key to test is to verify that the implementation of the design in software was done correctly. Without that, there could be test escapes and thus related field defects.

To summarize, DFMEA can help ensure that a test verifies the intended design.

If the DFMEA is completed during the requirements or early in the design phase, the scenarios captured in the DFMEA are used as inputs to help plan thorough and efficient test plans and test cases. During the test phase, the error and failure scenarios are tested to enhance product robustness. Test failures can be evaluated to determine if a design fault exists, and if so, serve as a lessons learned input to improve future DFMEAs (Fig. 15.11 and Fig. 15.12).

15.10 THE DFMEA TEAM

DFMEA is fundamentally a team effort, and therefore should only be performed with the key members of the feature team.

The DFMEA team should consist of a small group of experts from various disciplines—typically four to six people, but the exact team size should be dictated by the scope of the DFMEA. For the DFMEA to be effective, the knowledge, participation, and candor of the participants are essential. DFMEA can also be a team-building technique, as well as a technical tool for product innovation and problem solving.

The DFMEA team leader is responsible for:

- Leading the DFMEA sessions (setting up meetings, agenda, action items, posting DFMEA spreadsheets work-in-progress),
- Ensuring DFMEA team participation (experts from test, development, architecture, system engineering, etc.),
- Ensuring the required team members are present (confirm ability of each team member to meet at the scheduled time). A major key to the success of DFMEA is the brainstorming productivity and energy released when all key stakeholders are present.

Initially, the DFMEA process expert or facilitator may be invited to help guide the session as needed. The expert can bring tremendous insight to the team, reinforce confidence in the way the session is being conducted, and help nudge the team back on track if needed to help ensure an optimal, productive session. However, the process expert should resist the temptation to micromanage the session, identifying

FUNCTIONALITY	FAILURE MODE	EFFECT / IMPACT OF FAILURE	CAUSE OF FAILURE	CONTROLS/PREVENTION	TEST CASE ID	RPN Ranking S x O x D	Revised RPN	RECOMMENDED ACTIONS
100 User session support	Concurrency access response time >2 seconds.	Slow GUI Response time	Performance overload	None	FMEA-Test Case 8721-1	75	30	Display hour glass perhaps. Focus on load and stress testing
20 User session support single server config	FM Server failure. Users cannot log-in. All users impacted. No access to EMS	No access to the system	FM Se… Failure	…d to ensure auto …t on failure		36	36	Need to ensure auto reboot on failure
100 User session support	Multiple user access within a short period of time (burst mode)	Slow response to user access.	Perfor… overload	…U Utilization alarm raised?	FMEA-Test Case 8721-2	210	84	Determine if a CPU alarm is raised. Add performance test case
100 User session support	Cardinality exceedance. What happens if we exceed 100 users? Performance impact? Need to block session?	System overload. Slow response. Possible unpredictable behavior	Too many users	No Controls in place	FMEA-Test Case 8721-3	252	28	Limit session count to max of 100. Add check to Fault Manager. Make this value configurable in the system properties file.

Scenario requires a test case.

Figure 15.11 Mapping Scenarios to Test Cases

SCENARIO ID	FUNCTIONALITY	FAILURE MODE	EFFECT / IMPACT OF FAILURE	CAUSE OF FAILURE	CONTROLS/PREVENTION	TEST CASE ID	Severity (S)	Revised Severity	Occurrence (O)	Revised Occurrence	Detection (D)	Revised Detection	RPN Ranking S x O x D	Revised RPN	RECOMMENDED ACTIONS
1	80 User session support - 3k/5k	Concurrency access - response time > 3 seconds.	Slow GUI Response time	Performance overload	None	FMEA-FR10311-1	5 - moderate effect	5 - moderate effect	5 - Occasional Failures	3 - Relatively few Failures will occur	3 - High likelihood to detect Failure Mode	2 - Very High likelihood to detect Failure Mode	75	30	Display hour glass perhaps. Focus on load and stress testing
2	20 User session support - single server config	FM Server failure. Users cannot log-in. All users impacted. No access to EMS	No access to the system	FM Server HW Failure	Need to ensure auto reboot on failure		9 - Very Serious Effect	9 - Very Serious Effect	2 - Remote Chance of Failure	2 - Remote Chance of Failure	2 - Very High likelihood to detect Failure Mode	2 - Very High likelihood to detect Failure Mode	36	36	Need to ensure auto reboot on failure
3	80 User session support - 3k/5k	Multiple user access within a short period of time (burst mode)	Slow response to user access.	Performance overload	CPU Utilization alarm raised?	FMEA-FR10311-3	7 - Major Effect	7 - Major Effect	6 - Moderate number of Failures	4 - Few Failures will occur	5 - Moderate Likelihood to detect Failure Mode	3 - High likelihood to detect Failure Mode	210	84	Determine if a CPU alarm is raised. Add test case to performance test
4	80 User session support - 3k/5k	Cardinality exceedance. What happens if we exceed 80 users? Performance impact? Need to block session? Additional 20% before blocking perhaps.	System overload. Slow response. Possible unpredictable behavior	Too many users	No Controls in place	FMEA-FR10311-4	7 - Major Effect	7 - Major Effect	6 - Moderate number of Failures	2 - Remote Chance of Failure	6 - Moderately low Likelihood to detect Failure Mode	2 - Very High likelihood to detect Failure Mode	252	28	Limit session count to max of 80. Add check to MotoFM. Make this value configurable in the EMS properties file.
5	10 PM/CSL simultaneous search sessions	Number of sessions exceeded. Max searches limited. User feedback message indicating number of searches has been exceeded.	Additional sessions not allowed. No warning or grace on this limit.	Too many sessions	We have a limit already in place	FMEA-FR10311-5	7 - Major Effect	7 - Major Effect	6 - Moderate number of Failures	2 - Remote Chance of Failure	6 - Moderately low Likelihood to detect Failure Mode	2 - Very High likelihood to detect Failure Mode	252	28	Need to enforce user limit. Add check to Moto FM. Do we need a warning prior to reaching the limit? Who would get the message?
6	30 CLI user support	Number of sessions exceeded. Performance impact.	System overload. Slow response. Possible unpredictable behavior	Too many users	No Controls in place	FMEA-FR10311-6	7 - Major Effect	7 - Major Effect	6 - Moderate number of Failures	2 - Remote Chance of Failure	6 - Moderately low Likelihood to detect Failure Mode	2 - Very High likelihood to detect Failure Mode	252	28	Limit session count to max of 30. Make this value configurable in the EMS properties file.
7	30 ITNCM user sessions support	Number of sessions exceeded. Performance impact.	System overload. Slow response. Possible unpredictable behavior	Too many users	No Controls in place	FMEA-FR10311-7	7 - Major Effect	7 - Major Effect	6 - Moderate number of Failures	2 - Remote Chance of Failure	9 - Very Remote Likelihood to detect Failure Mode	2 - Very High likelihood to detect Failure Mode	378	28	Limit session count to max of 30. Add check to CM. Make this value configurable in the EMS properties file.
8	10 PM/CSL simultaneous search sessions	memory exhaustion (heap space) may occur resulting in currently information displayed being truncated	If this limit is exceeded, memory exhaustion (heap space) may occur. Currently information displayed is truncated.	Too many users	No Controls in place		7 - Major Effect	7 - Major Effect	6 - Moderate number of Failures	3 - Relatively few Failures will occur	9 - Very Remote Likelihood to detect Failure Mode	3 - High likelihood to detect Failure Mode	378	63	Limit the amount of information displayed based on memory exhaustion detection. Avoid corrupted information displayed. Check needed also to avoid possible crash.

Figure 15.12 Snapshot of a Completed DFMEA

every minor flaw. This will result in the brainstorming session participants worrying more about the process and making mistakes rather than the core goal: identifying problems early in the development cycle. Minor issues can more effectively be dealt with after the session is over by discussion with the DFMEA lead.

Optimize the use of the brainstorming session. Prompt each team member for input, writing down each idea, probing for additional scenarios or issues. It is not important to fill out every column during the brainstorming session. In fact, momentarily stopping the brainstorming session while attempting to fill out various fields, greatly detracts from the brainstorming session and momentum is lost. Capture the key ideas as they flow during the session; some of the other fields can be filled out during subsequent DFMEA sessions.

15.11 DFMEA ADVANTAGES AND DISADVANTAGES

Advantages

- Aligns product design with customer expectations.
- Finds problems before the system or product is built. "Do it right the first time."
- Reduces cost of poor quality.
- Captures engineering knowledge and fosters innovation and collaborative and collective ownership.
- Provides historical documentation for future reference to aid in the analysis of field failures.
- Reduces the possibility of the same kind of failure in future projects.
- Emphasizes problem prevention.
- Minimizes last minute product changes and associated high cost.
- Catalyst for teamwork and idea exchange.
- Identifies conceivable failures and their effects on operational success.
- Lists potential failures and the relative magnitude of their effects.
- Provides prioritized risk-reducing corrective actions,
- Provides support for risk-based test plan creation.
- Drives the creation of action plans to improve the design and eliminate potential defects early in the project life cycle.

Disadvantages

- The required detail makes the process time consuming. The initial brainstorming for some complex features or functional areas may span several weeks.
- Assumes the causes of problems are all single event in nature (combinations of events are captured as a single initiating event).
- Requires open communication and cooperation of the chosen participants to be effective.
- Depends on recruiting the right participants.
- Without the follow-up sessions and disciplined completion of identified actions, DFMEA will not be as effective.

15.12 LIMITATIONS OF DFMEA

DFMEAs are limited by the following:

- *Time.* The DFMEA brainstorming sessions require time, and for DFMEAs with larger and more complicated scopes, investments in multiple sessions are required.
- *Expert Participation.* The key participants are almost always limited in the time they have available to focus on the critical DFMEA tasks. Without the correct experts meeting together, the effectiveness of the DFMEA is drastically reduced.
- *System Architecture.* An accurate understanding of the requirements, system architecture, and design is important. If the design is immature or ill-defined or the key participants are not well-versed in the design or functionality being analyzed, the number of unique scenarios that can be uncovered will be limited.
- *Failure Rate or Occurrence Information.* For new systems, accurate failure rate data and frequency of occurrence can be difficult to obtain. However, for the DFMEA, exact values are not required. Techniques to estimate failure rates and the occurrences are described in this chapter and in later chapters.

15.13 DFMEAs, FTAs, AND RELIABILITY ANALYSIS

Two important tools for reliability analysis are DFMEA and fault tree analysis (FTA).

DFMEA aids in producing block-diagram reliability analysis, aids in producing maintenance manuals, and is often used for fault management design. DFMEA identifies design weaknesses and provides tools for assessing failure severity and the adequacy of fault detection.

FTA, on the other hand, starts with a system failure affecting system reliability and identifies critical component failures occurring separately or coupled with other events resulting in system failure. The FTA shows the relationship between these events using logic gates.

FTA and DFMEA can complement each other, as illustrated in Figure 15.13. With FTA, we start with the top event of interest: system outage greater than 30 seconds, for example. We then decompose this event into particular subsystems or component failures that could cause this top event. The subsystems or components are decomposed into lower level components until the lowest level of decomposition is reached. This becomes our root cause events.

Conversely for DFMEA, we start with a set of failure modes and determine how these events affect the components, subsystem, system, and the effect on the end user.

FTA is better suited for "top-down" analysis. We identify what the top failure event is and then decompose this event, drilling down until we get to a basic event that is decomposed no further. When used as a "bottom-up" tool, DFMEA can complement FTA by identifying additional causes and failure modes, resulting in top-level symptoms.

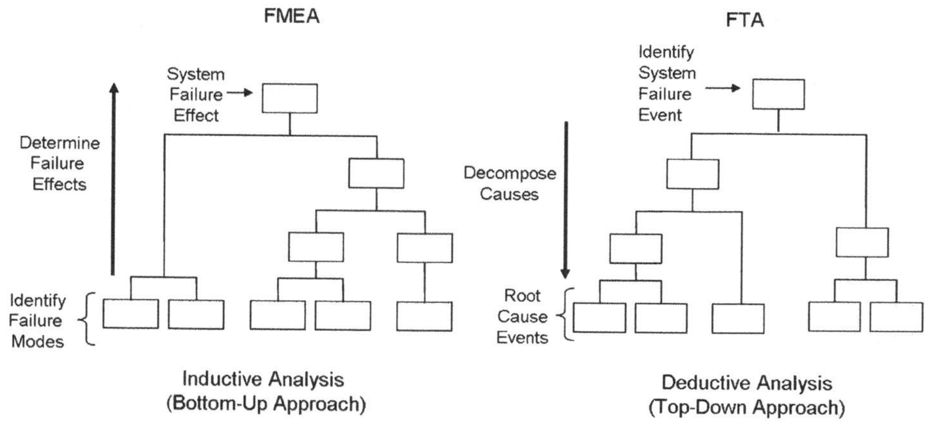

Figure 15.13 DFMEA and FTA Comparison

15.13.1 Example: Wireless Controller System Availability

In this example, a large integrated communication system with several redundancy features to support high availability (Five 9s) is examined. Many triggers could result in failure requiring a failover to the redundant components. In addition, the redundancy scheme itself was complicated and subject to failures as yet uncovered. Due to the complexity of the redundancy scheme and the critical importance of preventing a complete failure in the field, a DFMEA was conducted. The DFMEA looked at historical data from systems already deployed and the complexity of the failures and failover mechanisms.

During the brainstorming sessions, an exhaustive list of problems was identified and an availability model was developed to incorporate the failure mode scenarios. These failures were prioritized using the RPN score, a threshold determined, and the high risk items were addressed using the action plan. As additional field data became available, the model was updated and enhanced to improve its ability to predict failures and system reliability. The updated model helped to improve the accuracy of the occurrence and severity scores. This in turn drove the next set of changes to improve availability.

Some of the changes included new detection mechanisms and error handling associated with other system events, such as initialization, software upgrade, and system reconfiguration.

The system improvement updates were phased in over several releases spanning approximately one year, resulting in system availability that exceeded the original Five 9s requirement.

The key tools employed that allowed the achievement of high availability was the DFMEA and the FTA-based availability model. These two tools complemented each other and provided the appropriate accuracy and focus to achieve the availability goals.

15.14 SUMMARY

Learning from each failure after it occurs in the lab or out in the field is both costly and time consuming. If the failure is reported by a customer, customer dissatisfaction leading to loss of current customers and future business is a very plausible scenario. The root cause of the problem must be identified before any fix can be attempted. Some problems, as engineers will attest, can be notoriously hard to debug. Once the problem has been identified, the first solution may be too risky, and a less optimal fix proposed initially, with a more complete solution provided in a later software release.

Failures need to be prioritized according to how serious their consequences are, how frequently they occur and how easily they can be detected. A DFMEA does this and also documents current knowledge and actions about the risks of failures for use in continuous improvement.

DFMEA provides a systematic method of identifying a large number of potential failures using thought experiments. DFMEAs provide risk mitigation, in both product and process development phases. Each potential failure is considered for its effect on the product or process or customer and, based on the risk level, actions are determined to reduce, mitigate, or eliminate the risk prior to software implementation and deployment. The DFMEA drives a set of actions to prevent or reduce the severity or likelihood of failures, starting with the highest priority ones. Each failure scenario removed during the product design phase provides large dividends and cost–benefits for the company when a more robust and reliable product is deployed.

CHAPTER 16

Fault Tree Analysis

16.1 WHAT IS FAULT TREE ANALYSIS?

Fault tree analysis (FTA) is a systematic method of modeling different failures that can occur within a given system or component in a top-down hierarchical manner. These failures are associated using logic gates, which define how different failures interact with each other and impact the top level node. So in essence, each basic node in a fault tree represents a failure, and each intermediate node represents an association of these failures. Since this is a tree of failures or faults, FTA is an excellent and intuitive tool for modeling system availability.

When creating a fault tree, we start with a major failure as the top node and start identifying failures that trigger the top node event. The decomposition continues until we reach a set of basic (atomic) events. What is considered as a basic event depends on the scope of the analysis. Typically, a basic event should be "controllable" by the design team or organization, that is, the engineer can alter the characteristic of the basic event to influence the top-level event. For example, if we are performing a processing board failure analysis, we could consider failures of resistors, capacitors, memory chips, and so on. If we are working at a system level using hardware components from a third-party vendor, the basic event can then be only the manifestation of the hardware failure and how the system could recover from that failure.

There are two types of fault trees: qualitative and quantitative. A qualitative tree represents just the failures and their associations and does not include any quantitative data—it is an association of failure logic. A quantitative fault tree includes data that quantifies the failure characteristics of each node. For example, a node can be characterized by MTBF, MTTR, and failure probability.

Building a fault tree requires detailed knowledge of the system architecture and is usually an iterative process. Failures identified during creation of a fault tree can be used to enhance/correct the system architecture and design and thus increase the robustness of the system.

Designing High Availability Systems: Design for Six Sigma and Classical Reliability Techniques with Practical Real-Life Examples, First Edition. Zachary Taylor and Subramanyam Ranganathan.

A fault tree contains two basic elements:

1. *Events.* Events that characterize the failure. These can be basic or atomic events or composite events.
2. *Logic Gates.* Logic gates that associate the events into different groups.

16.2 EVENTS

An event is typically a fault or a failure that triggers a logic gate to fire. From a systems perspective, an event could be loss of power, or a drop in capacity, or reset of a card. Different types of events are represented in a Fault Tree, a few of which we will consider in this chapter.

16.2.1 Basic Event

A basic event is usually an atomic event; that is, an event that is typically not decomposed any further in the fault tree. For example, a power supply failure could be a basic event when modeling a complex system. However, if we are modeling the power supply itself as the subject of the FTA, an oscillator failure within the power supply, for example, may become a basic event. The level of granularity of decomposition typically depends on the system at hand, the scope of coverage, controllability of the design, and also the extent of detail necessary. It could also depend on what level of information is available and how critical that subsystem is with respect to the complete system.

Figure 16.1 is a symbolic representation of a basic event using the commercial FTA tool Relex from Windchill.

16.2.2 Repeated Event

When building a fault tree, it is quite possible that the same event impacts several branches of the tree simultaneously. These are modeled as *repeated* events. For

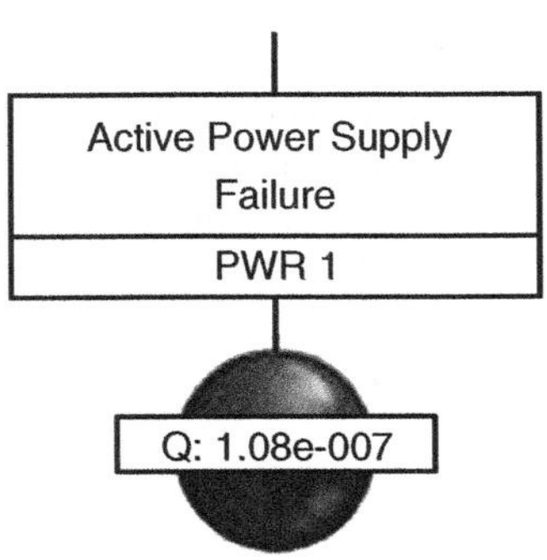

Figure 16.1 Basic Event

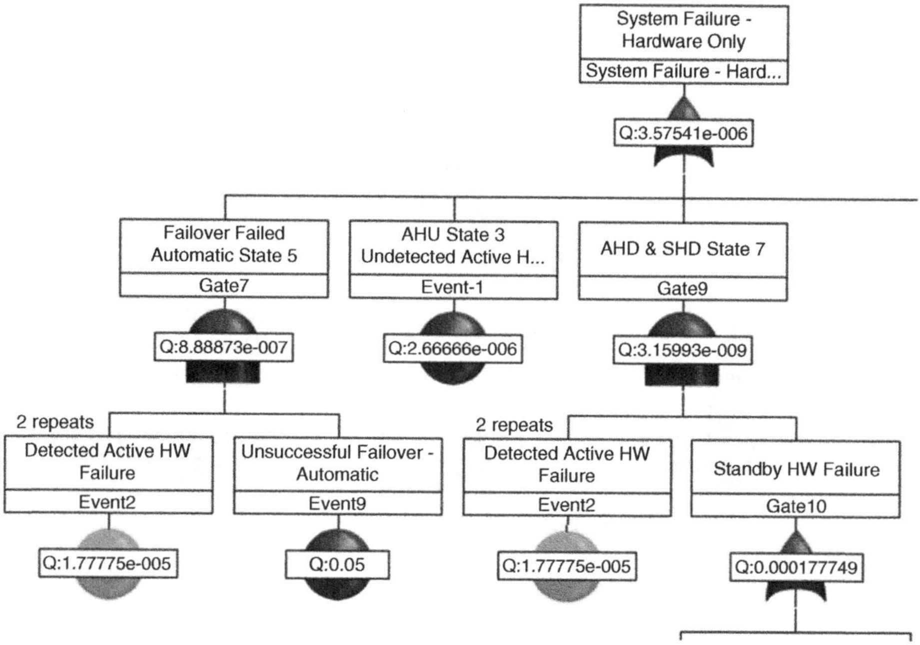

Figure 16.2 Repeated Event

example, if there are two devices drawing power from the same supply and if that power supply fails, it impacts both devices. In that case, power supply failure will be a repeated event. Repeated events represent the same instance of the failure in different parts of the fault tree. Figure 16.2 illustrates the use of repeated events (Detected Active HW Failure) in a fault tree.

Other types of events, such as conditional events and spare events, are described in the literature and the reader is encouraged to explore these additional extensions of the fault tree.

16.3 LOGIC GATES

Logic gates are an integral part of a fault tree. They define the relationships between different failure events all the way up to the top level event in a fault tree. Typically *OR* and *AND* logic gates are commonly used. These are also referred to as static gates and characterize basic digital logic.

OR Gate. This defines the logic where just one input event must be true in order for the output of the gate to be true (Fig. 16.3).

AND Gate. This defines the logic where all input events must be true in order for the output of the gate to be true (Fig. 16.4).

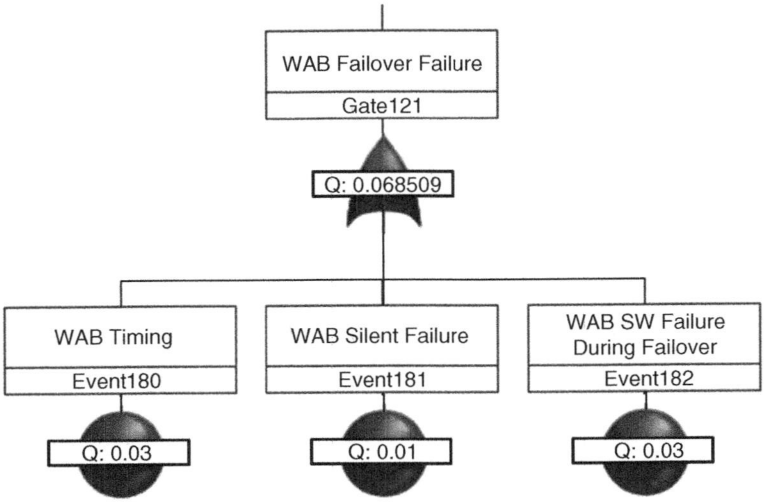

Figure 16.3 OR Gate with Three Inputs

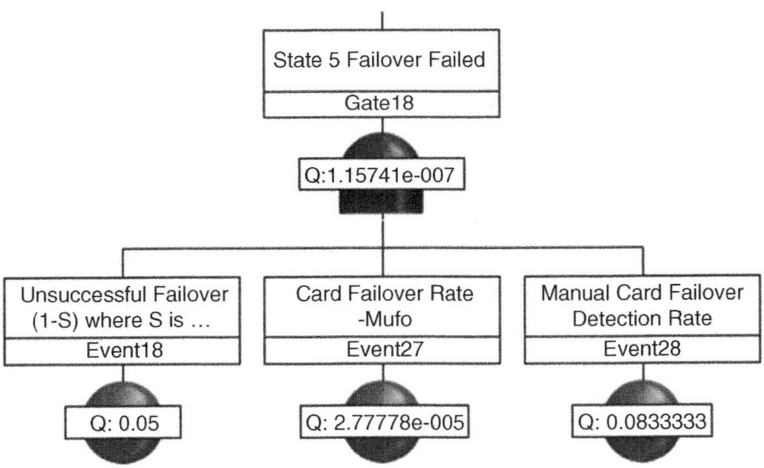

Figure 16.4 AND Gate with Three Inputs

Transfer Gate. A transfer gate is just a connector that connects logical blocks of a fault tree. It helps the architect to organize a fault tree across different pages or functional levels. A transfer gate is represented as a triangle as shown in Figure 16.5. The complement of the transfer gate is also shown in Figure 16.5.

Other special purpose gates, such as PAND (Priority AND), are also supported by tools such as Relex. Timing, sequence dependencies, and/or state information is

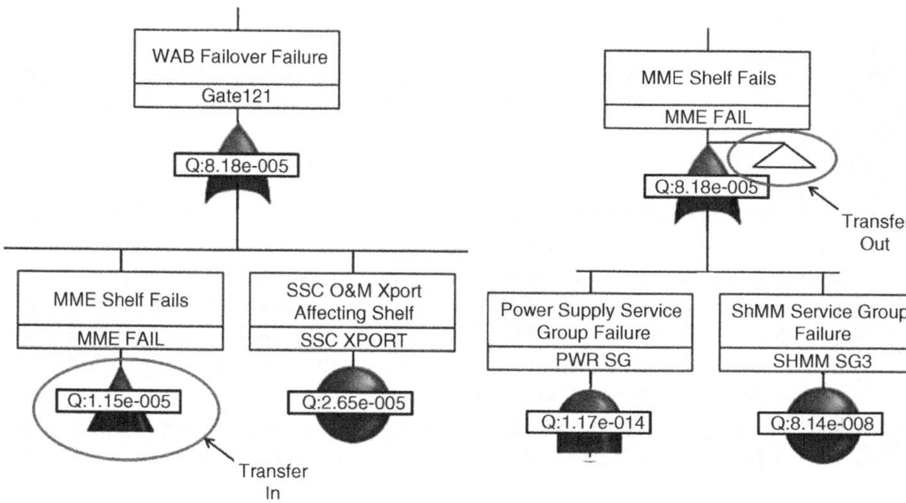

Figure 16.5 Transfer Gates

built in to these specialized gates. This is similar to using a high-level programming language or a macro to get a function accomplished that could be more complex to be realized by just using basic gates. However, all fault trees presented in this book have been built using basic AND/OR/Transfer gates. Readers interested in advanced topics on FTA may refer to the available literature.

16.4 CREATING A FAULT TREE

Having discussed the structure and components of a fault tree, we can now discuss the dynamics of its creation. To create a fault tree, there has to be some knowledge and understanding of the following:

1. The system architecture and functionality for which the fault tree is being created.
2. The top level failure that is being modeled.
3. The failures that can happen within the system and how they are interrelated to the top-level failure.

Let us consider a simple system that contains an active card (card A) and a standby card (card B). The system functions fully and normally if just one card is operating. When a failure occurs for card A, card B automatically takes over. This fault management function is built through a combination of hardware and software. There is seamless transition between the two cards whenever a failure of one card occurs and the other card is on standby. Also, A and B cannot be in the active state or standby state simultaneously. The cards can be manually fixed and in addition have some auto-recovery capability.

We now need to identify a failure that can become the root (top node) of the fault tree. This is the system failure. That is, both the active and standby cards fail, which render the system nonfunctional.

What are the failures that can lead to a system failure?

(a) Both the active and standby cards fail. The system can no longer function.

(b) The active card fails but the standby card fails to take over. Let us assume the standby card continues to operate in the standby mode and does not transition to the active role. This can happen due to a variety of reasons: the standby card lost communication with the active card, it failed to recognize that the active card has gone down, or it failed to transition to active state after recognizing that it needed to take over.

(c) Failover from the active to the standby card exceeds the upper spec timing limit, which results in the end user perceiving a system outage. In other words, the system becomes unavailable for a longer transient period of time than is acceptable to the customer due to the impact on the required system functionality.

Now if we start looking at the failures in terms of logic, we could say that the system failure occurs due to:

(a) Active failure **AND** standby failure

(b) Active failure **AND** standby does not take over at all,

(c) Active failure **AND** standby takes over after an unexpected long period of time (say 1 minute).

That is, a system failure occurs when either (a) **OR** (b) **OR** (c) happens (Fig. 16.6).

Since the same instance of "Active Card Failure" is associated with other failures, it becomes a repeated event.

We can apply the same logic to build more complex systems by digging into the failures and associating them using logic gates.

EXAMPLE 16.1

Let us consider a two parallel component availability calculation using FTA. Card 1 has an MTBF of 100,000 hours and an MTTR of 8 hours, whereas Card 2 has an MTBF of 50,000 hours and an MTTR of 16 hours. A full system failure occurs when both parallel components have failed.

What is the steady-state availability of the resulting system?

Solution

We first calculate the unavailability of each component (basic events). The system fails only if both components have failed, and thus the components are *ANDed* together in the fault tree. The unavailability of Card 1 is:

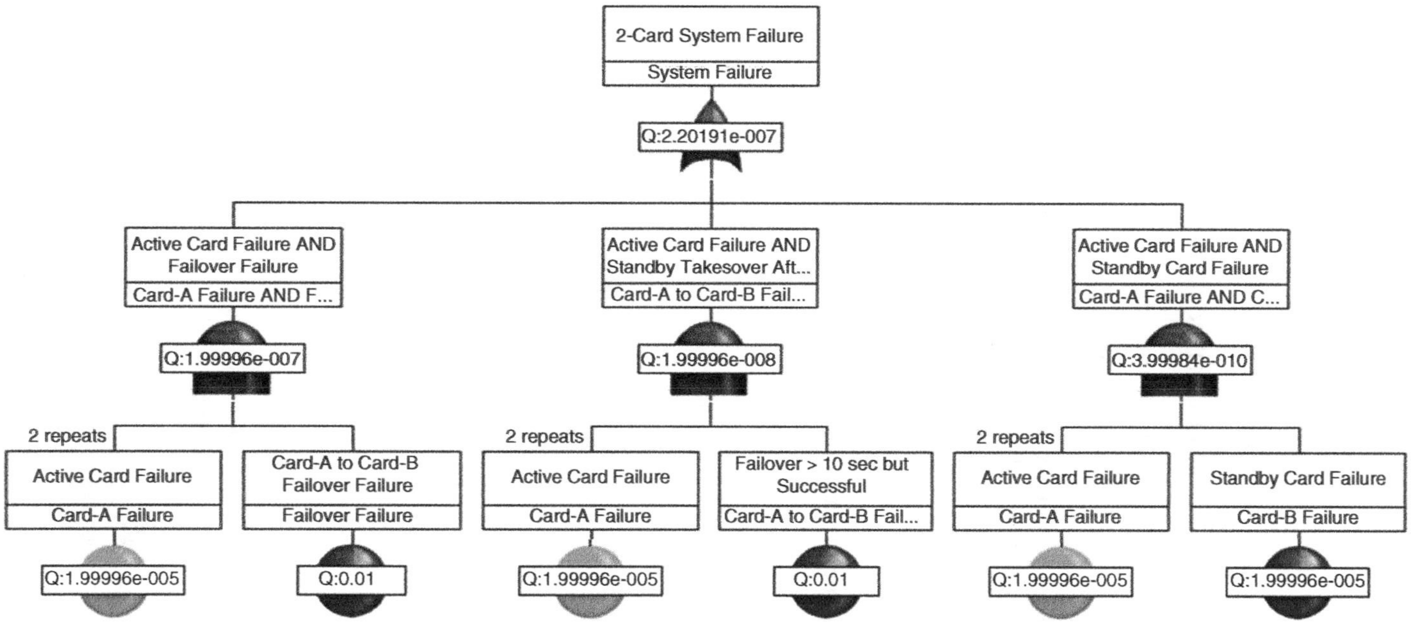

Figure 16.6 Fault Tree of an Active–Standby Pair

$$Q_1 = \text{MTTR}_1/\text{MTBF}_1$$
$$Q_1 = 8/100{,}000$$
$$Q_1 = 8.0\text{E-}5.$$

Likewise, the unavailability of Card 2 is:

$$Q_2 = \text{MTTR}_2/\text{MTBF}_2$$
$$Q_2 = 16/50{,}000$$
$$Q_2 = 3.2\text{E-}4.$$

We next construct the FTA diagram combining these basic events into the given redundancy configuration and calculating the unavailability of the system (Fig. 16.7).

From the diagram, the system availability is:

$$\begin{aligned} A_S &= 1 - Q_s \\ A_s &= 0.99999997. \end{aligned} \tag{16.1}$$

This gives the same result as obtained from the Markov calculation in Chapter 11, Section 11.4.11.

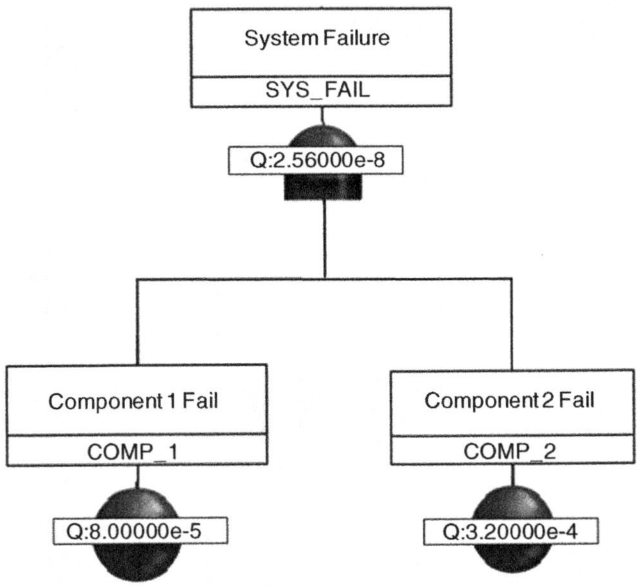

Figure 16.7 Two-Component Fault Tree Model

We note that this method is simpler than the Markov analysis and applies when we are interested in properties of the system in which we answer a binary question (system has failed/has not failed, component 1 is unavailable/available), and so on. In this case, FTA provides an intuitive and simple method to obtain these results. In addition, the fault tree itself provides insight into the nature of the system architecture in terms of how individual component failures are related to system failures.

16.5 FAULT TREE LIMITATIONS

Fault Trees have significant limitations, although some of these limitations have been overcome by commercial reliability software.

1. *Fault Severity.* Failures can have different impacts to the system; however, the top gate of a fault tree that combines many failures has only one severity.
2. *Failure Dependencies.* These are not captured in traditional fault trees. However, some commercial reliability software allows these dependencies to be captured using gates with built-in dynamic properties.
3. *Fault Propagation.* Failures may change the configuration of the system. The fault tree is static and does not change based on a failure trigger.
4. *Repair, Recovery, and Maintenance Actions.* Repairable systems are not captured in traditional fault trees. However, some software, such as Relex, allows you to insert a repair action and value inside the basic event as illustrated in the previous example. But even with repair actions captured, the modeling using fault trees becomes more cumbersome when the repair action and the failure event are separated by other events and dependencies.
5. *Fault Duplication.* A single event may trigger multiple fault tree branches in different parts of the fault tree. Some fault tree software allows fault duplication to be captured (example of repeated events in Relex).
6. *Degraded Modes.* FTA is not capable of easily addressing partial failures, degraded states, dynamic behavior over time, and "what-if" scenarios.

16.6 SUMMARY

FTA is a top-down reliability technique for deriving failure nodes that in combination with other failures can result in the occurrence of the top event (e.g., system outage).

FTA is an intuitive graphical approach that shows a clear relationship between faults and system failure. We applied FTA to the two-component nonrepairable model previously analyzed using Markov techniques. FTA has several significant limitations that we need to be aware of when applying this tool.

We will explore FTA in several examples and case studies in Chapters 18, 21, and 22.

CHAPTER 17

Monte Carlo Simulation Models

17.1 INTRODUCTION

Monte Carlo analysis is a simulation-based technique that enables quick analysis of system models. This technique can be used during requirements and design phases to visualize system behavior and predict system performance and reliability.

In this chapter, we consider Monte Carlo simulations with examples from Excel, Crystal Ball, and Matlab.

The Monte Carlo technique was first used by scientists working at the nuclear research lab in Los Alamos, New Mexico and named after the Monaco resort town famous for its casinos. Monte Carlo simulation has been used to model a wide variety of physical and conceptual systems. It is extensively used in financial planning, business, reliability engineering, and safety engineering for quantitative risk analysis and control.

In many engineering and scientific applications, Monte Carlo simulation is typically the method chosen for evaluating the impacts of parameter uncertainty on model predictions. This methodology allows a full mapping of the uncertainty in model inputs into the corresponding uncertainty in model outputs.

Monte Carlo simulation uses random sampling of either defined or unknown probability distributions representing uncertainty of events or variables of interest. The distributions are used to generate random inputs to a system transfer function which calculates the output values. Each set of samples and resultant outputs is called a *trial*. The simulation is repeated hundreds or thousands of times with independent random samples selected for each trial. The output values from each trial are stored, and after all trials have been run, the output values are used to create probability distributions. Monte Carlo models can walk through thousands of scenarios and generate predictions by taking randomness into account. It can show the extreme possibilities along with all possible consequences of most likely events. Thus, the simulation can tell us not only what could happen, but how likely it is to happen.

Designing High Availability Systems: Design for Six Sigma and Classical Reliability Techniques with Practical Real-Life Examples, First Edition. Zachary Taylor and Subramanyam Ranganathan.

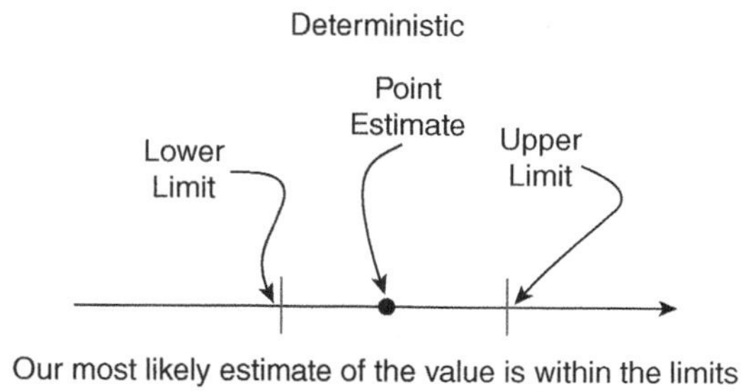

Figure 17.1 Deterministic Modeling

Assume we are given the task of designing a control system that must meet a set of tolerance requirements that are specified in terms of upper and lower specification limits. Based on the expected mean of the noise inputs to the system, we determine that the output response of the system falls well within our specification limits. We use the term noise loosely to refer to, among other things, variability in the true reliability of the components that comprise the system, unaccounted characteristics of the system that are not captured by the model, and other nontrivial influences. The mean of the noise represents a statistic or *point estimate* of the underlying variability of the noise. A point estimate is a single value or statistic calculated from a set of sample data selected from a population. This provides the most likely estimate of an unknown (fixed or random) parameter of the sampled population.

Using the point estimate as an input to the system, we calculate the output based on the system transfer function. The calculation is deterministic, that is, given any particular input, we can determine the exact output. Thus, using our point estimate of the noise, our model indicates that our system will meet the tolerance requirements as shown in Figure 17.1.

However, this result is only correct for the point estimate (the mean of the noise distribution). Since we know the noise or uncertainty itself is variable and we cannot predict the specific value of the noise, how likely is it that a particular value of this random variable will result in a scenario in which the system output no longer falls within the upper and lower limits? If we are able to characterize the noise using a known probability distribution, we can determine how likely it is that it will result in our system falling outside of our required tolerance. Figure 17.2 shows an example of the stochastic output of a system that has a 10% probability of not meeting the specification. We have gained additional information about the behavior of the system not revealed by the deterministic analysis using mean values. Since the system output has a quantifiable probability of being outside of the limits, this may affect our decision to proceed with this particular design.

Several approaches can be used to address uncertainty.

Single Point Estimation. Each uncertain variable within a model is assigned a "best guess" estimate. These values are used as inputs to predict system behavior. This is quick and easy to calculate, but provides a limited view of the system being modeled and could result in critical characteristics of the system being

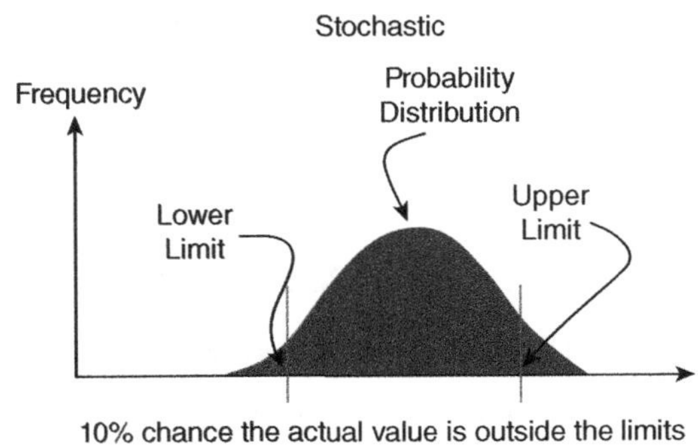

Figure 17.2 Stochastic Modeling

overlooked. A single point estimate, such as an average, may not be representative of the typical system behavior and does not provide the ranges of possible values and the likelihood of those values which can be critically important in determining whether the system meets minimum acceptable criteria. We should always keep in mind that a point estimate is a summary of data providing some view of the data but discarding other information that could be important.

Scenario Analysis. Several possible input values are manually chosen, such as the most-likely, best-case, and worst-case scenarios and used to determine the system behavior for these specific values. This increases our insight into the system behavior and provides a discrete set of possible outcomes, but does not provide the complete range of possibilities and the likelihood of occurrence of these outcomes. For example, is the worst-case scenario more likely than the best-case scenario? We cannot easily answer that question when only scenario analysis is employed.

What-If Situational Analysis. An incremental set or range of input possibilities is selected and for each of these inputs, the system behavior is determined. This approach provides more scenarios and a broader range of outputs, but the selection and calculation process is laborious and only reveals possibilities, not probabilities.

Monte Carlo Simulation. A random sampling of probability distribution functions are selected as model inputs to produce many possible outcomes. The results provide probabilities of the occurrence of the different outcomes.

Monte Carlo simulation provides a number of advantages over *deterministic*, or "single-point estimate" analysis:

- *Probabilistic Results.* Results show not only what could happen, but how likely each outcome is.
- *Graphical Results.* Output probability distribution graphs can be automatically generated using the output data collected from each trial.

- *Sensitivity Analysis.* With just a few cases, deterministic analysis makes it difficult to see which inputs impact the outcome the most. With Monte Carlo simulation, the relative contribution of each input to the output of interest can be readily calculated.
- *Scenario Analysis.* In deterministic models, it is very difficult to model different combinations of values for different inputs to see the effects of truly different scenarios. With Monte Carlo simulation based on thousands of trials, the specific set of input values that resulted in a range of output values of interest can be identified and analyzed.
- *Input Correlation.* In Monte Carlo simulation, interdependent relationships between input variables can be modeled, and the effects of these relationships on the outputs can be determined.

By using probability distributions, we attempt to capture the full range of possible values that can be expected for an input event and the relative likelihood of that value occurring. These probability distributions can be an excellent technique for describing uncertainty in the occurrence of an event.

Monte Carlo simulation can be used on a wide range of complex systems, including multi-input, multi-output, and multi-function systems and for system models that are too complex to derive a general set of equations for the solution. We could mathematically derive the output probability distribution of a complex model based on input probability distributions and the system transfer function. However, this approach can be very difficult, and in many cases, no solution is possible. As we saw in the Markov analysis, deriving the deterministic dynamic state behavior for even a simple model can become increasingly difficult as the number of states increase. We look to tool support to help us with our engineering analysis when analyzing more complete system models that more accurately reflects the system being built.

Statistics can be derived from the resultant output distributions that are close to the theoretical values based on the principle of the law of large numbers (the larger the sample, the more certain that the sample mean will be a good estimate of the system mean).

For large complex systems, the Mean Time between Failures (MTBF) or Mean Time to Repair (MTTR) may be difficult to determine using an analytical approach. By executing a Monte Carlo simulation over thousands of trials for a defined system model, we can estimate these reliability parameters from the simulation output. The simulation is run and the time between failures for each trial is captured. This is repeated n times and the estimated MTBF calculated:

$$\overline{\text{MTBF}_s} = \frac{1}{n}\sum_{j=1}^{n} \text{TBF}_j. \tag{17.1}$$

The precision of the estimate can be obtained by calculating the sample standard deviation:

$$\sigma_{\text{MTBF}} = \sqrt{\frac{1}{n-1}\sum_{j=1}^{n}(\overline{\text{MTBF}} - \text{TBF}_j)}. \tag{17.2}$$

Steps for Monte Carlo Simulation:

1. Define the relationship (transfer function) between one or more inputs and the outputs of interest.
2. Define a statistical distribution for each input.
3. Select a random value from each distribution.
4. Calculate the output value(s) based on the random values chosen for each input and store these values,
5. Repeat Steps 3 and 4 until a sufficient number of trials have been run (typically thousands of trials).
6. Plot each output histogram and calculate desired output statistics.

17.2 SYSTEM BEHAVIOR OVER MISSION TIME

We can use the Monte Carlo technique to generate random failure times from each component's exponential failure distribution. The overall system reliability is then obtained by simulating system operation over some period of time (mission time) and calculating the probability distribution of system reliability behavior. Note that the system reliability may vary significantly depending on the mission time chosen.

Starting with a Markov state diagram, we specify the point estimates and the exponential probability distribution for each of the reliability parameters that are of interest.

Component failure events are generated from random samples of the component failure distributions. Additionally, scheduled events (e.g., preventive maintenance actions), and conditional events (events initiated by the occurrence of other events, such as repair actions) may also be included to create a simulated lifetime scenario for the system.

The accuracy of the model depends on many factors including:

1. The level of detail captured in the model.
2. The correctness of the MTBF/MTTR estimates.
3. The validity of the constant failure rate and repair rate assumptions.
4. The number of trials simulated.
5. The mission time.

17.3 RELIABILITY PARAMETER ANALYSIS

With Monte Carlo simulation, we can analyze variations in our assumptions regarding MTBF and MTTR values. When these variations are injected in the model and thousands of simulation trials are executed, we can extract information on the likelihood of different values for the overall system availability and reliability. The Monte Carlo technique is used to generate random values of MTBF and MTTR from a failure distribution assumption for each component's failure and repair. The distributions are determined by the analysis based on the estimated most likely values for each component's MTBF and MTTR. The overall system availability distribution is then obtained.

Starting with the Markov state diagram of a particular configuration of interest, the next step is to specify the point estimate and the probability distribution for each of the parameters that are of interest.

The Monte Carlo simulation models require the inputs to be a probability distribution instead of a single number. The normal and triangular distributions are common choices. Historical data (nonparametric distributions), if available, can also be used instead of parametric distributions.

How do you interpret some of the properties of distributions? The mean (also called the first moment) represents the average value. The variance (also called the second moment) represents the associated risk, the third moment represents the distribution's skewness, and the fourth moment measures the peakedness.

17.3.1 Sensitivity Analysis

Sensitivity analysis is used to evaluate the relative influence of component availability on the availability of the entire system. This information can be used to identify critical components in the system that adversely affecting system availability, identify weaknesses in the architecture and design, and quantify how specific changes impact availability. Armed with this information, the designers can focus their efforts on those components that yield the highest impact, that is, make changes that have the biggest bang for the buck.

Sensitivity analysis is usually generated as part of a model's output as a result of Monte Carlo simulation. Sensitivity analysis represents the percentage each factor (little x's) in the model impacts the variability of the big-Y. For example, if we assume $Y = f(x1, x2, x3)$ as a model. The sensitivity analysis will give us the percent impact of each factor $x1$, $x2$, and $x3$ on the output Y. Let us say $x1$'s impact is 50%, $x2$'s impact is 30%, and $x3$'s impact is 20%. This analysis is usually helpful to tighten or reduce the variation of the output by controlling the variation of the inputs. In this assumption, since $x1$ has the highest percentage impact on Y, controlling the variation in $x1$ will help reduce the variation of Y or the spread of the distribution of Y. Let us illustrate this through a real-life example.

Let us assume that our availability model contains different factors with MTBF, MTTR, failure probabilities, and so on. We run a Crystal Ball Monte Carlo simulation in which the MTBFs and MTTRs are sampled from input probability distributions for a trial and repeated over many trials. For the output, we obtain the probability distribution for Five 9s availability and the sensitivity analysis graph, given in Figure 17.3 and Figure 17.4, respectively.

The sensitivity analysis indicates that SSCHW has 55.1% impact on the system Five 9s. SSCTiming has a negative 31.3% impact on Five 9s. Since SSCHW has the highest impact, the availability engineer should turn his attention to the control of the variation of SSCHW in order to reduce the impact in the output. In this example, reduction of variation in the SSCHW not only reduced the output variation, but also resulted in a right shift of the mean of Five 9s. Figure 17.5. The new sensitivity analysis is shown in Figure 17.6. The highest impacting factors are now very different from what they were previously. This, in essence, is the usefulness of sensitivity analysis.

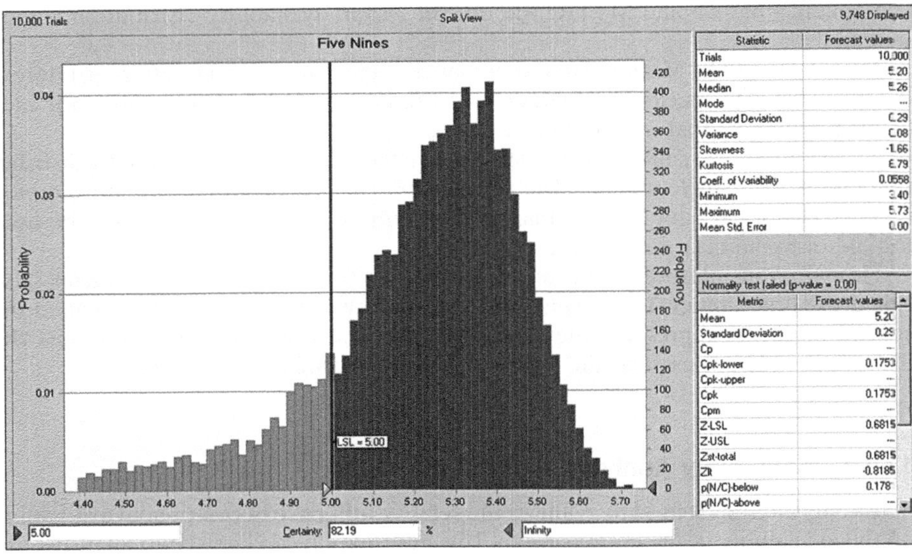

Figure 17.3 Availability Model: % Nines Output

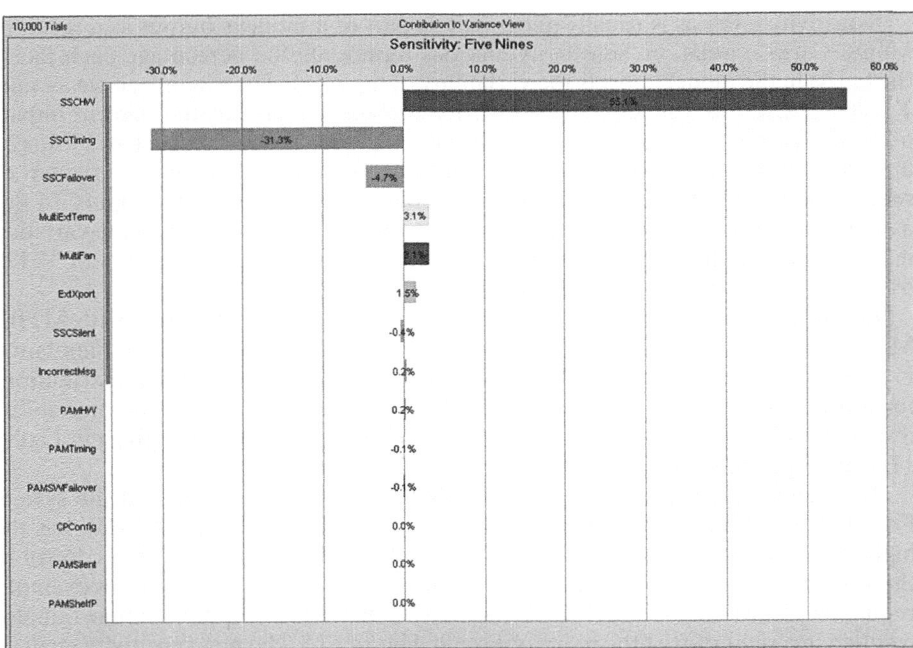

Figure 17.4 Sensitivity Analysis

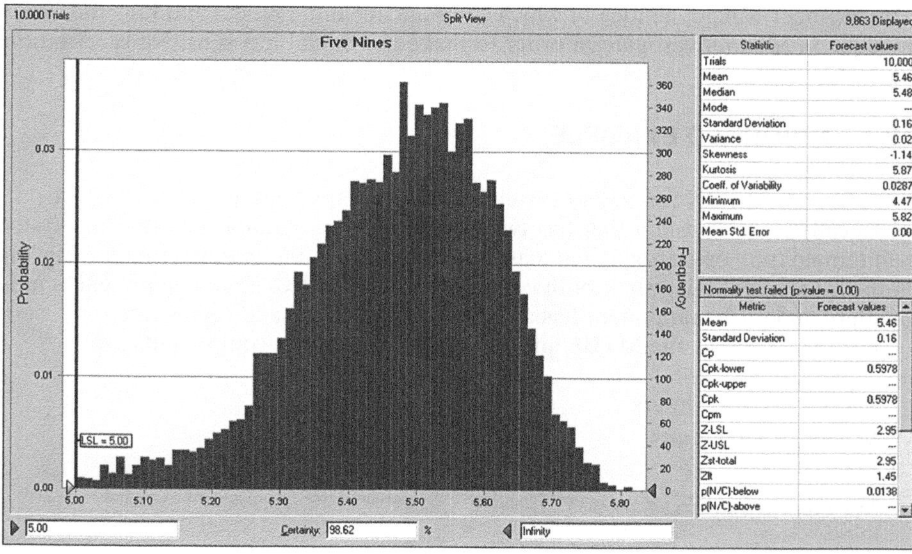

Figure 17.5 Revised Availability Model: % Nines Output

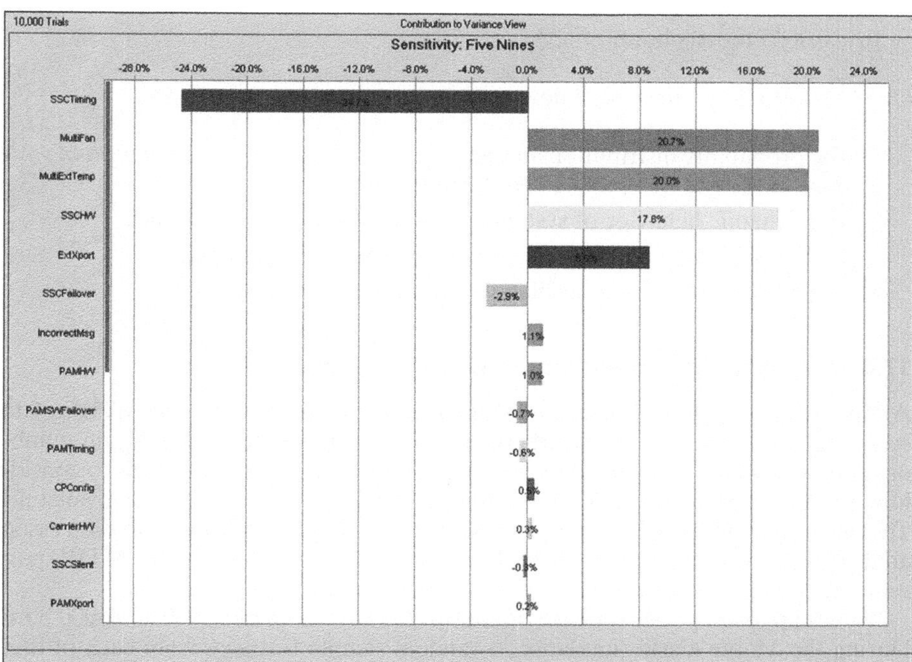

Figure 17.6 Revised Sensitivity Analysis

The engineer will get an idea of which components in the model and thus the system that she needs to investigate in order to make the model less sensitive to variations.

17.4 A WORKED EXAMPLE

Let us revisit the 2N redundant system introduced in Chapter 11, Section 11.4.

Recall this is a system that has two components in parallel such that the system will remain fully operational if at least one of the two components is functioning, that is, the system will fail only if both components have failed. Each component is independent and each component has a different failure rate and repair rate.

For this example, the MTBF and MTTR of each of the components are:

$MTBF_1 = 10{,}000$ hours
$MTBF_2 = 3000$ hours
$MTTR_1 = 8$ hours
$MTTR_2 = 16$ hours.

These values correspond to the following failure rates and repair rates:

$\lambda_1 = 0.0001$ failures/hour
$\lambda_2 = 0.000333$ failures/hour
$\mu_1 = 0.125$ repairs/hour
$\mu_2 = 0.0625$ repairs/hour.

Our goal is to determine the following:

1. The probability distribution and sensitivity of availability as a function of variations in the MTBF and MTTR of components.
2. The dynamic behavior of system availability as a function of time.
3. The steady state system availability using Markov techniques.
4. The steady state system availability using fault tree analysis.

17.4.1 Availability Sensitivity to MTBF Changes

Although we may have reasonable estimates of the availability and MTBF of the system we are building, the true MTBF will not be known until a significant number of systems have been deployed over a long period of time. We may want to consider how the uncertainty in the MTBF estimate of our components affects availability. The uncertainty in the MBTF estimate is captured as a probability distribution associated with the MTBF point estimate. The more uncertain we are of our MTBF point estimates, the larger the variance of our distributions.

We set up assumptions for each of the two MTBF values of 10,000 and 3000 hours. During the Monte Carlo simulation, a random sample is drawn from each of these probability distributions, and the resulting system availability and Five 9s values are calculated for each trial. Ten thousand trials are performed, and the resultant availability and Five 9s probability distributions are graphed.

The availability distribution represents the range of availability values that can be expected for any system with the median value roughly equal to the base case. In Crystal Ball, the *base case* is the initial value in an assumption, decision variable, or forecast cell prior to running the simulation. This initial value is our point MTBF estimate. The simulation results represent the likelihood of our fielded system achieving 5 9s given the variability in MTBF estimates.

We will calculate system availability by considering three different distributions for the MTBFs: uniform, triangular, and beta distributions.

17.4.2 Availability Sensitivity to Uniform MTBF Distribution

The uniform distribution implies that the "true" MTBF may be any value between the minimum and maximum values, and any of these values are equally likely. For component 1, our best guess or estimate is that the component has an MTBF of 10,000 hours. Let us assume our analysis or engineering judgment indicates that the smallest possible value of this MTBF could be one-fifth of our most likely point estimate that is, 2000 hours. We also estimate that the largest MTBF value could be 50% larger than our point estimate, that is, 15,000 hours. We are more concerned about how the lower MTBF affects availability. So as a rule of thumb, the lower MTBF minimum is more extreme than the maximum. This large variance between the minimum and maximum is intended to ensure that the worst-case possibilities are considered.

Likewise, for component 2, which has an MTBF of 3000 hours, we establish a uniform distribution with a minimum MTBF of 600 hours and a maximum MTBF of 4500 hours.

Using Crystal Ball, these two uniform MTBF distribution assumptions are assigned to the components and a forecast for the system availability in terms of percentages, and Five 9s probability distribution prediction is created (Fig. 17.7 and Fig. 17.8). The LSL for the figures is 0.99999 and 5.0, respectively, which results in a certainty value of 65.89%. This tells us that given our uncertainty in the MTBF estimates, we have a 66% chance of successfully achieving our Five 9s requirement.

Analyzing the Five 9s probability distribution as shown may provide a clearer picture on availability since the distribution is symmetric (mean, median, and base case are approximately the same). We have a 65.89% confidence of the system achieving Five 9s. If the system requirements indicate the variance in our predicted availability must be no less than four 9s or the certainty of meeting the average availability requirement must be at least 75%, then the 65.89% certainty value is not acceptable. We will need to take action to address the deficiency in our design by:

1. Increasing the MTBF, that is, increasing the reliability of our components
2. Decreasing the MTTR (smaller detect, failover, and repair times)
3. Decreasing the standard deviation (spread) of the availability probability distribution by reducing the variation in our MTBF estimates
4. Changing the system architecture to achieve the required availability.

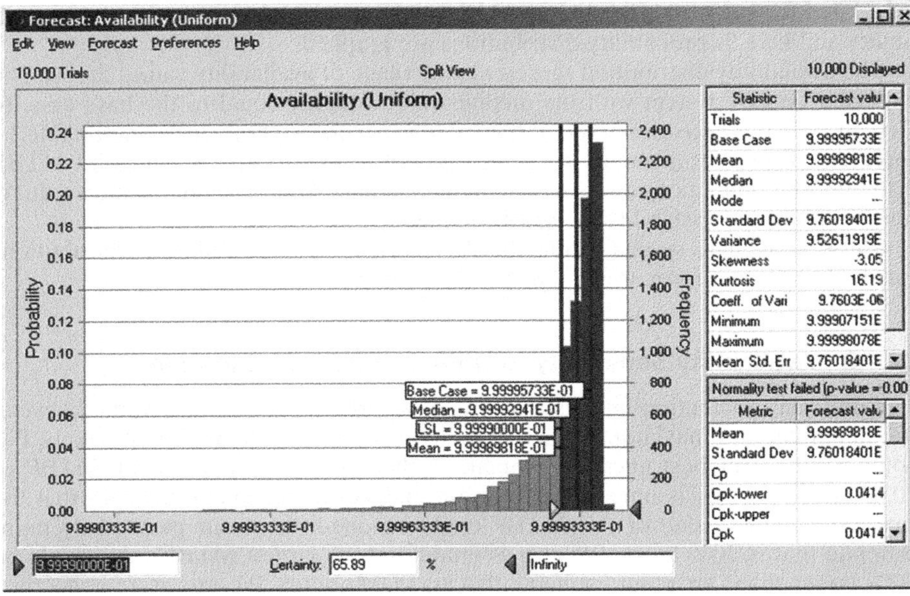

Figure 17.7 Availability Distribution: Uniform MTBF Assumptions

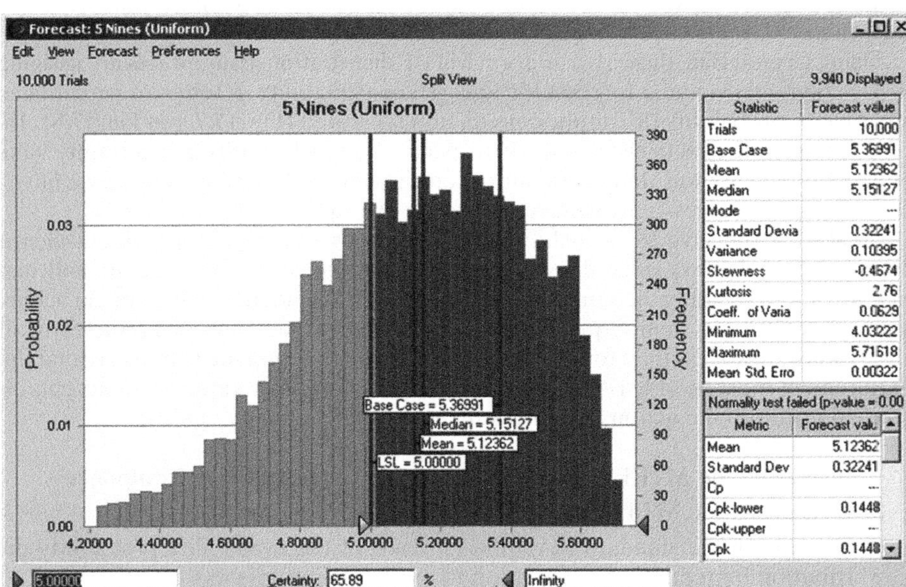

Figure 17.8 Five 9s Availability Distribution: Uniform MTBF Assumptions

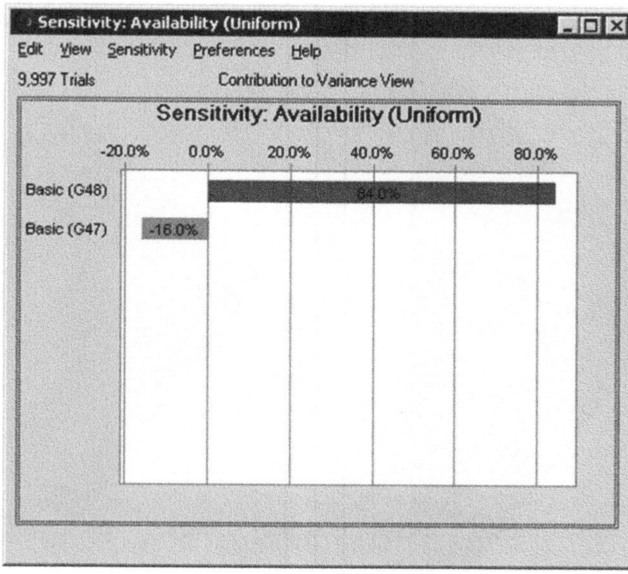

Figure 17.9 Five MTBF Sensitivity Analysis: Uniform MTBF Assumption

From the Five 9s distribution (Fig. 17.8), in the worst-case scenario (although unlikely), our average availability across all systems could be as low as 4.2 9s (99.9937% availability). For this scenario, the risk is too great that we would not meet the Five 9s requirement, and thus we will need to improve the system availability and/or reduce the uncertainty in our MTBF estimates—particularly the minimum values. The sensitivity chart (Fig. 17.9) shows the impact each component's MTBF has on system availability. From this chart, availability is most sensitive to variations in our estimate of the MTBF for component 2. In fact, 84% of the variation in availability is due to this component. Thus, we should focus our efforts on improving the minimum expected MTBF of component 2.

17.4.3 Availability Sensitivity to Triangular MTBF Distribution

Now what if we have reason to believe that our estimated MTBF values are the most likely values, and that other MTBF values are possible but decrease in likelihood the further we move away from the most likely value? One way to capture these assumptions is with a triangular probability distribution. For component 1, we select a triangular distribution and enter 2000 hours for the minimum value, 10,000 hours for the likeliest value, and 15,000 for the maximum value. No possible values exist outside the minimum and maximum values. For component 2, we select a triangular distribution with 600 hours for the minimum value, 3000 hours for the likeliest value, and 4500 hours for the maximum value (Fig. 17.10).

The resulting availability distributions are shown in Figure 17.11 and Figure 17.12.

Assumption: MTBF1

Triangular distribution with parameters:

Minimum	2,000.00
Likeliest	10,000.00
Maximum	15,000.00

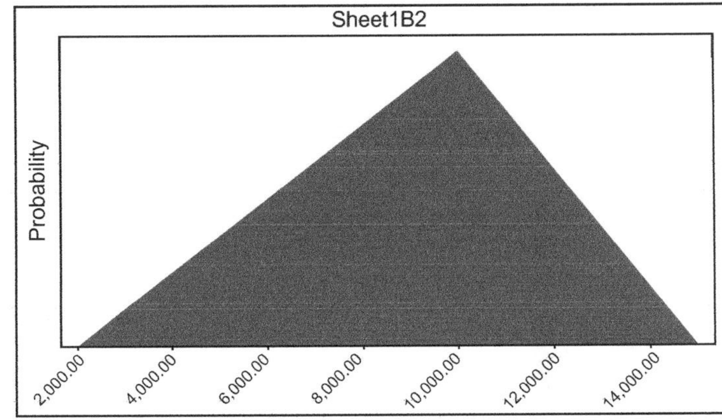

Assumption: MTBF2

Triangular distribution with parameters:

Minimum	600.00
Likeliest	3,000.00
Maximum	4,500.00

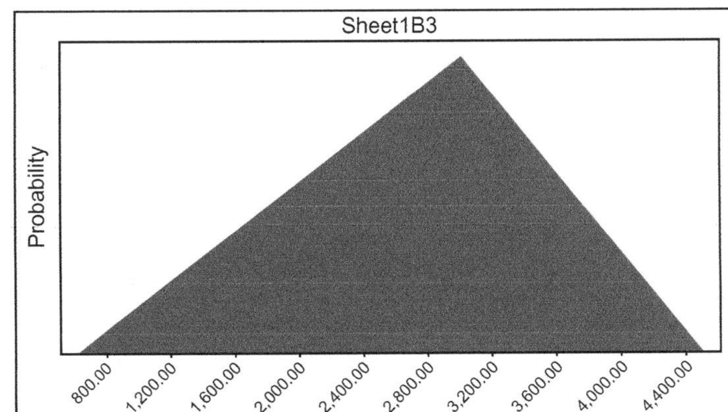

Figure 17.10 Triangular MTBF Assumptions

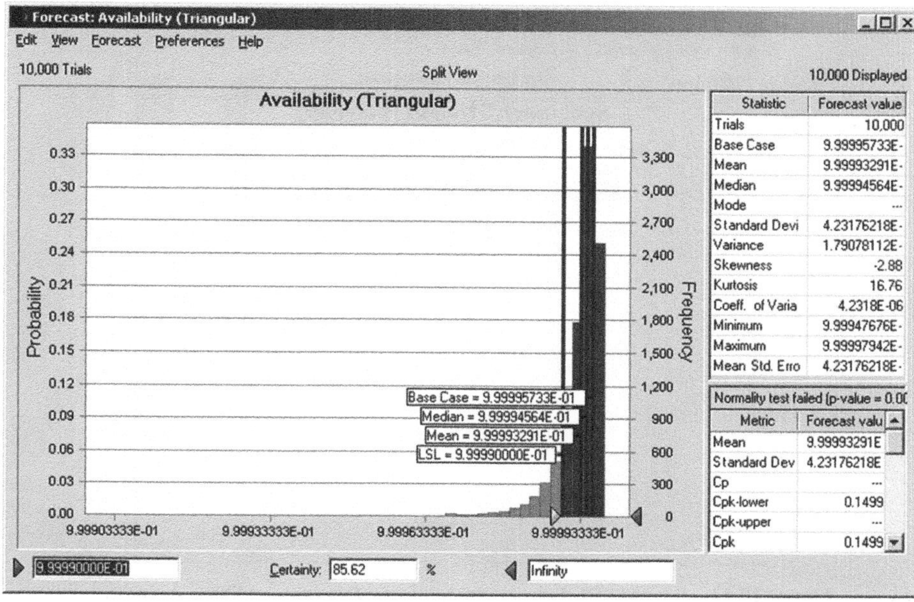

Figure 17.11 Availability Distribution: Triangular MTBF Assumptions

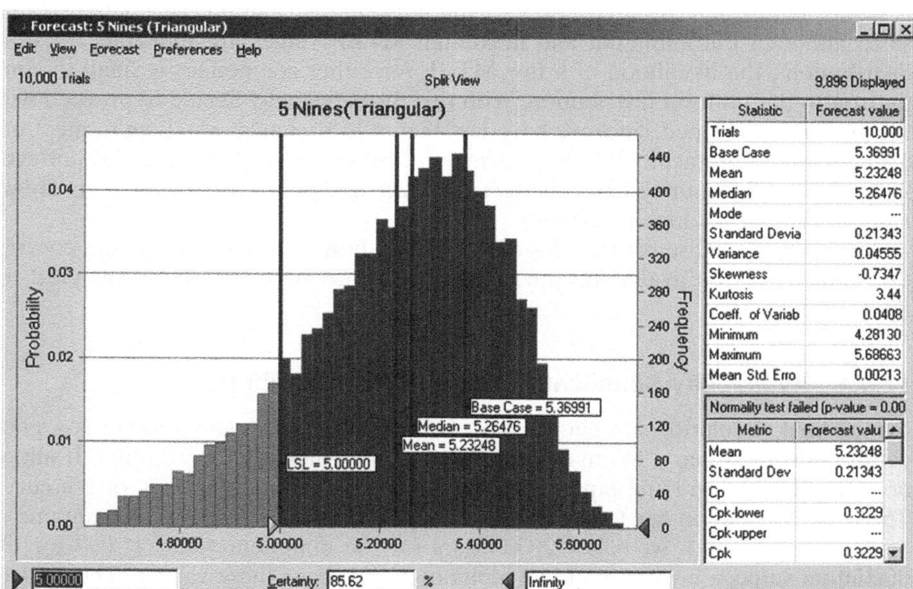

Figure 17.12 Five NINEs Availability Distribution: Triangular MTBF Assumptions

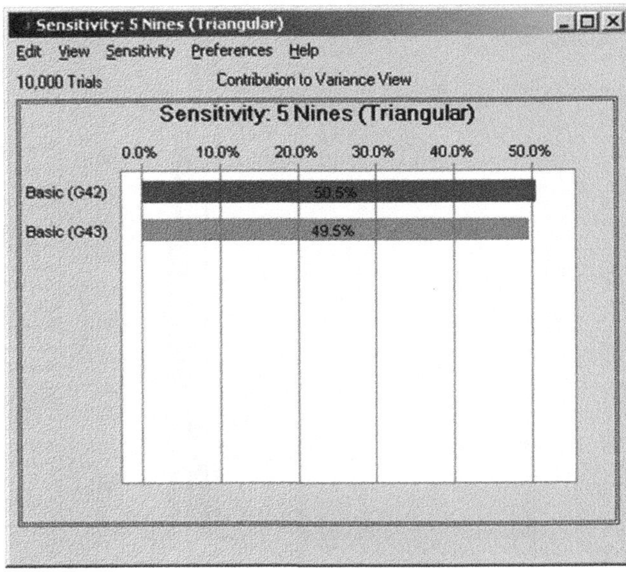

Figure 17.13 Five MTBF Sensitivity Analysis: Triangular MTBF Assumption

Notice the difference in the availability distributions of these figures in comparison with the uniform MBTF simulation. The availability distribution is skewed more to the right, and our likelihood of meeting our Five 9s requirement is now 85.62%. This is higher than the 65.89% certainty of the uniform distribution. This is reasonable, since although the minimum and maximum MTBF values are the same for both distributions, the likelihood of a low MTBF for either component is small in comparison to the uniform distribution. With this result, we may decide to proceed with the system as designed, since we have less than a 15% chance of failing to meet our availability requirement. Or if more stringent certainty requirements of 90% or even 95% are specified, our system must be reanalyzed and perhaps changed to ensure we meet Five 9s with a higher certainty.

Interestingly, the sensitivity chart (Fig. 17.13) shows that each subsystem contributes equally to the availability uncertainty and thus both should be analyzed for improvement.

17.4.4 Availability Sensitivity to Beta MTBF Distribution

In our final simulation, we choose a Beta distribution. This distribution is a good compromise between a Normal distribution and a Triangular distribution. It allows us to capture the gradual tapering of MTBF estimates on either side of the curve. The beta distribution has four parameters: alpha, beta, minimum, and maximum.

For component 1, we select 2000 hours for the minimum value, 15,000 for the maximum value, a beta of 3 and an alpha of 3. We chose these values to get a symmetric distribution. For component 2, we select 600 hours for the minimum value and 4500 hours for the maximum value (Fig. 17.14).

Assumption: MTBF1

Triangular distribution with parameters:

Minimum	2,000.00
Maximum	5,000.00
Alpha	3
Beta	3

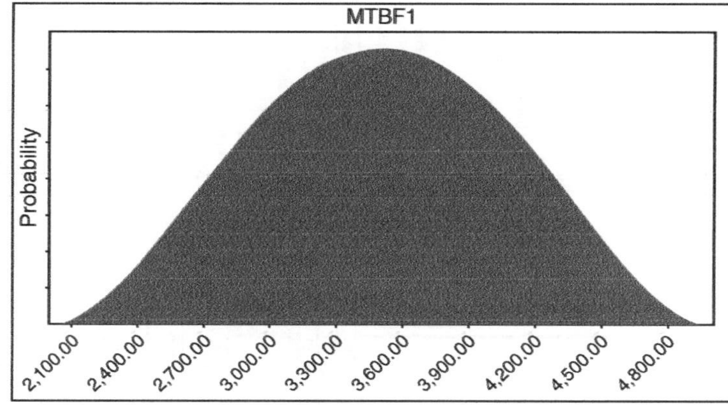

Assumption: MTBF2

Triangular distribution with parameters:

Minimum	600.00
Maximum	4,500.00
Alpha	3
Beta	3

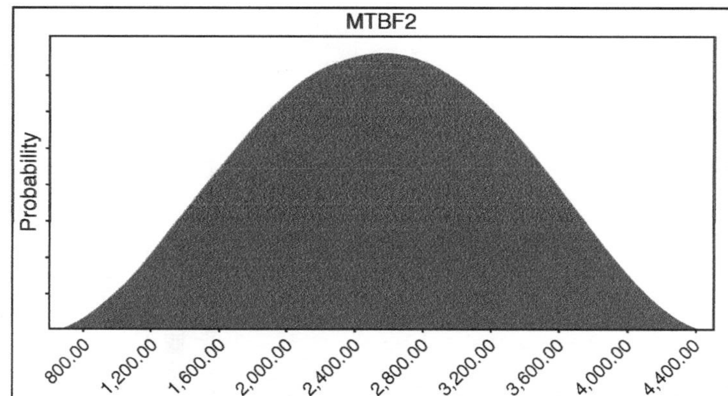

Figure 17.14 Beta MTBF Assumptions

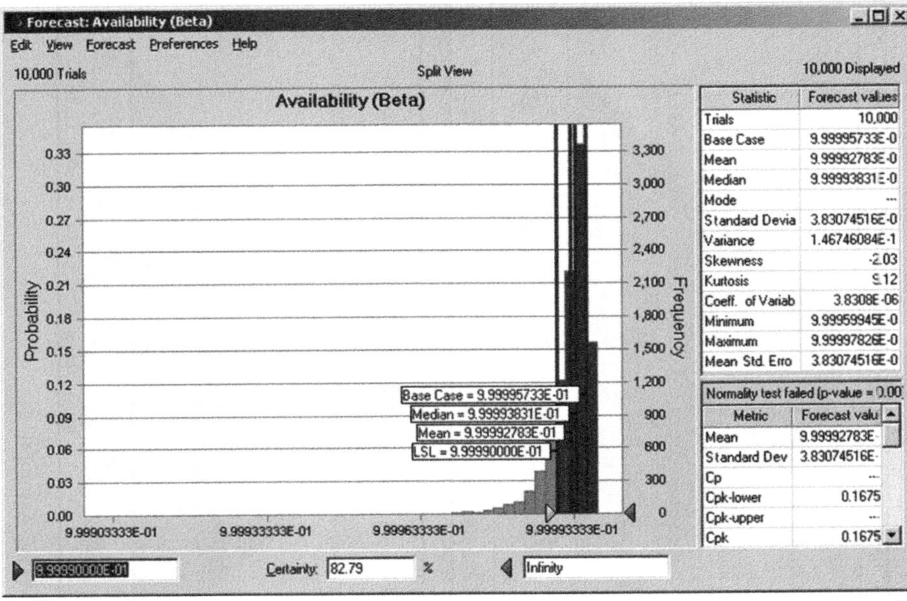

Figure 17.15 Availability Distribution: Beta MTBF Assumptions

The resulting probability distributions are shown in Figure 17.15 and Figure 17.16.

Notice the availability distributions of these figures are similar to the availability distributions for the triangular MBTF simulation. Our likelihood of meeting the Five 9s requirement is now 82.79% slightly worse than that of the triangular distribution. With this result, we may decide to proceed with the system as designed since we have less than an 18% chance of failing to meet our availability requirement.

The sensitivity chart (Fig. 17.17) shows that each component contributes equally to the availability uncertainty, and thus both should be analyzed for improvement.

Table 17.1 provides a comparison of the Five 9s across different MTBF distribution assumptions.

17.4.5 Availability for MTTR Variations

Now that we have considered variations in MTBF, let us consider variations in MTTR. One simulation approach is to use a normal distribution to represent the < MTTR. The most likely repair times would remain 8 and 16 hours for components 1 and 2, but we include a standard deviation of repair time. For this example, we chose for subsystem 1 a mean of 8 hours and a standard deviation of 2 hours, that is, 68% of the time, we would expect the repair to take between 6 and 10 hours. Similarly, for subsystem 2, we select a normal distribution with a mean of 16 hours and a standard deviation of 4 hours. The resulting Monte Carlo simulation forecasts are shown in Figure 17.18 and Figure 17.19. In these scenarios, the MTTR variations have minimal impact on meeting our availability requirements.

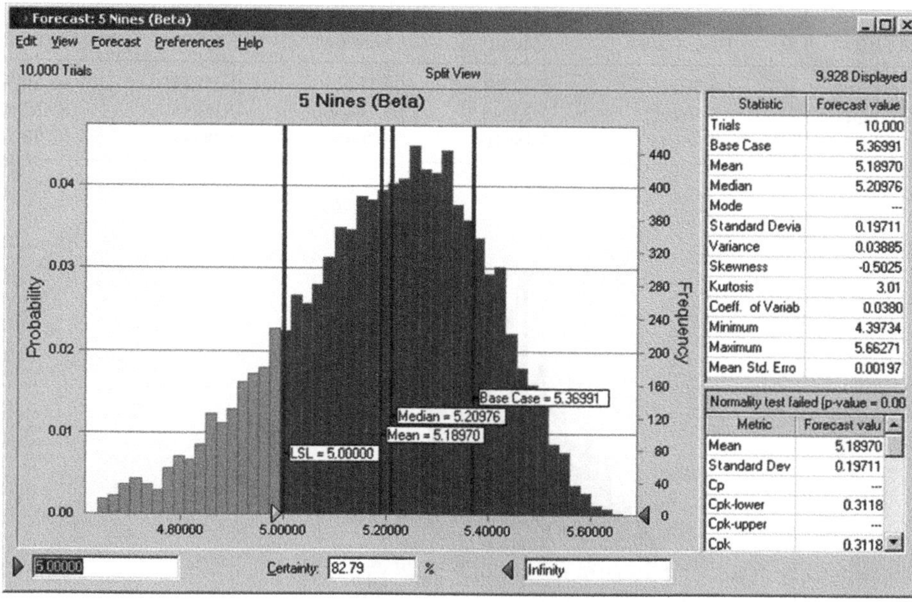

Figure 17.16 Five 9s Availability Distribution: Beta MTBF Assumptions

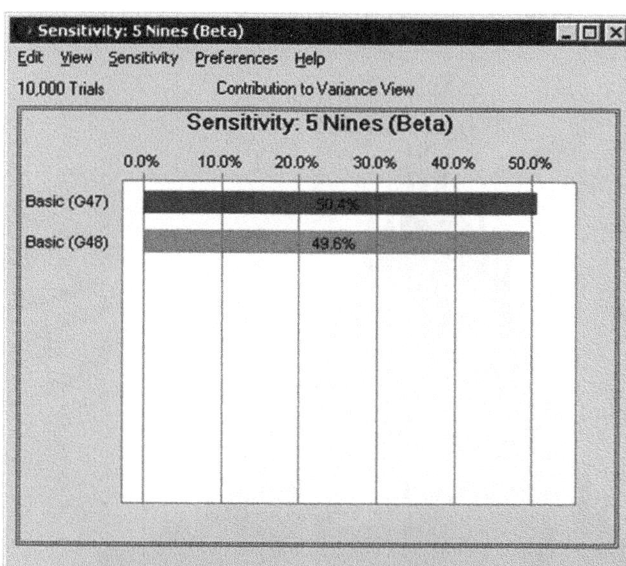

Figure 17.17 Five MTBF Sensitivity Analysis: Beta Distribution

TABLE 17.1 Comparison of Five 9s across Different MTBF Distributions

MTBF	Five 9s				
	Base Case	Mean	Median	LSL	Certainty
Uniform	5.36991	5.12362	5.15127	5.00000	65.89%
Triangular	5.36991	5.23248	5.26476	5.00000	85.62%
Beta	5.36991	5.18970	5.20976	5.00000	82.79%

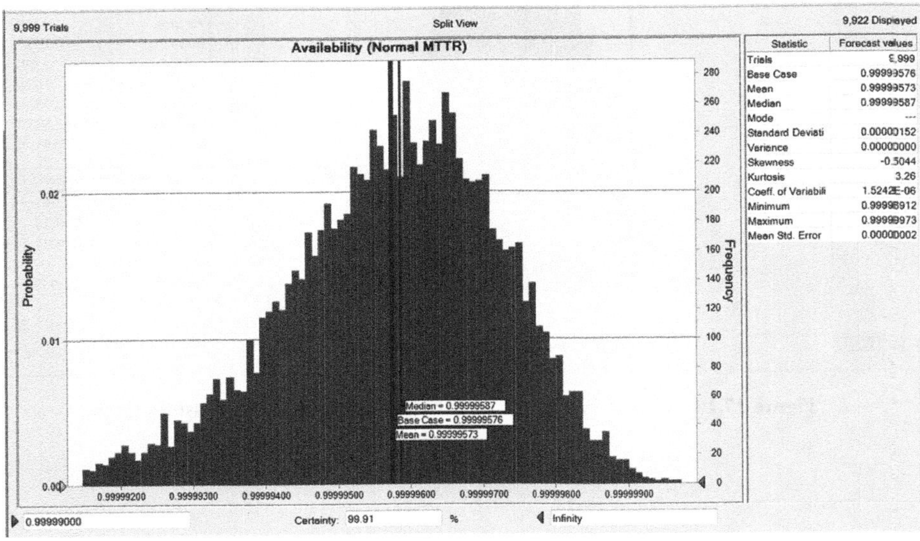

Figure 17.18 Availability Distribution: Normal MTTR Assumption

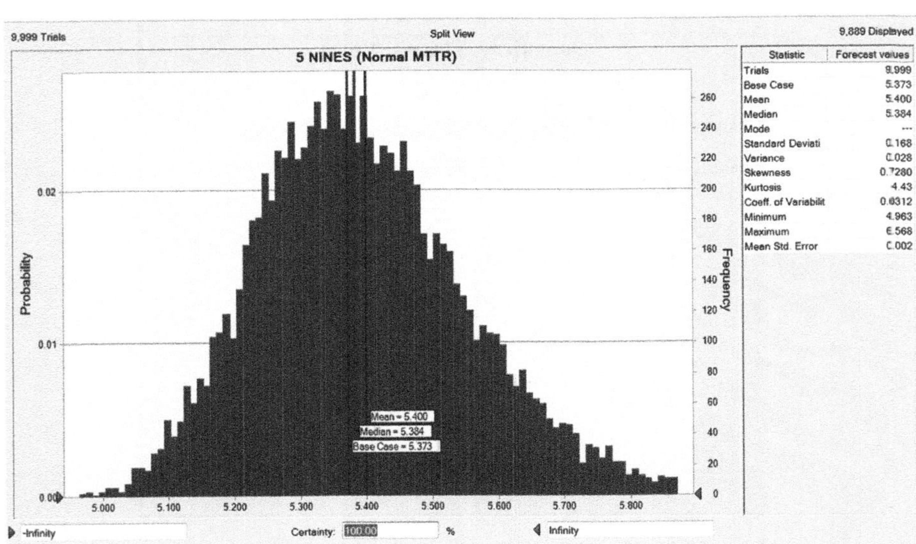

Figure 17.19 Five 9s Distribution: Normal MTTR Assumption

17.5 COMPONENT AND SYSTEM FAILURE TIMES USING MONTE CARLO SIMULATIONS

In the worked example in Section 17.4, we considered variations in MTBF and MTTR values, but did not take into account the system availability with respect to mission time. In this section, let us run Monte Carlo simulations to see how a system could behave based upon a Markov model.

Let us assume a two-component parallel system with repair, with the following parameters:

$MTBF_1 = 1000$ hours
$MTBF_2 = 300$ hours
$MTTR_1 = 8$ hours
$MTTR_2 = 16$ hours.

These values correspond to the following failure rates and repair rates:

$\lambda_1 = 0.001$ failures/hour
$\lambda_2 = 0.00333$ failures/hour
$\mu_1 = 0.125$ repairs/hour
$\mu_2 = 0.0625$ repairs/hour.

Each trial performs a simulation from $t = 0$ to $t =$ *mission time*. For each trial, we can determine the time to first system failure, the times to repair the system, the times individual components fail, and other reliability information.

Without access to tools such as Matlab and Crystal Ball, we can still create a Monte Carlo simulation using Excel. For two of these trials, that is, two missions, the times subsystems and system failures occur are graphed (Fig. 17.20 and Fig. 17.21). The

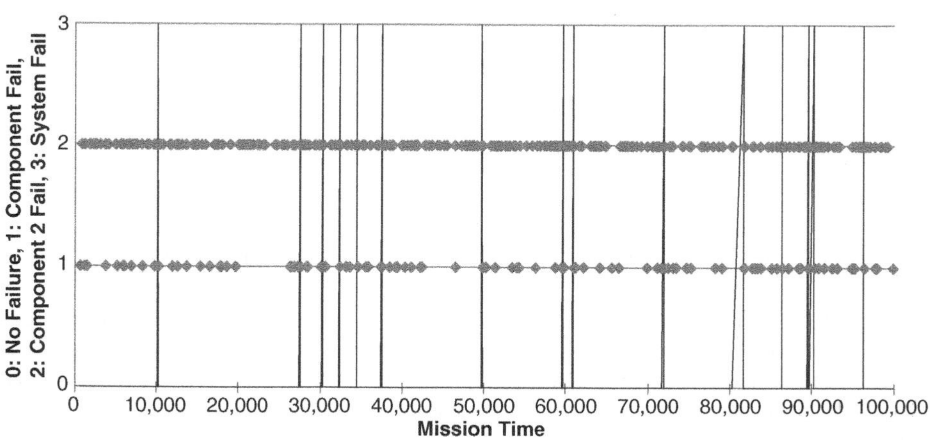

Figure 17.20 Monte Carlo Simulation of Component and System Failures (1)

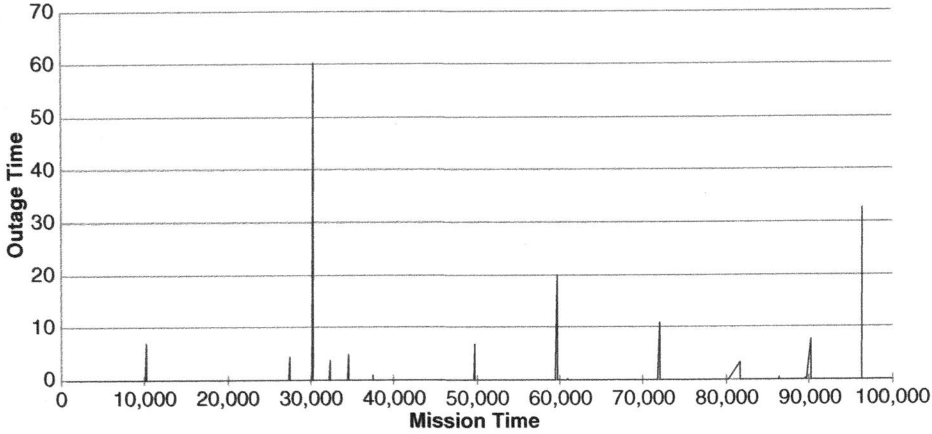

Figure 17.21 Monte Carlo Simulation of Outage Time versus Mission Time (1)

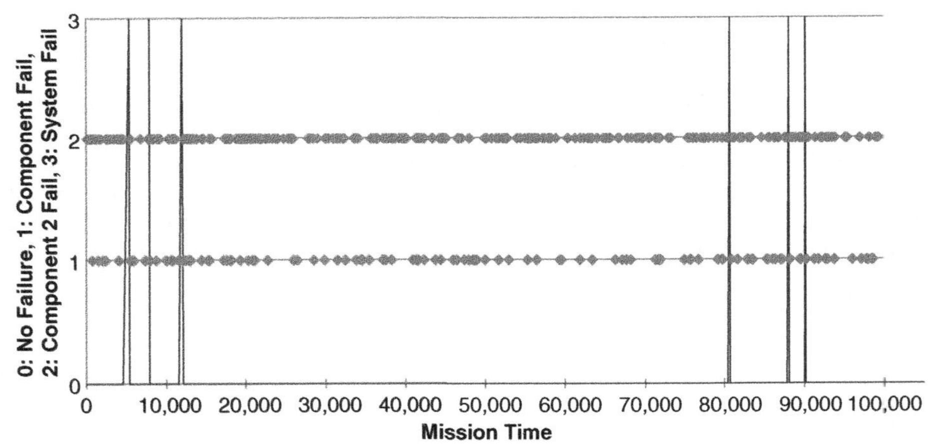

Figure 17.22 Monte Carlo Simulation of Component and System Failures (2)

duration of the system outages for these trials are plotted in Figure 17.22 and Figure 17.23.

This technique has important practical applications. For example, a network operator can predict outage probabilities during operation over a 1-year period and can take maintenance actions to prevent or minimize these outages. Using this information, the customer can minimize large revenue losses due to outages and ensure smooth operations.

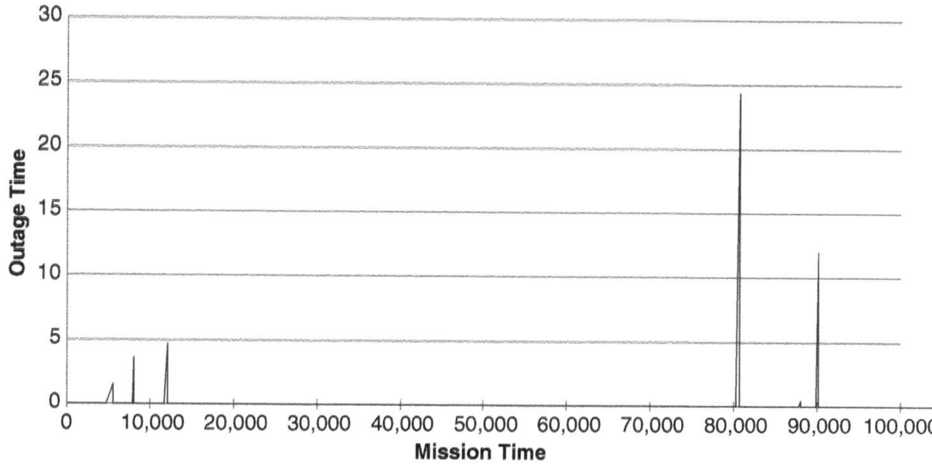

Figure 17.23 Monte Carlo Simulation of Outage Time versus Mission Time (2)

17.5.1 Simulation Algorithm

In Excel, we can generate a random number from an exponential distribution with a given mean value using the formula:

$$-\text{LN}(1-\text{RAND}())^{*}\ \text{MEANVALUE}. \tag{17.3}$$

The MEANVALUE is set to MTBF1, MTBF2, MTTR2, or MTTR1 depending on the parameter, we are simulating.

For example, `-LN(1-RAND())* MTBF1` creates a random time to fail for component 1.

We start by assuming both subsystems are operational and create a random time-to-fail value for each. For the first subsystem that fails, we determine the repair time using Equation (17.3) with the MEANVALUE set to the MTTR time of the first failed component. We then determine if the time to fail of the other component falls within the repair time interval. If so, we have a system failure since both components have failed. We continue the process identifying the events that occur along the time interval between 0 and the mission time and then graph the results.

The pseudo-code for this algorithm is shown in Figure 17.24.

17.6 LIMITATIONS OF USING NONTIME-BASED MONTE CARLO SIMULATIONS

Recall from our example with Crystal Ball, we did not use the exponential distribution to estimate the MTBF and MTTR values. The exponential distribution has been used throughout this book to determine the time a component may fail and the repair

```
Create a table of random calendar failure times for Component 1 and Component 2
Loop until Mission Time
TTF1(i)=TTF1(i) - LN(1-RAND())*MTBF1
TTF2(i)=TTF2(i) - LN(1-RAND())*MTBF2
End Loop

Run simulation
T=0
Loop until T=Mission Time
      If TTF1(idx1)<TTF2(idx2),
                  Comp Fail (i)
                  Clock Time at Fail(i)=TTF1(Idx1)
                  Repair Time(i) = Clock Time at Fail(i) - LN(1-RAND())*MTTR1
                  Idx1 = Idx1+1
                  If TTF2 < Repair Time(i)
                              System Fail(i) = 1
                  Else
                              System Fail(i) = 0
                  End
      Else
                  Comp Fail (i)
                  Clock Time at Fail(i)=TTF1(Idx1)
                  Repair Time(i) = Clock Time at Fail(i) - LN(1-RAND())*MTTR1
                  Idx1 = Idx1+1
                  If TTF2 < Repair Time(i)
                              System Fail(i) = 1
                  Else
                              System Fail(i) = 0
                  End
      End IF
End Loop
```

Figure 17.24 Component and System Failure Times Pseudo-code

rates. We may be tempted to use the exponential distribution to determine system behavior over time using Crystal Ball. However, the results that are obtained from this simulation may not reflect the behavior of the system over time. The architecture of the system must first be considered to determine how to interpret the results. When we vary the MTBFs and MTTRs and calculate availability for each trial, the resultant availability only represents the availability of the system at the time of the *first* failure and repair action. This is misleading, since the true availability of the system is calculated over the duration of the mission time. We can illustrate this through an example of a one-component two-state repairable system simulated using Crystal Ball and Matlab.

For the Crystal Ball simulation, the fail time and the repair time for each trial is obtained from sampling the respective exponential distribution probabilities. Let us take an example with the MTBF and MTTR with MTBF = 1000 hours and MTTR = 4 hours. The availability calculated in Crystal Ball is MTBF/(MTBF + MTTR). For 1000 trials, the availability distribution is shown in Figure 17.25.

Next, let us run the simulation using Matlab, where the availability is calculated based on the total outage hours and the total mission time. The outage time for each trial is:

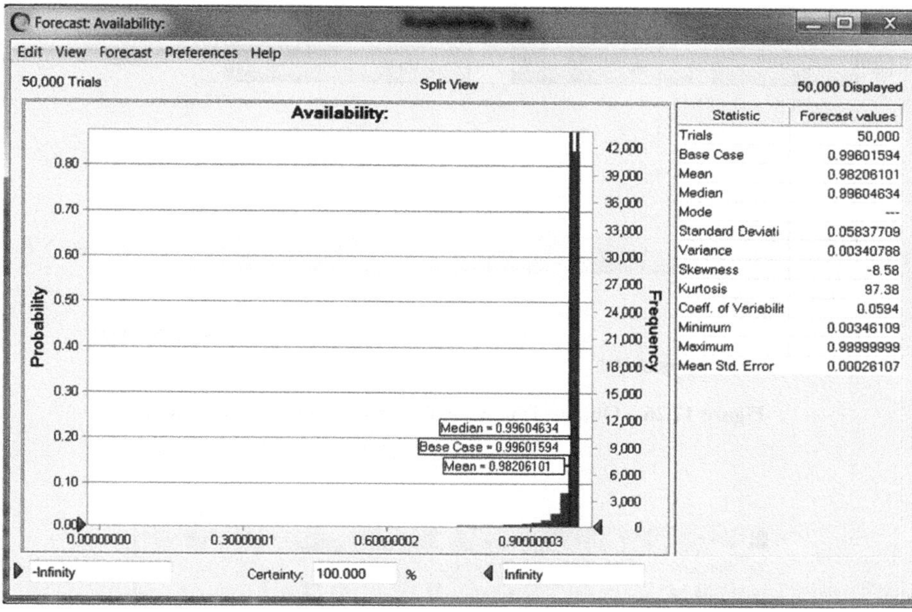

Figure 17.25 Monte Carlo Simulation of Availability: Exponential MTBF Assumptions

$$\text{Outage Time}_i = \sum_{j=1}^{n} \text{Downtime}_j.$$

For each trial, we could have 0, 1, or more down times over the mission time. We sum up all the down times to get the total outage time during the mission.

The average availability can be calculated as the average of sum of the availabilities of each trial:

$$\text{Average Availability} = \frac{\left(\sum_{i=1}^{\text{Number of Trials}} 1 - \frac{\text{Outage Time}_i}{\text{Mission Time}} \right)}{\text{Number of Trials}}. \tag{17.4}$$

An illustration of how the outage times are summed up across trials is shown in Figure 17.26. Notice the Crystal Ball simulation only covers the first fail and first repair events for a system that may experience several failures over its expected mission time. On the other hand, if the mission time is shorter than a randomly sampled MTBF, that failure is erroneously counted and added to the simulation results.

Using Matlab, (MarkovMissionTimeSim.m), the resulting availability prediction for 1000 trials over a mission time of 1 year is plotted in Figure 17.27.

Table 17.2 shows the statistics of the Monte Carlo availability simulations and compares the Crystal Ball simulation with the Matlab statistics at the time of first

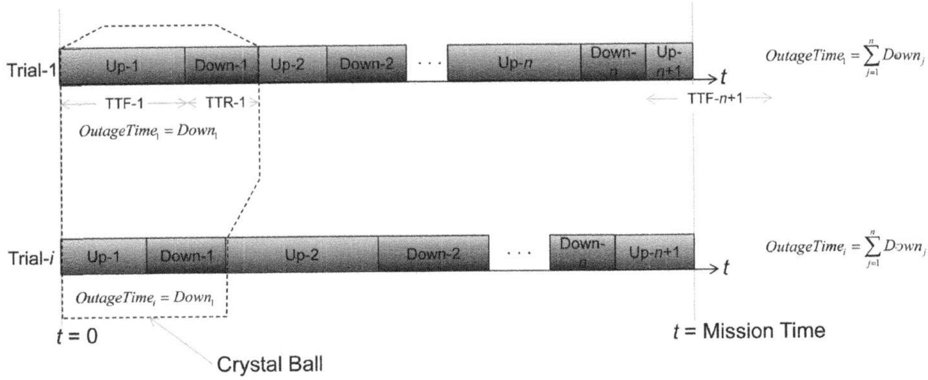

Figure 17.26 Outage Time Calculations for Simulated Systems

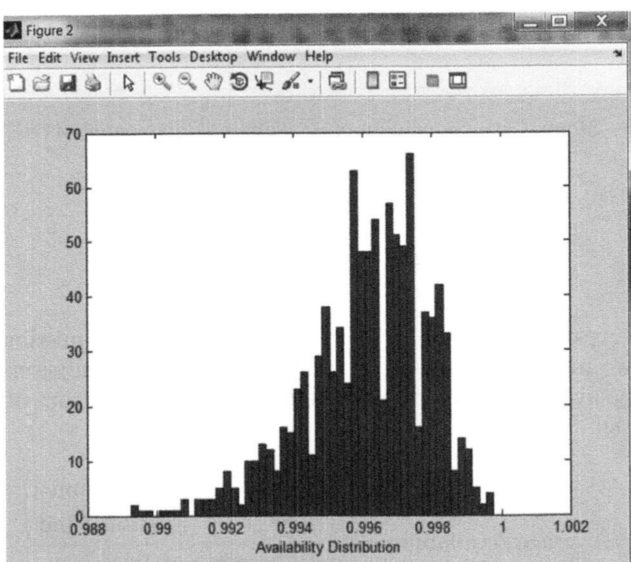

Figure 17.27 Average Availability Monte Carlo Simulation Using Matlab

failure and at the end of the mission time. Crystal Ball gives the average of the availabilities at the first time the system has failed and subsequently repaired. Notice the difference between the Matlab simulation and the Crystal Ball simulation. The Crystal Ball simulation prediction of the availability is incorrect because it does not take into account the notion of mission time. However, for 1000 trials, for the time of *first* failure and *first* repair, we obtain similar statistics (columns 1 and 2).

TABLE 17.2 Availability Statistics for Monte Carlo Simulation Models

	Crystal Ball	Matlab First Fail	Matlab Average Avail
Average	0.98172077	0.97946916	0.99610366
Median	0.99572791	0.99551905	0.99634700
Variance	0.00361206	0.00433239	0.00000317
Min	0.07768782	0.20000000	0.98926940
Max	0.99999227	0.99979960	0.99977170

Using Matlab to calculate the true average availability for the life-time or mission-time, the distribution statistics in the third column is obtained. Observe that the point availability using the equation MTBF/(MTBF + MTTR) = 0.99602. This value is close to the Matlab Monte Carlo calculated availability of 0.99610366, but orders of magnitude different than the Crystal Ball simulation (0.98172).

Now let us compare the variances. We see that the Crystal Ball variance is approximately 1000 times higher than the average availability distribution from Matlab. Why is that? In the Matlab model, the system can experience multiple failures and multiple repairs during the simulated mission time for each trial, which combine with the net effect of reducing the variation of the availability since we are averaging these values during each trial. For the Crystal Ball simulation, variations are injected only for the first Uptime and Downtime for each trial. If the first failure occurs early, the calculated availability of the system will be less than the true availability since the behavior of the system for the entire mission time is not considered. Thus, the minimum and maximum values are more extreme for the Crystal Ball simulation.

Mission time also affects reliability. If the mission time is short in comparison with the MTBF of the component, the reliability increases. If the mission time is longer in comparison with the MTBF of the component, the reliability decreases. Some vendors will provide very high reliability data for their products based on a predetermined mission time. If that mission time assumption is much smaller than the specified mission time for your product, then the reliability data provided by the vendor may be too high and hence should be adjusted to reflect the mission time.

17.7 SUMMARY

Monte Carlo simulation is a powerful technique used to characterize a range of system behavior based on assumed probability distributions of parameters associated with a model.

Monte Carlo simulation has been applied to a wide range of problems, including financial prediction. In terms of reliability modeling, we combine the probability distributions for many reliability parameters of interest in our model to predict the range of possible system availability and the likelihood of each outcome. These data is plotted for visual analysis. Based on these data, we can determine the likelihood of the minimum availability requirement of our system being met.

Sensitivity analysis shows us which parameters have the most impact on system availability. With this information, we can focus our design activities on optimizing/improving system behavior by improving the reliability of this parameter or minimizing their impacts.

Time-based Monte Carlo simulation shows us when outages are most likely to occur on average given a large number of systems being simulated.

A comparison of tools and simulation techniques helps us determine the benefit and limitations of each technique.

CHAPTER 18

Updating Reliability Estimates: Case Study

18.1 INTRODUCTION

This chapter provides a case study on the techniques for improving reliability parameter estimates by incorporating information from a product's field operations. We will explore an example from the telecom industry in which the initial availability model was created as part of system design. After the system was deployed, field failure data became available, and these data were used to revise the original reliability parameter estimates. The original reliability model was built using fault tree analysis. We show specifically how to incorporate new failure data into the original model to obtain a more accurate predictive model.

18.2 OVERVIEW OF THE BASE STATION CONTROLLER—DATA ONLY (BSC-DO) SYSTEM

Figure 18.1 provides a high level overview of the network architecture of a standard CDMA System with EVDO (EVolution—Data Only) capabilities. EVDO supplements the CDMA 1X voice-based system with high speed data capability. The BSC-DO is an enhanced IP-based system that offers faster processing, reduced foot print, high scalability, and capacity for CDMA network operators to provide high speed data services.

The BSC System has several call processing elements along with the element manager that assists with the O&M functionality and operator interface. There can be multiple BSCs in a network and each BSC manages many Base Transceiver Systems (BTS).

Figure 18.2 shows the commercial BSC-DO chassis in the lab. This system is commercially deployed in several North American markets. The BSC-DO system is made

Designing High Availability Systems: Design for Six Sigma and Classical Reliability Techniques with Practical Real-Life Examples, First Edition. Zachary Taylor and Subramanyam Ranganathan.

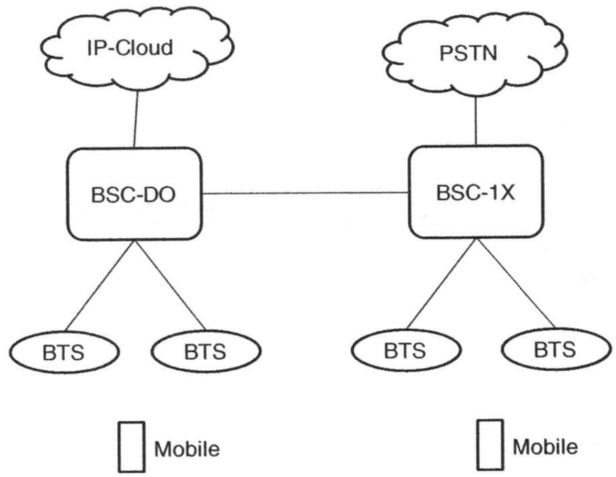

Figure 18.1 Simplified System Architecture

up of multiple shelves, and each shelf has a set of cards that handle Call Processing and O&M functions.

Figure 18.3 shows the logical functionalities within the system. These functional capabilities along with the hardware architecture are incorporated into a reliability model. The reliability model was built using fault tree analysis and Design Failure Modes and Effects Analysis (DFMEA).

18.3 DOWNTIME CALCULATION

Based upon experience with other systems deployed in the field that are similar to this system, the average system outage duration was estimated to be approximately 45 minutes. The outage time is defined as the time between the outage occurrence and outage resolution, that is, the system is fully functional again. For this system, the predicted availability based on design was 0.999985:

$$\begin{aligned}\text{Outage minutes per year} &= (\text{Unavailability}) \times (\text{Number of minutes per year}) \\ &= (1 - 0.999985) \times 365.25 \times 24 \times 60 \\ &= 7.889 \text{ minutes/year}\end{aligned}$$

$$\begin{aligned}\text{Frequency of outages} &= (\text{Average outage duration})/(\text{Outage minutes per year}) \\ &= (45 \text{ minutes})/(7.889 \text{ minutes/year}) = 5.7 \text{ years} \approx 68 \text{ months.}\end{aligned}$$

Given this outage time, we can expect on average, one outage every 68 months.

For this case study, three distinct failures resulted in three different outages with each outage directly impacting the availability of the BSC system. These outages occurred in three different geographic locations in North America.

Figure 18.2 BSC-DO System

18.3.1 Outage 1

The analysis indicated that the outage was triggered due to loss of communication between the active and standby systems. This fault was not detected either by the active or standby systems. Further drill-down revealed that the root cause of the failure was the active controller card hardware failure.

18.3.2 Outage 2

An analysis showed the following sequence of events (Fig. 18.4).

1. A failure resulted in a failover of the transport layer from the active controller card to the standby controller card. This resulted in loss of communication between the payload cards and the controller card.

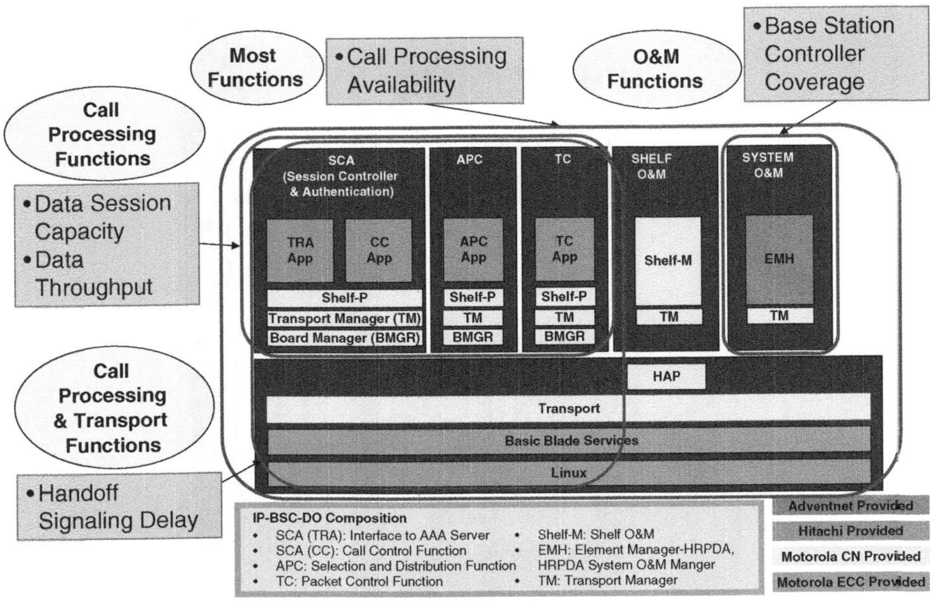

Figure 18.3 BSC-DO Functional Architecture

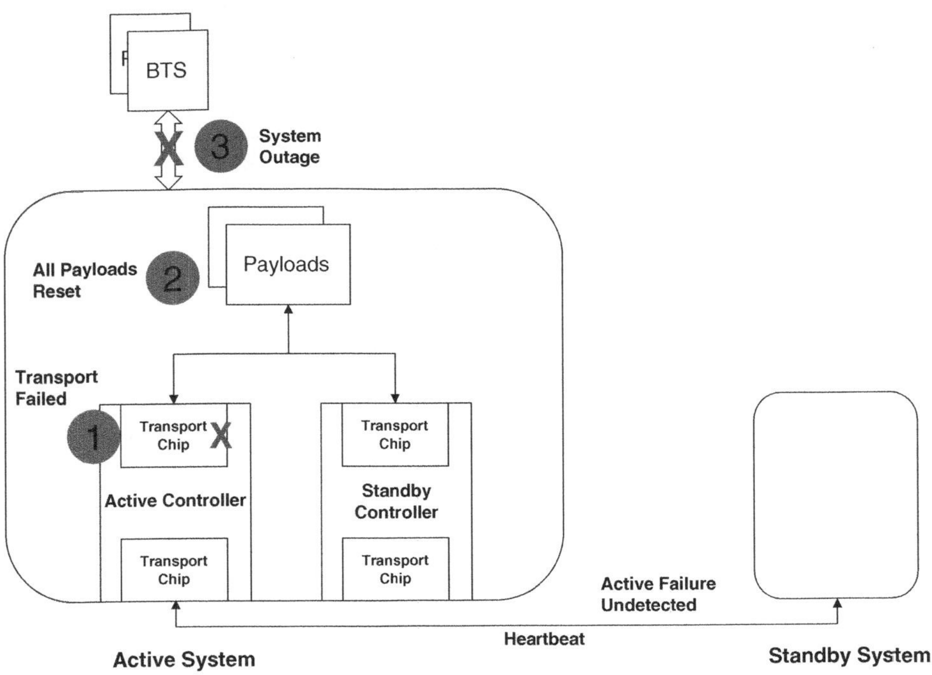

Figure 18.4 Sequence of Events Leading to Cage Outage

2. The payload cards automatically rebooted in an attempt to recover communication. Since the heartbeat mechanism was functional between the active and standby systems, the standby did not detect the failure.
3. A system outage occurred and the outage was manually detected.

The root cause of this problem was traced to an undetected transport hardware failure on the controller card.

18.3.3 Outage 3

A payload card had a machine check exception but no interrupt or other indication was generated (silent failure), resulting in the calls assigned to that payload card failing quickly. This outage lasted 34 minutes. Root cause analysis determined that the CPU did not generate a machine check interrupt that would have caused the payload card to reset. The fix was to add an exception signature in the system log and raise an alarm, which results in a reset of the payload card.

18.4 CALCULATING AVAILABILITY FROM FIELD DATA ONLY

In light of the outages, the following questions were considered:

1. How did these events affect the availability analysis?
2. What changes should be made to prevent an outage when these conditions occur and further improve the reliability of the system?

The analysis determined that by fixing these defects, the availability requirement of 99.9985 could be achieved.

The field outage data are used to calculate the system MTBF. At the time of the analysis, the BSC-DOs had been deployed in the field for 895.5 system months and three field outages had occurred. Thus the point estimate of the system MTBF and availability was:

$$\text{MTBF} = (\text{total system months})/(\text{number of outages})$$
$$\text{MTBF} = 895.5/3 = 298.5 \text{ months.}$$
$$\text{Availability} = 1 - (\text{MTTR}_S/\text{MTBF}_S)$$
$$\text{Availability} = 1 - ((45/60)/(298.5 \times 30 \times 24)) = 0.99999651.$$

Chi-square analysis was employed to provide availability estimates based on cumulative field data. For the chi-square analysis, the greater the field data, the higher the accuracy and confidence levels of the availability.

The upper and lower bound confidence factors were obtained from the chi-square distribution and multiplied by the MTBF point estimate to obtain the upper and lower MTBF bounds. The availability was then calculated.

TABLE 18.1 MTBF and Availability Summary

% Confidence Level	Lower Chi-Square Factor	Upper Chi-Square Factor	Lower MTBF Bound (system-months)	Lower Availability	Upper MTBF Bound (system-months)	Upper Availability
60	0.544	1.9543	162	99.9994	583	99.9998
80	0.449	2.7222	134	99.9992	813	99.9999
90	0.3869	3.6689	115	99.9991	1095	99.9999

18.5 ASSUMPTIONS BEHIND USING THE CHI-SQUARE METHODOLOGY

When the system was deployed, the constant failure rate assumption was applied to all components within the system reliability model. Using the techniques described in Chapter 12, the MTBFs were calculated.

Table 18.1 provides a summary of the calculated MTBF and availability for the deployed system.

The customer commitment for the BSC-DO System was 99.9985% availability. The field data and calculations showed that the customer requirement was exceeded.

18.6 FAULT TREE UPDATES FROM FIELD DATA

A second approach is to match field failures to specific failure events in the fault tree. Figure 18.5 is a snapshot of the first two failures (Outages 1 and 2) described in the previous sections. The arrow pointing to "SSC Xport Affecting Shelf" contains the estimate that we had before the system was deployed in the field. We updated this part of the model based on field data to increase the accuracy of the model and to predict system reliability.

The point MTBF estimate = Total number of system months of deployment of all systems divided by the number of transport failures = 895.5/2 = 448 months. We can use this number to calculate the unavailability. Unavailability = MTTR/MTBF = 2 hours/(448 × 30 × 24 hours) = 6.2e-06. The fix for the first two outages reduced the unavailability by 75%. The modified unavailability = 1.55e-06 (Fig. 18.6).

The third outage corresponded to PAM (Processor—Advanced Mezzanine ATCA blade) Failover Failure in the FTA (Fig. 18.7). The unavailability of this event was originally calculated to be 7.76e-7. The field data indicated an MTBF point estimate (Total system months)/(Number of outages due to "PAM Failover Failure") was 895 system months/1 outage = 895. The actual outage time for this failure was measured to be 34 minutes. Thus the unavailability was 34 minutes/895 system months = 8.789e-7.

The fault tree was updated taking in account the revised unavailability value as shown in Figure 18.8.

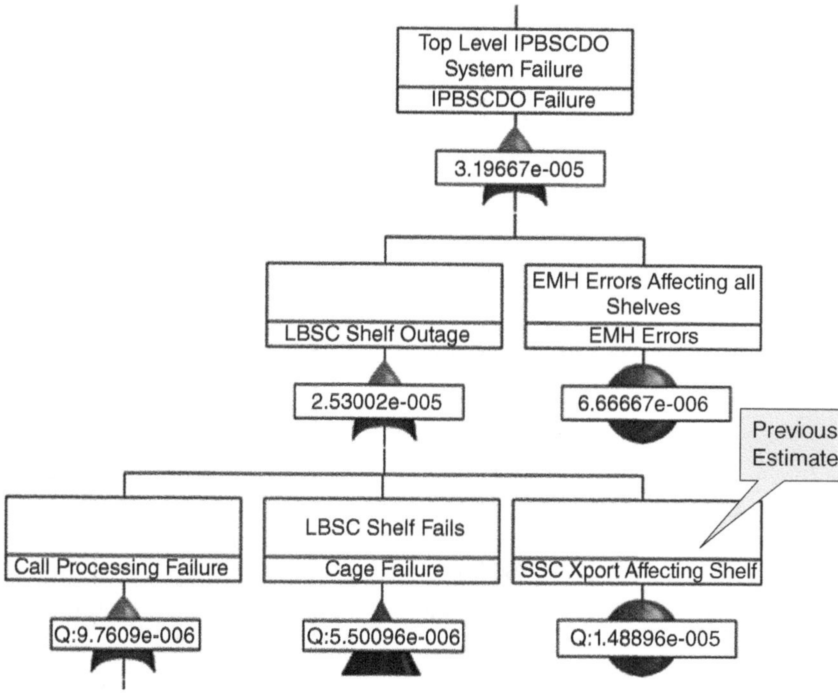

Figure 18.5 Fault Tree Estimate Before Field Data

However, this issue was also fixed through hardware and software changes, further reducing the unavailability of the "PAM Failover Failure and PAM Failure" event by 10%. The modified unavailability then becomes:

Unavailability = 8.789E-7 × 0.9

Unavailability = 7.9101E-7.

This new unavailability estimate replaced the previous estimate in the FTA, as shown in Figure 18.9.

After incorporating these numbers based on the fixes, the FTA yielded an increase in the overall availability by 20%.

The first row of Table 18.2 shows the overall BSC-DO availability before field data. The next two rows show the availability improvements based on the fixes implemented and deployed.

An additional exercise that can be performed is to analyze the hardware failures. The hardware failures can be tracked based upon the number and types of boards that have been returned for repair or replacement. The MTBF for these boards can be calculated in a similar fashion as we described previously: MTBF = number of system months of operation/number of failures per board type. We can then update the Fault Tree based on these numbers.

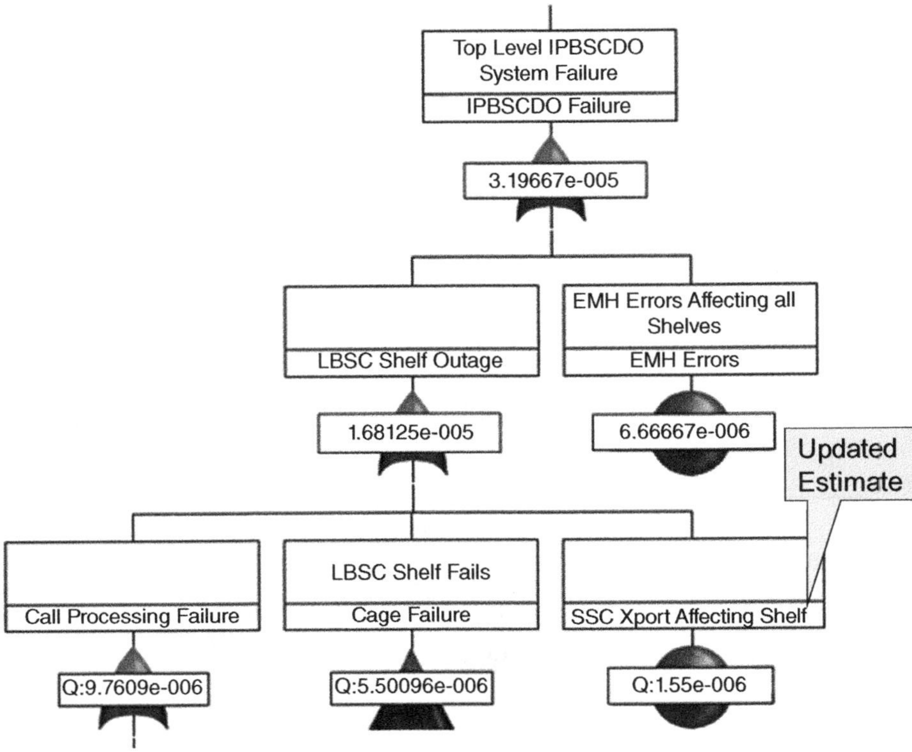

Figure 18.6 Fault Tree Estimate After Field Data

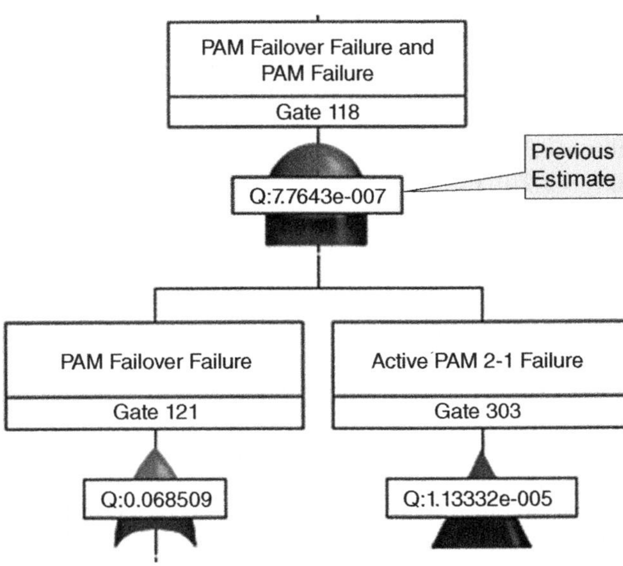

Figure 18.7 Fault Tree Estimate Before Field Data

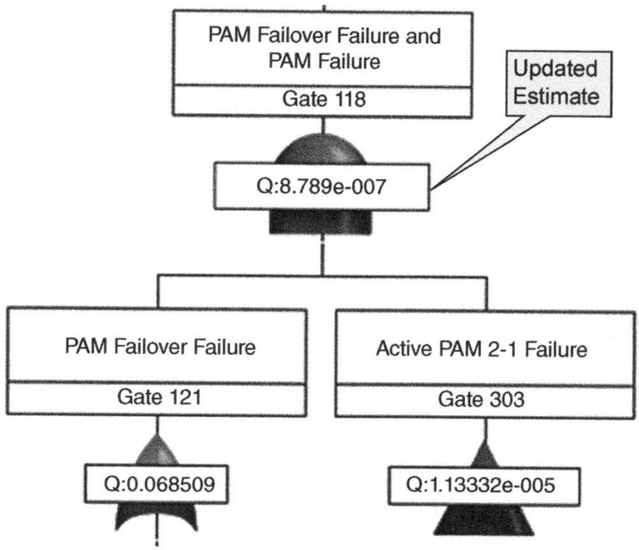

Figure 18.8 Fault Tree Estimate After Field Data

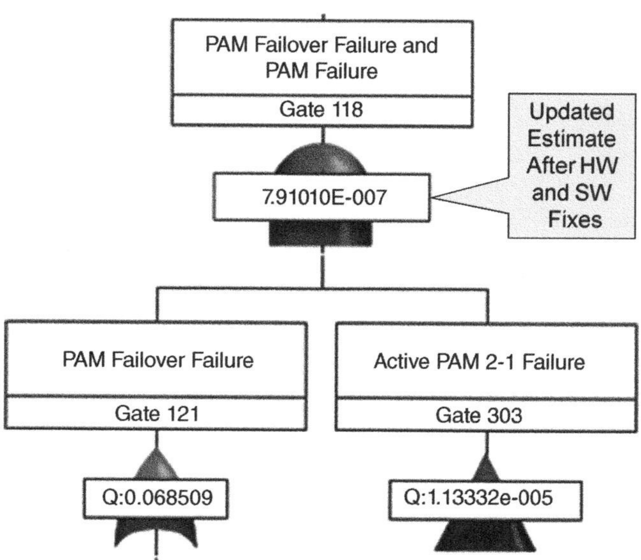

Figure 18.9 Fault Tree Estimate After Hardware and Software Fixes

TABLE 18.2 LBSC Shelf Availability Revisions with Field Data

	BSC Availability		
	Nines	% Avail	Delta %
Before field data	4.60	99.9975	N/A
Improved (First and second outages fixed)	4.67	99.9979	+14.5%
Improved (All three outage issues fixed)	4.70	99.9980	+20.7%

18.7 SUMMARY

This chapter described some practical techniques that can be used to analyze field data and revise the availability model and predictions. This analysis can be continued during the life of the system as and when more field data are obtained.

The availability analysis can be periodically updated based on potential new outages and/or additional field exposure. Even if we have zero outages over a long period of time, the system unavailability will not remain the same. This will actually increase the availability. On the contrary, if we have more outages, the overall availability of the system will decrease. The system availability is dynamic and must be updated as more information becomes available. It is a good practice to monitor and track system availability regularly to ensure that a system is healthy and operational at satisfactory levels.

CHAPTER 19

Fault Management Architectures

19.1 INTRODUCTION

Fault management is an integral part of any system design. It is this aspect of the design that makes or breaks system availability. When designing high availability systems, emphasis must be placed on various Fault management mechanisms that will absorb or eliminate different sources of faults.

What types of faults should be identified and modeled? Faults and the effect on the system of these faults can be identified through a Design Failure Modes and Effects Analysis (DFMEA). Although a DFMEA may identify hundreds of failure scenarios, failures can be grouped together based on the type, system impact, recovery action, and repair actions. Faults can be grouped into the following categories:

1. *Permanent Hardware Faults.* Hardware failures that cause the hardware component (card, server, external disk, etc.) to become permanently failed requiring the replacement of the failed component.
2. *Transient Software Faults*
   - (a) If the problem is isolated to a particular application or service, the fault management policy may result in a restart of the application. If restart fails, the FM policy is escalated to attempt other recovery actions.
   - (b) Reboot the operating system. Reboot the OS and perform a cold restart of all processes on the server or card.
3. *Operator-Initiated Faults.* Failures caused by operator error may be a nontrivial contributor to overall system failures. The failure rates for these errors are highly dependent on the user interface design of the system, that is, ease of use, and the training and skill level of the operator. The system architect should always consider the possibility of these errors and design the system to minimize them through the use of warnings, constraints on parameter editing, restricting

Designing High Availability Systems: Design for Six Sigma and Classical Reliability Techniques with Practical Real-Life Examples, First Edition. Zachary Taylor and Subramanyam Ranganathan.

user access (e.g., role-based access control [RBAC]), validating all changes entered by the operator, creating training material, online help manuals, better ergonomics, and so on.

4. *Design Faults.* Faults in software or hardware design. A programming mistake is a fault.
5. *Software Logic Errors.* Logic bugs, thread protection, performance, and so on.
6. *Environmental Faults.* Fires, floods, earthquakes, sabotage, and power failures.
7. *Intermittent Faults.* Faults that cause the system to oscillate between faulty and fault-free operation

Hence, it is important to design a functional and elegant Fault Management architecture considering different perspectives, such as reliability, operability, availability, and maintainability.

19.2 FAULTS, ERRORS, AND FAILURES

Definitions:

- *Fault.* A fault is the occurrence or the existence of an undesired state within a system.
- *Error.* An error is a variation from the nominal desired performance or operation of the system. An error is a manifestation of a fault. An error that deviates from the expected service may be tolerated for a specified period of time before it becomes a failure.
- *Failure.* A failure is the inability of the system to provide the specified or expected service. A failure is a manifestation of an error. An error can transition into a failure in due course.

According to IEC 50 (191), "failure is the event when a required function is terminated (exceeding the acceptable limits)," while a fault is "the state of an item characterized by inability to perform a required function excluding the inability during preventive maintenance or other planned actions, or due to lack of external resources." An error is a "discrepancy between a computed, observed, or measured value or condition and the true, specified or theoretically correct value or condition." An error is not yet a failure because it is within the acceptable limits of deviation from the desired performance (target value). An error is sometimes referred to as an incipient failure (e.g., see OREDA 2002; IEEE Std. 500).

According to NASA, a failure is "an unacceptable deviation from the design tolerance or in the anticipated delivered service, an incorrect output, the incapacity to perform the desired function." A fault is "[a] defect, imperfection, mistake, or flaw of varying severity that occurs within some hardware or software component of system. "Fault" is a general term and can range from a minor defect to a failure."

When a fault occurs, it creates a latent error that becomes effective when it is activated; otherwise, the fault is dormant, that is, the error is not manifested or detected in the system. The fault may exist early in the development cycle prior to deployment or shipping the product, for example, undetected software bugs.

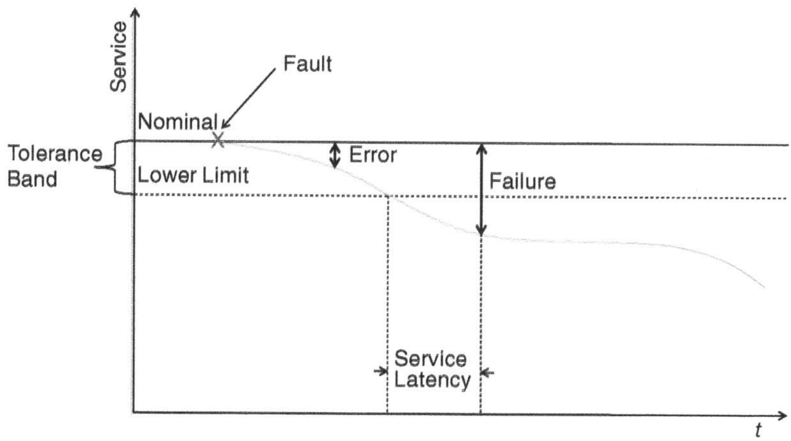

Figure 19.1 Differences between Faults, Errors, and Failures

The cause of an error is a fault. An activated error is a detectable discrepancy between the observed system operation or performance and the nominal expected condition caused by the fault. An error is the intermediate stage between a fault and a failure (Fig. 19.1).

If the error affects the system service, a failure occurs. An error is the manifestation of a fault in the system, and a failure is an error that degrades service outside of acceptable tolerance limits or results in a complete loss of service functionality.

When a system is not able to deliver correct service, a system or service failure occurs. The service failure deviates from the specified or expected behavior perceived by the user. The specified behavior may include performance thresholds and maximum service delay times. If the service degrades outside these bounds, the service is considered failed.

A software fault (also known as a software bug) is uncovered when an internal error occurs within the program. The software error may result in complete program failure or cause a different failure mode perceptible to the user. A software fault is the ultimate cause of the software failure.

The relationship among faults, errors, and failures is illustrated in Figure 19.1 and Figure 19.2.

When the fault manifests itself as an error or subsequently a failure, the troubleshooting team identifies the root cause and the fault is removed. The ability to identify and recreate the activation pattern of a fault is called *reproducibility*. In some cases, the root cause cannot be easily determined, and indirect methods, such as failure mitigation, are implemented. Examples of difficult-to-find faults include timing faults, resource contention faults, transient faults, and rare unreproducible scenarios.

When a failure occurs, it results in one or more failure effects. These effects describe how the performance or operation of the system is impacted by the failure. We previously discussed the relationships between failures, failure modes, and failure effects in Chapter 15. A failure of a system component causes an internal fault of the

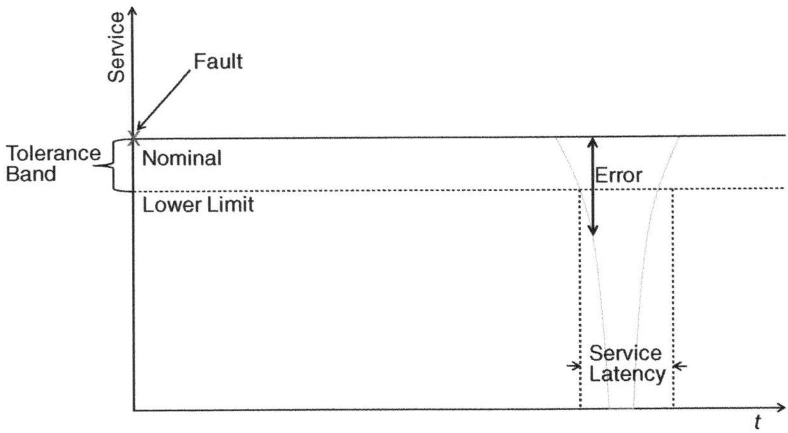

Figure 19.2 Differences between Faults, Errors, and Failures

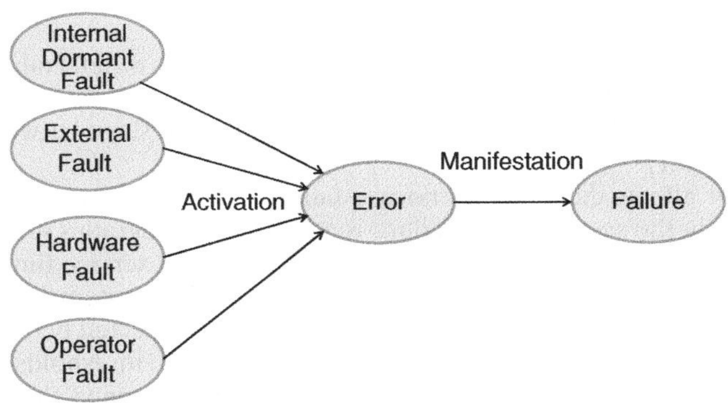

Figure 19.3 Fault Propagation

system, which may or may not result in system failure or may cause an external fault to another system that utilizes the system's service.

For example, for a 2N redundant system, if one of the components fail, the system (although it now has a fault), will remain operational since the remaining component provides full system functionality.

The fault propagation is illustrated in Figure 19.3.

19.2.1 Failure Analysis Considerations

Failures may be classified either by the operator or by the system based on severity levels, such as critical, major, and minor, depending on the impact of the failure to the service and other consequences resulting from this failure. These severity levels are dependent on the system and customer requirements. In some cases, the severity levels for each failure can be modified dynamically after a better understanding is obtained. Typically, an alarm is generated to notify the user of the failure and a severity is associated with the alarm.

Failures may be classified as:

1. *Fail-Operational.* A component has failed but the system remains fully operational.
2. *Fail-Safe.* The system has failed but does not result in injury of personnel or damage to the system.
3. *Catastrophic.* The system has failed and damage to equipment and/or injury to personnel occur.

19.2.1.1 Failure Independence

If a failure has a single cause and effect, that is, it does not result in secondary effects, the failure is independent. During reliability modeling, the initial simplifying assumption is that the failure is independent. The DFMEA helps to identify other effects of failures and dependencies. When these are identified, the mitigation of these effects should be considered since they reduce reliability.

19.2.1.2 Common Mode Failure

A common mode failure spans redundancy components or affects nonredundant parts of the system. Common mode failures may result in a failure of the system or degradation in system performance. The common mode failures should be identified using DFMEA or other techniques and eliminated or minimized.

19.2.1.3 Fault Isolation

Once a failure occurs, we would like to minimize its impact on the system. By isolating the failure to a single component within a redundant component framework, for example, we can prevent the failure from impacting the operation of the system. Redundancy is a powerful technique for reducing the probability of system failures.

19.3 FAULT MANAGEMENT DESIGN

The fundamental objective of Fault Management design is to keep a system alive and robust despite various occurrences and influences of faults. One of the modes of design is to have a protection group. For example, we design a system to have a robust failover mechanism such that the failure of one component in the system is covered by a redundant component. If that component has a hard failure, we should be able to detect it. We use mechanisms like heartbeat (keep-alive messages), checkpointing, NARQ, handshake messages, and so on. Any of those basic techniques will help identify a problem, and if it turns out that the problem is isolated to a processing card, then the protection mechanism will ensure that the failover to the redundant component takes place.

An area of analysis in the architecture when we estimate the reliability parameters is the *detectability* of the problem. How likely is it that when we have a problem, we will be able to quickly identify the problem? The detection mechanism that is part of fault management should be designed to pick up those failures.

In addition to detectable failures (*observability*), systems have undetected or "silent" failures that should be accounted for as part of reliability prediction. In situations where the repair action is to reset the card, the MTTR for the card will typically be several minutes (*controllability*). Once the system is aware of the problem, automatic repair and recovery actions can be taken. However, if this fails, and the card must be manually replaced, the MTTR could be several hours. So the more robust the fault management architecture is coupled with an optimal maintainability policy, the higher the availability of the system.

Auto recovery is another key ingredient of Fault management design. It is helpful for transient software failures, such as buffer overflows, memory leaks, locks that have not been released, random occurrence of fault or a state that is unstable that causes the system to lock up, and so on. For these types of behavior, generally, a reboot of the card may resolve the problem.

Other fault management procedures include checkpointing and rollback in software, and robust install and upgrade procedures.

A dedicated hardware line can trigger a reboot independent of the software failure. In situations where the software failure is not responsive to external commands, we'd have to look at independent paths to increase the chances of recovering that card. Other methods include an internal watchdog time-out that will trigger a card reboot. The more repair actions we have built into the system, the more robust the system becomes. There is of course a trade-off of the number of such mechanisms that we build into a system. We do not want to overdesign these mechanisms because we get to a point where we start to have diminishing returns.

Another key aspect for the customer to operate the system and have high availability is for the alarm and notification mechanisms to be clear, precise, and fast such that there is minimal effort in understanding the problem and how it can be fixed. If an alarm is vague, the interpretation of what action needs to be taken may not be clear. There may be no documentation to say whether the system is functional, degraded, or failed. The system engineer/architect must always consider the maintainability of the system as part of system design.

19.4 REPAIR VERSUS RECOVERY

We need to keep in mind the difference between a repair action and a recovery action.

A repair action is a separate activity in which the failed component is automatically repaired by the system (e.g., software component restart) or manually repaired (e.g., failed board replaced).

On the other hand, a recovery action is employed when a failed component has been identified and we need to mitigate the impact of this failure on the system. For example, with redundancy architecture employed, the failed component may be bypassed and a redundant component brought into service.

Generally, as soon as a failure is detected in one component, we would want to execute a recovery action first prior to repair. The purpose is to isolate this failure and to either prevent this failure from affecting the system or minimize the impact to the system. Once we have isolated this failure and it is no longer within the system's functional boundary, we can focus on repair action. As an example, if we have a critical call processing application executing and the application does not communicate over a specific interface say for 5 seconds, then calls may start to fail. A possible requirement would be to attempt to bring into service a redundant component in less than 5 seconds after a failure has occurred, to ensure that calls are not dropped. The repair action can then be initiated.

19.5 DESIGN CONSIDERATIONS FOR RELIABILITY MODELING

When beginning a new development project, the analyst determines the scope of the reliability analysis. She will usually have some idea of the system requirements and functional architecture, and typically will have some knowledge of the system hardware and software that will be considered. The first step is to identify the components to be modeled. A component can range from a capacitor on a circuit board when considering a new hardware board design or a server blade when designing a new back-end server for an internet web site.

For each of these components, we create a reliability estimate using one of the techniques illustrated in this text. This initial estimate will likely change over time—sometimes radically, as we obtain more information. With the components and their reliability data identified, we can then perform an initial DFMEA to identify failure modes of these components and the effects of these failures on the system. (Practically, a DFMEA may also be performed immediately after identifying the components). Armed with this information, we can create reliability models based on existing system architecture or we may create multiple models that in turn help drive the system architecture.

Start with the simplest model, that is, make simplifying assumptions, such as failover times are instantaneous and 100% reliable. Group similar components into an aggregate component, ignore higher reliability component. This will allow the designer to quickly determine whether the system can meet its reliability goals. If the model fails, add redundancy, monitoring, maintenance assumptions, and so on that help move you closer to achieving your goal. Once the goal has been achieved, we can refine the simplified model to more effectively mirror the system being designed and gain more confidence in meeting the requirements.

With the more mature model, we focus in on areas of the model that contribute more to the unreliability and unavailability of the system. We can perform a more detailed DFMEA to identify and correct weaknesses in the architecture and improve some of our reliability estimates. The DFMEA will also help identify the degraded modes of operation and optimal techniques to ensure system robustness during these conditions. Actions taken could include replacing components with more reliable components, enhancing automatic repair techniques, and providing more informative alarming (reduces manual repair time).

19.6 ARCHITECTURE TECHNIQUES TO IMPROVE AVAILABILITY

How do we improve availability?

1. For high availability systems, we select high reliability hardware and operating systems.

2. Use or implement more reliable software components. It is important to study the reliability characteristics of all the components in the system. (Increasing overall MTBF.)
3. Improve the user interface and complementing this with operator training. By doing so, we increase overall system availability given that we decrease the number of opportunities for operator error.
4. Decrease repair time. If an operator intervention is required for repair, decreasing repair times will increase the availability of the system. (Decreasing overall MTTR.)
5. Add automatic repair and recovery: task restarts, board resets, error checking and correction techniques, and so on.
6. Implement redundancy. One of the primary characteristics of a highly available system is the use of redundancy. Redundant hardware and software components can greatly increase the overall availability of the system assuming that most of the common mode failures have been removed. Not every component can be made redundant; however, the idea is to maximize availability by minimizing single point failure modes.

19.7 REDUNDANCY SCHEMES

At the start of a project, the reliability goals and requirements are initially at a high level, for example, the system shall achieve Five 9s availability. The initial redundancy configuration selection should be as simple as possible. If this design meets our requirements, we need explore no further. Only when requirements are not met should additional redundancy and more reliable components be considered.

Different types of redundancy are employed:

1. *Hardware Redundancy.* Hardware is duplicated such that failure of any particular subsystem or component will not cause an outage.
2. *Information Redundancy.* Corrupt data are detected and automatically corrected by adding checksum information.
3. *Communication Link Redundancy.* Redundant transport links allow communication path failures to be bypassed by taking an alternate path. High availability processing cards and systems have transport redundancy built-in by the manufacturer.
4. *Software Redundancy.* Different software is created such that failure in one software program does not affect the other software program. This technique is rarely used due to costs.

Some common redundancy schemes are described below.

19.7.1 Three-Component Majority Voting (Triplex-Duplex)

In this configuration, we have three nonrepairable components each with a failure rate of λ. In State 1, all the three components are operational (Fig. 19.4). The outputs

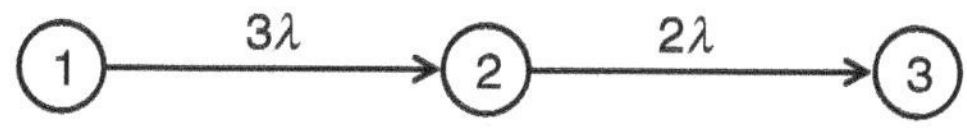

Figure 19.4 Three-Component Majority Voting (Triplex-Duplex)

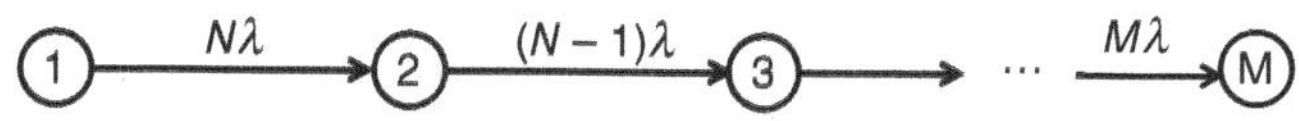

Figure 19.5 *M* of *N* Voting Redundancy

of each of these three components are compared with each other and if one of the outputs disagrees with the other two, that component is no longer considered operational and the system moves to State 2. If the output between those two remaining components does not match, the system is then considered as failed and moves to State 3.

The components are considered nonrepairable if they cannot be automatically repaired or manually replaced during the mission time. Many safety-critical systems make use of this redundancy schemes. Examples include fly-by-wire flight control systems, nuclear power plant control equipment, defense industry, and medical equipment.

19.7.2 *M* of *N* Voting

Here, we generalize the previous system to apply to *M* parallel redundant components in which a minimum of *N* components is required for operation. Each time a component fails, that is, no longer provides the correct output, it is removed from the system. This system allows multiple failures to be tolerated until we reach the minimum required set of components (Fig. 19.5).

Let us compare this type of architecture for different levels of redundancy (Fig. 19.6). The higher the number of components, the greater is the system reliability. As we can see, regardless of the level of redundancy, the system will almost certainly fail prior to 3500 hours. If we look at mission time = 800 hours, the two out of two system has only a 20% chance of working, the two out of three system has around 40% chance of working, the two out of four system has around 60% chance of working, and finally the two out of five system has approximately 75% chance of being operational. The mission time is critical to selecting the configuration required. In this example, if the mission time exceeds 3500 hours, none of these solutions will be acceptable. Also note that as mission time increases, the differences across all system configuration types diminish.

19.7.3 Cold Standby

A single active component provides system functionality. When the active component fails, the cold standby component is powered up and replaces the failed component. Since the standby component is not powered up initially, it cannot fail. When the system is in State 2, the standby component is powered up and has a failure rate equivalent to the failed unit (Fig. 19.7).

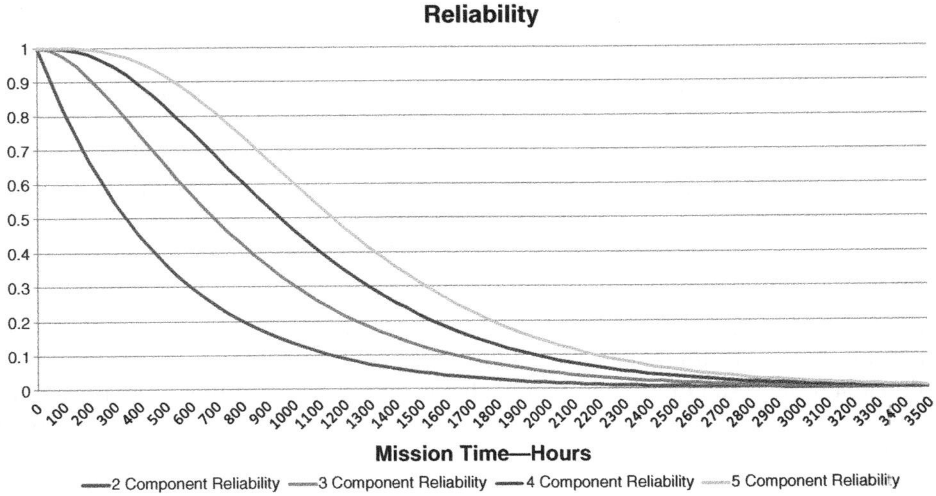

Figure 19.6 Two Out of *N* Reliability

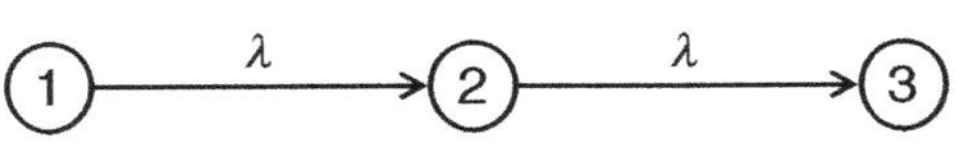

Figure 19.7 Cold Standby

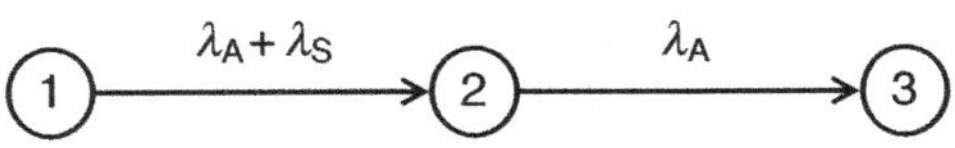

Figure 19.8 Hot Standby

19.7.4 Hot Standby

For a hot standby system, the standby component that protects the active component is powered on. The failure rate λ_S may be different from that of the active component λ_A, since it may not be executing the active load. If the active or the standby fails, the system moves to State 2. Once in State 2, we have only one component operating, and if that component fails, we move to State 3. If the active fails at State 1, this model assumes perfect switching and zero downtime to transition the standby to the active. If the standby fails, then the active component is not impacted (Fig. 19.8).

19.7.5 Hot Standby with Switch Failure

Let us extend the hot standby system by incorporating an imperfect switch. There is some probability that the switching mechanism will fail; and if it fails, the system

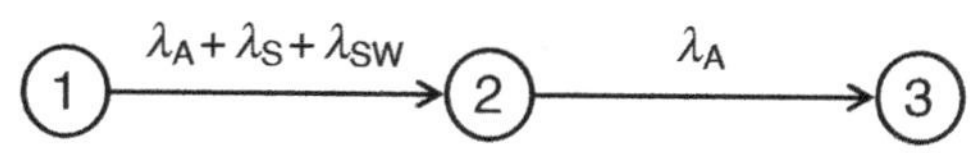

Figure 19.9 Hot Standby and with Imperfect Switch Failure

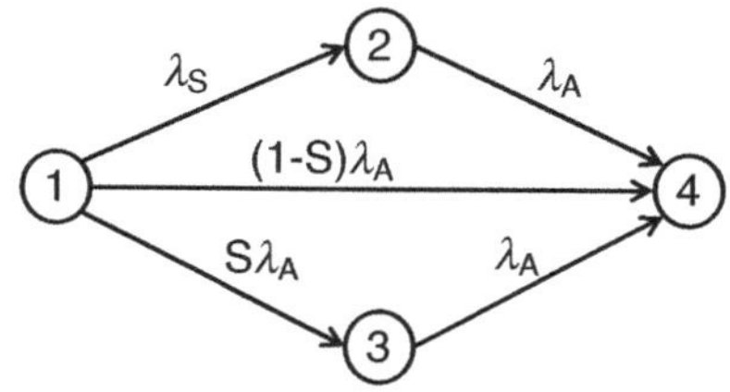

Figure 19.10 Hot Standby with Successful Switch Probability

moves to State 2. If the switch fails, at the rate of λ_{SW}, the standby component cannot be used, and thus the system will fail if the active component also fails (Fig. 19.9).

19.7.6 Hot Standby with Successful Switch Probability

Another way to model a standby with an imperfect switch is to consider the proportion of time, the switch "S" action is successful using a value between 0 and 1 (Fig. 19.10). 1 indicates that the switching is always successful. 0 indicates that the switching is never successful. If the switching is not successful, the system goes from State 1 to State 4 directly. If the switching is successful, the system moves to State 3 and the newly active component will continue to operate until it fails. If the standby fails in State 1, the system moves to State 2 and the active component continues to be operational.

When the active component fails and the switching mechanism is not successful, the system immediately fails although a functional redundant component is still available.

An example comparison of the reliability of these standby models is shown in Figure 19.11.

19.7.7 2N Redundancy

For a 2N redundant system, two equivalent redundant components are operating in parallel. A 2N redundant system is similar to a hot standby system. The redundant units actively check the operation of each other. Upon detection that the active component is not working, the redundant backup unit will take over (failover). The failure rate of the standby unit may be the same as the active unit, since both are the same hardware. However, different software is executing on the standby unit, so the failure rates of the software may be different. A 2N redundant system is modeled the same as a hot standby (Fig. 19.8). If the failure rates of the redundant units are the same, then $\lambda_S = \lambda_A$.

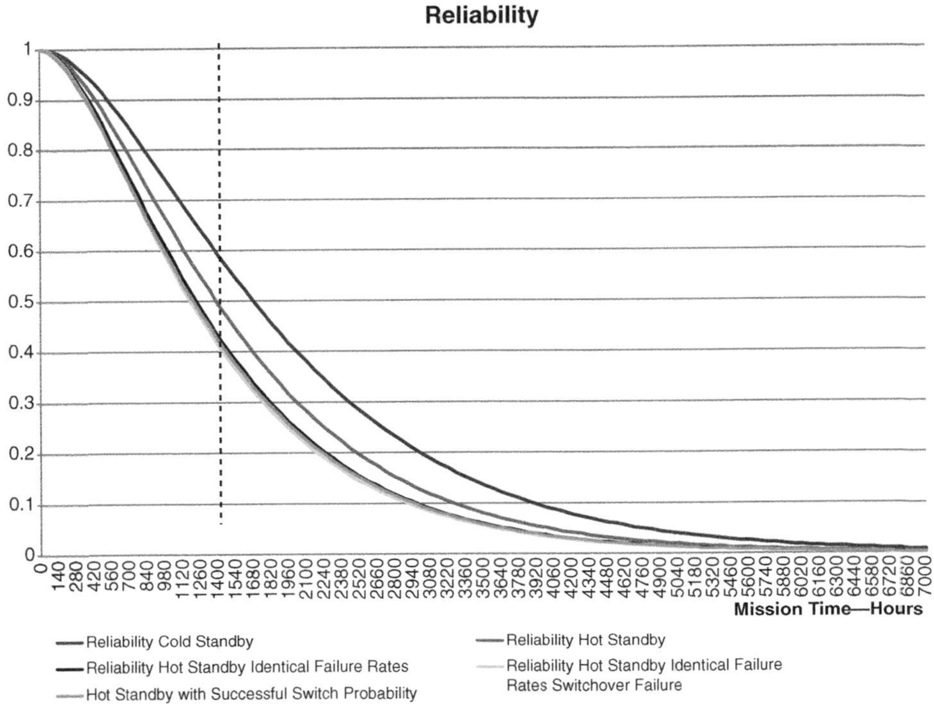

Figure 19.11 Standby Reliability

Other terminology is also used, such as 1 + 1 and 1:1. In general, these are the same as a 2N redundancy scheme, but are more commonly used in communication systems where two redundant communication links are being managed. If a problem is detected on the redundant communication link, the standby link takes over. Some data may also be processed on the redundant link, either fully redundant with the active link or partial. Communication between the two nodes is critical to determine if one node has failed or not.

One difference between a 1:1 and a 1 + 1 system is the failback mechanism. Although no universally adopted standard exists that specifies the failback behavior, for a 1:1 redundancy configuration, generally after the problem has been resolved with the failed system or link, the system switches back to its original configuration. This is important in some systems where the routing of main and alternative links is critical, more capacity is available on the active link, or operability and maintainability are drivers.

To gain efficiencies in bandwidth, a 1:1 system can be designed for low priority traffic to be routed when in normal operation (i.e., more bandwidth is available for being carrying traffic). During a switchover of the active to the standby the low priority traffic is dropped to ensure full bandwidth is available for the high priority "protected channel" traffic. After reverting back to the original configuration, the low priority channel is once again available and the low priority traffic resumes.

For a 1 + 1 or 2N redundancy configuration on the other hand, failback may not be automatic. The system will remain with the backup unit functioning as the active indefinitely. The rationale for this behavior is to reduce the probability of inducing a failure by the failback process itself. More importantly, in many systems, such as servers or communications equipment, switching between nodes is imperfect. Some outage may occur during the switching even with no failure present. So when managing these networks, if the default configuration is to be maintained, the switchback will typically occur in a maintenance window to minimize traffic disruptions. In other cases, the physical component that is assigned the active role is not as important, so the active role can "float" between the two physical devices.

19.7.7.1 Split Brain One common problem with a symmetric 2N redundant configuration is determining which component has failed. If Server A fails to respond to Server B's heartbeat, Server B may declare Server A as nonfunctioning and automatically take over the system functionality. However, Server A may also detect no heartbeats from Server B and assume Server B is dead. This leads to resource contention and more problems as each server assumes full access to shared services, such as disk drives, logical IP addresses, and other services.

A couple of ways exist to avoid this:

1. *Alternate Communication Paths.* If alternate paths exist, then we may be more confident that our peer server is truly dead if it fails to respond on both paths. The paths should be sufficiently diverse to minimize the possibility of a common mode failure affecting both paths simultaneously. For example, one path may be a fiber channel link while another path may pass through an external router.
2. *Adding Another Voter.* With 2N redundancy, we need a tiebreaker, that is, a third independent voter. An external disk array or an internal controller card can provide that vote. Quorum is achieved when a component receives at least two votes: one from the component and the other from the tie-breaker device. Several commercial high availability software packages, including Oracle, Redhat, and SAS, use these techniques.

Once the winner has been declared, fencing off access to shared services is performed to ensure no contention issues arise from the failed system.

19.7.8 Hot Standby with Successful Switch Probability, Repairable System

Let us consider the same model (Fig. 19.10), but add repair capability to the model. If the active fails, then it is automatically repaired at the repair rate μ_A, and if the standby fails, it is repaired at the repair rate μ_S. In addition, for some percentage of the active failures, the system does not failover to the standby correctly and the system transitions to State 5. In this state, the system must be manually repaired at a repair rate μ_M (Fig. 19.12).

This type of design is typical in the telecom industry.

Both States 4 and 5 are failed states. If the active fails and the failover is not successful, the system enters State 5. Since the active has failed and the standby did not

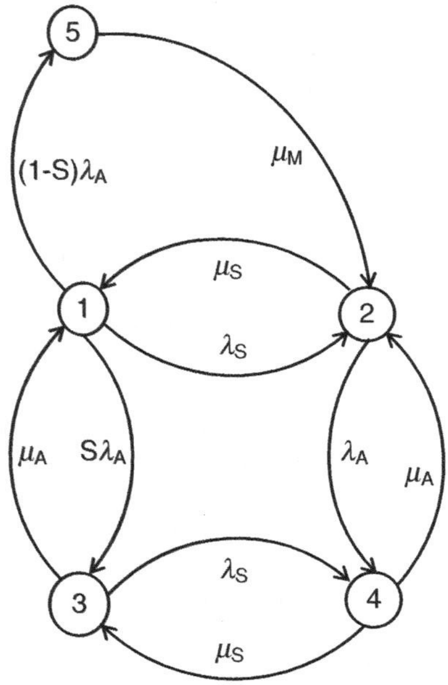

Figure 19.12 Hot Standby Repairable System with Successful Switch Probability

take over the active role, we will assume both are unusable. Either the original active is repaired or the standby is brought in service as the active with a rate of repair equal to μ_M. This will bring the system to State 2 where the system is operational but the standby system must be repaired.

EXAMPLE 19.1

We are analyzing a 2N redundant system consisting of two payload blades in a chassis. The failure rates and repair rates for the cards are the same:

$MTBF_A = 1000$ hours
$MTBF_S = 1000$ hours
$MTTR_A = 8$ hours
$MTTR_S = 8$ hours
$MTTR_M = 8$ hours.

Using the system model in Figure 19.12, determine the steady-state availability of the system with different values of failover probability success and compare with a 2N redundant system with perfect switching.

Solution

The failure rates and repair rates:

$\lambda_A = 0.001$ failures per hour
$\lambda_S = 0.001$ failures per hour
$\mu_A = 0.125$ repairs per hour
$\mu_S = 0.125$ repairs per hour
$\mu_M = 0.125$ repairs per hour.

This system can be simulated by a commercial tool, such as Relex. However, in this example, we will use Matlab (Markov_Static_Model.m). The model is captured in Matlab, and the steady-state probabilities are calculated. The availability is the sum of the steady state probabilities of being in States 1–3.

The availability for different values of S is shown in Table 19.1.

The change in system availability with respect to the failover success probability S is depicted in Figure 19.13.

Using the same models we analyzed in Chapter 11, for a 2N redundant system with the same failure and repair rates, we obtain:

$$A = 0.999937.$$

We see this is the same value as the last row in the table. With perfect switching, our availability is the highest and exceeds 4 nines.

For a single component system with the same failure and repair rates, we obtain:

$$A = 0.992063.$$

This availability is essentially the same as the first row in the table in which the failover never succeeds. If failover does not work, the system is reduced to a single repairable component.

TABLE 19.1 Availability for Different Failover Success Probabilities

S	Availability	Unavailability	Five 9s
0	0.9921875	0.00781250	2.10721
0.1	0.9929542	0.007045763	2.152072
0.2	0.9937228	0.006277214	2.202233
0.3	0.9944932	0.005506848	2.259097
0.4	0.9952653	0.00473466	2.324711
0.5	0.9960394	0.003960645	2.402234
0.6	0.9968152	0.003184797	2.496918
0.7	0.9975929	0.002407113	2.618504
0.8	0.9983724	0.001627587	2.788456
0.9	0.9991538	0.0008462136	3.07252
1.0	0.9999370	0.00006298816	4.200741

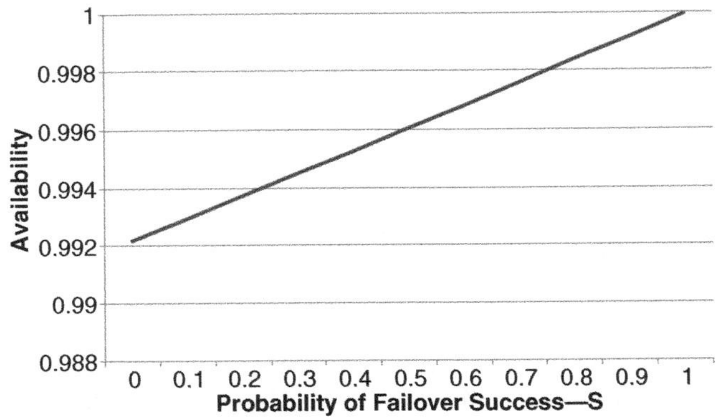

Figure 19.13 Availability versus Failover Probability

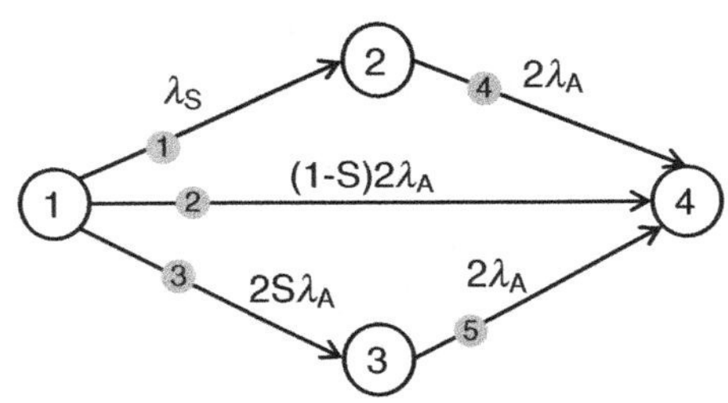

Figure 19.14 Hot Standby and Two Active Components

19.7.9 Hot Standby and Two Active Components

This is an $N + 1$ configuration with $N = 2$. In this configuration, we have one standby component protecting multiple active components. The failure of any active component will result in the standby taking over. This model is very similar to the previous model; however, the failure rate of this system is higher since we have an additional active component that can fail (Fig. 19.14).

EXAMPLE 19.2

It is interesting to observe the transition probabilities for each state in the system. We create a Markov simulation using Matlab (dynMarkovSim.m) with the following parameters:

$MTBF_A = 1000$ hours
$MTBF_S = 1000$ hours
$S = 0.95$.

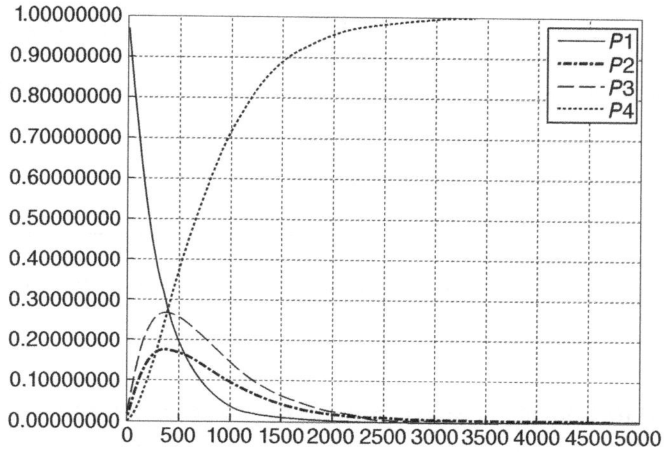

Figure 19.15 Dynamic Change in Availability for 5000 hours

What are the dynamic state probabilities over a mission time of 5000 hours?

Solution

Figure 19.16 shows the dynamic behavior for the four states of the system. The system starts out fully functional with a probability of 1 in State 1 and decreases exponentially until the mission time of 5000 hours where the system is most likely failed. State 4 shows how the probability of the system failing increases from 0 to an almost certain chance of failing at 5000 hours. The intermediate States 2 and 3 also have dynamic properties with a probability of being in those states peaking at around 300 hours (Fig. 19.15).

19.7.10 Load Sharing

Another popular configuration is load sharing. In this scheme, the system is overprovisioned such that if any one component fails, the workload can be distributed among the remaining components with no impact to system performance (Fig. 19.16). In this example, we have three components in which the work is equally distributed. If one of these components fails, the system moves to State 2. State 2 is a fully operational state with the remaining two components. If either one of these remaining components fail, the system is still functional albeit in a degraded mode.

For example, a web server that contains three backend servers in which the requests are equally distributed in a round-robin fashion. The failure of any one server will not impact the performance or operation of the system. If a subsequent server fails, the responses may be much slower than the specified requirement; however, the transactions will generally be successful.

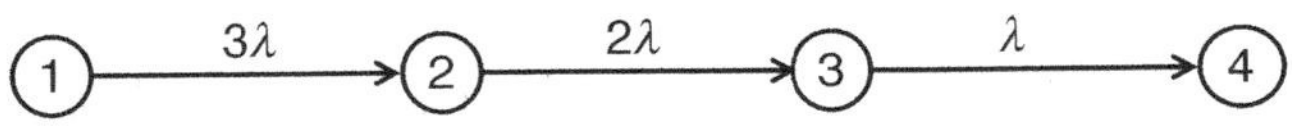

Figure 19.16 Load Sharing, and Degraded Operation, Failure Rate Independent of Load

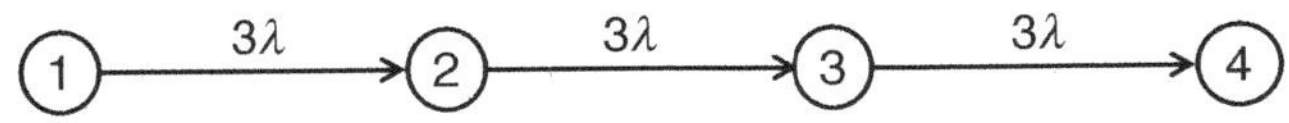

Figure 19.17 Load Sharing, and Degraded Operation, Failure Rate Proportional to Load

19.7.11 Load Sharing, Failure Rate Proportional to Load

Taking the same example as above, we now consider the impact of the transaction load on the failure rate of the system. If the failure rate is proportional to traffic, then after the failure of one component, the load of the failed server is spread equally across the remaining two servers. Thus, the failure rate increases by 50% on the two remaining servers giving us a 3λ failure rate between State 2 and State 3. If another server fails and all of the load of that server must be moved to the remaining server, that is, the remaining server does three times the work, then the failure rate of that server becomes 3λ (Fig. 19.17).

19.7.12 *N*:1 Redundancy

N:1 redundancy aka *N*:1 protection means that there is 1 protection channel for *N* working channels. If there is a failure on one of the *N* working channels, that traffic gets switched to the protection channel. Once the problem clears the protection switch reverts and low priority traffic can again be carried over the shared protection channel. The protection provided is often reduced as the physical cables, fibers, etc. are generally physically routed via the same path. If the protection channel is routed through different paths than the working path(s), the redundant network uptime potential is greater, that is, traffic protection is provided as well as equipment protection.

19.7.13 *N* + 1 Redundancy

$N + 1$ and $N + M$ redundancy ensures system availability in the event of single component failure. The *N* components have at least one independent backup component or multiple backup components (*M*). The configuration generally operates in an active/passive or active/standby mode since the backup components, although actively participating in fault management and redundancy management, do not actively participate in a service providing role during normal operation. The level of transparency (disruption to system availability) during failover is dependent on a specific solution, though some degradation or outage of varying lengths will occur during the failover process followed by full restoration of the system services.

Examples of $N + 1$ redundancy:

- Connecting devices (servers, and so on) in dual switch SAN (storage area network) fabrics employ a discrete path to each switch. Only one path is active at any given time; resiliency is provided by the availability of an additional path if the active path becomes unavailable.
- Server cluster power supplies.
- A set of processing blades in a chassis providing the same service. The service requests are distributed evenly across the blades. The standby blade provides a checkpointing and monitoring service for the other blades. Upon failure, the recovery of full system functionality is expedited by the checkpointed files used during the failover.

Similar to our discussion on 1 + 1 or 2N redundancy, all components of the redundant configuration are the same. Failure of one of the active components will result in the standby blade for that redundant group taking over the services and acting in the role of the failed active blade. The failed blade, once recovered, automatically assumes the standby role. If a subsequent failure occurs, the new standby will take over the functionality of this failed active, and so on. Thus, the standby card has no fixed position and "floats" randomly within the resource group.

The standby card can optionally be fixed to a specific location. For this configuration, a failback is required after the failed active card returns to service. This failback can be automatic or manual depending on the requirements of the customer.

19.7.14 Escalating Fault Recovery

Let's consider a robust system in which escalating failure recovery actions are employed (Fig. 19.18). If a software application on our system fails, we can attempt to restart that application assuming the time to restart has a minimal impact on system operation. If the application restart is successful (S_{RST}), the system returns to fully functional operation. The failover time is longer than the application restart time, so our first line of action is to restart the application. If the failover is not successful ($1 - S_{FO}$), the system fails and enters State 4. If the restart fails, the system moves to State 3, at which point a failover to a redundant component is attempted.

EXAMPLE 19.3

Figure 19.19 illustrates the reliability of a system with the following parameters: $\lambda_{SW} = 0.001$, $\mu_{SW} = 60$, $\mu_{FO} = 12$, $S_{FO} = 0.95$, and $S_{RST} = 0.9$. It also compares the system reliabilities for different values of failover success rate (S_{FO}) and reset success rates (S_{RST}). When the failover and reset success rates are zero, the reliability drops dramatically compared with higher values of those success rates.

19.8 SUMMARY

To improve the reliability and availability or our system, fault management techniques should be considered. When a fault occurs, can our system automatically detect this problem and then mitigate its impact on the system?

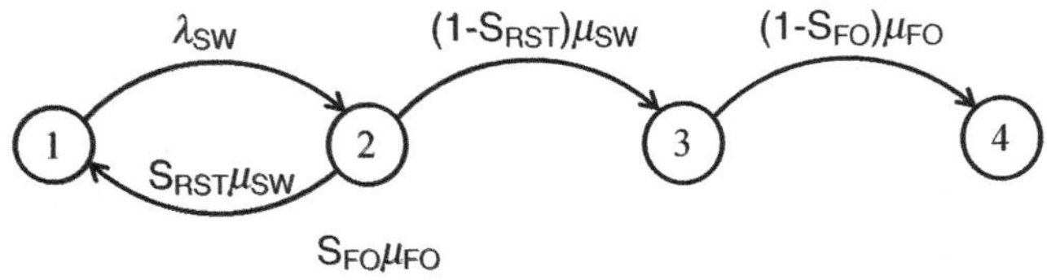

Figure 19.18 Escalating Fault Recovery

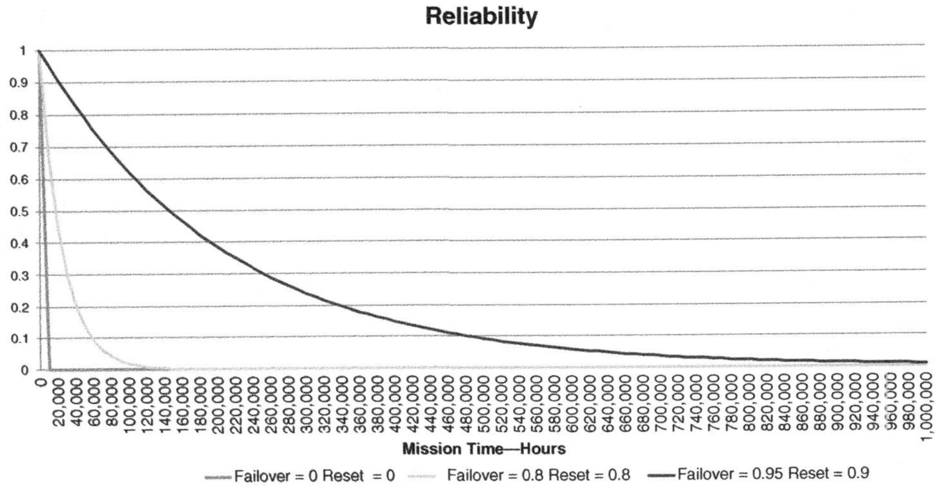

Figure 19.19 Reliability with Escalating Fault Recovery

Fault coverage, time to detect, redundancy schemes, recovery time, and robustness of the system in the event of failures are important considerations. This chapter focused on these aspects of fault management.

Redundancy is one of the most powerful design techniques for minimizing or eliminating the impact of failures on the overall system behavior. We discussed a variety of models used in the fields of aerospace and telecommunications engineering. We compared the system effects of redundancy with repair and redundancy without repair. In certain mission-critical systems, such as fly-by-wire aircraft flight control systems, the failed components cannot be repaired or replaced until after the mission has been completed.

Comparisons of the various techniques, including several examples were provided.

CHAPTER 20

Application of DFMEA to Real-Life Example

20.1 INTRODUCTION

In Chapter 15, the Design Failure Modes and Effects Analysis (DFMEA) technique was described. With this background information, we will now consider a case study from the telecom industry that will illustrate how DFMEA can be applied in a practical manner with significant improvements to the availability and overall quality of the product. The case study focuses on the Base Station Controller—Data Only (BSC-DO) system previously introduced in Chapter 18. The system architecture consists of multiple cages and cards within each cage. The DFMEA case study focuses on the cage failover functionality, which has a 2 + 1 configuration (two active cages and one standby cage) as shown in Figure 20.1. Upon failure of an active cage, the standby cage will take over the functionality of the active cage.

20.2 CAGE FAILOVER ARCHITECTURE DESCRIPTION

The first two cages are active and the third cage is the standby cage. If an active cage fails, we want to ensure that the standby cage will take over all the functionality of the failed cage. As part of the failure detection, each cage monitors the other cages using a heartbeat mechanism. If the heartbeat is not received within a specified time interval, the standby cage will declare that the active cage is no longer functioning. The cage failure could occur for several reasons, such as a power failure, a transport issue, a faulty software upgrade, and so on. The failover should happen automatically that is, the standby cage will take over all the functionality of the active cage, and all logical IP addresses will be reassigned to the standby cage. The system will remain in this configuration until the operator forces a failback to the previously active cage. Another interesting point is that calls during failover and calls during failback will

Designing High Availability Systems: Design for Six Sigma and Classical Reliability Techniques with Practical Real-Life Examples, First Edition. Zachary Taylor and Subramanyam Ranganathan.

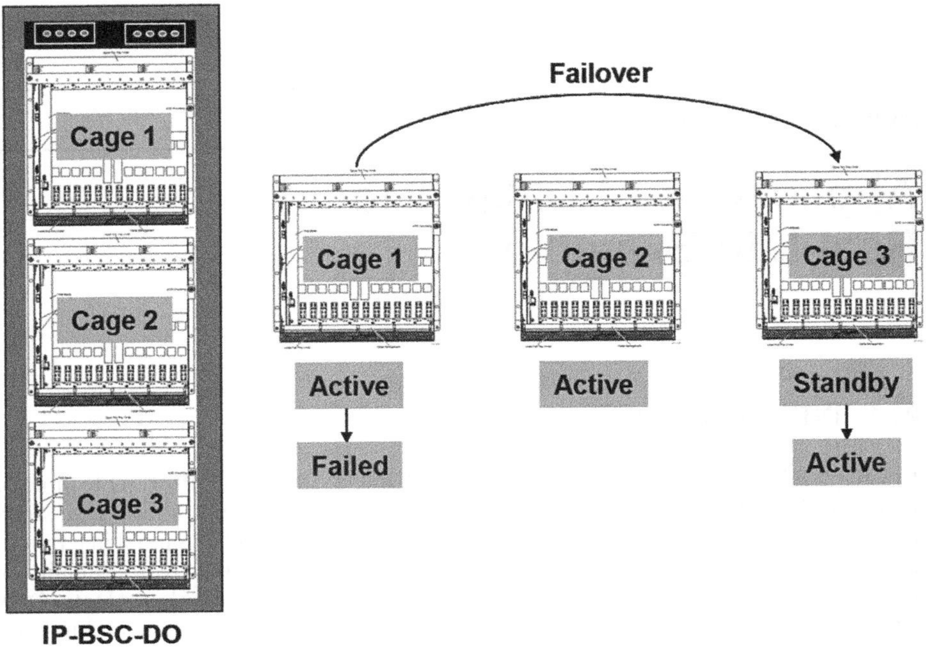

Figure 20.1 BSC Cage Failover

be dropped. The failback is designed only to be operator initiated to minimize call impacts, that is, performed during the maintenance window when call traffic is low or other system repair activities are being performed.

The required window for a successful failover is ≤30 seconds. The failover cannot be initiated by the operator but is designed to occur only when a failure that impacts the cage occurs. Even if the operator decides to reset an active cage, the command will be intercepted and the cage failover will be prevented.

The DFMEA was specifically focused on cage failover to identify all failure modes that were significant. The DFMEA identified 24 unique failure scenarios, and each of those was assigned a Risk Priority Number (RPN). A pareto of the RPNs was performed, and the highest RPNs were addressed first as they corresponded to the highest risks. The general rule of thumb applied to this project was that RPNs ≥ 100 should be addressed as high risk RPNs. To follow up with the mitigation actions, a change request was assigned to each of the high risk RPNs and analyzed for risk mitigation actions. Fixes were identified as part of the DFMEA itself and then implemented. Figure 20.2 shows the RPN pareto for the 24 failures that were identified. The issues were mapped to requirements, design, and code corresponding to phases in the development life cycle.

Figure 20.3 shows the improvements identified by the DFMEA sessions. The drop in the revised RPNs shown in the figure indicates significant drop in the technical risks. Some of the RPNs were assigned to a subsequent release for various reasons.

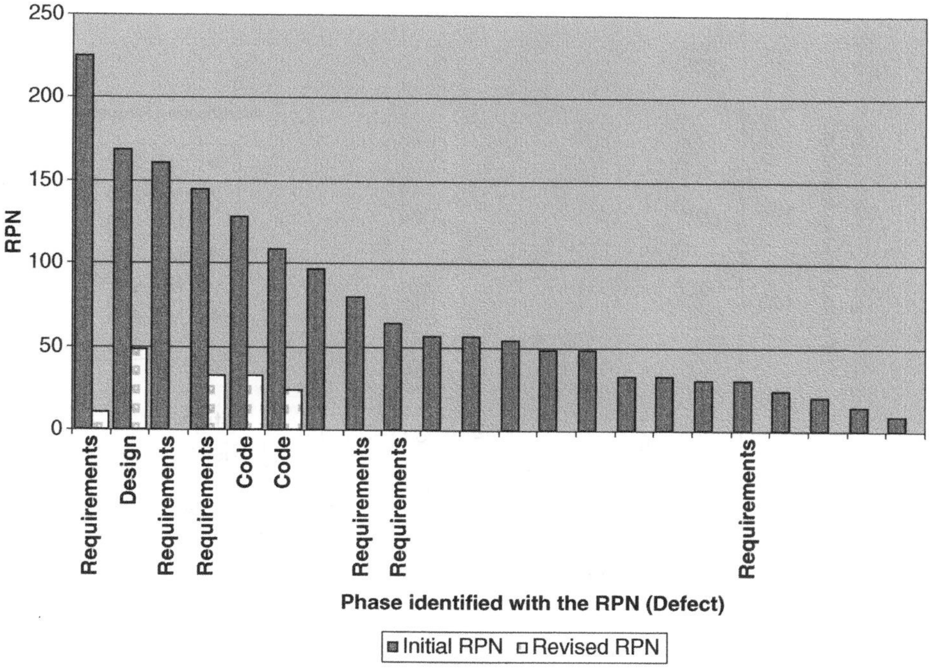

Figure 20.2 Cage Failover DFMEA RPN Pareto

Project Management determined which release should contain a particular fix based on other dependencies, such as staffing and resource constraints, etc.

20.3 CAGE FAILOVER DFMEA EXAMPLE

As an example, let us drill down into one of the DFMEA entries for Cage Failover. Each entry in the DFMEA is given a unique identifier. We consider a failure scenario in which a cage failover occurs when no failure exists. The standby cage is specifically designed to protect an active cage when a cage failure occurs, so part of the brainstorming activity is to look at false failure signatures. What if the cage failover occurs when no cage failure exists? How can this be addressed? One of the triggers could be an operator action. There may be a certain set of commands that an operator could issue that might inadvertently cause a cage failover. Each cage has redundant resources. For example, the cage controller has 2N redundancy, and if one of the cage controller cards is disabled, the other cage controller will take over. The operator is allowed to disable these cards for maintenance. If a failure occurs on an active cage controller, the redundant card will take over, and the operator can administratively "lock" the failed card and attempt to isolate, diagnose, and repair it. The operator

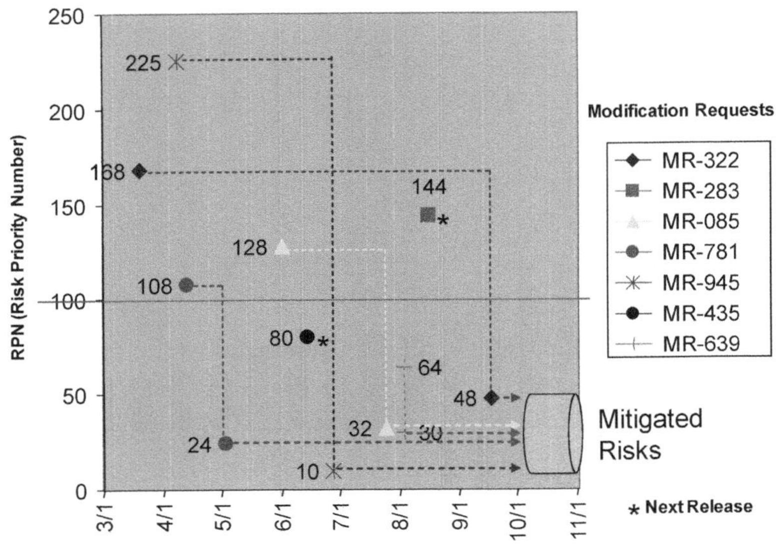

Figure 20.3 DFMEA Risk Mitigation

may reset the card to attempt recovery. If the reset is successful, the card will return to service after initialization.

A particular failure mode uncovered in this DFMEA: The operator could issue a reset for an active controller card when the redundant controller card is disabled. When both cards are disabled, the cage controller capability is unavailable—communication between that cage and the external environment is disrupted. The standby cage that is monitoring the active cage will not receive heartbeats and will declare the active cage failed and take over the functionality of that cage.

If this occurs, all calls will be cleared for 30 seconds. This is a flaw in the design. The severity is set to "Moderate Effect" because this will not result in a long outage. The likelihood of this happening (Occurrence) is "Medium," since the operator can issue this command at any time. Detection ranking is set to "High" since the operator would be informed of a cage failover after issuing the Reset Command.

So to summarize the failure scenario: The operator resets one controller card when the other controller card is already disabled or failed, thus triggering a failover inadvertently.

The DFMEA action plan was to prevent the operator from resetting the last enabled controller card when the system is operational. The revised Severity ranking is still moderate because the effect of this failure is the same. The occurrence of this event was prevented and thus the revised Occurrence score went down significantly after the change was made to the software. The revised Detection score became Very High because if the operator issues this command, an immediate feedback is provided as a pop-up warning, indicating that the operator is not allowed to lock or reset the last enabled controller card.

TABLE 20.1 Example of Risk Reduction for a Failure Mode

HW/SW/ UI/System	Failure Mode	Failure Mode Details (Notes)	Effect	Severity
System	Failover erroneously triggered by operator action	Call processing resources assigned to active cage removed. No conflict with standby cage (newly active)	Calls dropped and no new calls established for up to 30 seconds. Loss of cage redundancy protection.	5—Moderate effect

Revised Severity	Cause of Failure	Occurrence	Revised Occurrence	Current Process Controls
5—Moderate effect	Operator action: Reset of active controller when no standby exists. The design does not support this action. It does support reset of the entire cage, i.e., no false failover during a cage reset.	9—Very high	1—Unlikely	None

Detection	Revised Detection	RPN	Recommended Actions	Action Tracking ID	Revised RPN
5—Medium	2—Very high	225	Do not allow reset of the last enabled controller	MR-945	10

HW, hardware; SW, software; UI, user interface.

The root cause of this problem was traced to requirements as there was no requirement that explicitly indicated that the operator is prevented from locking or resetting the last enabled controller card. The original RPN was 225. The revised RPN after the fix was put in was reduced to 10 (Table 20.1).

Table 20.2 provides examples of the failure modes that resulted in software design changes along with the initial and revised RPN scores.

20.4 DFMEA SCORECARD

The DFMEA scorecard indicates the progression of risk of a given feature or functionality across the product life cycle. The top 10 RPN scores are selected, and the

TABLE 20.2 Examples of Software Changes due to Cage Failover DFMEA

Issue	Initial RPN	Design Change	Revised RPN
Cage Failover exceeds 30 seconds—transport layer overload due to broadcast storm	168	Implement a spanning tree algorithm in software that will detect the broadcast loop and perform recovery actions	48
Failed cage remains active because of Duplicate IP Addresses—database synchronization problem	144	Standby cage migration software design to include logic to arbitrate with active cage and relinquish control if necessary	32
Failover erroneously triggered by operator action—will cause dropped calls and prevent new call originations for a time interval	225	Design controls built into software—do not allow reset of the last enabled controller card	10
Call Processing (CP) Error on Standby Cage—all CP resources lost because it received invalid Service Node Index (SNI)	128	Modify cage initialization software procedures to set valid SNI attributes so that they are used during standby transition	32

corresponding high, low, and average RPN scores are tracked at each development phase gate to ensure that the technical risks are proactively mitigated. Table 20.3 indicates that the technical risk was reduced by around 55% between requirements and design phases.

20.5 LESSONS LEARNED

1. Some of the problems that were identified as part of the DFMEA resulted in code changes. This implied that if the DFMEA had been performed earlier in the life cycle, then those problems may have been fixed prior to developing the code. We know that the cost of fixing a problem increases dramatically as we progress from requirements to design, code, test, and finally deployment.
2. Participation of key stakeholders is critical. If key subject matter experts are not available, then the analysis is incomplete. For example, several problems associated with the transport layer were not identified because of the unavailability of a transport layer expert. It is best to have everyone in the same meeting rather than trying to discuss with the expert offline. One of the important aspects of performing a DFMEA is the interaction between the experts.
3. RPN ranking is a tremendous help in prioritizing problems and following up on actions. It may seem a little ambiguous to assign ratings from 1 to 10. For example, how would you differentiate a Severity 7 from a Severity 8? Each severity should have a specific description. If we consistently apply the same

TABLE 20.3 DFMEA Scorecard

New Unique and Difficult Feature	Critical Parameter	DFSS Method	System Engineering Owner	Requirements Top 10 RPNs Average	Requirements Top 10 RPNs Min	Requirements Top 10 RPNs Max	Design Top 10 RPNs Average	Design Top 10 RPNs Min	Design Top 10 RPNs Max
Base Station Controller Commercial	Call Processing Availability—Cage Failover	DFMEA	Joe Smith	123	56	225	56	30	96

DFSS, Design for Six Sigma.

criteria for all of the DFMEAs, the ranking will be more objective. These rankings provide a better indication of the relative risk of problems rather than the alternative of relying on just gut instinct.

4. The DFMEA complements existing requirements analysis and design techniques that are already in place. The DFMEA should be integrated into the existing product development process, and more importantly, integrated into the mindset of the product engineering teams to make it effective.
5. Precise definitions and an understanding of the DFMEA fields by all participants are important to keep the focus on the analysis.
6. Choose the right tool. Look at the tools in light of what best works for the team. Many commercial tools are available to support DFMEA—some tools may work better than others, and perhaps a simple spreadsheet will prove sufficient. In our analysis, Excel spreadsheets were used for the DFMEAs. Pick the tool that works well for the team.

20.6 SUMMARY

This chapter provided a case study of a DFMEA for an actual telecommunications project. As a result of conducting this DFMEA, design faults were identified that otherwise may not have been discovered. Design changes were introduced to prevent these faults from being discovered after product introduction. Using the initial and revised RPNs, the design improvements can be quantified. The lessons learned from this project were identified and used to improve the effectiveness and efficiency of the subsequent DFMEAs for this project.

CHAPTER 21

Application of FTA to Real-Life Example

21.1 INTRODUCTION

In Chapter 15, fault tree analysis (FTA) was introduced. With this background information, we will now consider a case study from the telecom industry that will illustrate how FTA can be applied in a practical manner with resulting improvements to the overall quality of the product. The case study focuses on the Base Station Controller—Data Only (BSC-DO) system previously introduced. We provide a step-by-step approach to create the model, introduce how we use different tools for different purposes along the way, and ultimately estimate the Five 9s for the system.

21.2 CALCULATING AVAILABILITY USING FAULT TREE ANALYSIS

Before creating the fault tree, we identify the top event, which is the BSC-DO system failure. This failure is assumed to contribute to the system outage. Next, we identify the types of failures at the high level that result in a system failure. Critical parameters are identified, and the one we will focus on is the "Call Processing Service Availability" critical parameter.

Call processing service unavailability occurs when the system has a loss of call processing capacity greater than 10% for a duration of at least 5 seconds. We use tools such as Design of Experiments, Regression, and Monte Carlo Simulations along with Sensitivity Analysis to identify the mathematical models and significant factors that impact call processing availability. The sensitivity analysis allows us to understand how the changes in the parameters associated with the system affect the overall call processing availability.

The fault tree is made up of a several hundred nodes representing hardware, software, physical, and logical components. Figure 21.1 identifies the top-level fault tree hierarchy for the overall system.

Designing High Availability Systems: Design for Six Sigma and Classical Reliability Techniques with Practical Real-Life Examples, First Edition. Zachary Taylor and Subramanyam Ranganathan.

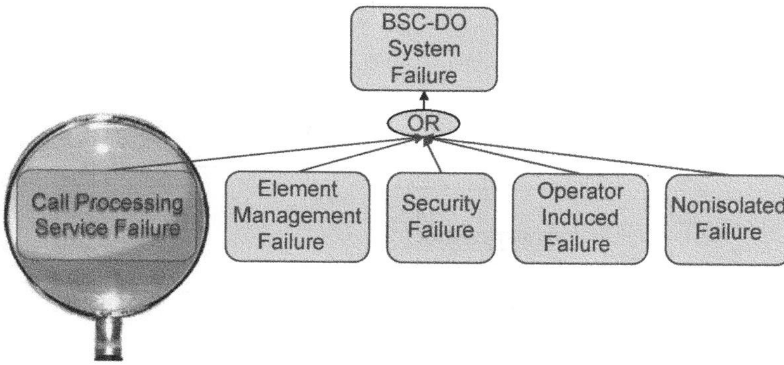

Figure 21.1 Top-Level Fault Tree

Nonisolated failures refer to failures in the system that cannot be attributed to one of the other categories. Security and Operator failures can impact call processing because there can be an outage due to a malicious security attack (especially in IP-based systems) or an inadvertent action on the part of the operator. For this analysis, we consider the system impact based on the failure of any component that affects call processing service.

The Mean Time between Failures (MTBF), Mean Time to Repair (MTTR), and failure probabilities are used for calculations. The MTBF information can be obtained from the hardware datasheets typically from the manufacturer's specifications for the hardware when available. Another technique is to look at similar cards or similar equipment from the same manufacture or other manufacturers. Finally, we can look at actual reliability data gathered from other projects that have used similar hardware. For software, we obtain reliability estimates based on past experiences, complexity, existing software ported to the platform, software maturity, organizational software maturity levels, and so on. Using the information we have gathered, the software component MTBFs are created. After the system is deployed to the field, these reliability estimates are updated. Once our assumptions have been made, the data gathered, and the fault tree created, we can identify high risk nodes based on the model and target them for additional analysis.

21.3 BUILDING THE BASIC EVENTS

If a hardware can be replaced in the field that is, cards, disk drives, power supplies, fans, and so on, each of these "Field Replaceable Units" should be represented as a separate component in the fault tree. If the hardware component fails, regardless of the specific cause of the component failure, we are only interested in this level of analysis whether the board is fully functioning, degraded, or failed. Looking at the transfer gate "Carrier 1 Hardware Failure" shown in Figure 21.2, this gate represents all the possible modes of failures that result in that carrier card failing.

For software, we can aggregate each software component executing on a specific card. All possible failure modes of a particular application are identified in the Design

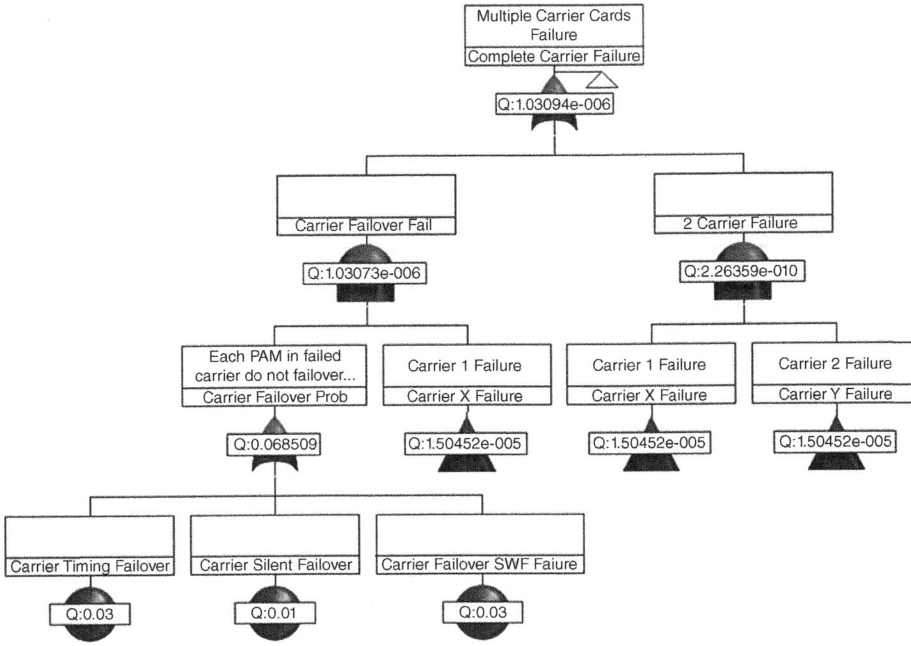

Figure 21.2 Hardware and Software Failure Nodes

Failure Modes and Effects Analysis (DFMEA) and contribute to the software failure rate.

We also consider other physical components, such as fans, power supplies, chassis, and so on. The maturity of the system architecture or design alternatives will drive specific components being modeled. Beginning with the initial system architecture, observe how potential failures can affect the overall system. This analysis will drive architecture decisions, such as redundancy configuration and fault management. FTA should be started as early as possible to facilitate a more iterative approach to system design. The improvement of the system architecture is driven by the iterative FTA. Fault trees can expose potential weak links in a system architecture that can be rectified during the design itself.

The fault tree can be decomposed into different subsystems for individual analysis as needed. These individual models may then be combined into a larger system level fault tree. In Figure 21.2, we see how basic events interact to create a higher level node.

In Relex, each basic event is characterized by probability, MTBF, corrective time, etc. as shown in Figure 21.3.

21.4 BUILDING THE FAULT TREE

The BSC-DO architecture includes separate redundancy schemes for different resource groups. While building the fault tree, failures related to the failover

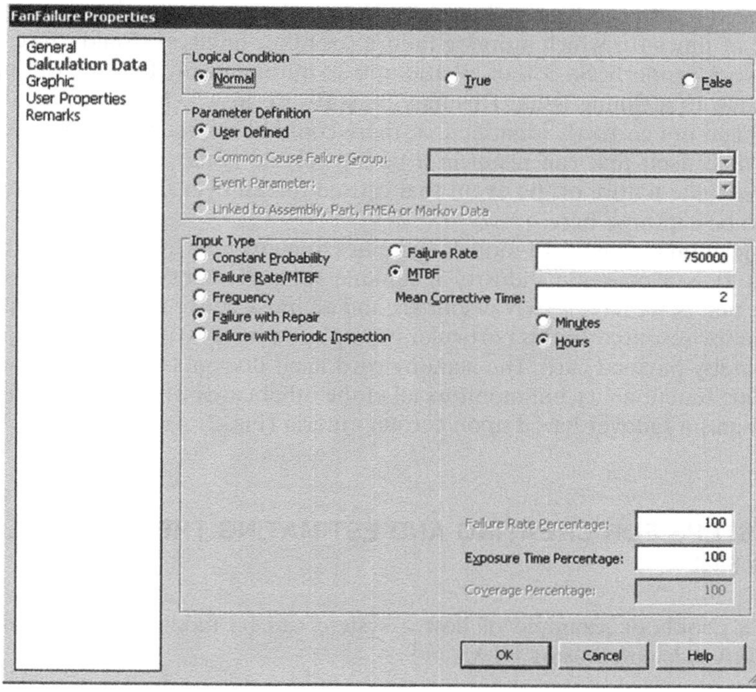

Figure 21.3 Example of Characterizing an Atomic Node

functionality were considered because failovers are like moving parts in an architecture. Given that the PAM card has failed, what is the probability that a successful failover will occur?

We express failover failures in terms of percentage of failovers that are not successful. Failures that prevent successful failover can be classified under the following categories:

- *Timing Failure* (3%). Timing issues that prevent the failover from occurring within the required time. For example, the detection of the failure is delayed until after an outage has already occurred.
- *Silent Failure* (1%). Failures on the failed card that are not detected by the fault management software. For example, 1% of the time, when a failover is required, it does not occur, since the failure that drives the failover is not detected. Eliminating silent failures in any system architecture is critical, and FTA helps achieve this objective.
- *Failover Failure* (3%). The failures are detected properly and detected within the required time, but a logic error in the failover mechanism results in the failover not completing successfully. For this case study, the chance of a failover failure is 3%.

We need to use engineering judgment and past data to determine the appropriate number of times for which a timing failure (or other failures) would cause failovers not to work properly. So 3% of all attempts to automatically failover will be unsuccessful due to a timing issue. This may be software in which failure signatures are delayed and not correctly identified, or there could be an issue with the actual hardware switch itself that can result in a timing related issue. Anything that could be sensitive to the nature of the event that caused the failover from a time perspective needs to be captured here.

This general technique of modeling applies to many high availability system architectures. It is a common industry technique to implement an $N + 1$ redundancy scheme, that is, we have 1 to N resources, and all those active resources are protected by 1 standby resource. In this particular example, we have 10 payload cards, protected by 1 standby payload card. The standby card itself does not provide any active call processing functionality, but monitors all of the other cards within the resource group and initiates a failover based upon certain criteria (Fig. 21.4).

21.5 STEPS FOR CREATING AND ESTIMATING THE AVAILABILITY USING FTA

Here's a cookbook example of how a system can be modeled, simulated, and its availability calculated using FTA:

1. Create a fault tree using a modeling tool, such as Relex. It is very useful to have a graphical representation of the fault tree to help easily identify the relationships between the components and how they drive the overall availability (top event).
2. Identify/estimate all the failure rates for each individual modeled component (basic event).
3. Once the tree is complete, it is relatively easy to identify the high risk failure nodes for further analysis. The purpose of this step is to create a mathematical relationship between the high risk nodes and the top level node in the fault tree.
4. Transfer failure rate data for each node into the availability calculation spreadsheet. This is useful for more in-depth analysis.
5. Create a Design of Experiments (DOE) using the failure nodes data (optional)
    - Use Minitab to create the experiment.
    - The resulting regression equation can then be used as the output from the Monte Carlo simulation.
6. Run experiment in Relex and populate the results in Minitab (optional).
    - Obtain the regression equation.
    - The equation for availability is determined from Minitab.
    - Use the regression equation to set up a Monte Carlo simulation.
7. Set up a Monte Carlo simulation using Crystal Ball and inject the necessary variation for each node. Run the Monte Carlo Simulation.
    - Review the probability density curve for availability.
    - Analyze the capability metrics—especially the Cpk value.

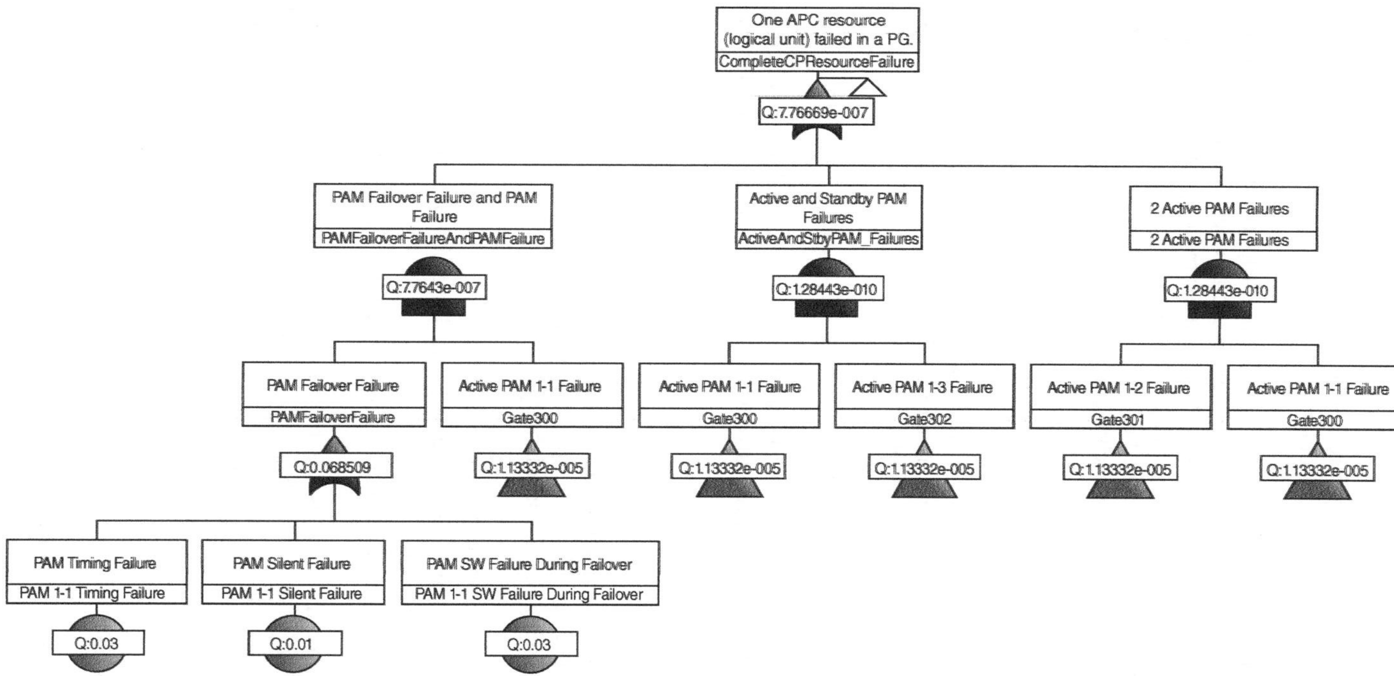

Figure 21.4 Example of Building a Fault Tree

Step 1: Creating the fault tree

Once we have some idea of the preliminary system architecture, we can begin creating the fault tree. The fault tree modeling is an iterative process that both drives and is driven by the system architecture.

Using the Relex tool we take a look at the properties of each one of the events (Fig. 21.5). A numeric value and short text is used to describe the event. These events can be listed in a spreadsheet (Table 21.1). For example, 104 will be the event ID. You can use a Relex filter to create and apply the same failure component attributes across multiple branches in the tree. The fault tree shows the relationships between the basic events, based on the system architecture and how these events drive the overall system availability.

Step 2: Estimating failure events

Table 21.1 shows a snapshot of the worksheet that contains all basic events in a tabular format. Each event has a unique ID, along with other characteristics, such as MTBF, MTTR, and Probability of Failure.

Step 3: Identifying high-risk nodes

Figure 21.6 is a snapshot of the fault tree in tabular form. We identified failover failures from the five different resource groups as the key events that drive call processing service availability.

Step 4: Actual gate calculations

The actual fault tree gate unavailability calculations are shown in Table 21.2. We have three different ways of calculating the overall failure rates of the system, and thus the Call Processing Service availability:

(a) The summation formula
(b) Esary Proschan method (upper and lower bounds are applied for the gate probabilities) and
(c) Exact probability calculations (provided by the Relex tool).

We can move from the Relex tool directly to the spreadsheet and create many different "*what if*" scenarios that can be quickly calculated and then analyzed in the spreadsheet. The additional value of using the spreadsheet is that we can go beyond our initial four failover nodes that we started with in the analysis.

Figure 21.7 shows the relationship between the fault tree events and the spreadsheet events.

Step 5: Setting up the design of experiments

If we do not have a mathematical relationship between Call Processing Failure and different card failovers, a DFSS design of experiments can be setup to derive that

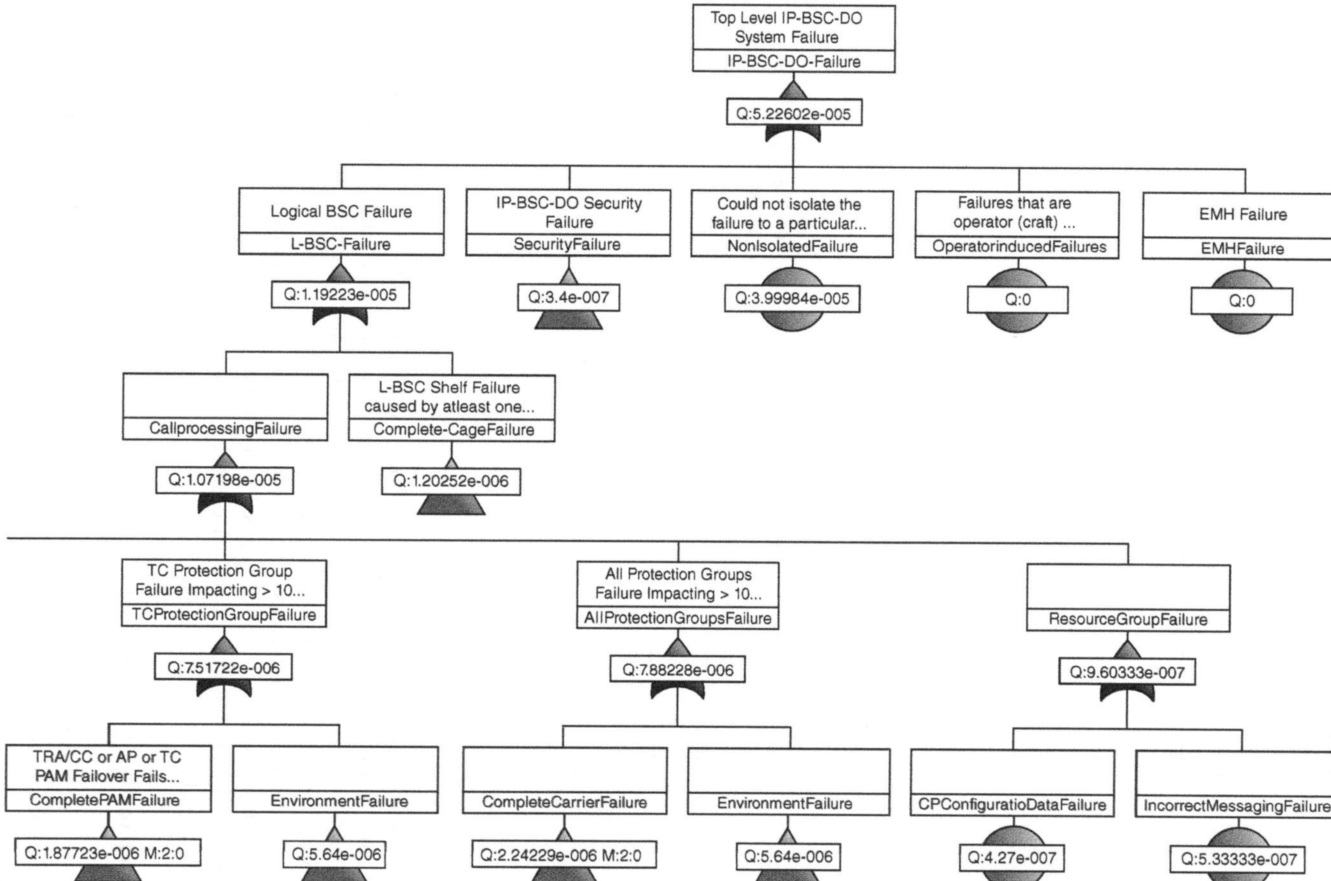

Figure 21.5 Top-Level Fault Tree Using Relex Tool

TABLE 21.1 Basic Failure Events

	A	B	C	D	E	F
1						
2		ID	Device	MTBF	MTTR (hours)	Probability
3			EVENTS			
4		1	Carrier HW Failure	550000	2	3.6364E-06
5		2	Carrier SW Failure	200000	2	0.00001
6		104	FRU Timing Failure			0.03
7		105	FRU Silent Failure			0.01
8		106	FRU SW Failover Failure			0.03
9		7	CP Config Data Failure	937500	0.5	5.3333E-07
10		8	Incorrect Message Failure	937500	0.5	5.3333E-07
11		9	2 Or More Fan Failures	750000	2	2.6667E-06
12		10	Temp Ext Env Multi FRU	750000	2	2.6667E-06
13		12	SAM SW Failure	200000	2	1.00000E-05
14		13	SAM HW Failure	1300000	2	1.5385E-06
15		19	Temperature Env Single FF	750000	2	2.6667E-06
16		20	Voltage Env Single FRU	750000	2	2.6667E-06
17		26	PAM HW Fail	1500000	8	5.3333E-06
18		27	PAM App Fail	50000	0.1	2.00000E-06
19		28	PAM BMGR Fail	100000	0.1	1.00000E-06
20		29	PAM Transport Fail	50000	0.1	2.00000E-06
21		30	PAM Shelf-P Fail	100000	0.1	1.00000E-06
22		32	PAM Stby App Fail	50000	0.1	2.00000E-06
23		43	HAP SW Failure	50000	2	4.00000E-05
24		44	Internal SSC Xport Fail	50000	2	4.00000E-05
25		45	Mishandled External Xport	50000	2	4.00000E-05
26		46	SSC HW Fail	1200000	2	1.66667E-06
27		47	O&M SW Fail	100000	2	2.00000E-05
28		52	Active Power Supply Fail	2600000	2	7.69231E-07
29		116	Cage Failover SW Failure			0.125

relationship. Minitab offers a framework to create design of experiments. For this experiment, we define three different levels—the base unavailability and the other two levels corresponding to +40% and −40% of the base. The percentages chosen for the other two levels are based on the confidence that we have on the original base value. Minitab also generates a run order for the experiments. The Relex tool was used to simulate these different runs. Each run had a particular set of values for the different failover failures. In this case, since we have four factors, we did a two-level full factorial experiment (Table 21.3).

Step 6: Run the experiment

For the levels corresponding to the assumed variations, the experiment is run using the Relex tool to generate the Call Processing availability values for different combinations of the selected nodes.

Table 21.4 illustrates the output of the different runs of Call Processing Service Availability and the corresponding Five 9s. A regression analysis is then performed

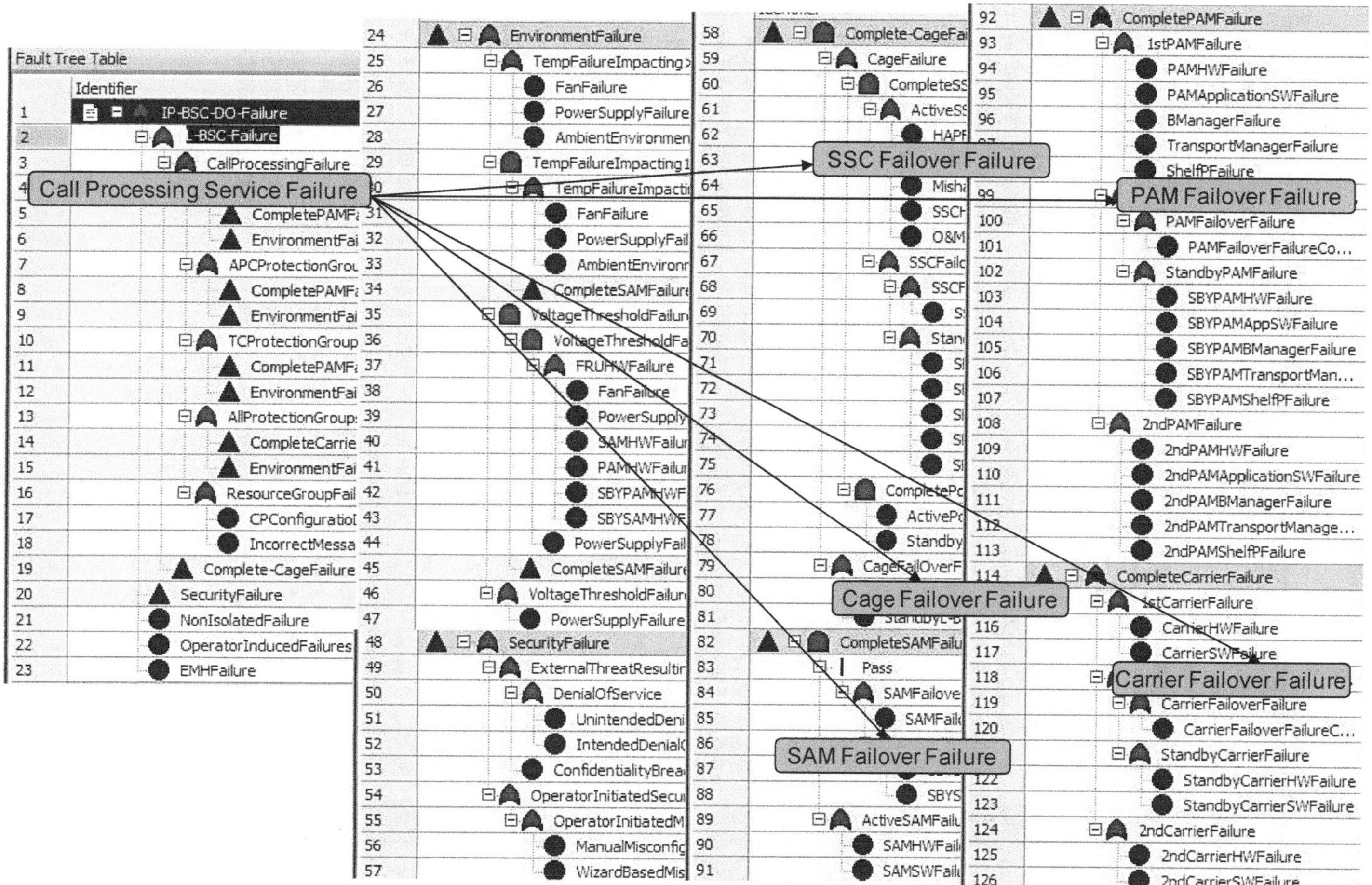

Figure 21.6 Identifying High Risk Nodes

TABLE 21.2 Calculating Probabilities Using Three Different Methods

2	ID	Device	Gate Logic	Summation	Easy Prochan	Exact (Relex)	E
30		GATES					
31	3	Carrier Failure	1 AND 2	1.36364E-05	1.36363E-05	1.36362E-05	
32	4	2 Carrier Failure	3 AND 3	1.85950E-10	1.85949E-10	1.85946E-10	
33	107	FRU Failover Error	104 OR 105 OF	7.00000E-02	6.85090E-02	6.85090E-02	
34	5	Carrier Failover Failure	3 AND 107	9.54545E-07	9.34211E-07	9.34203E-07	
35	6	Complete Carrier Failure	4 OR 5	9.54731E-07	9.34397E-07	9.34377E-07	
36	14	Active SAM Failure	12 OR 13	1.15385E-05	1.15384E-05	1.15383E-05	
37	15	Standby SAM Failure	11 OR 13	1.15385E-05	1.15384E-05	1.15383E-05	
38	16	Both SAMs Fail	14 AND 15	1.33136E-10	1.33136E-10	1.33133E-10	
39	114	SAM Failover Fail	107 AND 15	8.08E-07	7.90E-07	7.90E-07	
40	18	Complete SAM Failure	16 OR 114	8.07825E-07	7.90621E-07	7.90604E-07	
41	21	Temp Fail > 1 FRU	9 OR 10	5.33333E-06	5.33333E-06	5.33331E-06	
42	22	Temp and SAM Fail	18 AND 19	2.15420E-12	2.10832E-12	2.10827E-12	
43	23	Volt and SAM Fail	18 AND 20	2.15420E-12	2.10832E-12	2.10827E-12	
44	24	Full Power Supply Fail	52 AND 53	5.91716E-13	5.91716E-13	5.91715E-13	
45	115	Volt Failure > 1 FRU	24	5.91716E-13	5.91716E-13	5.91715E-13	
46	25	Environment Failure	21 OR 22 OR :	5.33334E-06	5.33333E-06	5.33332E-06	
47	31	PAM Failure	26 OR 27 OR	1.13333E-05	1.13333E-05	1.13332E-05	
48	33	STBY PAM Failure	26 OR 32 OR	1.13333E-05	1.13333E-05	1.13332E-05	
49	34	Active and Stby PAM Fail	31 AND 33	1.28444E-10	1.28443E-10	1.28443E-10	
50	35	2 Active PAM Fail	31 AND 31	1.28444E-10	1.28443E-10	1.28443E-10	
51	103	PAM Failover Error	104 OR 105 OF	7.00000E-02	6.85090E-02	6.85090E-02	
52	36	PAM Failover Fail	103 AND 31	7.93333E-07	7.76432E-07	7.76430E-07	
53	37	SCA, TC, or SCA Resource	34 OR 35 OR	7.93590E-07	7.76689E-07	7.76690E-07	
54	41	Resource Group Failure	7 OR 8	1.06667E-06	1.06667E-06	1.06670E-06	
55	42	Call Processing Failure	37 OR 37 OR	9.73551E-06	9.66443E-06	9.66434E-06	
56	48	Active SSC Fail	43 OR 44 OR	1.41667E-04	1.41659E-04	1.41654E-04	
57	49	Standby SSC Fail	48	1.41667E-04	1.41659E-04	1.41654E-04	
58	50	Active and Stby SSC Fail	48 OR 49	2.00694E-08	2.00673E-08	2.00659E-08	
59	51	SSC Failover Fail	113 AND 48	9.91667E-06	9.70493E-06	9.70458E-06	
60	53	Standby Power Supply Fail	52	7.69231E-07	7.69231E-07	7.69231E-07	
61	54	Power Supply Fail	52 AND 53	5.91716E-13	5.91716E-13	5.91715E-13	
62	55	Complete SSCs Failure	50 OR 51	9.93674E-06	9.72500E-06	9.72327E-06	
63	56	Cage Failure	54 OR 55	9.93674E-06	9.72500E-06	9.72233E-06	
64	57	MultiCage Failure	56 AND 56	9.87387E-11	9.45756E-11	9.45422E-11	
65	58	Cage Failover Failure	56 AND 114	1.24209E-06	1.21563E-06	1.21549E-06	
66	59	Complete Cage Failure	57 OR 58	1.24219E-06	1.21572E-06	1.21549E-06	
67	11	STBY SAM SW Failure	12	1.00000E-05	1.00000E-05	1.00000E-05	
68	60	LBSC Failure	42 OR 59	1.09777E-05	1.08801E-05	1.08798E-05	
69	113	SSC Failover Failure	104 OR 105 OF	7.00000E-02	6.85090E-02	6.85090E-02	

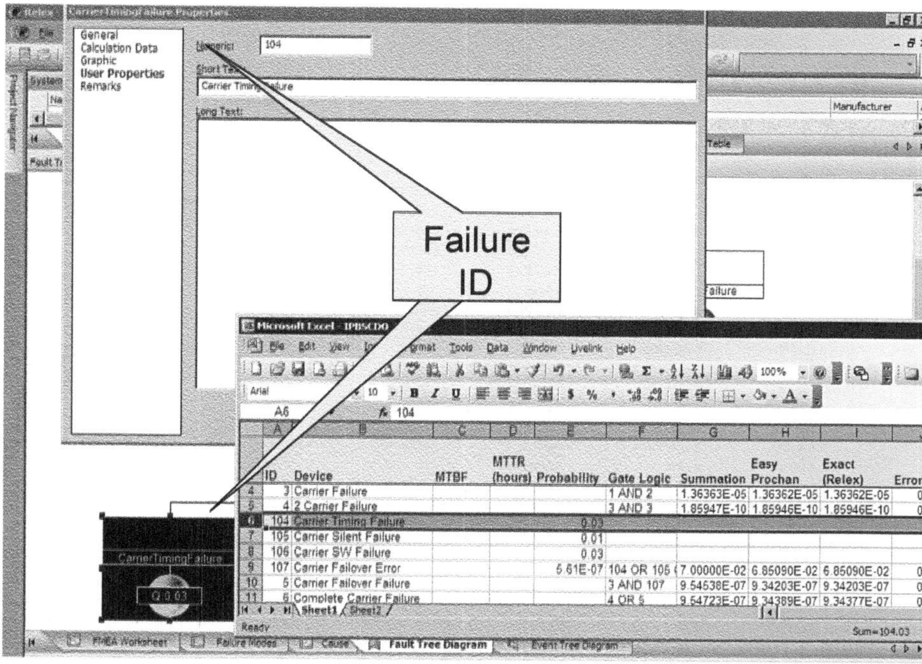

Figure 21.7 Relationship between the Initial Basic Event Spreadsheet and the Fault Tree

TABLE 21.3 Design of Experiments in Minitab

	C1 StdOrder	C2 RunOrder	C3 CenterPt	C4 Blocks	C5 CageFailover	C6 PAMFailoverFailure	C7 CarrierFailoverFailure	C8 SSCFailoverFailure	C9 SAMFailoverFailure
1	1	1	1	1	0.0000007	0.0000005	0.0000006	0.0000058	0.0000011
2	2	2	1	1	0.0000017	0.0000005	0.0000006	0.0000058	0.0000005
3	3	3	1	1	0.0000007	0.0000011	0.0000006	0.0000058	0.0000005
4	4	4	[illegible]	1	0.0000017	0.0000011	0.0000006	0.0000058	0.0000011
5	5	5	[illegible]	1	0.0000007	0.0000005	0.0000013	0.0000058	0.0000005
6	6	6	1	1	0.0000017	0.0000005	0.0000013	0.0000058	0.0000011
7	7	7	1	1	0.0000007	0.0000011	0.0000013	0.0000058	0.0000011
8	8	8	1	1	0.0000017	0.0000011	0.0000013	0.0000058	0.0000005
9	9	9	1	1	0.0000007	0.0000005	0.0000006	0.0000134	0.0000005
10	10	10	1	1	0.0000017	0.0000005	0.0000006	0.0000134	0.0000011
11	11	11	1	1	0.0000007	0.0000011	0.0000006	0.0000134	0.0000011
12	12	12	1	1	0.0000017	0.0000011	0.0000006	0.0000134	0.0000005
13	13	13	1	1	0.0000007	0.0000005	0.0000013	0.0000134	0.0000011
14	14	14	1	1	0.0000017	0.0000005	0.0000013	0.0000134	0.0000005
15	15	15	1	1	0.0000007	0.0000011	0.0000013	0.0000134	0.0000005
16	16	16	1	1	0.0000017	0.0000011	0.0000013	0.0000134	0.0000011
17	17	17	0	1	0.0000012	0.0000008	0.0000009	0.0000096	0.0000008

Experiment Number

TABLE 21.4 Minitab Experiment

Editor Tools Window Help Assistant

Output data entered in DOE table

C4	C5	C6	C7	C8	C9	C10	C11
Blocks	CageFailover	PAMFailoverFailure	CarrierFailoverFailure	SSCFailoverFailure	SAMFailoverFailure	CallProcessingServiceFailure	Five9s
1	0.0000007	0.0000005	0.0000006	0.0000058	0.0000011	0.0000072	5.14127
1	0.0000017	0.0000005	0.0000006	0.0000058	0.0000005	0.0000082	5.08697
1	0.0000007	0.0000011	0.0000006	0.0000058	0.0000005	0.0000078	5.10519
1	0.0000017	0.0000011	0.0000006	0.0000058	0.0000011	0.0000088	5.05498
1	0.0000007	0.0000005	0.0000013	0.0000058	0.0000005	0.0000080	5.09851
1	0.0000017	0.0000005	0.0000013	0.0000058	0.0000011	0.0000089	5.04902
1	0.0000007	0.0000011	0.0000013	0.0000058	0.0000011	0.0000086	5.06569
1	0.0000017	0.0000011	0.0000013	0.0000058	0.0000005	0.0000096	5.01962
1	0.0000007	0.0000005	0.0000006	0.0000134	0.0000005	0.0000072	5.14126
1	0.0000017	0.0000005	0.0000006	0.0000134	0.0000011	0.0000082	5.08697
1	0.0000007	0.0000011	0.0000006	0.0000134	0.0000011	0.0000078	5.10518
1	0.0000017	0.0000011	0.0000006	0.0000134	0.0000005	0.0000088	5.05497
1	0.0000007	0.0000005	0.0000013	0.0000134	0.0000011	0.0000077	5.11561
1	0.0000017	0.0000005	0.0000013	0.0000134	0.0000005	0.0000086	5.06425
1	0.0000007	0.0000011	0.0000013	0.0000134	0.0000005	0.0000086	5.06568
1	0.0000017	0.0000011	0.0000013	0.0000134	0.0000011	0.0000096	5.01962
1	0.0000012	0.0000008	0.0000009	0.0000096	0.0000008	0.0000084	5.07620

to obtain a regression equation that mathematically characterizes the system availability in terms of the originally assumed five factors (Fig. 21.8).

Step 7: Monte Carlo Simulation from Crystal Ball tool

We use Crystal Ball to set up the Monte Carlo simulation for the regression model that we obtained from the DOE and regression analysis. We inject variances for each factor in the model for Crystal Ball to simulate. The injected variance is driven by our confidence in the original data. The lower the confidence in the data, the higher the injected variance becomes (and vice versa). Though "SSCFailoverFailure" was not significant in the regression equation, we took that into account in the Monte Carlo simulation, just to confirm with the sensitivity analysis; thus, we have four terms in the assumption (Table 21.5).

The Monte Carlo Simulation is run to get the probability distribution for Call Processing Availability (Fig. 21.9) along with the Sensitivity Analysis (Fig. 21.10). The associated capability metrics (Cpk) are also reviewed to establish the design margin.

21.6 SUMMARY

This chapter provided a case study of FTA for a telecommunications project. We noted a variety of factors could cause a system outage—including events outside the system itself, such as environmental or security issues. We examined one important aspect of system failure—call processing failure and identified issues or faults that could result in this failure.

Fractional 5 factorial DOE

Factors :	5	Base Design :	5, 16
Runs :	17	Replicates :	1
Blocks :	none	Center pts (total) :	1

Display order : Run order
Display Units : Uncoded

Factors and Their Uncoded Levels

Factor	Name	Low	High
A	Cagefailover	0.0000007	0.0000017
B	PAMFailoverFailure	0.0000005	0.0000011
C	CarrierFailoverFailure	0.0000006	0.0000013
D	SSCFailoverFailure	0.0000058	0.0000134
E	SAMFailoverFailure	0.0000005	0.0000011

Responses and Models

Responses : CallProcessingServiceFailure
Terms : A B C D

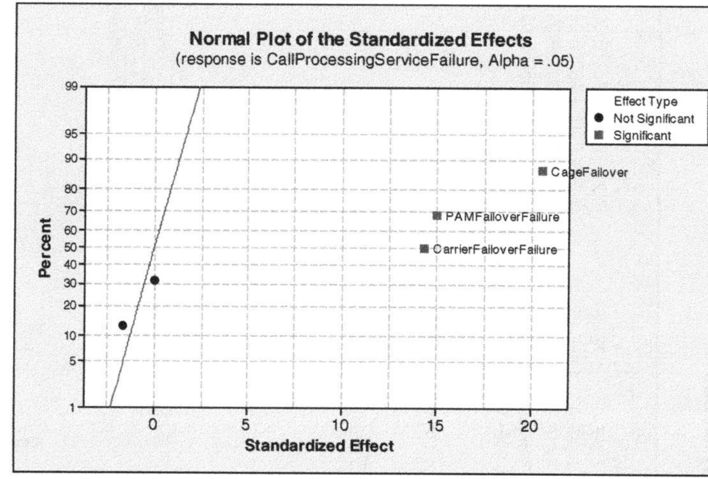

R–Sq = 98.73% R–Sq(pred) = 96.82% R–Sq(adj) = 98.16%

Estimated Coefficients for CallProcessingServiceFailure using data in uncoded units

Term	Coef
Constant	5.52569E–06
CageFailover	1.00415
PAMFailoverFailure	1.13153
CarrierFailoverFailure	0.895048
SSCFailoverFailure	–0.0100666
SAMFailoverFailure	–0.0000097

Regression Equation

Call_Processing_Failure =
5.53E-06
+ 1.132 * PAM_Failover_Failure
+ 0.895 * Carrier_Failover_Failure
+ 1.004 * Cage_Failover_Failure

Figure 21.8 Regression Model from DOE Results

TABLE 21.5 Setting Up the Model in Crystal Ball

A16

	A	B	C	D	E	F	G	H
1			-40%	+40%		-40%	+40%	
2	CageFailoverFailure	1.2155E-06	7.29E-07	1.7E-06		7.2929E-07	1.7017E-06	
3	PAMFailoverFailure	7.7643E-07	4.66E-07	1.09E-06		4.6586E-07	1.0870E-06	
4	CarrierFailoverFailure	9.3420E-07	5.61E-07	1.31E-06		5.6052E-07	1.3079E-06	
5	SSCFailoverFailure	9.7046E-06	5.82E-06	1.36E-05		5.8227E-06	1.3586E-05	
6	Constant	0.00E+00	0.00E+00	0.00E+00				
7								
8								
9	CallProcessingServiceFailure	1.03E-05	6.17E-06	1.44E-05				
10								
11								
12	Five9's	4.98804996	5.209899	4.841922				
13	Lower Limit	5						

Assumptions

Forecast

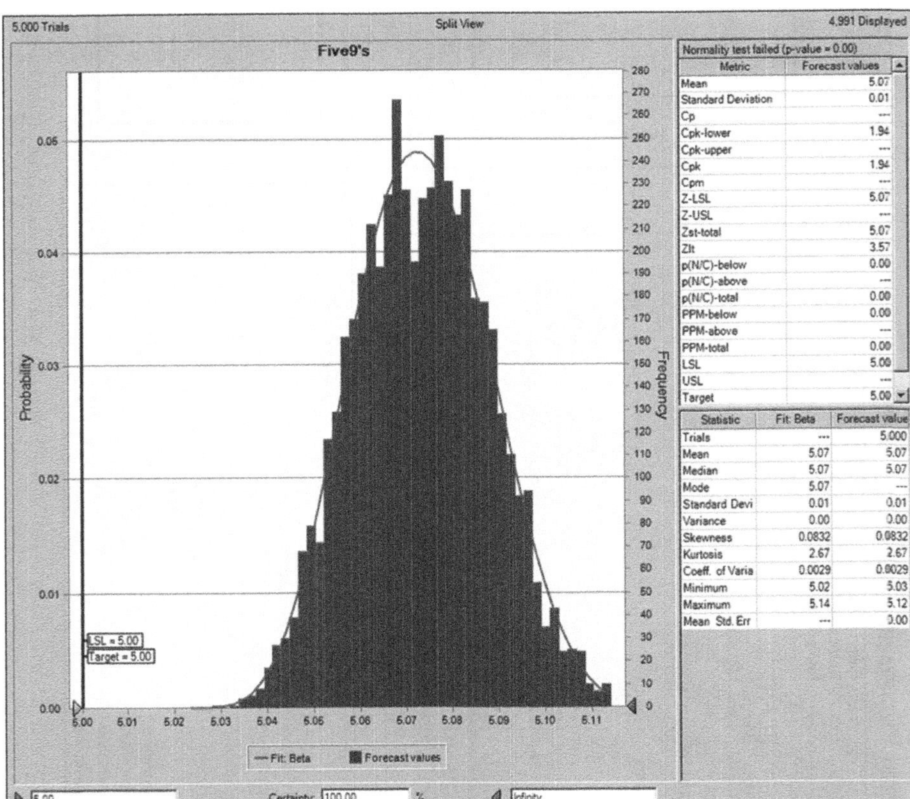

Figure 21.9 Monte Carlo Simulation Results

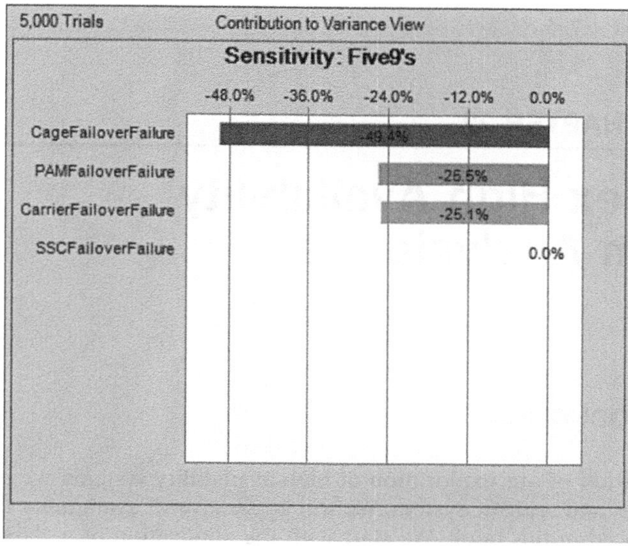

Figure 21.10 Sensitivity Analysis

A good approach is to initially capture the fault tree in a modeling tool, such as Relex, and then transfer these data and relationships to a spreadsheet for easier manipulation.

We identified the high risk nodes and used DOE and Monte Carlo Analysis to determine the impact of these nodes on the system. This simulation results show how likely it is the system would achieve its Five 9s availability requirement. The sensitivity analysis showed which failures have the largest impact on the system. Armed with this information, the design team can focus on the components that have the biggest impact on availability. By addressing these items, we greatly improve the probability of meeting our availability requirements.

CHAPTER 22

Complex High Availability System Analysis

22.1 INTRODUCTION

As a culmination of our exploration of high availability systems, we now consider a more complex and realistic system. We will apply several availability techniques that were introduced in this book. We start with the application of a classical technique, Markov analysis, to a real-life system, and then apply Design for Six Sigma (DFSS) techniques, such as fault tree analysis and Monte Carlo simulations. The objective is to illustrate how these techniques can be synergized to obtain a better understanding of the predictions, as well as to show how they are also interrelated. We compare the outputs of the dynamic Markov modeling along with static fault tree analysis. Since Markov models quickly become complicated with the explosion of states, we illustrate how such a complex model can be simplified using certain underlying assumptions.

We consider a 2N redundant system with both hardware and software components. Only one card is active at any time and the other card is a hot standby. The standby takes over as the active card either due to a failover or due to an operator command. We model the states of the two cards from both hardware and software perspectives. We first look at the hardware state machine followed by the software state machine—each considered independently. Then we combine the two state machines into a single Markov model and discuss some of the findings.

22.2 MARKOV ANALYSIS OF THE HARDWARE COMPONENTS

Figure 22.1 illustrates the state transitions associated with the two card system.

The values of Mean Time between Failures (MTBF), Mean Time to Repair (MTTR), and other model parameters can be obtained through different estimation techniques described in the previous chapters (Table 22.1).

Let us trace through a few state transitions.

Designing High Availability Systems: Design for Six Sigma and Classical Reliability Techniques with Practical Real-Life Examples, First Edition. Zachary Taylor and Subramanyam Ranganathan.

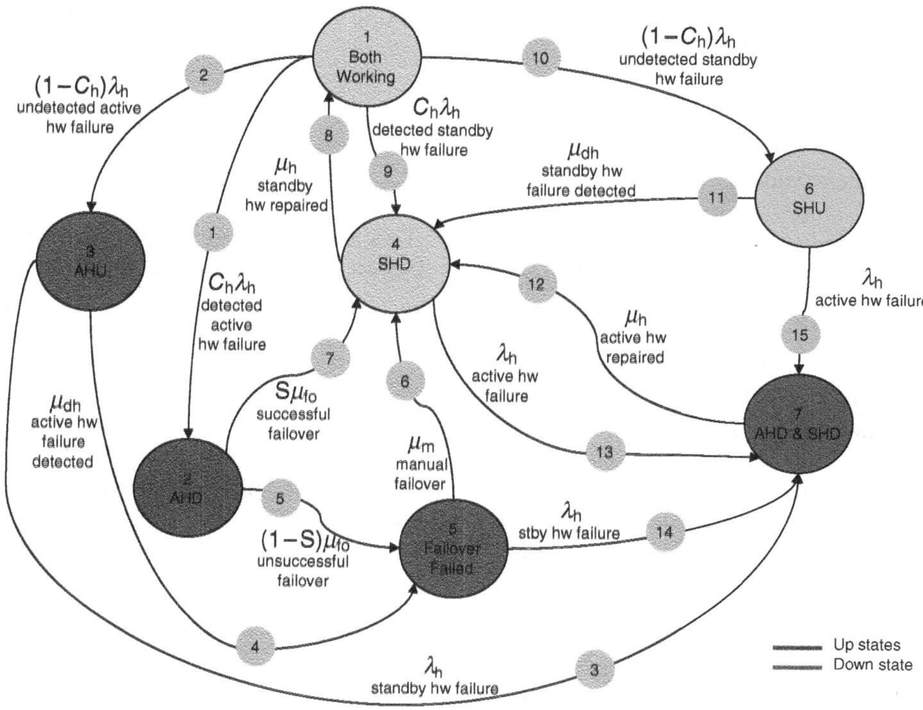

Figure 22.1 Controller System: Hardware State Diagram

TABLE 22.1 Hardware Markov Model Parameters

Symbol	Description	Calculation
C_h	% of hardware failures that are detected	Number of detected failures/total number of failures
S	% of successful failovers	Number of successful failovers/total number of failovers
λ_h	Card hardware failure rate	1/MTBF
μ_h	Card hardware repair rate	1/MTTR
μ_s	Card reset rate	1/Card reset time
μ_{dh}	Manual card failure detection rate	1/Manual card failure detection time
μ_m	Manual card failover rate	1/Manual card failover time. (This includes the time required to manually detect a failure and initiate a successful failover. Thus, it could be of the order of a few minutes to a few hours.)
μ_{fo}	Card Failover rate	1/Card failover time

Normal operation consists of one active and one standby card at any given time. Let us assume that:

Card 1 = Active
Card 2 = Standby
State = State 1.

While in State 1, let us assume that Card 1 fails and the failure is detected by the system. In this case, the system transitions to State 2, where Card 2 is still in the standby state.

Card 1 = Failed Hardware
Card 2 = Standby
State = State 2.

If a failure occurs and failover is successful, Card 2 becomes the new active card. This transitions the system to State 4 from State 2.

Card 1 = Failed Hardware
Card 2 = Active
State = State 4.

From State 1, however, if an undetected failure occurs, the System transitions to State 3, in which Card 1 has failed but the system is not aware of this failure and Card 2 continues to be in standby mode.

Card 1 = Failed Hardware
Card 2 = Standby
State = State 3.

The System remains in this state until the failure is eventually detected manually (transitions to State 5).

Card 1 = Failed Hardware, Failure Detected Manually
Card 2 = Standby
State = State 5.

From State 5, the system transitions to State 4 if the manual failover is successful.

Card 1 = Failed Hardware, Failure Detected Manually
Card 2 = Active
State = State 4.

From State 5, the system transitions to State 7 if the manually initiated failover is *not* successful.

Card 1 = Failed Hardware, Failure Detected Manually
Card 2 = Failed Hardware
State = State 7.

While in State 3, if the standby card also fails, the system transitions to State 7 where both cards have failed.

Card 1 = Failed Hardware, Undetected
Card 2 = Failed Hardware
State = State 7.

From State 7, after one of the cards is repaired and becomes active, the system transitions to State 4. Note that State 4 implies that one of the cards is still failed. In this case, let us assume that Card 1 was repaired and became active.

Card 1 = Active
Card 2 = Failed Hardware
State = State 4.

From State 1, if the standby card, Card 2, has an undetected failure, the system transitions to State 6. The system is operational because the active card, Card 1, is still functioning as an active card.

Card 1 = Active
Card 2 = Failed Undetected
State = State 6.

From State 1, if the active card fails and the failure is detected, and the failover is successful, the standby card, Card 2 becomes active (State 4).

Card 1 = Failed Hardware
Card 2 = Active
State = State 4.

Note the other state transitions as illustrated in Figure 22.1. Though theoretically there can be $2^7 = 128$ transitions, not all of the transitions are applicable, and only those that contribute significantly to failure behavior are considered.

Each *transition* is modeled as a "rate" that is, the rate of flow of state transition probability from one state to the other. For example, transition from State 1 to State 3 is represented as $(1 - C_h)\lambda_h$. This is because $(1 - C_h)$ corresponds to the percentage of hardware failures that are not detected, and λ_h is the hardware failure rate. Similarly, the transition from State 1 to State 4 is represented as $C_h\lambda_h$.

In the steady-state analysis, the probability of transitioning from one state to another is a constant. In addition, the sum of the transitions *from* one state is equal to the sum of the transitions *to* that same state.

So for State 1, the transition equation is:

$$P(\text{State } 1)[(1-C_h)\lambda_h + C_h\lambda_h + C_h\lambda_h + (1-C_h)\lambda_h] = P(\text{State } 4)(\mu_h).$$

Similarly, we can come up with steady-state equations for all the other states. Since we have seven states, the system must be in one of those 7 states at any given time. The sum of the probabilities of all the states must be equal to 1.

$$P(\text{State } 1) + P(\text{State } 2) + \cdots + P(\text{State } 7) = 1.$$

As described in Chapter 9, we put these equations in canonical form and solve for the steady-state probabilities for each state.

Note that this model represents only the hardware availability. In the next section, we will consider software state transitions and the corresponding software availability.

EXAMPLE 22.1

Consider the 2N-redundant hardware controller, modeled as shown in Figure 22.1 with the following reliability parameters:

$MTBF_H =$ 500,000 hours
$MTTR_H =$ 8 hours
$MTTR_{DH} =$ 12 hours
$MTTR_{FO} =$ 10 seconds
$C_h = 0.9$.

Determine the Five 9s availability of the system.

Converting the above times to rates:

$\lambda_h = 0.000002$ failures per hour
$\mu_h = 0.125$ repairs per hour
$\mu_{dh} = 0.0833$ number of detected hardware failures/hour
$\mu_{fo} = 360$ number of failovers/hour.

We create the model using a software program and enter the transition rate parameters. Executing the Matlab program Markov_SS_Model.m, a snapshot of which is given in Figure 22.2. The generated steady-state probabilities and sojourn time for each state along with overall system availability is:

```
Controller System  -  HW Model
State Probability          Minutes/Year
1     9.999630e-001        525940.527856
2     4.999815e-009        0.002630
3     2.399854e-006        1.262227
4     3.199882e-005        16.830097
5     2.179867e-007        0.114652
```

```
6    2.399854e-006       1.262227
7    5.922641e-010       0.000312
Availability: 9.999974e-001
FIVE 9s:5.581130
```

The steady state availability of this system is 5.58 9s.

```
File Edit Text Go Cell Tools Debug Desktop Window Help
S=0.99;
Ch = 0.9;
MTBFh = 500000; %Mean Time Between Board Failures
MTTRh = 8; %Mean Time to Repair Board
MTTDh = 12; %Mean time to detect undetected hardware failures
MTFO = 10/3600; %Mean time for succcessful failover
Muh = 1/MTTRh;
Lamh=1/MTBFh;
Mudh = 1/MTTDh; %Rate of detecting undetected hardware failures
Mufo = 1/MTFO; %failover rate (failovers/hour)
Mum = 1;%Manual failover

Rate(1)= Ch*Lamh;
Rate(2)= (1-Ch)*Lamh;
Rate(3)= Lamh;
Rate(4)= Mudh;
Rate(5)= (1-S)*Mufo;
Rate(6)= Mum;
Rate(7)= S*Mufo;
Rate(8)= Muh;
Rate(9)= Ch*Lamh;
Rate(10)= (1-Ch)*Lamh;
Rate(11)= Mudh;
Rate(12)= Muh;
Rate(13)= Lamh;
Rate(14)= Lamh;
Rate(15)= Lamh;

FailStates(2)=1;
FailStates(3)=1;
FailStates(5)=1;
FailStates(7)=1;
NextStates(1,1)=2;
NextStates(1,2)=3;
NextStates(3,3)=7;
NextStates(3,4)=5;
NextStates(2,5)=5;
NextStates(2,7)=4;
NextStates(1,9)=4;
NextStates(1,10)=6;
NextStates(4,8)=1;
NextStates(6,11)=4;
NextStates(6,15)=7;
NextStates(7,12)=4;
NextStates(4,3)=7;
NextStates(5,6)=4;
NextStates(5,14)=7;
CompSensitivity.m | calcStateTimes.m | calcMTBTrans.m | secondode.m* | SSStateMatrix.m | SteadyStateProb.m
```

Figure 22.2 Matlab Simulation for HW Markov State Machine

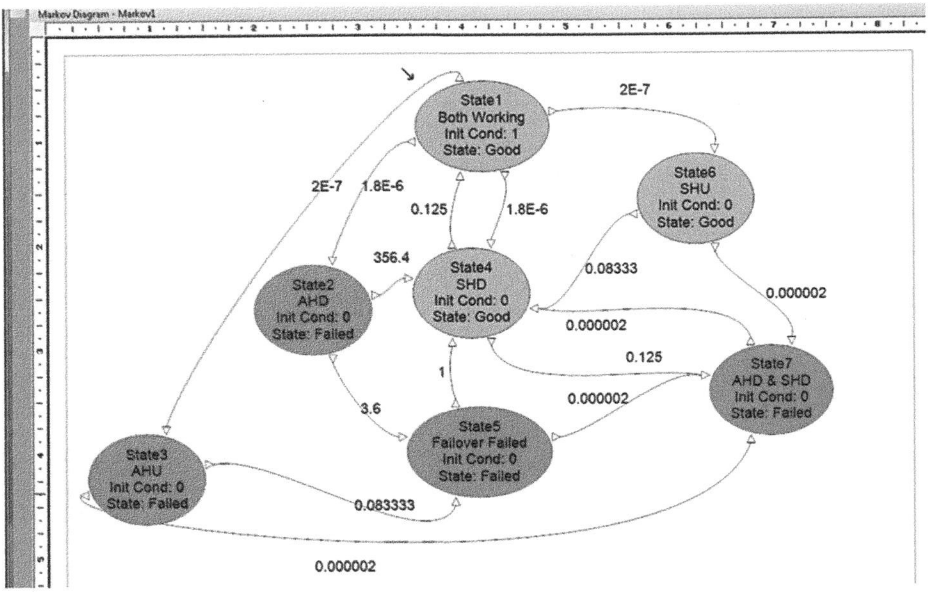

Figure 22.3 Controller System: Hardware State Diagram (Relex)

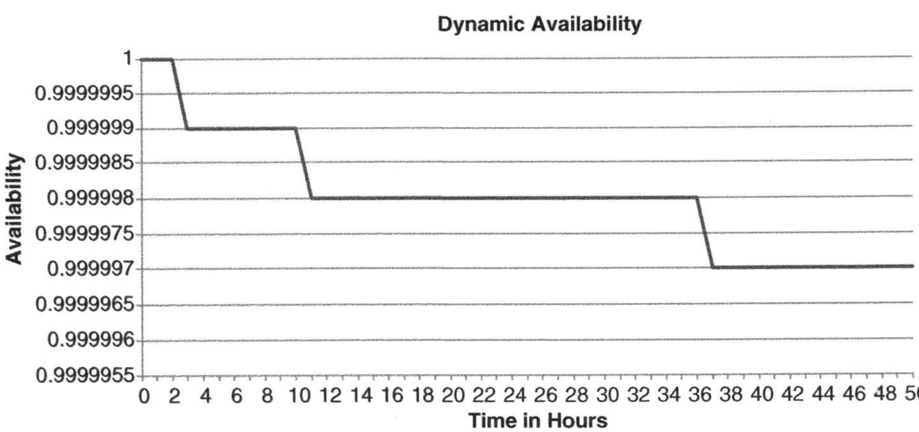

Figure 22.4 Controller System: Dynamic Availability for Hardware

Another method is to create this simulation using the Relex Markov tool. The state diagram is entered graphically into Relex. Using the hardware state diagram shown in Figure 22.1, we create the state machine in Relex (Fig. 22.3).

As we see from the simulation output (Fig. 22.4), the steady-state hardware system availability is 0.999997, corresponding to 5.52 nines.

22.3 BUILDING A FAULT TREE FROM THE HARDWARE MARKOV MODEL

Using the Markov model from Figure 22.1, we can derive the equivalent fault tree wherein the failure states represent the logic gates (Fig. 22.5).

The fault tree information can be captured in a spreadsheet as shown in Table 22.2.

Using this spreadsheet and Crystal Ball, we can run the Monte Carlo simulation to calculate the availability of the hardware-only system. The resultant availability forecast output is shown in Figure 22.6. In addition, the sensitivity analysis is shown in Figure 22.7.

Matlab provides another way to show sensitivity analysis. With Matlab, we can plot the parameter changes and the resultant outage impact as shown in Figure 22.8. A notable difference from Crystal Ball Sensitivity output is that herein, we can see the impact of varying each individual factor on the output system availability, in this case represented as Outage Minutes. The top left chart, illustrates how by changing S, the probability of Successful Failover, impacts the Outage Minutes. The red line corresponds to the outage minutes for a value of $S = 95\%$ (the baseline value). For example, $S = 70\%$ corresponds to an outage of approximately 1.75 minutes, as shown in the top left graph of the figure.

22.4 MARKOV ANALYSIS OF THE SOFTWARE COMPONENTS

Now let us model the software components of the system. Figure 22.9 depicts the software state transitions associated with the two card system.

The software Markov model parameters are given in Table 22.3.

The states are traced using the same method described in the hardware Markov model.

So for State 1, for example, the sum of all transitions out of State 1 is equal to the sum of all transitions into State 1. This transition equation is:

$$P(\text{State 1})[(1-C_s)\lambda_{sw} + C_s\lambda_{sw} + C_s\lambda_{sw} + (1-C_s)\lambda_{sw}] = P(\text{State 4})(\mu_{sw}).$$

Similarly, we can come up with steady-state equations for all of the other states as well. Since we have seven states, the system should be in one of those seven states at any given time, and the sum of probabilities of being in all the states must be equal to 1:

$$P(\text{State 1}) + P(\text{State 2}) + \cdots + P(\text{State 7}) = 1.$$

As described in Chapters 9–11, we put these equations in canonical form and solve for the steady-state probabilities for each state. The solution to these equations can be obtained using a tool, such as Matlab or Relex.

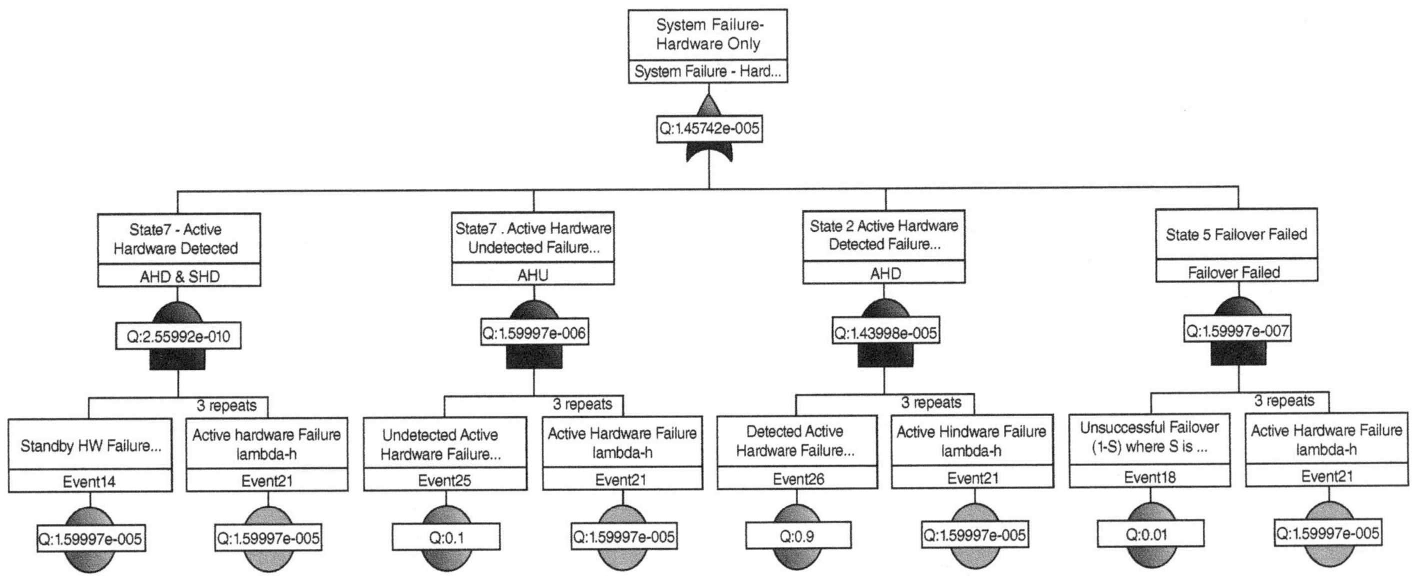

Figure 22.5 Fault Tree Derived from the HW Markov Model

TABLE 22.2 Setting Up the Monte Carlo Simulation for the Fault Tree

Monte Carlo Simulation of Fault Tree

Device	MTBF (hours) or Probability	MTTR (hours)	Probability	Markov Equivalent	Distribution	Summation
EVENTS						
Active HW Failure	500000	8			Normal	1.59997E-05
Standby HW Failure	500000	8			Normal	1.59997E-05
Undetected HW Failure	0.1		0.1	$(1\text{-}C_h) = 1\text{-}0.9 = .1$	Normal	0.1
Unsuccessful Failover	0.05		0.05	$(1\text{-}S) = 0.05$	Normal	0.05
Probability of Detecting a Hardware Failure	0.9		0.9	$C_h = 0.9$	Normal	0.9
Card Failover Rate	2.78E-05		2.78E-05	μ_{fo}	Normal	2.78E-05
Manual Card Failover Detection Rate	0.0833333		0.0833333	μ_{dh}	Normal	0.0833333
GATES						
State 7—Active Hardware Detected Failure & Standby Hardware Detected Failure				$\lambda_h * \lambda_h * \lambda_h * \lambda_h$	N/A	6.55318E-20
State-3 Active Hardware Undetected Failure				$(1\text{-}C_h) * \lambda_h$	N/A	1.59997E-06
State 2 Active Hardware Detected Failure				$(1\text{-}S) * C_h * \lambda_h$	N/A	1.43998E-05
State 5 Failover Failed				$(1\text{-}S) * \mu_{fo} * \mu_{dh}$	N/A	1.16E-07
OUTPUT						
System Failure (Including Failover Time)—Hardware Only					N/A	1.61E-05
Five Nines—System Failure (Including Failover Time)—Hardware Only					N/A	4.792756628

Markov Variables	Value
C_h: % of Hardware failures that are detected	0.9
S: % of Successful Failovers	0.95
λ_h: Card Hardware Failure Rate	1.59E-05 MTTR/(MTB F+MTTR)
μ_{fo}: Card Failover Rate	0.0000277
μ_{dh}: Manual Card Failure Detection Rate	0.0833333

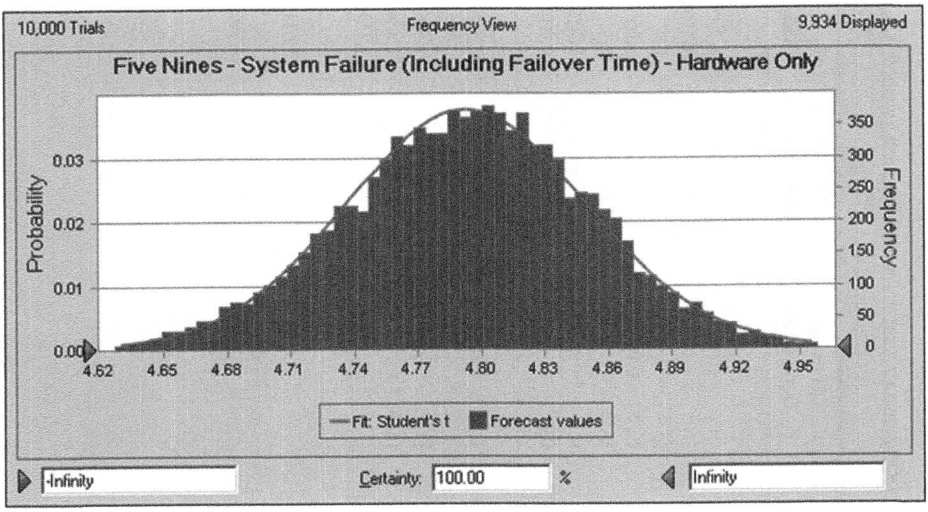

Figure 22.6 Simulation Output from Crystal Ball

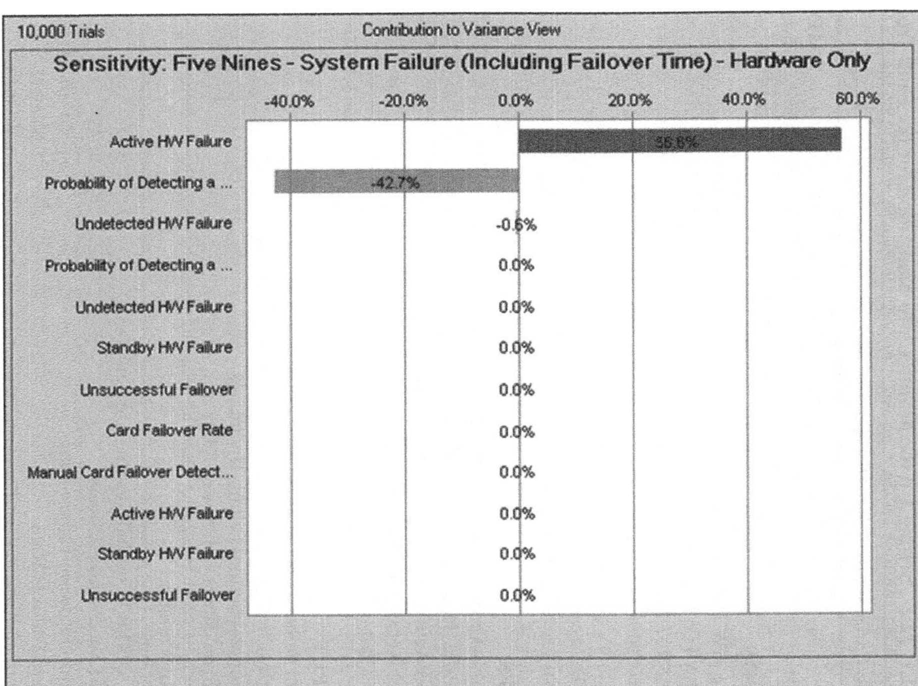

Figure 22.7 Sensitivity Analysis from Crystal Ball

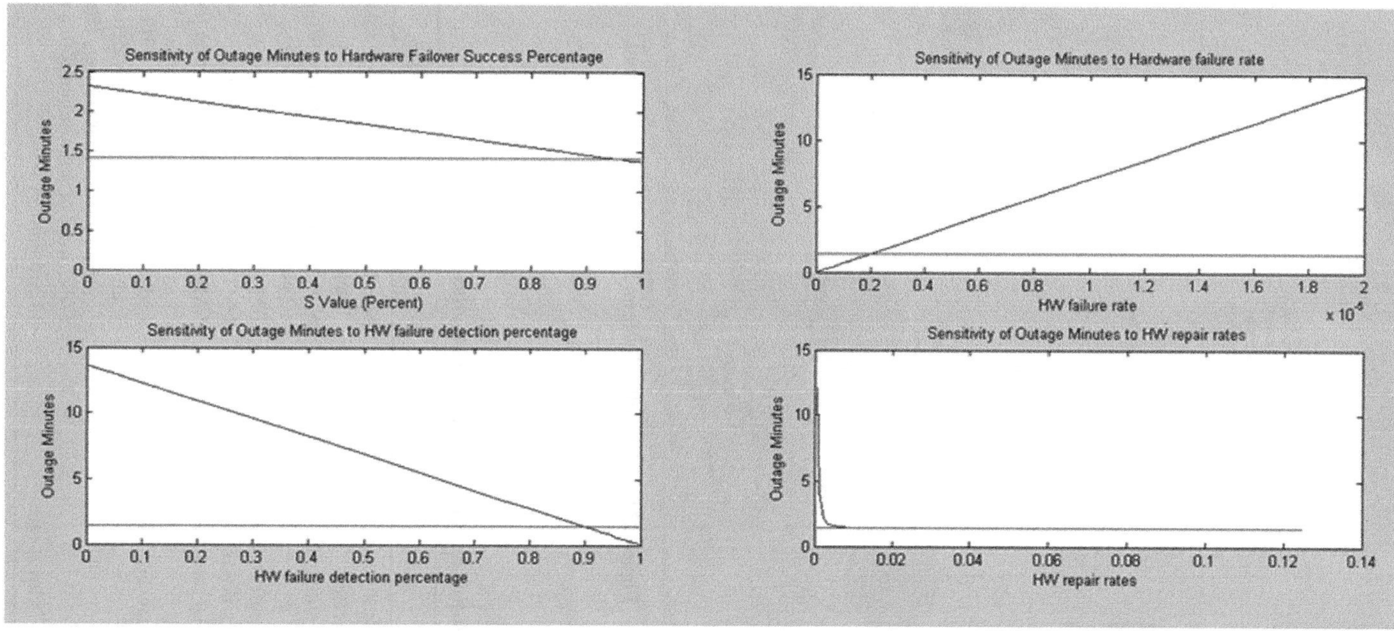

Figure 22.8 Sensitivity Analysis from Matlab

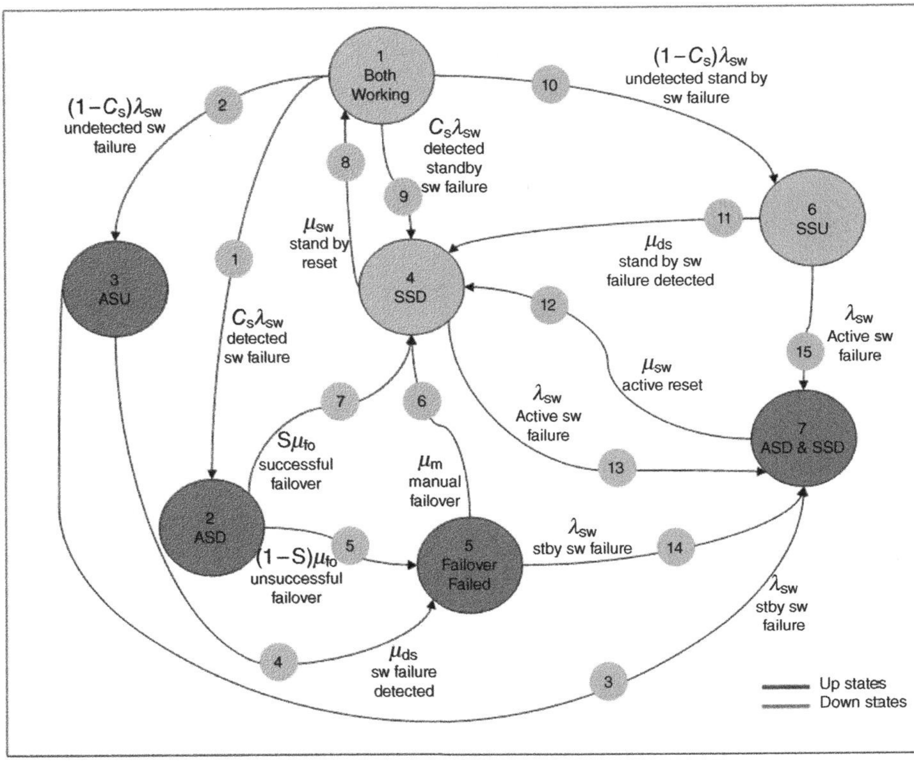

Figure 22.9 Controller System: Software State Diagram

TABLE 22.3 Software Markov Model Parameters

Symbol	Parameter	Description
C_s	% of software failures that are detected	Number of detected failures/total number of failures
S	% of successful failovers	Number of successful failovers/total number of failovers
λ_{sw}	Card software failure rate	1/MTBF
μ_{sw}	Card software repair rate	1/MTTR (=card reset time)
μ_{ds}	Manual card failure detection rate	1/Manual card failure detection time
μ_m	Manual card failover rate	1/Manual card failover time. (This includes the time required to manually detect a failure and initiate a successful failover. Thus it could be of the order of a few minutes to a few hours.)
μ_{fo}	Card failover rate	1/Card failover time

EXAMPLE 22.2

Determine the software availability based on the model in Figure 22.5 with the following reliability parameters:

$MTBF_{sw}$ = 50,000 hours
$MTTR_{sw}$ = 2 minutes
$MTTD_{sw}$ = 12 hours (Mean Time to Detect an undetected software failure)
MTFO = 10 seconds (Mean Time for one Successful Failover)
$C_s = 0.91$
$S = 0.99$.

Converting the above times to rates:

$\lambda_{sw} = 0.00002$ software failures per hour
$\mu_{sw} = 30$ software repairs per hour
$\mu_{dsw} = 0.083333$ detected software failures/hour
$\mu_{fo} = 360$ failovers/hour
$\mu_m = 1$ manual failover/hour.

Using the Matlab program Markov_SS_Model.m, a snapshot of which is given in Figure 22.10, the following output is generated:

```
Controller System - SW Model
State Probability                  Minutes/Year
1      9.999534e-001               525935.515029
2     5.055320e-008                0.026589
3     2.159381e-005                11.357481
4     1.333271e-006                0.701247
5     1.981436e-006                1.042156
6     2.159381e-005                11.357481
7     3.100155e-011                0.000016
Availability: 9.999764e-001
FIVE 9s:4.626613
```

22.5 MARKOV ANALYSIS OF THE COMBINED HARDWARE AND SOFTWARE COMPONENTS

If we combine both the hardware and software Markov models, the number of states will grow significantly (Fig. 22.11). Although the diagram looks intimidating, we can tame it relatively easily with the aid of reliability software tools.

```
S=0.99;
Cs = 0.91;
MTBFsw = 50000; %Mean Time Between Board Failures
MTTRsw = 2/60; %Mean Time to reset software
MTTDsw = 12; %Mean time to detect undetected software failures
MTFO = 10/3600; %Mean time for succcessful failover
Musw = 1/MTTRsw;
Lamsw=1/MTBFsw;
Mudsw = 1/MTTDsw; %Rate of detecting undetected software failures
Mufo = 1/MTFO; %failover rate (failovers/hour)
Mum = 1;%Manual failover

Rate(1)= Cs*Lamsw;
Rate(2)= (1-Cs)*Lamsw;
Rate(3)= Lamsw;
Rate(4)= Mudsw;
Rate(5)= (1-S)*Mufo;
Rate(6)= Mum;
Rate(7)= S*Mufo;
Rate(8)= Musw;
Rate(9)= Cs*Lamsw;
Rate(10)= (1-Cs)*Lamsw;
Rate(11)= Mudsw;
Rate(12)= Musw;
Rate(13)= Lamsw;
Rate(14)= Lamsw;
Rate(15)= Lamsw;

FailStates(2)=1;
FailStates(3)=1;
FailStates(5)=1;
FailStates(7)=1;
NextStates(1,1)=2;
NextStates(1,2)=3;
NextStates(3,3)=7;
NextStates(3,4)=5;
NextStates(2,5)=5;
NextStates(2,7)=4;
NextStates(1,9)=4;
NextStates(1,10)=6;
NextStates(4,8)=1;
NextStates(6,11)=4;
NextStates(6,15)=7;
NextStates(7,12)=4;
NextStates(4,3)=7;
NextStates(5,6)=4;
NextStates(5,14)=7;
```

Figure 22.10 Matlab Simulation for SW Markov State Machine

EXAMPLE 22.3

Let us consider a numeric example for the combined state machine. In addition to the parameters defined in the HW-only and SW-only state machines, we introduce a parameter μ_{md} which represents the rate of manual detection of a problem. All other parameter values are the same as in the previous two examples.

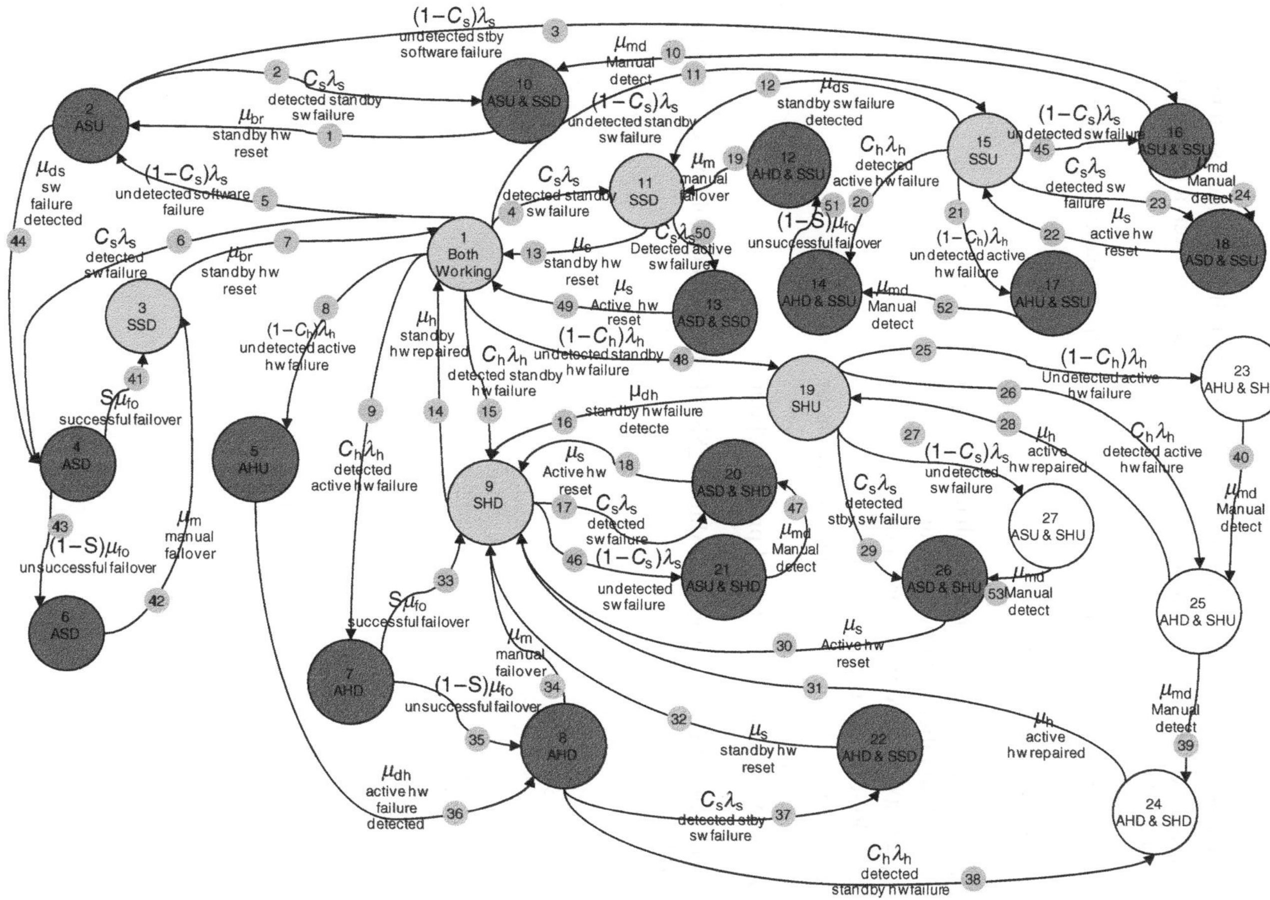

Figure 22.11 Controller System: Combined Hardware and Software State Diagram

$$\mathrm{MTTD_{md}} = 1.2 \text{ hours.}$$

1. Determine the steady-state Five 9s availability of the combined system.
2. Using Monte Carlo simulation, plot state transitions over a mission time of 10,000 hours.

Combining this with our previous rates, the complete parameter list is:

$\lambda_h = 0.000002$ failures per hour
$\mu_h = 0.125$ repairs per hour
$\mu_{dh} = 0.0833$ number of detected hardware failures/hour
$\mu_{fo} = 360$ number of failovers/hour
$\lambda_{sw} = 0.00002$ software failures per hour
$\mu_{sw} = 30$ software repairs per hour
$\mu_{dsw} = 0.083333$ detected software failures/hour
$\mu_m = 1$ manual failover/hour
$\mu_{md} = 0.8333$ manually detected problems/hour.

Using the Matlab program Markov_SS_Model.m, the following output is generated:

```
Controller System - Combined HW SW Model
State Probability          Minutes/Year
1       9.999129e-001      525914.168829
2       2.159812e-005      11.359746
3       3.333043e-006      1.753047
4       5.555071e-008      0.029217
5       2.399791e-006      1.262194
6       1.999826e-007      0.105183
7       4.999564e-009      0.002630
8       2.179766e-007      0.114647
9       3.199721e-005      16.829253
10      7.199365e-011      0.000038
11      3.333033e-006      1.753042
12      4.319520e-011      0.000023
13      1.011020e-011      0.000005
14      1.199867e-011      0.000006
15      2.159760e-005      11.359473
16      4.665137e-010      0.000245
17      5.183424e-011      0.000027
18      7.199208e-011      0.000038
19      2.399192e-006      1.261879
20      1.066574e-010      0.000056
21      6.911398e-010      0.000364
22      6.611958e-013      0.000000
23      5.758061e-012      0.000003
24      1.849369e-011      0.000010
```

```
25      2.303224e-011        0.000012
26      7.997307e-012        0.000004
27      5.182255e-011        0.000027
Unavailability: 2.447805e-005
Availability: 9.999755e-001
FIVE 9s:4.611223
```

For the second part of the problem, we make use of a Matlab script (markovMissionTimeSim.m). The result of one execution trial of this tool is shown in Figure 22.12. Notice the system transitions to State 11, four times, and State 4, three times, over the 10,000-hour period. These transitions occur during an active and standby software failure, respectively. This is expected since the software failure rates are 10 times higher than the hardware failure rates.

We also simulate this using the Relex Markov tool. The state machine in Figure 22.11 was input into Relex (Fig. 22.13) and the dynamic availability graphed (Fig. 22.14).

22.6 TECHNIQUES FOR SIMPLIFYING MARKOV ANALYSIS

One of the disadvantages of using Markov state machines is state explosion. Beyond relatively simple models, the number of states to be modeled can grow dramatically. As we have seen from the previous example of the combined hardware and software controller model, the total number of states rose to 27. To make the analysis more

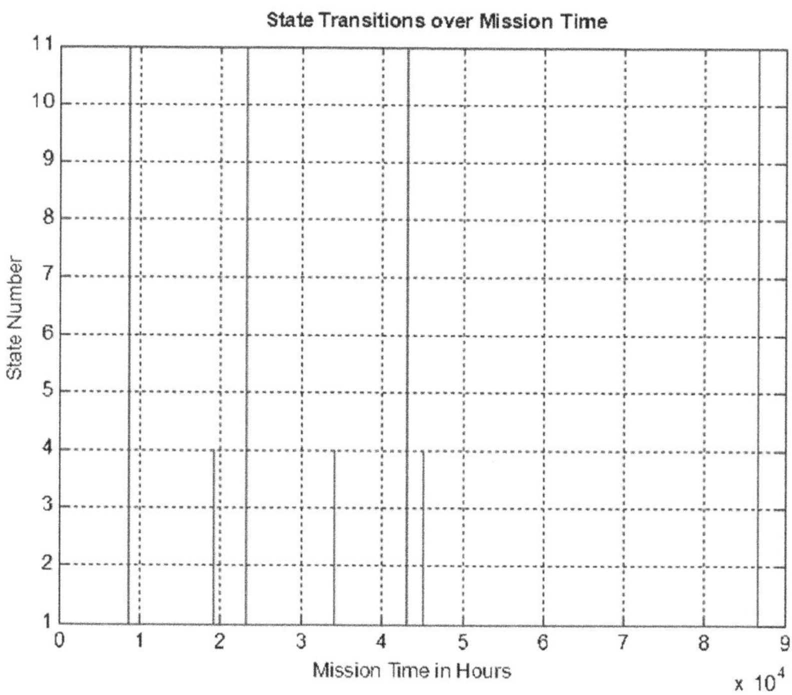

Figure 22.12 Matlab Simulation for Combined Markov State Machine: Transitions over Time

manageable and easier to understand, we use a variety of techniques to simplify the models with little loss in overall accuracy. If any parameter in the Markov Model becomes insignificant in comparison to others, we can remove that parameter and its associated state transitions from the model.

One technique is to combine states that (1) provide little additional information or (2) the transition probabilities between the states are insignificant when compared with other state transitions.

Another technique is to modularize the system by having multiple state machines and combining these state machines through a fault tree model. We will consider both techniques.

Referring to Figure 22.9, we can combine States 2 and 5 if the failover time is insignificant. If the failover time is insignificant or does not affect our availability requirements, we can remove the transition from State 2 to State 4 and only consider the failover failed state when we have an active failure. The simplified model is shown in Figure 22.15 and the equivalent simplified Fault Tree is shown in Figure 22.16.

The Matlab output of the simplified state machine along with the original state machine output is given below. Observe that the availability and unavailability are virtually identical. This simplified analysis may be sufficient for our requirements. The value of this simplification becomes more evident as the state space grows larger.

```
>> Markov_SS_Model (9,0)
Controller System - HW Model
State Probability          Minutes/Year
1      9.999630e-001       525940.527856
2      4.999815e-009       0.002630
3      2.399854e-006       1.262227
4      3.199882e-005       16.830097
5      2.179867e-007       0.114652
6      2.399854e-006       1.262227
7      5.922641e-010       0.000312

Unavailability: 2.623432e-006
Availability: 9.999974e-001
FIVE 9s:5.581130

>> Markov_SS_Model (12,0)
Controller System - Simplified HW Model
State Probability          Minutes/Year
1    9.999814e-001         525950.207737
2    2.179911e-007         0.114655
3    2.399898e-006         1.262250
4    1.599970e-005         8.415203
5    0.000000e+000         0.000000
6    2.943936e-010         0.000155
Unavailability: 2.618183e-006
Availability: 9.999974e-001
FIVE 9s:5.582000
```

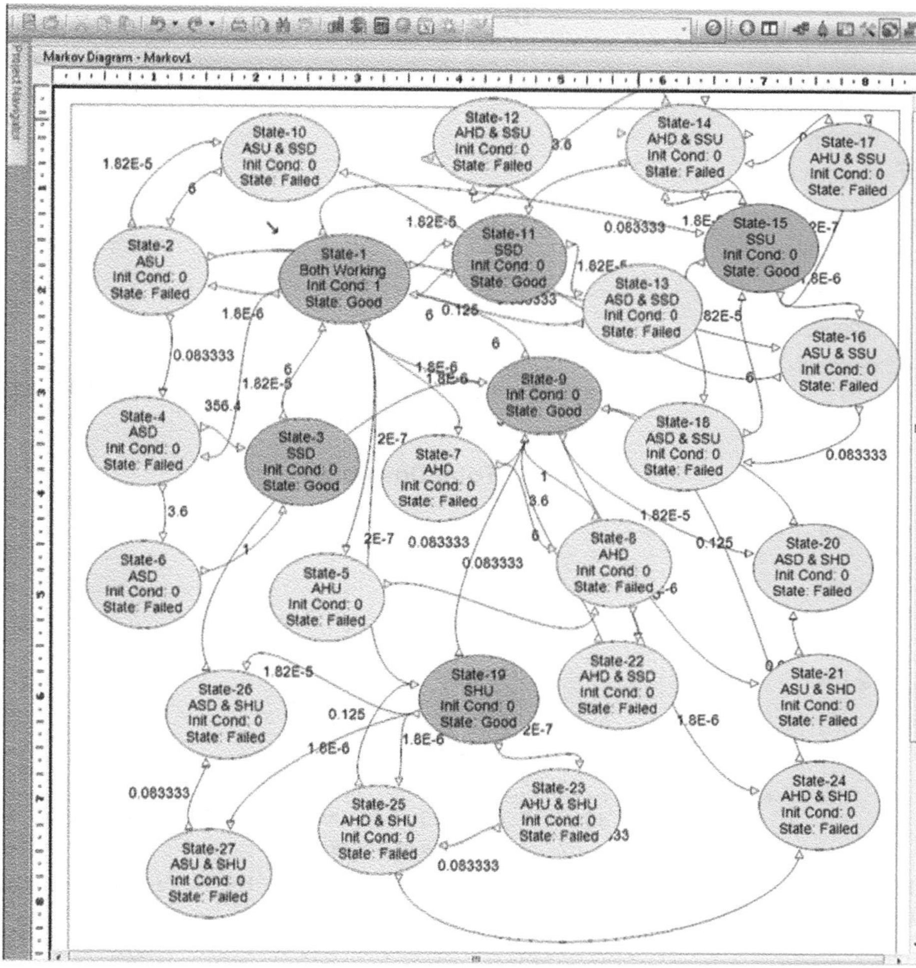

Figure 22.13 Controller System: Hardware and Software State Diagram (Relex)

The second technique is to modularize the state machines and combine them using fault trees. To illustrate this, let us look at a two-state Markov model for hardware and a two-state Markov model for software. The system is good in State 1 and failed in State 2. To combine both the hardware and software Markov models into one Markov model, we create a four-state combined model that has both hardware and software as follows:

State 1: HW and SW Working
State 2: HW Failed, SW Working

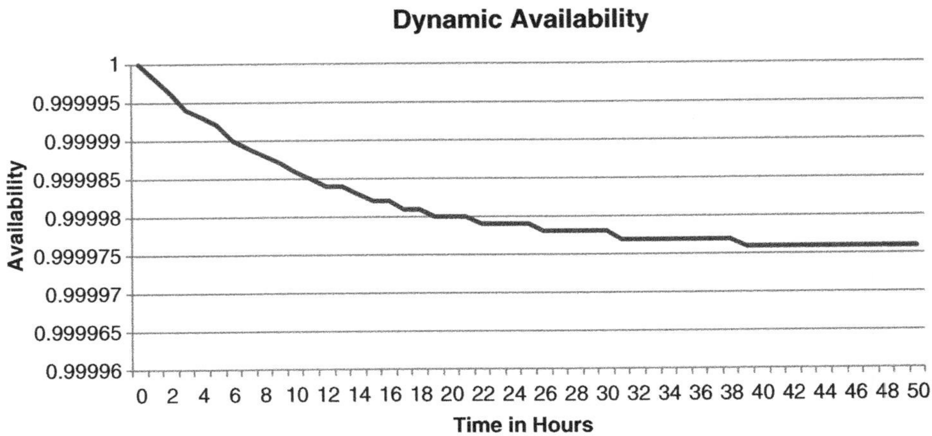

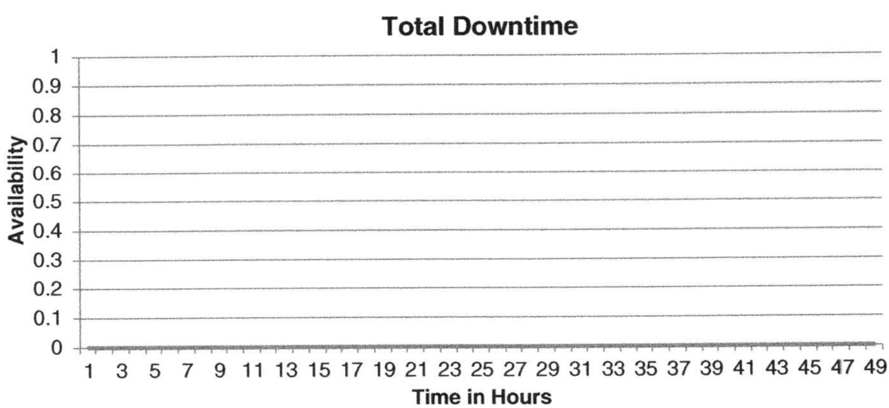

Figure 22.14 Controller System: Dynamic Availability for Hardware and Software Combined

State 3: SW Failed, HW Working

State 4: Both HW and SW have failed.

Instead of creating a four state model, we could combine the original two state models using fault tree combinational logic. All the good states are AND-ed, and all the bad states are OR-ed. The illustration of the combined model is show in Figure 22.17.

These models are equivalent if we are only concerned with the probability of being in a good state or a failed state. In other words, if we are not interested in whether the hardware failed or the software failed, but we are interested in the overall system availability, we can then combine them using the fault tree logic.

A comparison of the outputs of the two types of models is shown below:

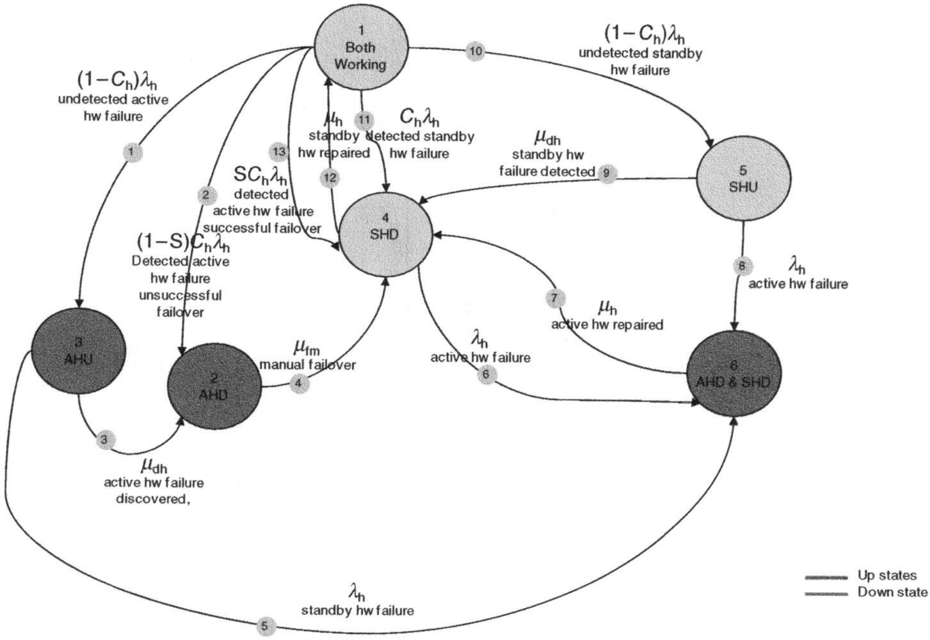

Figure 22.15 Controller System: Hardware State Diagram—Simplified

```
Markov_SS_Combined_Model ('FTA', 15,16,1)
State  Probability          Minutes/Year
1      9.930388e-001        522298.686210
2      6.961202e-003        3661.313790
Unavailability: 6.961202e-003
Availability: 9.930388e-001
FIVE 9s:2.157316
```

```
>> Markov_SS_Model (17,0)
Simple 4-state Repairable Combined HW & SW Model
State  Probability          Minutes/Year
1      9.930388e-001        522298.686210
2      1.986078e-003        1044.597372
3      4.965194e-003        2611.493431
4      9.930388e-006        5.222987
Unavailability: 6.961202e-003
Availability: 9.930388e-001
FIVE 9s:2.157316
```

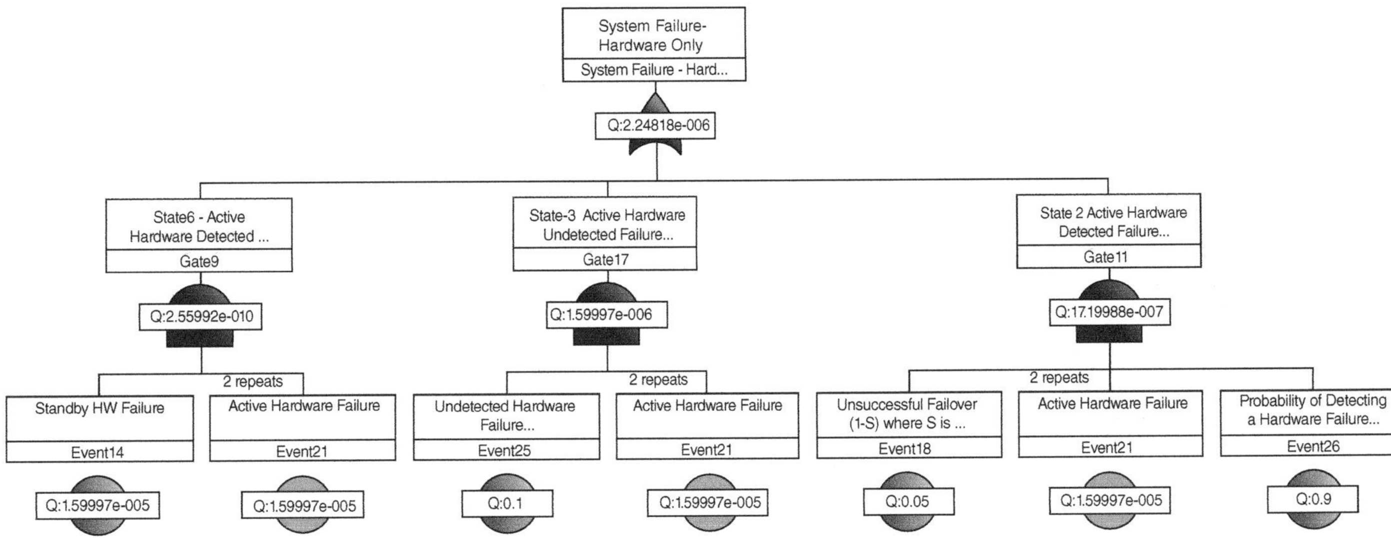

Figure 22.16 Equivalent Fault Tree Analysis for the Simplified HW System

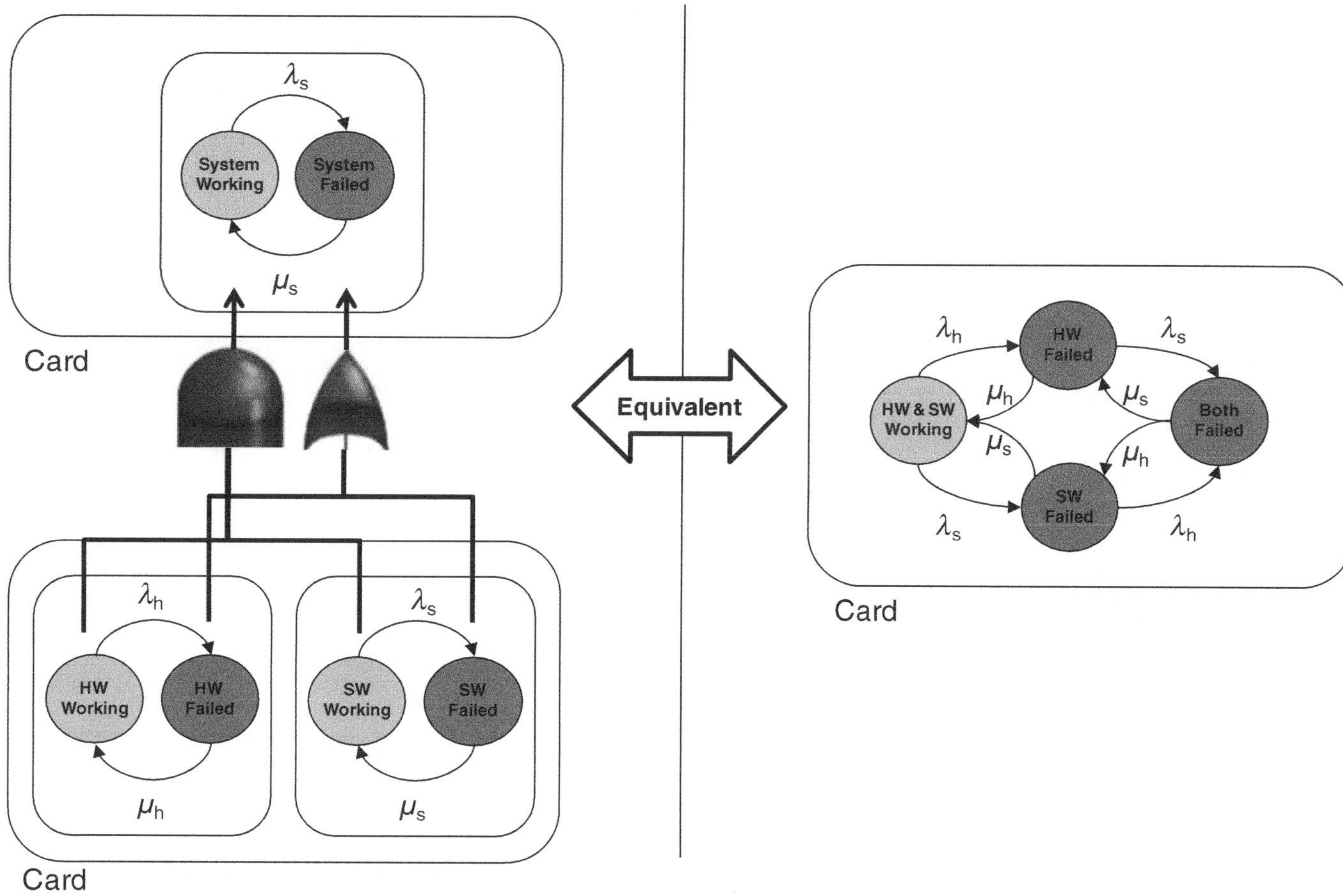

Figure 22.17 Markov Simplification Using FTA

TABLE 22.4 Setting Up the Monte Carlo Simulation for the Fault Tree for the Simplified HW System

Monte Carlo Simulation of Fault Tree

Device	MTBF (hours) or Probability	MTTR (hours)	Probability	Markov Equivalent	Distribution	Summation
EVENTS						
Active HW Failure	500000	8			Normal	1.59997E-05
Standby HW Failure	500000	8			Normal	1.59997E-05
Undetected HW Failure	0.1		0.1	$(1-C_h) = 1-0.9 = .1$	Normal	0.1
Unsuccessful Failover	0.05		0.05	$(1-S) = 0.05$	Normal	0.05
Probability of Detecting a Hardware Failure	0.9		0.9	$C_h = 0.9$	Normal	0.9
GATES						
State 6—Active Hardware Detected Failure & Standby Hardware Detected Failure				$\lambda_h * \lambda_h$	N/A	2.55992E-10
State-3 Active Hardware Undetected Failure				$(1-C_h) * \lambda_h$	N/A	1.59997E-06
State 2 Active Hardware Detected Failure				$(1-S) * C_h * \lambda_h$	N/A	7.19988E-07
OUTPUT						
System Failure—Hardware Only					N/A	2.32022E-06
Five Nines—System Failure—Hardware Only					N/A	5.634471045

Markov Variables	Value
C_h: % of Hardware failures that are detected	0.9
S: % of Successful Failovers	0.95
λ_h: Card Hardware Failure Rate	1.59E-05 MTTR/(MTB F+MTTR)

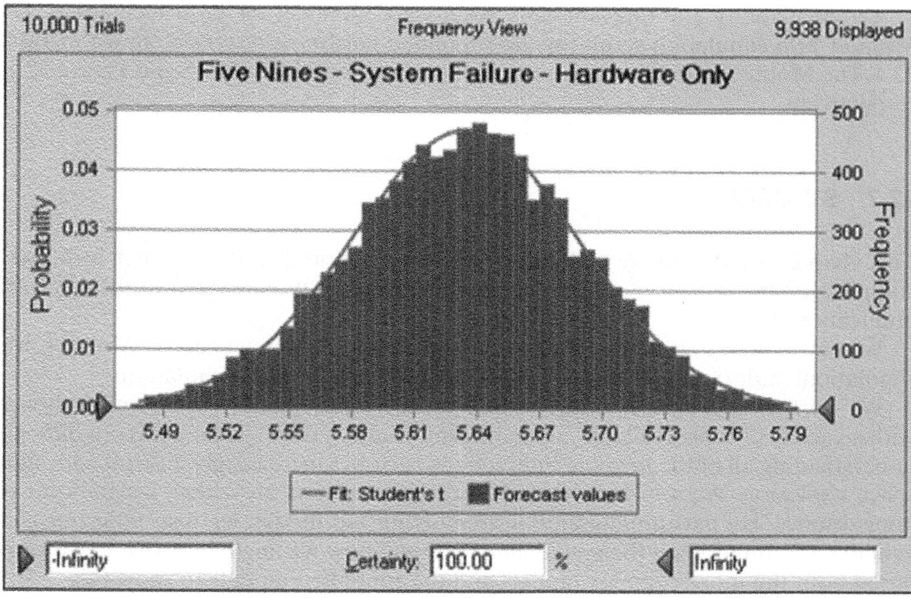

Figure 22.18 Simulation Output Calculating Five 9s

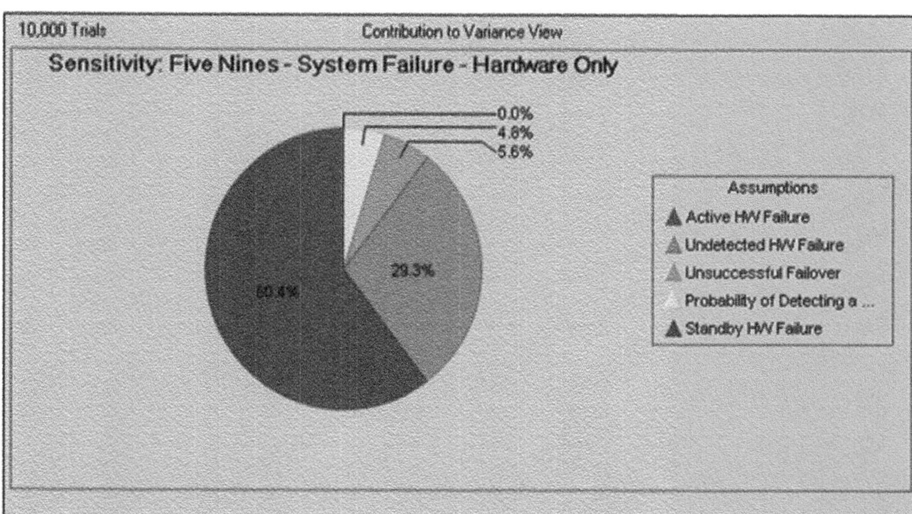

Figure 22.19 Sensitivity Chart Affecting Five 9s

The availability and unavailability results are identical for both techniques. The Monte Carlo simulation setup for the simplified hardware system is shown in Table 22.4. The Monte Carlo Simulation results are shown in Figure 22.18 and Figure 22.19 respectively.

22.7 SUMMARY

This chapter applied many techniques we have explored in this chapter to a more complicated but practical system architecture frequently employed in the telecommunications industry.

We considered a variety of techniques—including Markov analysis, Monte Carlo simulation, fault tree analysis, sensitivity analysis, and Matlab simulation.

We saw firsthand how state space explosion in Markov analysis can occur. Although with a variety of tools available, extracting information from the model is not difficult, understanding all of the possible interactions becomes increasingly difficult with state space explosion. We see that many of these interactions are either insignificant or nonexistence. By partitioning and/or simplifying the model, we lose little but gain valuable information on the system behavior and how we might improve it. We compared both the more complex models and the simplified models to show this.

With the approach to solving complex analysis problems provided in this chapter, as well as the many other techniques for solving reliability problems and improving reliability explored in this text, the reliability analyst will be able to successfully tackle a wide variety of reliability analysis and design problems. These techniques applied optimally will help contribute to the improvement of the reliability of the systems and products the analyst encounters.

REFERENCES

[1] National Institute of Standards and Technology. http://www.itl.nist.gov/div898/handbook/apr/section4/apr451.htm (accessed May 15, 2013).

[2] Hayt, W.H., Jr. and Kemmerly, J.E. *Engineering Circuit Analysis*. McGraw Hill, 1993.

[3] El-Haik, B. and Al-Aomar, R. *Simulation-Based Lean Six-Sigma and Design for Six-Sigma*. John Wiley & Sons, 2006.

[4] Trivedi, K.S. *Probability and Statistics with Reliability, Queuing, and Computer Science Applications*, 2nd edition. John Wiley & Sons, 2001.

[5] QuEST Forum. TL 9000 Telecom Quality Management System. http://www.tl9000.org/ (accessed May 15, 2013).

[6] Pham, H. (ed.). *Handbook of Reliability Engineering*. Springer, 2003.

[7] Birolini, A. *Reliability Engineering: Theory and Practice*, 6th edition. Springer, 2010.

[8] Dinesh Kumar, U., Crocker, J., Chitra, T., and Saranga, H. *Reliability and Six Sigma"*. Springer, 2006.

[9] Ranganathan, S. and Redzic, C. Applying DFSS for third-party intensive systems. SSIRI, IEEE, 2009.

[10] Maass, E. *DFSS for Hardware and Software*. Prentice Hall, 2009.

[11] McDermott, R.E., Mikulak, R.J., and Beauregard, M.R. *The Basics of FMEA*, 2nd edition. Productivity Press, 2009.

[12] Bauer, E., Zhang, X., and Kimber, D.A. *Practical System Reliability*. John Wiley & Sons, 2009.

[13] Ash, C. *The Probability Tutoring Book*. Institute of Electrical and Electronics Engineers, Inc., 1993.

[14] Viniotis, Y. *Probability and Random Statistics for Electrical Engineers*. McGraw-Hill, 1998.

[15] Pukite, J. and Pukite, P. *Modeling for Reliability Analysis*. IEEE Press, 1998.

[16] Miller, I. and Freund, J.E. *Probability and Statistics for Engineers*. Prentice-Hall, Inc., 2010.

[17] Triola, M.F. *Elementary Statistics Using Excel*, 2nd edition. Pearson Education, Inc., 2009.

[18] Ross, S.M. *Introduction to Probability Models*, 10th edition. Academic Press, 2009.

[19] Hogg, R.V. and Tanis, E.A. *Probability and Statistical Inference*, 8th edition. Prentice Hall Inc., 2008.

[20] Utas, G. *Robust Communications Software*. John Wiley & Sons, 2005.

[21] Taylor, J.R. *An Introduction to Error Analysis*, 2nd edition. University Science Books, 1997.

[22] Hsu, H. *Probability, Random Variables, & Random Processes*. McGraw-Hill, Inc., 2010.

Designing High Availability Systems: Design for Six Sigma and Classical Reliability Techniques with Practical Real-Life Examples, First Edition. Zachary Taylor and Subramanyam Ranganathan.

[23] Papoulis, A. and Unnikrishna Pillai, S. *Probability, Random Variables, and Stochastic Processes*, 4th edition. The McGraw-Hill Companies, Inc., 2002.

[24] Yates, R.D. and Goodman, D.J. *Probability and Stochastic Processes*, 2nd edition. John Wiley & Sons, Inc., 2005.

[25] Howard, R.A. *Dynamic Probabilistic Systems, Semi-Markov and Decision Processes*. Dover Publications, Inc., 2007.

[26] Golnaraghi, F. and Kuo, B.C. *Automatic Control Systems*. Prentice-Hall, Inc., 2010.

[27] Hsu, H.P. *Signals and Systems*. McGraw-Hill, Inc., 2010.

[28] Johnson, R. and Kuby, P. *Elementary Statistics*. Brooks/Cole, 2010.

[29] Weiss, N.A. *Introductory Statistics*. Addison-Wesley Publishing Company, Inc., 1982.

[30] Laprie, J.C., Dependable computing and fault tolerance: Concepts and terminology. Fifteenth International Symposium on Fault-Tolerant Computing. Ann Arbor, MI, 1985.

[31] Gray, J. A census of tandem system availability between 1985 and 1990. Technical Report 90.1 (Part Number 33579). Tandem Computers Inc., January 1990.

[32] Voas, J., Charron, F., McGraw, G., Miller, K., and Friedman, M. Predicting how badly "good" software can behave. *IEEE Software*, 14(4):73–83, 1997.

[33] Lyu, M.R.-T. Software reliability theory. Encyclopedia of Software Engineering, 2002.

[34] Hennessy, J.L. and Patterson, D.A. *Computer Architecture—A Quantitative Approach*, 3rd edition. Morgan Kaufmann, 2011.

[35] Grottke, M. and Trivedi, K.S. Fighting bugs: Remove, retry, replicate, and rejuvenate. *Computer Communications*, 40(2):107–109, 2007.

[36] Mullen, R. The lognormal distribution of software failure rates: Origin and evidence. *Proceedings of the 9th IEEE International Symposium on Software Reliability Engineering*. November 1998.

[37] Garg, S., Van Moorsel, A., Vaidyanathan, K., and Trivedi, K.S. A methodology for detection and estimation of software aging. page 283, IEEE Computer Society, Los Alamitos, CA, USA, 1998.

[38] Gokhale, S.S. and Trivedi, K.S. Analytical models for architecture-based software reliability prediction: A unification framework. *IEEE Transactions on Reliability*, Vol. 55:No. 4–December, 2006.

[39] Yacoub, S., Cukic, B., and Ammar, H. Scenario-based analysis of component-based software. *Proceedings of the Tenth International Symposium on Software Reliability Engineering*. Boca Raton, FL, November 1999.

[40] Gokhale, S. Accurate reliability prediction based on software structure. *Proceedings of IASTED Conference on Software Engineering and Applications (SEA 03)*. Marina Del Rey, CA, November 2003.

[41] Kececioglu, D. *Reliability and Life Testing Handbook*. Prentice Hall, 1993.

[42] Hoyland, A. and Rausand, M. *System Reliability Theory: Models and Statistical Methods*. John Wiley & Sons, 1994.

[43] Nelson, W. *Applied Life Data Analysis*. John Wiley & Sons, 1982.

[44] O'Connor, P. *Practical Reliability Engineering*. John Wiley & Sons, 2003.

[45] Romeu, J.L. Reliability estimations for exponential life. RIAC START, Volume 10. http://src.alionscience.com/ (accessed May 15, 2013).

[46] Tijms, H.C. *A First Course in Stochastic Models*. John Wiley & Sons, 2003.

[47] Klinger, D.J., Nakada, Y., and Menéndez, M.A. *AT&T Reliability Manual*. AT & T Bell Laboratories, 1999.

[48] Dovich, R.A. *Reliability Statistics*. ASQ Quality Press, 1990.

[49] Rubinstein, R.Y. and Kroese, D.P. *Simulation and the Monte Carlo Method*. John Wiley & Sons, 2007.

[50] Vose, D. *Risk Analysis: A Quantitative Guide*. John Wiley & Sons, 2008.

[51] Stamatis, D.H. *Failure Mode and Effect Analysis: FMEA from Theory to Execution*. ASQ Quality Press, 2013.

[52] Stamatis, D.H. *The OEE Primer: Understanding Overall Equipment Effectiveness, Reliability, and Maintainability*. Auerbach Publications, 2010.

[53] Yang, G. *Life Cycle Reliability Engineering*. John Wiley & Sons, 2007.

[54] Savage, S. L. *The Flaw of Averages: Why We Underestimate Risk in the Face of Uncertainty*. John Wiley & Sons, 2012.

[55] Mann, N., Schafer, R.E., and Singpurwalla, N.D. *Methods for Statistical Analysis of Reliability and Life Data*. John Wiley & Sons, 1974.

[56] Vesely, W. NASA fault tree handbook with aerospace applications version 1.1. August 2002.

INDEX

Note: Page numbers in *italics* indicate figures; those in **bold**, tables.

Designing High Availability Systems: Design for Six Sigma and Classical Reliability Techniques with Practical Real-Life Examples, First Edition. Zachary Taylor and Subramanyam Ranganathan.

Printed and bound by CPI Group (UK) Ltd, Croydon, CR0 4YY
17/09/2026
14964286-0006